Leeds Trinity
University College

LIBRARY

This book is due for return on or before the last date stamped below

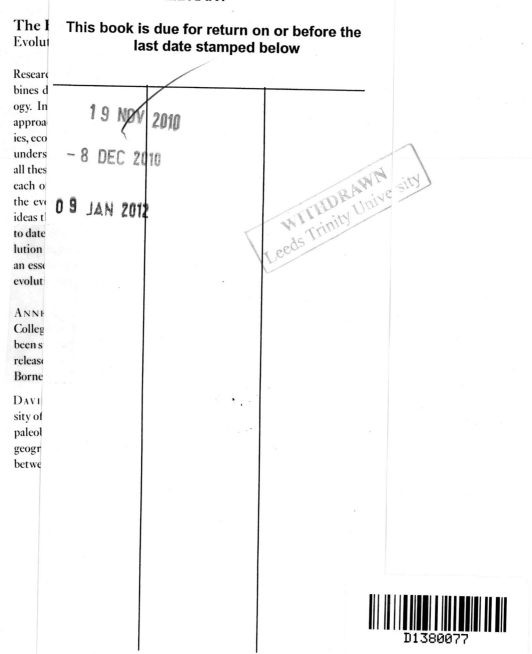

1 9 NOV 2010

- 8 DEC 2010

0 9 JAN 2012

The I
Evolut

Resear
bines d
ogy. In
approa
ies, eco
unders
all thes
each o
the ev
ideas t
to date
lution
an ess
evolut

ANNF
Colleg
been s
releas
Borne

DAVI
sity of
paleol
geogr
betwe

D1380077

The Evolution of Thought

Evolutionary Origins of Great Ape Intelligence

Edited by

Anne E. Russon
Department of Psychology, Glendon College, York University, Toronto

David R. Begun
Department of Anthropology, University of Toronto, Toronto

CAMBRIDGE
UNIVERSITY PRESS

CAMBRIDGE UNIVERSITY PRESS
Cambridge, New York, Melbourne, Madrid, Cape Town, Singapore, São Paulo

Cambridge University Press
The Edinburgh Building, Cambridge CB2 8RU, UK

Published in the United States of America by Cambridge University Press, New York

www.cambridge.org
Information on this title: www.cambridge.org/9780521783354

© Cambridge University Press 2004

First published 2004
This digitally printed version 2007

A catalogue record for this publication is available from the British Library

Library of Congress Cataloguing in Publication data
The evolution of thought : evolutionary origins of great ape intelligence / edited by Anne E. Russon
and David R. Begun.
   p.   cm.
Includes bibliographical references.
ISBN 0 521 78335 6
1. Apes – Evolution.   2. Apes – Psychology.   3. Animal intelligence.   I. Russon, Anne E.   II. Begun, David R.
QL737.P96E83   2004
155.7 – dc22   2003058668

ISBN 978-0-521-78335-4 hardback
ISBN 978-0-521-03992-5 paperback

Contents

Contributors

DAVID R. BEGUN
Department of Anthropology
University of Toronto
Toronto, ON M5S 3G3, Canada
begun@chass.utoronto.ca

JOANNA BLAKE
Department of Psychology
York University
4700 Keele St
North York, ON M3J 1P3, Canada

RICHARD W. BYRNE
School of Psychology
University St. Andrews
St. Andrews, Fife KY16 9JU, Scotland, UK
rwb@st-andrews.ac.uk

MARK FLINN
Department of Anthropology
107 Swallow Hall
University of Missouri
Columbia, MO 65211, USA
FlinnM@missouri.edu

DANIEL L. GEBO
Department of Anthropology
Northern Illinois University
De Kalb, IL 60115-2854, USA
dgebo@niu.edu

KEVIN D. HUNT
Department of Anthropology, SB 130
Indiana University
Bloomington, IN 47405, USA
kdhunt@ucs.indiana.edu

JAY KELLEY
Department of Oral Biology, College of Dentistry
University of Illinois at Chicago
801 South Paulina St.
Chicago, IL 60612-7213, USA
jkelley@uic.edu

LÁSZLÓ KORDOS
The Geological Institute of Hungary
H-1143 Budapest
Stefánia út 14, Hungary
kordos@mafi.hu

CAROL E. MACLEOD
Department of Anthropology
Langara College
Vancouver, BC V5Y 2Z6, Canada
caroleli@sfu.ca

SUE TAYLOR PARKER
Department of Anthropology
Sonoma State University
Rohnert Park, CA 94928, USA
parker@sonoma.edu

RICHARD POTTS
Director, Human Origins Program
Department of Anthropology
National Museum of Natural History
Smithsonian Institute
Washington, DC 20560-0112, USA
Potts.Rick@NMNH.SLEDU

SIGNE PREUSCHOFT
Living Links, Yerkes Primate Center
Emory University
954 North Gatewood Road
Atlanta, GA 30329, USA
Present address: Haydnstraße 25, 44147 Dortmund, Germany

CAROLINE ROSS
School of Life & Sport Sciences
University of Surrey
London, UK
c.ross@roehampton.ac.uk

ANNE E. RUSSON
Department of Psychology
Glendon College, York University

2275 Bayview Ave.
Toronto, ON M4N 3M6, Canada
arusson@gl.yorku.ca

MICHELLE SINGLETON
Department of Anatomy
Midwestern University
555 31st Street
Downers Grove, IL 60515, USA
msingl@midwestern.edu

CAREL P. VAN SCHAIK
Biological Anthropology and Anatomy
Duke University
Box 90383 Durham, NC 27708-0383, USA
vschaik@acpub.duke.edu

CAROL V. WARD
Department of Anthropology
Department of Pathology and Anatomical Sciences
107 Swallow Hall

University of Missouri
Columbia, MO 65211, USA
WardCV@missouri.edu

DAVID P. WATTS
Department of Anthropology
Yale University
P.O. Box 208277
New Haven, CT 06520-8277, USA

JUICHI YAMAGIWA
Laboratory of Human Evolution Studies
Faculty of Science, Kyoto University
Sakyo-ku, Kyoto, 606-8502, Japan
yamagiwa@jinrui.zool.kyoto-u.ac.jp

GEN YAMAKOSHI
Graduate School of Asian and African Studies
Kyoto University
Sakyo-ku, Kyoto, 606-8502, Japan
yamakosh@jinrui.zool.kyoto-u.ac.jp

Preface

This book arose from three realizations. First, there is an important need for good models of great ape cognitive evolution. Studies of comparative primate cognition over the last two decades increasingly show that all great apes share a grade of cognition distinct from that of other nonhuman primates. Their cognition appears to be intermediate in complexity between that of other nonhuman primates and humans, so it offers the best available model of the cognitive platform from which human cognition evolved. Understanding the position of the great apes is then essential to understanding cognitive evolution within the primate order and ultimately, in humans. Second, existing reconstructions of the evolutionary origins of great ape cognition are all in need of revision because of advances in research on great ape cognition itself, on modern great ape adaptation, and on fossil hominoids. Third, developing an accurate picture of the evolutionary origins of great ape intelligence requires bringing together expertise from a highly diverse range of fields beyond modern great ape cognition. Essential are current understandings of the brain, life histories, social and ecological challenges, and the interactions among them in both living and ancestral hominids.

We therefore assembled a team of contributors with expertise spanning the topics currently recognized as relevant to cognitive evolution in the great ape lineage, with the aim of piecing together the most comprehensive picture possible today. We asked all our contributors to explore the implications of their realm of expertise for cognition and cognitive evolution. We are grateful to all of them for their willingness to embark on this enterprise and for sticking with the sometimes trying process of fitting this broad range of material together. The product is a compilation of our contributors' views on adaptations relevant to cognition in the great ape lineage and our attempt to integrate their material into a coherent picture. Our sense is that a coherent picture does emerge. That contributors working from very different perspectives often voiced similar conclusions adds to our sense that this picture has considerable substance.

We do not presume that our reconstruction will close the book on the evolutionary origins of great ape cognition. Although we covered most if not all of the major issues currently recognized as important in the evolution of great ape mentality, the breadth of the material involved means that our coverage is inevitably brief. Further, our contributors pointed to additional factors in need of consideration and there remain vast areas of importance that have been little researched or that are still crying for evidence. This picture will undoubtedly change as understanding improves. Our hope is that this collective work will contribute to filling the need for good models of the evolutionary origins of great ape intelligence and at the same time spur efforts to improve our picture where it proves lacking.

1 · Evolutionary reconstructions of great ape intelligence

Psychology Department, Glendon College of York University, Toronto

INTRODUCTION

Research increasingly shows great apes surpassing other nonhuman primates in their mentality, achieving abilities traditionally considered uniquely human. Importantly, the cognitive capacities that distinguish them include rudimentary symbolic processes, in the sense of processes that operate on the basis of mental images rather than direct sensory-motor phenomena. Although this view does not represent consensus among experts (e.g., Tomasello & Call 1997), many well-respected researchers now accept this interpretation of the empirical evidence (e.g., Byrne 1995; Langer & Killen 1998; Parker & McKinney 1999; Parker, Mitchell & Boccia 1994; Parker, Mitchell & Miles 1999; Russon, Bard & Parker 1996; Savage-Rumbaugh, Shanker & Taylor 1998; Whiten & Byrne 1991; Wrangham *et al.* 1994).

If great apes are capable of symbolic cognitive processes, views of symbolism as having evolved within the human lineage are incorrect. Implications for understanding cognitive evolution within the primates are complex and important. First, neither the landmark significance of symbolism to cognition nor its importance in understanding the evolution of higher primate cognition is diminished by this revision. What is altered is timing. Symbolic cognition shifts from an achievement of the human lineage to a foundation for it. Second, reconstructions of the conditions leading to the evolution of symbolic processes remain important, but existing reconstructions lose much of their weight because they focus on conditions linked with the divergence of the human lineage. If symbolic processes are the joint province of humans and great apes, ancestral large hominoids are their probable evolutionary source. At this vastly different point in time and probably in space,

a very different set of conditions likely affected them. Finally, what is unique to the human mind must be re-evaluated.

This volume aims to reconstruct the evolutionary origins of great ape intelligence. This is not the first such reconstruction; over half a dozen have been developed over the last 25 years, primarily by scholars of cognition (e.g., Byrne 1997; Byrne & Whiten 1988; Parker & Gibson 1977, 1979; Parker 1996; Povinelli & Cant 1995; Russon 1998). While their expertise on issues of cognition is undisputed, their navigation and rendition of evidence and debate in the other key areas can be less sound. Many extant reconstructions, for instance, rely on outdated or flawed views of modern great ape anatomical or behavioral adaptations, sociality, ecology, or ancestry (for discussion, see Byrne 1997, 2000; Russon 1998). Reconsidering the evolutionary origins of great ape intelligence is well worth undertaking at this time. The accumulation of empirical evidence is generating better models of cognitive processes in living great apes. The body of knowledge on the behavioral, anatomical, social, and ecological traits of living great apes is affording increasingly reliable identification of potentially conservative traits. A recent upsurge of interest in hominoid evolution occasioned by significant fossil finds and increasingly sophisticated molecular taxonomic methods enormously improves the prospects for honing in on the critical pieces of the ancestral hominoid picture that concern cognition.

To orient our attempt, this first chapter revisits existing reconstructions of cognitive evolution that implicate the great apes. Aims are to highlight why and where evolutionary reconstructions of great ape cognition are in need of revision, the factors potentially at play, and our approach to developing a new reconstruction.

Copyright Anne Russon and David Begun 2004.

RECONSTRUCTING GREAT APE COGNITIVE EVOLUTION

Reconstructing the events responsible for the evolution of great ape cognition entails, in part, a logic that links cognitive capacity with observable physical features (e.g., Byrne 2000; Parker & McKinney 1999). Failure to adhere to this logic undermines many existing reconstructions. Very briefly, great ape cognition requires powerful, sophisticated brains. Whatever the reasons for their evolution, such large brains support increasingly complex behavior. While it is very difficult to establish whether there was direct selection in ancestral great apes for more complex forms of behavior, it is likely that once attained, this capacity was used to ecological and social advantage. So, if common forms of complex behavior can be identified in living great apes that distinguish them from other nonhuman primates, then these behaviors and their putative cognitive, anatomical, ecological, and social correlates may represent conservative traits that owe to common ancestry. Once such a suite of characters is identified, it should be possible to infer related aspects of behavior in ancestral great apes, the ancestral conditions that could have favored them, and the cognitive processes that evolved to govern them.

Reconstructions of cognitive evolution in the primates have further been guided by their own set of premises. First, enhancements to primate cognition are presumed to have been adaptive, i.e., achieving greater behavioral flexibility by enhanced cognition was directly selected for, not an incidental byproduct (e.g., Byrne 1995; Byrne & Whiten 1988; Gibson 1993; Povinelli & Cant 1995; Parker 1996). Brain enhancements that are fortuitous luxuries are unlikely to be maintained or even to occur because brain tissue is especially costly energetically (Aiello & Wheeler 1995; Armstrong 1983). Second, modern cognition (abilities, development, functions), as expressed in natural habitats, is taken as a good proxy for ancestral precursors.

HUMAN COGNITIVE EVOLUTION

Reconstructions of human cognitive evolution regularly borrow the great apes to define the primitive intellect from which human intelligence diverged and upon which it built. I review three recent models to illustrate how scholars of human cognitive evolution have tended to portray great ape cognition and the problems so occasioned.

Donald

Donald (1991, 1993, 2000), a neuropsychologist, modeled the human mind as evolving from the ancestral, prehominin condition in three cognitive transformations, wherein cognitive and cultural evolution are deeply and fundamentally interdependent. These transformations are founded on new memory representations because cognitive systems that are culturally dependent cannot replicate without systems for storing collective knowledge. Donald's starting point is an "episodic" culture in the common great ape–human ancestor, based primarily on modern chimpanzees. From this evolved "mimetic" (*Homo erectus*), "mythic" (*Homo sapiens*), then "theoretic" (modern human) cultures.

Critical to great apes are episodic and mimetic cultures, taken to represent great ape cognition, modern and ancestral, and the step beyond. In positioning great apes as "episodic," Donald characterizes their cognition as governed by procedural memory: able to store perceptions of events but poor at episodic recall, having little voluntary access to episodic memories without environmental cues. This would leave great apes unable, voluntarily, to shape and modify their own actions or to access their stored representations, so unable to invent gestures, mimes, and signs to communicate or to practice their skills systematically. Their experience would be an episodic lifestyle governed by the present. The "mimetic" cultures that followed, enabled by voluntary retrieval of stored memories independent of environmental cues, would surmount this episodic inability. This allows individuals to take voluntary control over their own output, including voluntary rehearsal and refinement, and mimetic skills like pantomime, reenactive play, self-reminding, imitative learning, and proto-pedagogy; in effect, it allows using their bodies as communication devices to act out events in quasi-symbolic form.

Critics have already shown that "episodic" underestimates great apes. Great apes' capabilities include the episodic recall and the voluntary control over motor output essential to mimesis (Byrne 1997; Byrne & Russon 1998; Matsuzawa 1996; Russon 1998; Schwartz & Evans 2001), bringing them close to the mimetic minds attributed to *Homo erectus* (Byrne pers.

commun.; Mitchell & Miles 1993; Parker & McKinney 1999).

Donald (2000) now accepts that great apes achieve more complex cognition, symbolic skills included, but discounts their importance on the grounds that they represent individual versus collective representational systems (i.e., symbolic cultures). He attributes many of great apes' most impressive achievements (e.g., language, stone tool making) to the transformative powers of human cultural rearing environments which, he believes, can transform them into "superprimates" by exploiting cognitive potential that has remained untapped for millions of years. This position is also disputable. Taï Forest chimpanzees use two gestures with shared collective meanings, leaf-clipping and knuckle-knocking, that verge on collective symbolic representations (Boesch 1996). That human enculturation induces higher than normal cognitive abilities in great apes is not well established and the claim has been contested on several fronts (Parker & McKinney 1999; Russon 1999b; Suddendorf & Whiten 2001).

Cosmides and Tooby

Cosmides and Tooby (1992), evolutionary psychologists, proposed that human cognition evolved through cognitive "modules" biologically designed to address the particular adaptive problems that ancestral humans encountered in their environment of evolutionary adaptedness, taken to be hunter–gatherer lifestyles in Pleistocene environments. Language, theory of mind, spatial relations, and tool use are among the modules proposed. Supposedly, these modules are "content rich," pre-fitted with knowledge relevant to the Pleistocene problems these hunter–gatherers faced, and have changed little since because too little time has passed to allow further evolutionary modification.

Limitations to this model have been pointed out. Mithen (1996) argued that modularity of this sort does not reflect what humans really do, mix and match their thinking. Byrne (2000) identified flaws in the logic and evidence of "adaptation to the Pleistocene." Hominins did change and diverge in the Pleistocene. Human ancestors pursued a lifestyle close to living hunter–gatherers (e.g., large animal hunting, fire, living shelters) only from about 40 000 years ago, too recently to have shaped human cognitive evolution. Finally, human traits offering evidence of evolutionary origins (e.g., infanticide,

homicide, mating systems) long predate hominins in the primates. Traits proposed as significant in human cognitive evolution almost certainly have much longer evolutionary histories than this model allows. Neglecting evidence on modern great apes and other primates leaves this model without a credible point of departure.

Mithen

Mithen (1996), an archeologist, proposed four "acts" in human cognitive evolution. Act 1 opened 6 Ma with ancestral great apes, Act 2 at 4–5 Ma with ancestral hominins, Act 3 at 1.8 Ma with *Homo erectus*, and Act 4 at 100 000 years ago with modern humans. Like others, Mithen uses living great apes, especially chimpanzees, to represent the cognitive capacities existing at the ancestral great ape–human divide.

Mithen assumes a fundamentally modular cognitive architecture (after Cosmides & Tooby 1992), and a recapitulationist position, that the sequence of developmental stages can be read as re-iterating the phylogenetic sequence of ancestral adult forms. Within this framework, he proposes three phases of cognitive evolution based on children's cognitive development (after Karmiloff-Smith 1992): generalized intelligence, specialized intelligences, and cognitive fluidity. Generalized intelligence comprises a suite of general-purpose, associative-level learning and decision-making mechanisms used in all domains to modify behavior in light of experience (e.g., trial and error learning, stimulus enhancement). Specialized intelligences are biologically designed modules for specific problem domains, operating in virtual isolation of one another. Three are proposed: social (for social interaction and mind-reading), natural history (for understanding the natural world, especially biology), and technical (for manufacturing, manipulating, and throwing stone and wooden artifacts). Cognitive fluidity is achieved by interconnecting specialized intelligences, allowing them to work together by enabling the flexible flow of knowledge and ideas among them.

Mithen portrays ancestral great ape cognition, Act 1, at the interface between phases 1 and 2: equipped with generalized intelligence, a social intelligence, and an incipient natural history intelligence (for resource distribution) that generated capacities comparable to those of other haplorhines but *somewhat* more powerful. Act 2 added further modularization, Act 3 added a language module that connected with the social but not technical or natural history modules (which remained isolated from each other), and Act 4 broke down barriers between modules to allow cognitive fluidity.

Many experts portray great ape cognition very differently. To illustrate, Mithen attributed chimpanzees' tool and foraging expertise to general intelligence, i.e., associative learning, whereas substantial evidence exists of their using rudimentary symbolism and hierarchization (e.g., Byrne 1995; Matsuzawa 2001; Parker & McKinney 1999; Russon 1998, 1999a; Suddendorf & Whiten 2001). He also claimed great apes show domain isolation because they miss opportunities at the social–foraging interface, like failing to learn foraging skills socially or use material culture to serve social strategies, whereas considerable evidence shows they use social learning in acquiring foraging skills (Byrne & Byrne 1993; Parker 1996; Russon 1999b; van Schaik & Knott 2001; van Schaik, Deaner & Merrill 1999; van Schaik *et al.* 2003; Whiten *et al.* 1999) and use tools socially (Boesch & Boesch-Achermann 2000; Goodall 1986; Ingmanson 1996; Peters 2001).

Summary

While most reconstructions of human cognitive evolution recognize the hominids as defining the cognitive platform from which hominins diverged and their evolutionary context, all suffer from underestimating that cognitive platform and therefore, from misidentifying the evolutionary conditions involved.

RECONSTRUCTIONS OF PRIMATE COGNITIVE EVOLUTION

Reconstructions of cognitive evolution within the primate order tend to fall into two categories, social and ecological, according to the type of selection pressure promoted as most influential, and to presume that influences operate in similar fashion across the order as a whole or at least across the haplorhines.

Social intelligence

The suggestion that primates' complex social lives shaped the evolution of their intellect can be traced to Jolly (1966), Kummer (1967), and Humphrey (1976). Tripartite relations, maneuvers to influence powerful individuals and potential allies, and tactical deception are among the facets of primate sociality singled out as cognitively complex. If communicative signals were selected for the signaler's competitive advantage more than for honest exchange (Krebs & Dawkins 1984), spiraling evolutionary arms races could have occurred, first to improve schemes for outwitting competitors (favoring abilities for agonistic cooperation and perhaps for generating misleading signals), then for dupes to enhance their abilities to detect honest information behind misleading signals. Such reasoning spawned the influential Machiavellian Intelligence hypothesis on the nature and evolution of primate cognition (Byrne & Whiten 1988). Cooperative advantages gained via social reciprocity, tallying favors exchanged, recognizing and categorizing conspecifics by family membership, etc. are also potential selection pressures in primate cognitive evolution (Cheney & Seyfarth 1990; de Waal 1996).

The social intelligence hypothesis argues that the social pressures on primates are more complex than the ecological pressures typically proposed as prime movers of cognitive evolution, range size and frugivory. Social problems present highly changeable information from changing animate partners, sensory input from diverse modalities, and multiple individual and social attributes. Social cognition must operate on this multifaceted information in parallel; ecological cognition, supposedly, faces a much lighter parallel mental load (Barton & Dunbar 1997). Accordingly, social pressures were the primary forces shaping primate cognitive evolution.

Dunbar and his colleagues have been major proponents of this hypothesis. They consistently find that their index of intelligence (neocortical ratio, the size of the neocortex relative to basic brain structures) correlates with indices of social complexity (group size) but not ecological complexity (range size or day journey length, adjusted for body size) in species where individuals live in intensely social groups rather than simple aggregations. Correlations hold within primates (within haplorhines, between strepsirhines and haplorhines, perhaps between haplorhines and hominins: Barton 1996; Dunbar 1992, 1995, 1998), within carnivores, and within cetaceans (Kudo & Dunbar 2001). They conclude that in such taxa, cognitive capacities constrain the number of individuals that can co-exist in one social group (Barton & Dunbar 1997; Dunbar 1992, 1998). This work is problematic with respect to primate cognitive evolution for at least two reasons. First, social complexity depends on factors beyond group size, such as group structure, group organization, and the range and subtlety of members' behavior (Byrne 1999, and see Parker, Chapter 4, van Schaik, Preuschofr & Watts, Chapter 11, this volume). Second, these studies and accounts concern cognition as a constraint on sociality, not sociality as a selection pressure for cognitive enhancement, so they say little about cognitive evolution (Parker & McKinney 1999).

Concerning great ape cognitive evolution, five issues deserve mention. (1) Most social activities promoted as cognitively complex (e.g., tripartite relations, tallying social exchange) occur in many haplorhines so they require only the cognitive capacities of monkeys, not the advanced capacities distinctive of great apes. Possible exceptions include high-level tactical deception (Byrne & Whiten 1997), consolation (de Waal & Aureli 1996) and symbolic communication (Boesch 1996; Savage-Rumbaugh et al. 1996). (2) Studies of group size–neocortex size correlations have included Pan and Gorilla but not the orangutan, who is large-brained and semi-solitary (Dunbar 1992, 1998). (3) Social intelligence proponents probably underestimate the ecological complexities facing great apes. Great apes' "technical" skills for obtaining difficult foods bear witness to the complexity of these ecological pressures (Byrne & Byrne 1991; Byrne, Corp & Byrne 2001; Russon 1998, 2003; Stokes 1999; Yamakoshi & Sugiyama 1995), and these pressures are multifaceted in arboreal or competitive conditions. These technical capacities are also relegated to evolutionary side effects under the social intelligence hypothesis, which fits poorly with the sense that they are central to great ape adaptation (Byrne 1997). (4) These social complexity measures do not reflect impressions that sociality is more complex in great apes than other haplorhines (e.g., Byrne 1995, 1997; Parker & McKinney 1999). (5) Close analysis of group–brain size correlations suggests cognitive differences between great apes and other haplorhines (Dunbar 1993; Kudo & Dunbar 2001), with great apes seeming to use more "computing power" than monkeys to manage the same number of relationships (Dunbar 1998). In other words, group size does not completely account for differences in cognitive power between haplorhines and great apes. While this hypothesis has been valuable in identifying the complex social pressures facing primates, it has offered little to reconstructing the evolution of a distinctive great ape cognition.

Ecological hypotheses

Diet

Diet, frugivory in particular, is the ecological pressure most often linked to the evolutionary enhancement of primate cognition. Foods distributed unpredictably in time and space or over large supplying areas, dietary diversity, and diets that rely on foods that are difficult to obtain have all been promoted as setting a selective premium on high intelligence (Clutton-Brock & Harvey 1980; Galdikas 1978; Gibson 1986; Menzel 1978; Menzel & Juno 1985; Milton 1981, 1988; Parker 1978; Parker & Gibson 1977; Wrangham 1977).

Fruit is especially patchy in spatial and temporal distribution compared with foliage, so frugivory could promote abilities like memory, spatial reasoning, or cognitive maps (Milton 1981, 1988). Two sympatric New World monkeys, frugivorous spiders and folivorous howlers, support this prediction: spiders have greater relative brain size, larger home ranges, and a more protracted dependency/learning period (Milton 1988). Frugivory also correlates positively with brain size in haplorhines although the effect is much smaller than group size (Barton 1996; Byrne 1997), as well as in bats, rodents, insectivores, and lagomorphs (Milton 1988). Diversifying the diet to include protein- and fat-rich foods may be responsible for large day ranges in frugivorous primates, chimpanzees included, rather than searching for ripe fruit (Hladik 1975).

The main problem with this broad view of dietary niche for reconstructing great ape cognitive evolution is that it does not distinguish great apes from other haplorhines. Although all great apes retain basically frugivorous diets and monkeys evolved greater capacities for folivory, some monkeys and the lesser apes have diets similar to those of great apes. Dietary pressures distinctive to great apes are more likely to be found in specific dietary features. Foods that are difficult to obtain, for instance, have often been proposed as selection pressures favoring the enhancement of great ape cognition, to enable the complex techniques needed to obtain them. Pressures stem from food defenses that pose "technical difficulties," like embeddedness, toxicity, or antipredator behavior in animal prey (e.g., Byrne 1997; Boesch & Boesch-Achermann 2000; Hladik 1977; Parker & Gibson 1977). Such defenses are common in foods that supplement fruits in great apes, especially fallback foods needed in periods of fruit scarcity. Both embeddedness and technical difficulty have inspired hypotheses on the evolution of a distinctively "great ape" cognition (Byrne 1997; Parker & Gibson 1977).

Ranging

Two ranging patterns might underpin evolutionary enhancements to primate cognition, range size and terrestriality. Increased range size could favor enhanced memory and cognitive maps (e.g., Clutton-Brock & Harvey 1980). Terrestrial life could have exerted selection pressures because it increases predation risks. Primates' preferred evolutionary response to predation appears to have been increased social group size, which could then have been the direct pressure for enhanced intelligence (e.g., Dunbar 1992; van Schaik 1983). Range size is a function of diet, body size, and group size, however. Gorillas that eat more fruit have longer daily travel distances than those that eat less (Yamagiwa 1999); frugivorous spider monkeys have larger home ranges than folivorous howlers (Milton 1988); and larger groups are likely to have to travel farther than smaller ones to fulfill their food needs. Accordingly, links between range size and cognition may owe to these underlying parameters. Further, neither range size nor terrestriality distinguishes great apes from other haplorhines, so neither can account for the evolution of great ape cognition.

Summary

Primate-focused reconstructions are unsatisfying as reconstructions of great ape cognition because they do not distinguish great apes from other haplorhines. They are valuable in offering broader views of evolutionary pressures affecting great apes as primates, the range of ecological and social pressures worth exploring in greater depth, and the haplorhine pattern from which they differ. Limits to these hypotheses do, however, illustrate the need to determine what promoted the great apes' evolutionary divergence from other halorhines in their cognition.

RECONSTRUCTIONS OF GREAT APE COGNITION

Some reconstructions address the evolution of a distinctive great ape cognition, considering that cognitive evolution within the primate order probably involved three major grade shifts, not the two shifts typically portrayed (strepsirhine to haplorhine, haplorhine to hominin) (Byrne 1997; Byrne & Whiten 1997). The third shift, intervening temporally between them, is from most haplorhines to hominids (great apes and humans), with all hominids showing greater cognitive sophistication.

Most of these hypotheses were stimulated by Parker and Gibson's (1977) extractive foraging model, which singled out great apes and cebus for their "intelligent" tool using abilities. Several constitute revisions of earlier reconstructions, provoked by inconsistent findings. Most are synthetic, in that they propose a suite of selection pressures acting in concert, or sequentially, to produce the distinctive mentality characteristic of all living great apes.

Extractive foraging

Parker and Gibson (1977, 1979; Gibson 1986) hypothesized that seasonal reliance on embedded foods and prolonged ontogeny shaped hominid cognitive evolution. Ancestral great apes faced selection pressures imposed by omnivorous diets with seasonal reliance on embedded foods like hard-shelled fruits and nest-building insects. Embedded foods demand extractive foraging techniques; when needed seasonally, they favored the evolution of flexible techniques assisted by "intelligent" tool use (i.e., tool users understand the causal dynamics involved; Parker & Potí 1990), which require enhanced cognitive abilities. Reliance on tool-assisted extractive foraging favored prolonging ontogeny because foraging independence requires complex skills. These complex skills require advanced cognitive abilities, so prolonging ontogeny helped immatures by extending parental support and cognitive development. Extending dependency increased pressures on caregivers, especially mothers, by interfering with further reproduction, and favored the ability for imitation to speed offsprings' acquisition of foraging skills. They hypothesized that intelligent tool use evolved independently in *Cebus* for similar reasons.

Valuable features of this model include the effort to identify specific dietary features that distinguish the hominids and the incorporation of prolonged ontogeny, a life history parameter that distinguishes hominids from other haplorhines. Prolonged ontogeny extends cognitive development and parental support into the juvenile period in great apes (Parker & McKinney 1999).

Imitative abilities in particular emerge near the onset of juvenility, when the most complex facets of foraging skills are likely being acquired.

This hypothesis remains prominent although several limitations are recognized. (1) Great apes surpass *Cebus* cognitively so if extractive foraging explains their common intelligent tool use, additional factors are needed to explain great apes' greater cognitive power. (2) Whether seasonal extractive foraging affects great apes differently than other haplorhines is unclear; baboons, for instance, are omnivorous seasonal extractive foragers but do not show comparable cognition (e.g., Byrne 1997). (3) Singling out embeddedness neglects other food defenses requiring equally complex techniques, such as spines, toxins, distasteful exudates, and digestive inhibitors (e.g., Byrne 1997; Russon 1998). (4) Intelligent tool use may not qualify as synapomorphic in great apes relative to other haplorhines. It is absent from the vast majority of wild great ape populations (van Schaik *et al.* 1996) and in the two species where it can be habitual, orangutans and chimpanzees, it is rare in most (orangutan) or some (chimpanzee) populations (Chapman & Wrangham 1993; van Schaik & Knott 2001; Wrangham *et al.* 1993). (5) These complications impose two additional assumptions on this hypothesis, both open to question: living chimpanzees best represent the common great ape ancestor in diet and foraging strategy, and intelligent tool use characterized the common ancestor but was subsequently lost or reduced in most descendants. (6) Intelligent tool use is not itself a cognitive process, but an expression of means–ends cognition. Means–end cognition also generates manipulative foraging techniques and cognitively, great apes' manipulative techniques are very similar to their tool-assisted ones (Byrne & Byrne 1991; Byrne *et al.* 2001; Stokes & Byrne 2001; Matsuzawa 2001; Russon 1998; Yamakoshi & Sugiyama 1995; and see Byrne, Chapter 3, Yamakoshi, Chapter 9, this volume). Great apes' intelligent tool use could reflect means–ends cognitive processes that evolved for other purposes and were subsequently recruited for tool use. (7) If tool-assisted extractive foraging qualifies as a cognitive adaptation in great apes then so should cooperative hunting. It too is an important contributor to foraging success, primarily in chimpanzees (Boesch & Boesch-Achermann 2000). (8) This hypothesis has difficulty explaining the wealth of cognitive enhancements that great apes show beyond foraging, especially in the social domain.

Apprenticeship

Parker (1996) extended the extractive foraging model to propose that what evolved in great apes was an apprenticeship system wherein cognitive capabilities depend on rich social input during development. Apprenticeship, here, means guided participation in shared activities of a routine nature (Rogoff 1992). Parker proposed the co-evolution of a suite of interrelated cognitive abilities in hominids – imitation, intelligent tool use, self-awareness, demonstration teaching – that enabled immatures to acquire the tool-assisted extractive foraging skills essential and unique to their clade and that relieved maternal pressures by boosting offsprings' capacities to acquire this expertise.

The particularly valuable feature of this model is that it integrates social and ecological hypotheses: it situates sophisticated cognitive abilities for social transmission at the heart of the evolutionary enhancements that characterize great ape cognition, portrays social and ecological abilities working together rather than in isolation, and envisions cognitive enhancements as achieved through changes to ontogeny. This set of social and physical abilities occurs as an interrelated cluster in living great apes (Mitchell 1994), supporting the suggestion that they evolved as an interrelated suite to support tool-assisted extractive foraging. That social input is essential to great apes' cognitive development and acquisition of ecological skills is amply supported, although not restricted to tool skills (e.g., Parker & McKinney 1999; Tomasello & Call 1997).

As a derivative of the extractive foraging hypothesis, however, this model faces the same limits associated with exclusive concentration on tools and extractive foraging. Further, even the extended suite of cognitive abilities hypothesized to have evolved in response to these selection pressures neither covers nor generates the full range of cognitive advantages that great apes show over other haplorhines.

Arboreal clambering

Povinelli and Cant (1995) proposed large body size combined with arboreal travel as the selection pressures that favored evolutionary enhancements to intelligence in the common ancestor of great apes. They argued that arboreal travel pressures acting on extremely large-bodied primates favored the cognitive capacity for a self-concept, in particular the self as a causal agent, to allow individuals to figure the effects of their own body weight into their arboreal travel. Modern orangutans model the last common ancestor, their arboreal travel problems model ancestral selection pressures, and their clambering mode of arboreal locomotion expresses their cognition (clambering is primarily suspensory, orthograde locomotion that employs all four limbs in irregular fashion to grasp and hold multiple supports). This meshes with impressions that orangutans' cognitive prowess is most evident in arboreal locomotion (e.g., Bard 1993; Chevalier-Skolnikoff, Galdikas & Skolnikoff 1982; MacKinnon 1974). Povinelli and Cant identified *Oreopithecus* as a highly suspensory fossil hominid exemplifying this lifestyle.

This hypothesis is valuable in bringing attention to the intellectual challenges of arboreal travel for large-bodied primates and in incorporating the fossil record, but several limitations are recognized. (1) It applies to orangutans but not clearly to other great apes or their common ancestor. Not all great apes rely on arboreal travel. Neither was the ancestral hominid condition clearly arboreal: orangutans' postcranial adaptations for arboreal locomotion are recently derived, they differ substantially from those of the other living great apes, and the ancestral condition *vis-à-vis* arboreality is ambiguous (e.g., Begun 1992; Martin & Andrews 1993; Moyà-Solà & Köhler 1993; Pilbeam 1996, Tuttle & Cortright 1988). (2) It argues for the evolution of a single cognitive ability, self-concept, so it does not explain the broad range of abilities seen in great apes and their generally high level, i.e., their cognitive systems. (3) It considers only selection pressures for a self-concept, but construes self-concept as dependent upon a generalized cognitive capability, mental representation, i.e., recalling to mind or "re-presenting" mental codes for entities and simple object relations in the absence of their normal sensory and motor cues. Enabling a self-concept, then, either required evolving the generalized capacity for mental representation or tapped a pre-existing representational capacity; either scenario requires further explanation. (4) It is not certain that ancestral hominids had brains large enough for such cognitive abilities. The large hominid that Povinelli and Cant suggest may have faced such arboreal pressures, *Oreopithecus bambolii*, had an unusually small brain (Harrison & Rook 1997), not the large brain associated with sophisticated abilities like self-concept and mental representa-

tion. (5) How to test this hypothesis empirically remains a puzzle.

Technical intelligence

Byrne (1997, 1999, 2000; Stokes & Byrne 2001) argued that what sets great apes apart from other haplorhines are numerous "technical" problems exacerbated by their exceptionally large body size. Significant among them for their cognitive challenge are foraging, ranging, arboreal locomotion, and nest building. Large size aggravates foraging problems for great apes, so they need greater foraging efficiency and rely more heavily on high-quality, physically defended foods (e.g., embedded). Large size probably also increases the difficulty of ranging, arboreal travel (per Povinelli & Cant 1995), and finding secure sleeping sites. Ancestral hominids would have faced similar selection pressures for improved foraging, aggravated by large body size, slow and inefficient locomotion ("brachiation"), and dietary constraints (unspecialized guts, no cheek pouches). These pressures favored solutions of greater complexity and efficiency. The unique evolutionary solution of the hominids was to organize voluntary behavior hierarchically. Cognitively, hierarchization involves reorganizing and refining cognitive structures into multi-leveled programs. It brings abilities like mental representation, planning and insight to cognition and increased speed and efficiency to behavior. It affects cognition generally, so it could have evolved in response to any of these problems. Payoffs are most evident in foraging-related activities but because hierarchization is generalized, it brings matching payoffs to social cognition such as understanding social partners as active agents with intentions.

This hypothesis accounts for the complex "technical" skills unique to the great apes and for the cognitive difficulties that even gorillas, the most folivorous great apes, face in foraging. In proposing cognitive advances that were generalized, it provides an explanation for cognitive enhancements across domains, as products of this overall increase in cognitive power. Others also single out cognitive hierarchization (e.g., Gibson 1990, 1993; Matsuzawa 2001; Russon 1998), which has been shown in food processing techniques in chimpanzees, gorillas, and orangutans (Byrne & Byrne 1991; Byrne *et al.* 2001; Russon 1998; Stokes & Byrne 2001) but not vervets (Harrison 1996). This hypothesis may, however, invite the same criticism launched at the social intelligence hypothesis: enhancements to social cognition are thereby relegated to automatic side effects, which sits poorly with the obvious advantages they provide and ignores the possibility of adaptive advances to social cognition. Also not incorporated are several factors known to affect great ape cognition (e.g., prolonged ontogeny) and the interplay among critical factors (e.g., effects of large body size on sociality, interactions between technical and social pressures).

Arboreal foraging

Russon (1998) reconsidered the suite of selection pressures proposed to have shaped great ape cognitive evolution – large size, difficult diet, prolonged ontogeny, arboreal travel – then revisited existing reconstructions. The main revision concerned arboreality. Arboreal travel has been advocated as a cognitive selection pressure, but arboreal foraging may be more important. Arboreality clearly complicates the cognitive challenges of obtaining difficult foods, at least in orangutans and chimpanzees (Russon 1999b; Stokes & Byrne 2001). Arboreal foraging, as a hypothesis, blends and extends technical intelligence and apprenticeship models: "technical" difficulties associated with a difficult dietary niche, large body size, and prolonged ontogeny all imposed cognitive selection pressures on ancestral great apes; and arboreality exacerbated foraging pressures. It argues for centralized hierarchization as a key underpinning for great ape cognition and for development is a critical factor in moderating ecological pressures and cognitive capabilities. What is valuable here is the attempt to generate a reconstruction that integrates the suite of plausible selection pressures, all proposed cognitive advances, and current evidence. Like the technical intelligence hypothesis, however, arboreal foraging suffers from relegating advances in social cognition to side-effects.

DISCUSSION

This overview emphasizes the need to revise reconstructions of great ape cognitive evolution. With evidence and opinion increasingly recognizing a distinct "great ape" cognition, reconstructing cognitive evolution in the primates, from the whole of the order to modern humans, first and foremost requires the incorporation of more accurate representations of great ape cognition. In particular, many existing reconstructions have not

differentiated the great apes from other haplorhine primates so they have underestimated great ape capacities, especially for symbolic processes. Recognizing primitive symbolism as the province of the hominids obligates substantial revisions of reconstructions of human cognitive evolution. Reconstructions of great ape cognitive evolution suffer similar problems, typically owing to considering sets of problem-specific cognitive abilities that fall short of representing the full cognitive breadth and complexity that great apes express. The few models that could account for great apes' full range of cognitive advances do so by proposing the emergence of generalized processes, such as mental representation or hierarchization, that enhanced cognitive capacities across the board.

Concerning selection pressures, many of those currently proposed to have favored evolutionary enhancements to primate cognition are not unique to hominids. Other primates have societies as complex, diets as diverse, seasonal, patchy, or embedded, and ranges as large, terrestrial, or arboreal as great apes, and great apes themselves vary on most of these. Explaining a unique great ape cognition requires at least one selection pressure on cognition, or an interaction among several pressures, that uniquely affected their common ancestor. In that context, ecological pressures currently appear to be the most likely to have shaped great ape cognitive evolution although social pressures may yet be shown to have had an important influence. Most of the plausible pressures are in any case interrelated, making it likely that a set of pressures, interacting or acting in sequence, shaped their cognitive enhancements.

If no existing reconstruction meets current standards, all help show the way forward. More accurate and complete portrayals of great apes are needed in virtually every facet of this exercise: modern great ape cognition, modern great ape adaptation, and great ape evolutionary history. We need accurate characterizations of the capacities and processes that distinguish great ape cognition from that of other nonhuman primates, of modern humans and, to the degree possible, of ancestral hominins. We need better understanding of modern great ape adaptation, especially the biological substrate that supports their cognition (e.g., the brain, life histories) and the social and ecological challenges to their cognition (e.g., diet, locomotion, habitat), as bases for establishing what roles their advanced cognitive capabilities play. In some cases, evidence on modern great ape adaptation is the only available basis for inferring shared ancestral traits and conditions. Finally, we need better representations of great apes' evolutionary history, both the traits of ancestral hominids and the conditions in which they lived – representations that are especially difficult to construct, given the incomplete evidence available.

Even with the incomplete material that has been woven into evolutionary scenarios, the difficulty of incorporating all the factors likely to be relevant and of representing the balance among them is increasingly evident. Accurate reconstruction may well require unraveling the effects of multiple pressures, including identifying which were fundamental and which represent compensations, determining which operated as constraints and which opened adaptive opportunities, and establishing the sequence of pressures and cascading effects. It remains to be seen whether the evolutionary record will eventually yield answers to these questions.

THE CURRENT VOLUME

This volume works toward developing the most comprehensive reconstruction of great ape cognitive evolution possible today, by assembling and integrating opinion from experts in each of the disciplines with evidence to offer on this issue – specialists in great ape cognition, behavior, ecology, sociality, and anatomy as well as paleontologists expert in the study of corresponding ancestral hominid traits. Contributors were asked to discuss their area of expertise with attention to implications for great ape cognition and its evolution.

We used existing reconstructions of primate cognitive evolution to guide our choice of topics. These suggest a variety of abilities that may represent cognitive adaptations along with modern and phylogenetic features that may underpin variation in cognitive capacities within the primate order (e.g., diet, range size, social complexity, terrestriality–arboreality). Several of these features, singly or in concert, assume distinct qualities in the hominids and so could underlie distinct forms of cognition in that clade. Whether any of these abilities constitute cognitive adaptations and whether any of these features qualifies as a direct pressure favoring evolutionary enhancements to cognition, all are useful in suggesting the major dimensions along which the hominids are distinct from other haplorhine primates that may somehow be tied to their cognitive capacities.

Correspondingly, we organized this volume into three parts, which address (1) what distinguishes great ape cognition, (2) what features of behavior, anatomy, sociality, and ecology characterize living great apes as a clade and show strong links to their cognition and (3) the corresponding conditions in the common ancestral hominid. The first part offers an overview of the cognitive capacities that characterize modern great apes and distinguish them from other nonhuman primates, to establish what intellectual phenomena may require evolutionary explanations different from those that apply to all haplorhine primates. The second and third parts assemble and assess evidence on ecological, social, behavioral, and anatomical factors linked with these distinctive cognitive phenomena in living and ancestral large hominids. Our aims are first, to assess whether the factors proposed could be linked with enhanced cognition in the ways portrayed by existing reconstructions, and second, to explore other factors and/or interactions among them that may have contributed to that cognition. Our final chapter attempts to integrate this material into a coherent, overall picture of the evolutionary origins of great ape cognition.

REFERENCES

Aiello, L. & Wheeler, P. (1995). The expensive tissue hypothesis. *Current Anthropology*, **36**, 199–221.

Armstrong, E. (1983). Metabolism and relative brain size. *Science*, **220**, 1302–4.

Bard, K. A. (1993). Cognitive competence underlying tool use in free-ranging orang-utans. In *The Use of Tools by Human and Non-Human Primates*, ed. A. Berthelet & J. Chavaillon, pp. 103–13. Oxford: Clarendon Press.

Barton, R. A. (1996). Neocortex size and behavioral ecology in primates. *Proceedings of the Royal Society of London, Series B*, **263**, 173–7.

Barton, R. A. & Dunbar, R. I. M. (1997). Evolution of the social brain. In *Machiavellian Intelligence II: Extensions and Evaluations*, ed. A. Whiten & R. W. Byrne, pp. 240–63. Cambridge, UK: Cambridge University Press.

Begun, D. R. (1992). Miocene fossil hominoids and the chimp–human clade. *Science*, **257**(2058), 1929–33.

Boesch, C. (1996). Three approaches for assessing chimpanzee culture. In *Reaching into Thought: The Minds of the Great Apes*, ed. A. E. Russon, K. A. Bard &

S. T. Parker, pp. 404–29. Cambridge, UK: Cambridge University Press.

Boesch, C. & Boesch-Achermann, H. (2000). *The Chimpanzees of the Taï Forest: Behavioural Ecology and Evolution.* Oxford: Oxford University Press.

Byrne, R. W. (1995). *The Thinking Ape: Evolutionary Origins of Intellect in Monkeys, Apes, and Humans.* Oxford: Oxford University Press.

(1997). The technical intelligence hypothesis: an alternative evolutionary stimulus to intelligence? In *Machiavellian Intelligence II: Extensions and Evaluations,* ed. R. W. Byrne & A. Whiten, pp. 289–311. Cambridge, UK: Cambridge University Press.

(1999). Human cognitive evolution. In *The Descent of Mind: Psychological Perspectives on Hominid Evolution,* ed. M. C. Corballis & S. E. G Lea, pp. 71–87. Oxford: Oxford University Press.

(2000). Evolution of primate cognition. *Cognitive Science,* 24(3), 543–70.

Byrne, R. W. & Byrne, J. M. E. (1991). Hand preferences in the skilled gathering tasks of mountain gorillas (*Gorilla gorilla beringei*). *Cortex,* 27, 521–46.

(1993). Complex leaf-gathering skills of mountain gorillas (*Gorilla g. beringei*): variability and standardization. *American Journal of Primatology,* 31, 241–61.

Byrne, R. W. & Russon, A. E. (1998). Learning by imitation: a hierarchical approach. *Behavioral and Brain Sciences,* 21, 667–721.

Byrne, R. W. & Whiten, A. (ed.) (1988). *Machiavellian Intelligence: Social Expertise and the Evolution of Intellect in Monkeys, Apes, and Humans.* Oxford: Clarendon Press.

(1997). Machiavellian intelligence. In *Machiavellian Intelligence II: Extensions and Evaluations,* ed. A. Whiten & R. W. Byrne, pp. 1–23. Cambridge, UK: Cambridge University Press.

Byrne, R. W., Corp, N. & Byrne, J. M. E. (2001). Estimating the complexity of animal behaviour: how mountain gorillas eat thistles. *Behaviour,* 138, 525–57.

Chapman, C. A. & Wrangham, R. W. (1993). Range use of forest chimpanzees of Kibale: implications for the understanding of chimpanzee social organization. *American Journal of Primatology,* 31(4), 263–73.

Cheney, D. L. & Seyfarth, R. M. (1990). *How Monkeys See the World: Inside the Mind of Another Species.* Chicago, IL: University of Chicago Press.

Chevalier-Skolnikoff, S., Galdikas, B. M. F. & Skolnikoff, A. (1982). The adaptive significance of higher

intelligence in wild orangutans: a preliminary report. *Journal of Human Evolution,* 11, 639–52.

Clutton-Brock, T. H. & Harvey, P. H. (1980). Primates, brains and ecology. *Journal of the Zoological Society of London,* 190, 309–21.

Cosmides, L. & Tooby, J. (1992). Cognitive adaptations for social exchange. In *The Adapted Mind: Evolutionary Psychology and the Generation of Culture,* ed. J. Barkow, L. Cosmides & J. Tooby, pp. 163-228. Oxford: Oxford University Press.

de Waal, F. B. M. (1996). *Good Natured: The Origin of Right and Wrong in Humans and Other Animals.* Cambridge, MA: Harvard University Press.

de Waal, F. B. M. & Aureli, F. (1996). Consolation, reconciliation, and a possible difference between macaques and chimpanzees. In *Reaching into Thought: The Minds of the Great Apes,* ed. A. E. Russon, K. A. Bard & S. T. Parker, pp. 80–110. Cambridge, UK: Cambridge University Press.

Donald, M. W. (1991). *The Origins of the Modern Mind: Three Stages in the Evolution of Culture and Cognition.* Cambridge, MA: Harvard University Press.

(1993). Précis of *Origins of the Modern Mind: Three Stages in the Evolution of Culture and Cognition. Behavioral and Brain Sciences,* 16, 737–91.

(2000). The central role of culture in cognitive evolution: a reflection on the myth of the "isolated mind". In *Culture, Thought, and Development,* ed. L. P. Nucci, G. B. Saxe & E. Turiel, pp. 19–38. Mahwah, NJ: Lawrence Erlbaum Publishers.

Dunbar, R. I. M. (1992). Neocortex size as a constraint on group size in primates. *Journal of Human Evolution,* 20, 469–93.

(1993). Coevolution of neocortical size, group size and language in humans. *Behavioral and Brain Sciences,* 16(4), 681–735.

(1995). Neocortex size and group size in primates: a test of the hypothesis. *Journal of Human Evolution,* 28, 287–96.

(1998). The social brain hypothesis. *Evolutionary Anthropology,* 6(5), 178–90.

Galdikas, B. M. F. (1978). Orangutans in hominid evolution. In *Spectrum: Essays Presented to Sutan Takdir Alisjahbana on his Seventieth Birthday,* ed. S. Udin, pp. 287–309. Jakarta: Dian Rakyat.

Gibson, K. R. (1986). Cognition, brain size, and the extraction of embedded food resources. In *Primate Ontogeny, Cognition, and Social Behaviour,* ed.

J. G. Else & P. C. Lee, pp. 93–103. Cambridge, UK: Cambridge University Press.

(1990). New perspectives on instincts and intelligence: brain size and the emergence of hierarchical mental constructional skills. In *"Language" and Intelligence in Monkeys and Apes: Comparative Developmental Perspectives*, ed. S. T. Parker & K. R. Gibson, pp. 97–128. New York: Cambridge University Press.

(1993). Tool use, language and social behavior in relationship to information processing capacities. In *Tools, Language and Cognition in Human Evolution*, ed. K. R. Gibson & T. Ingold, pp. 251–69. Cambridge, UK: Cambridge University Press.

Goodall, J. (1986). *The Chimpanzees of Gombe: Patterns of Behavior*. Cambridge, MA: Harvard University Press.

Harrison, K. E. (1996). Skills used in food processing by vervet monkeys, *Cercopithecus aethiops*. Unpublished doctoral dissertation, University of St. Andrews, St. Andrews, Scotland.

Harrison, T. & Rook. L (1997). Enigmatic anthropoid or misunderstood ape? The phylogenetic status of *Oreopithecus bambolii* reconsidered. In *Function, Phylogeny, and Fossils: Miocene Hominoid Evolution and Adaptations*, ed. D. R. Begun, C. V. Ward & M. D. Rose, pp. 326–62. New York: Plenum Press.

Hladik, C. M. (1975). Ecology, diet, and social patterning in Old and New World primates. In *Socioecology and Psychology of Primates*, ed. R. H. Tuttle, pp. 3–35. The Hague: Mouton Publishers.

(1977). Chimpanzees of Gabon and chimpanzees of Gombe: some comparative data on the diet. In *Primate Ecology: Studies of Feeding and Ranging Behaviour in Lemurs, Monkeys, and Apes*, ed. T. H. Clutton-Brock, pp. 481–501. New York: Academic Press.

Humphrey, N. K. (1976). The social function of intellect. In *Growing Points in Ethology*, ed. P. P. G. Bateson & R. A. Hinde, pp. 303–17. Cambridge, UK: Cambridge University Press.

Ingmanson, E. (1996). Tool-using behavior in wild *Pan paniscus*: social and ecological considerations. In *Reaching into Thought: The Minds of the Great Apes*, ed. A. E. Russon, K. A. Bard & S. T. Parker, pp. 190–210. Cambridge, UK: Cambridge University Press.

Jolly, A. (1966). Lemur social behaviour and primate intelligence. *Science*, 153, 501–6.

Karmiloff-Smith, A. (1992). *Beyond Modularity: A Developmental Perspective on Cognitive Science*. Cambridge, MA: MIT Press.

Krebs, J. R. & Dawkins, R. (1984). Animal signals: mind reading and manipulation. In *Behavioral Ecology: An Evolutionary Approach*, 2nd edn., ed. J. R. Krebs & N. B. Davies, pp. 340–401. Oxford: Blackwell Scientific.

Kudo, H. & Dunbar, R. I. M. (2001). Neocortex size and social network size in primates. *Animal Behaviour*, **62**(4), 711–22.

Kummer, H. (1967). Tripartite relations in Hamadryas baboons. In *Social Communication among Primates*, ed. S. A. Altmann, pp. 63–72. Chicago, IL: University of Chicago Press.

Langer, J. & Killen, M. (ed.) (1998). *Piaget, Evolution, and Development*. Mawhah NJ: Lawrence Erlbaum Associates.

MacKinnon, J. (1974). The behaviour and ecology of wild orangutans (*Pongo pygmaeus*). *Animal Behaviour*, **22**, 3–74.

Martin, L. & Andrews, P. (1993). Renaissance of Europe's ape. *Nature*, **365**(7), 494.

Matsuzawa, T. (1996). Chimpanzee intelligence in nature and in captivity: isomorphism of symbol use and tool use. In *Great Ape Societies*, ed. W. C. McGrew, L. F. Marchant & T. Nishida, pp. 296–209. Cambridge, UK: Cambridge University Press.

(2001). Primate foundations of human intelligence: a view of tool use in nonhuman primates and fossil hominids. In *Primate Origins of Human Cognition and Behavior*, ed. T. Matsuzawa, pp. 3–25. Tokyo: Springer.

Menzel, E. W. Jr. (1978). Cognitive mapping in chimpanzees. In *Cognitive Processes in Animal Behavior*, ed. S. H. Hulce & W. K. Honig, pp. 375–422. Hillsdale, NJ: Erlbaum.

Menzel, E. W. Jr. & Juno, C. (1985). Social foraging in marmoset monkeys and the question of intelligence. In *Animal Intelligence*, ed. L. Weiskrantz, pp. 145–58. Oxford: Clarendon Press.

Milton, K. (1981). Distribution patterns of tropical plant foods as an evolutionary stimulus to primate mental development. *American Anthropologist*, **83**, 534–58.

(1988). Foraging behaviour and the evolution of primate intelligence. In *Machiavellian Intelligence*, ed. R. W. Byrne & A. Whiten, pp. 285–305. Oxford: Clarendon Press.

Mitchell, R. W. (1994). Primate cognition: simulation, self-knowledge, and knowledge of other minds. In *Hominid Culture in Primate Perspective*, ed. D. Quiatt & J. Itani, pp. 177–232. Nivet, CO: University of Colorado Press.

Mitchell, R. W. & Miles, H. L. (1993). Apes have mimetic culture. *Behavioral and Brain Sciences*, **16**(4), 768.

Mithen, S. (1996). *The Prehistory of the Mind*. London: Thames & Hudson.

Moyà Solà, S. & Köhler, M. (1993). Recent discoveries of *Dryopithecus* shed new light on evolution of great apes. *Nature*, **365**(7), 543–5.

Parker, C. E. (1978). Opportunism and the rise of intelligence. *Journal of Human Evolution*, **7**, 596–608.

Parker, S. T. (1996). Apprenticeship in tool-mediated extractive foraging: The origins of imitation, teaching and self-awareness in great apes. In *Reaching into Thought*, ed. A. E. Russon, K. A. Bard & S. T. Parker, pp. 348–70. Cambridge, UK: Cambridge University Press.

Parker, S. T. & Gibson, K. R. (1977). Object manipulation, tool use and sensorimotor intelligence as feeding adaptations in cebus monkeys and great apes. *Journal of Human Evolution*, **6**, 623–41.

(1979). A developmental model of the evolution of language and intelligence in early hominids. *Behavioral and Brain Sciences*, **2**, 364–408.

Parker, S. T. & McKinney, M. L. (1999). *Origins of Intelligence: The Evolution of Cognitive Development in Monkeys, Apes, and Humans*. Baltimore, MD: Johns Hopkins.

Parker, S. T. & Potí, P. (1990). The role of innate motor patterns in ontogenetic and experiential development of intelligent use of sticks in cebus monkeys. In *"Language" and Intelligence in Monkeys and Apes*, ed. S. T. Parker & K. R. Gibson, pp. 219–46. New York: Cambridge University Press.

Parker, S. T., Mitchell, R. W. & Boccia, M. L. (ed.) (1994). *Self-Awareness in Animals and Humans*. New York: Cambridge University Press.

Parker, S. T., Mitchell, R. W. & Miles, H. L. (ed.) (1999). *Mentalities of Gorillas and Orangutans in Comparative Perspective*. Cambridge, UK: Cambridge University Press.

Peters, H. H. (2001). Tool use to modify calls by wild orang-utans. *Folia Primatologica*, **72**(4), 242–4.

Pilbeam, D. (1996). Genetic and morphological records of the Hominoidea and hominid origins: a synthesis. *Molecular Phylogenetics and Evolution*, **5**(1): 155–68.

Povinelli, D. J. & Cant, J. G. H. (1995). Arboreal clambering and the evolution of self-conception. *Quarterly Review of Biology*, **70**(4), 393–421.

Rogoff, B. (1992). *Apprenticeship in Thinking*. New York: Oxford University Press.

Russon A. E. (1998). The nature and evolution of intelligence in orangutans (*Pongo pygmaeus*). *Primates*, **39**, 485–503.

(1999a). Orangutans' imitation of tool use: A cognitive interpretation. In *The Mentalities of Gorillas and Orangutans*, ed. S. T. Parker, R. W. Mitchell, & H. L. Miles, pp. 119–145. Cambridge, UK: Cambridge University Press.

(1999b). Naturalistic approaches to orangutan intelligence and the question of enculturation. *International Journal of Comparative Psychology*, **12**(4), 181–202.

(2003). Developmental perspectives on great ape traditions. In *Towards a Biology of Traditions: Models and Evidence*, ed. D. Fragaszy & S. Perry. Cambridge, UK: Cambridge University Press (in press).

Russon, A. E., Bard, K. A. & Parker, S. T. (ed.) (1996). *Reaching into Thought: The Minds of the Great Apes*. Cambridge, UK: Cambridge University Press.

Savage-Rumbaugh, S., Shanker, S. G. & Taylor, T. J. (1998). *Apes, Language, and the Human Mind*. New York: Oxford University Press.

Savage-Rumbaugh, E. S., Williams, S. L., Furuichi, T. & Kano, T. (1996). Language perceived: *paniscus* branches out. In *Great Ape Societies*, ed. W. C. McGrew, L. F. Marchant & T. Nishida, pp. 173–84. Cambridge, UK: Cambridge University Press.

Schwartz, B. L. & Evans, S. (2001). Episodic memory in primates. *American Journal of Primatology*, **55**, 71–85.

Stokes, E. J. (1999). Feeding skills and the effect of injury on wild chimpanzees. Unpublished PhD dissertation, University of St. Andrews, St. Andrews, Scotland.

Stokes, E. J. & Byrne, R. W. (2001). Cognitive abilities for behavioral flexibility in wild chimpanzees (*Pan troglodytes*): the effect of snare injury on complex manual food processing. *Animal Cognition*, **4**, 11–28.

Suddendorf, T. & Whiten, A. (2001). Mental evolution and development: evidence for secondary representation in children, great apes, and other animals. *Psychological Bulletin*, **127**(5), 629–50.

Tomasello, M. & Call, J. (1997). *Primate Cognition*. New York: Oxford University Press.

Tuttle, R. H. & Cortright, G. W. (1988). Positional behaviour, adaptive complexes, and evolution. In *Orangutan Biology*, ed. J. H. Schwartz, pp. 311–30. Oxford: Oxford University Press.

Yamakoshi, G. & Sugiyama, Y. (1995). Pestle-pounding behaviour of wild chimpanzees at Bossou, Guinea: a newly observed tool-using behaviour. *Primates*, **36**, 489–500.

van Schaik, C. P. (1983). Why are diurnal primates living in groups? *Behaviour*, **87**, 120–44.

van Schaik, C. P. & Knott, C. (2001). Geographic variation in tool use in *Neesia* fruits in orangutans. *American Journal of Physical Anthropology*, **114**, 331–42.

van Schaik, C. P., Ancrenaz, M., Borgen, G., Galdikas, B., Knott, C. D., Singleton, I., Suzuki, A., Utami, S. S., Merrill, M. (2003). Orangutan cultures and the evolution of material culture. *Science*, **299**, 102–5.

van Schaik, C. P., Deaner, R. O. & Merrill, M. M. (1999). The conditions for tool use in primates: implications for the evolution of material culture. *Journal of Human Evolution*, **36**, 719–41.

van Schaik, C. P., Fox, E. A. & Sitompul, A. F. (1996). Manufacture and use of tools in wild Sumatran orangutans: implications for human evolution. *Naturwissenschaften*, **83**(4), 186–8.

Whiten, A. & Byrne, R. W. (1991). The emergence of metarepresentation in human ontogeny and primate phylogeny. In *Natural Theories of Mind*, ed. A. Whiten, pp. 267–82. Oxford: Basil Blackwell.

Whiten, A., Goodall, J., McGrew, W. C., Nishida, T., Reynolds, V., Sugiyama, Y., Tutin, C. E. G., Wrangham, R. W. & Boesch, C. (1999). Culture in chimpanzees. *Nature*, **399**, 682–5.

Wrangham, R. W. (1977). Feeding behaviour of chimpanzees in Gombe National Park, Tanzania. In *Primate Ecology: Studies of Feeding and Ranging Behaviour in Lemurs, Monkeys, and Apes*, ed. T. H. Clutton-Brock, pp. 508–38. London: Academic Press.

Wrangham, R. W., Conklin, N. L., Etot, G., Obua, J., Hunt, K. D., Hauser, M. D. & Clark, A. P. (1993). The value of figs to chimpanzees. *International Journal of Primatology*, **14**(2), 243–56.

Wrangham, R. W., McGrew, W. C., de Waal, F. B. M. & Heltne, P. G. (ed.) (1994). *Chimpanzee Cultures*. Cambridge, MA: Harvard University Press.

Yamagiwa, J. (1999). Socioecological factors influencing population structure of gorillas and chimpanzees. *Primates*, **40**, 87–104.

2 · Enhanced cognitive capacity as a contingent fact of hominid phylogeny

DAVID R. BEGUN
Department of Anthropology, University of Toronto, Toronto

INTRODUCTION

Relations among living great apes and humans have been worked out in recent years to the satisfaction of most researchers, from both the molecular and the morphological/fossil approaches (Begun 1999; Begun, Ward & Rose 1997; Page & Goodman 2001; Satta, Klein & Takahata 2000; Shoshani *et al*. 1996). It is now widely recognized that humans and *Pan* (chimpanzees and bonobos) are members of a clade (evolutionary lineage) to the exclusion of other living primates, and that among living apes gorillas are next most closely related, orangutans after that, and hylobatids (gibbons and siamangs) after that. The living great apes, humans, and their ancestral lineages then form a natural evolutionary group, the hominids (Table 2.1 and Figure 2.1). Where fossil hominids (Miocene to Pleistocene great apes and humans) fall within this framework is the subject of intense debate, but this question is not critical to the theme of this chapter.

Phylogenetic parsimony suggests that characteristics shared among all members of the hominid group probably evolved once in their common ancestor rather than repeatedly in the separate lineages of the Hominidae.[1] In some cases the symplesiomorphic (shared primitive) nature of characters shared among living hominids is supported by fossil evidence, such as brain size, body mass and rate of maturation (Begun & Kordos, Chapter 14, Gebo, Chapter 17, Kelley, Chapter 15, this volume). Among the shared characteristics that cannot be directly confirmed in the fossil record is great apes' distinctive intelligence. Most research, much of it discussed in this volume, converges on the conclusion that great apes are more intelligent than nonhominid primates. All great apes appear to have a more complex approach to the challenges of their environments than other primates, whether it involves complex social relations, communication, patterns of learning/teaching, or elaborate and/or strategic patterns of foraging (in this volume, see Bryne, Chapter 3, Parker, Chapter 4, Russon, Chapter 6, van Schaik, Preuschoft & Watts, Chapter 11, Yamagiwa, Chapter 12, Yamakoshi, Chapter 9). They are also known to outperform all other primates on a variety of cognitive tasks, in settings ranging from "semi-natural" to completely artificial (see many of the contributions to this volume).

Table 2.1. *A classification of hominoid genera discussed in this chapter*

Hominoidea
Proconsulidae
Proconsul
Hylobatidae
Hylobates
Hominidae
Homininae
Dryopithecus
Ouranopithecus
Pan
"Australopithecus"
Homo
Gorilla
Ponginae
Lufengpithecus
Sivapithecus
Pongo
Hominidae *incertae sedis*
Oreopithecus
Hominoidea *incertae sedis*
Morotopithecus

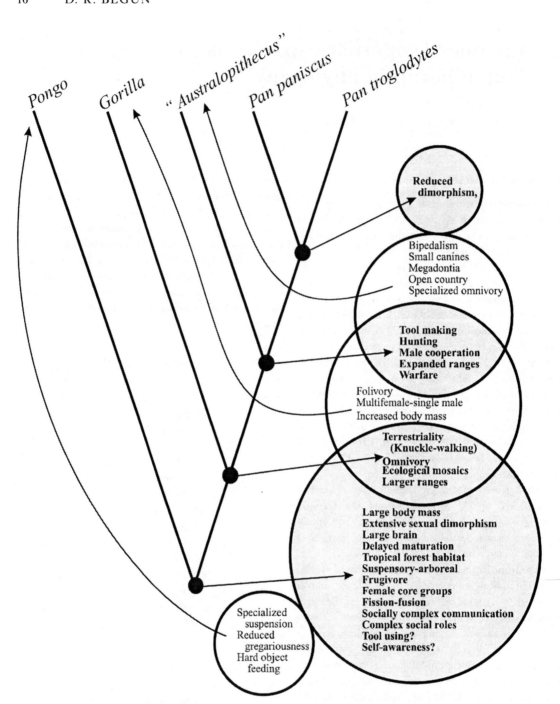

Figure 2.1. Cladogram and Venn diagram showing evolutionary relations among living hominids and "*Australopithecus*," and characters likely to have been present at specific nodes and in each terminal taxon. "*Australopithecus*" lacks many of the autapomorphies of modern humans, which facilitates comparison to other living hominids. It appears in quotes because it is probably a paraphyletic taxon in its current prevalent usage. Shaded areas enclose lists of characters of the LCA (last common ancestor) represented by the node connected to each arrow. Unshaded areas enclose lists of features characteristic of the terminal taxon to which each curved arrow leads. See text for discussion.

In this chapter I hope to outline the characters that are parsimoniously interpreted to have been present in the last common ancestor of the living Hominidae, given the pattern of relations we conclude to exist based on current lines of evidence. Detailed analyses of most of these characters are presented by other contributors to this volume. This exercise allows us to reconstruct aspects of the ancestral condition of hominid anatomy and behavior that have been related, in one way or another, to the evolution of great ape intelligence. It is difficult to test detailed aspects of this profile because the normal procedure involves the use of additional taxa to test hypotheses of character state transition sequences (the change from one character state in an ancestor to another in a descendant). There are no other extant hominid taxa to add to this analysis, and although some fossil hominids are reasonably well known, none preserves any direct evidence of their intelligence and in fact little of their biology is preserved. Some correlates of intelligence in hominids (body size, brain size, life history) can be reconstructed to some degree from fossils, and these are discussed elsewhere in this volume (Begun & Kordos, Chapter 14, Kelley, Chapter 15, Ward, Flinn & Begun, Chapter 18, this volume).

THE QUESTION OF HOMOLOGY

MacLarnon (1999) noted in the context of the reconstruction of behavioral evolution that the comparative analysis of shared features depends on the assumption that they are homologous. For the characters she is considering, she makes a strong case that the definition of homology must be "loosened" to incorporate both operational and phylogenetic homology, that is, characters that are indistinguishable functionally and morphologically (operational homology) and characters that are deduced cladistically to derive from commonality of descent (phylogenetic homology). Here, I am less concerned with the functional or behavioral significance of characters deduced to have been present in the hominid last common ancestor than in the logic of deducing their presence in the first place. For this, a requirement of phylogenetic homology is imperative, since we do not have the opportunity to operationalize homology in the past. There is no widely accepted method of establishing the presence of homology, and in cladistics homology is generally assumed a priori and falsified for specific characters by parsimony (Begun, in press). Until we have a

complete understanding of characters' states from the gene to the phenotype we will not be able to define similarities as homologies, and even then parallelism remains a possibility. In the case of hominids, for instance, we fail to falsify the hypothesis that high levels of sexual dimorphism and male–male competition are homologous in *Pongo* and *Gorilla*, but we do falsify the hypothesis that very low levels of sexual dimorphism and complex and intricate sexual social relations are homologous in bonobos and humans. As we learn more about the biology of behavior in great apes, and more about the fossil record of great apes and humans, we will be able to falsify more hypotheses of hominid behavioral homology and in so doing develop a more precise account of hominid behavioral evolution, that is, a more precise mapping of behavioral characteristics that are inherited from a common ancestor versus those that emerge inevitably from the interaction of phylogeny, ecology, and behavior.

A PROFILE OF THE LAST COMMON ANCESTOR (LCA) OF LIVING HOMINIDS

On the basis of the shared presence of characters among living hominids, and with some reference to better-known fossil hominids, we can propose a broadly defined hypothetical ancestral behavior and morphology of the LCA of living hominids (Figure 2.1).

Large body mass

All living great apes, and most if not all fossil great apes, have body masses exceeding 25 kg, which is large by primate standards (Smith & Jungers 1997; Begun & Kordos, Chapter 14, Gebo, Chapter 17, Potts, Chapter 13, Ward *et al.*, Chapter 18, this volume). The smallest living great apes are female *Pongo*, female *Pan troglodytes schweinfurthii*, and female *Pan paniscus*, all with mean body masses between about 33 and 36 kg, and minima around 26 kg (Smith & Jungers 1997; and author's notes from museum records). Even the minimum mass of 26 kg is larger than that of almost all other living primates, and is routinely reached or exceeded only by the largest male papionins (a tribe within the Cercopithecidae), *Papio anubis*, *Papio ursinus*, and *Mandrillus sphinx* (Smith & Jungers 1997). In no case, even among these largest of nonhominid primates,

18   D. R. BEGUN

do male *mean* body masses equal or surpass body mass means of the smallest female hominids.

Among fossil hominids, the smallest *Dryopithecus* (female *D. laietanus* and *D. brancoi*) were likely to have been smaller on average than the smallest living hominids, with the smallest females possibly weighing only about 20 kg (Begun & Kordos, Chapter 14, this volume; author's data). The smallest *Sivapithecus*, female *S. punjabicus*, may have been similar in body mass to the smallest living hominids, based on dental dimensions, although if this species was megadont (relatively large-toothed), it may have been closer to *Dryopithecus* in body mass. This is a definite possibility, given its dentognathic morphology (Singleton, Chapter 16, this volume). *Oreopithecus* is estimated to have been about 32 kg, close in body mass to the smallest female hominid means, but the specimen on which this estimate is based (IGF 11778, the famous "Florence skeleton" found by J. Hürzeler in 1958) is a male (Jungers 1987). Female *Oreopithecus* are estimated at about half that body mass. Even these comparatively small fossil hominids are larger than the majority of living primates. Finally, other clearly hominid taxa such as *Ouranopithecus* and *Lufengpithecus* are in the size range of large chimpanzees and small gorillas (de Bonis & Koufos 1994, 1997; Schwartz 1990, 1997; author's personal observations), as is *Morotopithecus*, though this taxon is less clearly a great ape (Gebo, Chapter 17, this volume; Gebo *et al.* 1997; MacLatchy *et al.* 2000; Pilbeam 1969). So, the ancestor of living hominids was almost certainly large compared with most primates.

Among hominoids more generally, both fossil and living, hylobatids (especially gibbons) are unusually small, suggesting that they may be phyletic dwarfs, that is, secondarily reduced in size compared with the ancestor they share with other hominoids. The living sister clade to the Hominidae is thus probably autapomorphic (uniquely derived) in body mass, and not representative of the primitive condition in hominids. A number of authors have suggested that body mass in hominids may be causally related to increased intelligence, either as a stimulus of selection for greater levels of self-awareness needed for safe arboreal locomotion (e.g., Gebo, Chapter 17, Hunt, Chapter 10, this volume) or as part of a broader phenomenon that led incidentally to the evolution of larger brains (e.g., Kelley, Chapter 15, Ross, Chapter 8, this volume). Ancestral conditions concerning body mass are then of considerable importance

in reconstructing the evolutionary origins of hominid cognition.

Strong sexual dimorphism

All hominids, including humans, are sexually dimorphic in body mass and certain aspects of skeletal morphology. However, living humans and bonobos are unusual among hominids in having low levels of sexual dimorphism. Even chimpanzees, which are about twice as dimorphic in body mass as humans, have low sexual dimorphism compared with most other hominids, fossil and living (Morbeck & Zihlman 1989; Zihlman 1984; Zihlman & Cramer 1978). Fossil humans (pre-*Homo*), *Gorilla*, and *Pongo* are all strongly sexually dimorphic in body mass and cranial anatomy, though fossil humans are less dimorphic in dental morphology (canine size and shape) than all great apes including all *Pan* (Jungers 1988; Kelley 1995a; Lockwood *et al.* 1996, 2000; McHenry 1988, 1992; Smith & Jungers 1997). All fossil great apes are also strongly sexually dimorphic (Begun 1994, 2002; Kelley 1995b, 1997, 2002; Kelley & Etler 1989; Kelley & Pilbeam 1986; Kelley & Qinghua 1991). Given this distribution, it is most likely that substantial (close to 2 : 1 male to female mean body mass) sexual dimorphism characterized the hominid LCA. Reduced sexual dimorphism, mainly in body mass, is autapomorphic in *Pan* and reduced dimorphism in skeletal anatomy is autapomorphic and homoplastic (acquired independently) in *Homo* and *Pan paniscus*. Although it is doubtful that the degree of sexual dimorphism has any direct relationship to intelligence, it may well be related to factors that are potentially related to the evolution of great ape intelligence, notably body mass and aspects of social and feeding behaviors (Bean 1999; Plavcan 1999; Rodman 1984; Temerin, Wheatley & Rodman 1984; and see Parker, Chapter 4, van Schaik *et al.*, Chapter 11, Yamagiwa, Chapter 12, this volume).

Large brain size

All living hominids have large brains in absolute dimensions and in some measures of relative size (Begun & Kordos, Chapter 14, this volume; Falk 1980, 1983, 1987; Hartwig-Scherer 1993; Harvey 1988; Holloway 1983, 1995; Kappelman 1996; Kordos & Begun 1998; Martin 1981, 1983, 1990; Martin & Harvey 1985; Potts, Chapter 13, this volume; Schultz 1936, 1941; Tobias 1971, 1975,

1983, 1995; Walker *et al.* 1983). Normalizing brain mass (controlling for overall body size) has proven to be a complex and vexing enterprise, with the result that different methods of accounting for the effects of overall body mass produce different results regarding measures of relative brain size (see Begun & Kordos, Chapter 14, this volume). However, it remains quite clear that no non-hominid primate of any size, including extremely large fossil cercopithecines well within the body mass range of living hominids, has a brain size even approaching that of any living hominid (Begun & Kordos, Chapter 14, this volume; Elton, Bishop & Wood 2001; Kordos & Begun 2001a; Martin 1993). It is probably the case that the exceptionally large body mass of all hominids compared with other primates obscures the truly large size of their brains in most normalization procedures.

In contrast, hylobatids have a brain mass that approximates that of other anthropoids of similar size (see references above). The encephalization quotient values (EQ, the most common method of normalizing for body mass differences) of the smallest hylobatids, gibbons, are the same as those of many hominids and higher than those of the largest hominids, male gorillas and orangutans. This is most likely due to the exceptionally large size of hominids and the secondarily reduced size of hylobatids. Gibbons have more hominid-like EQ values, while siamangs of twice the body mass have average monkey EQ values (Begun & Kordos, Chapter 14, this volume).

Direct evidence of large brain size in fossil hominids comes from two specimens of *Dryopithecus*, which provide absolute and relative brain size estimates in the range of small hominids (Begun & Kordos, Chapter 14, this volume; Kordos & Begun 2001a). Indirect evidence of a large brain comes from patterns of dental development and its relationship to life history and brain size in *Sivapithecus* (Kelley, Chapter 15, 1997, this volume) and the external dimensions of the frontal bone of *Lufengpithecus* (an undescribed, well preserved juvenile specimen, author's personal observations). *Proconsul*, a primitive hominoid, is said to be more encephalized than living monkeys, but this is based on a questionable association of brain size and body mass from two different individuals (and probably two different species) (Begun & Kordos, Chapter 14, this volume; Potts, Chapter 13, this volume; Walker *et al.* 1983).

In sum, fossil and living hominids are distinct from all other primates in their brain mass, both in absolute terms and in comparison with other primates of similar body mass. The connection between brain mass and intelligence seems intuitively obvious: all mammals that are considered to be more intelligent than their close relatives have larger brains, whether these are carnivores, cetaceans, strepsirhines, or anthropoids. The connection is actually a complex one. Brains are extremely metabolically expensive and there probably have to be adaptive payoffs to maintaining large ones. High intelligence with its associated adaptability is such a benefit.

Extended or delayed maturation

Hominids show delayed maturation, characterized by a lengthening of the period of skeletal and dental maturation, a delay in the onset of menarche, a lengthening of life span, and a diversity of other factors related to life history (Kelley, Chapter 15, Ross, Chapter 8, this volume). All living hominids are distinguished from other primates by a slower rate and/or an extended duration of growth processes leading to adulthood. Hylobatids mature more rapidly, and fossil hominids (*Sivapithecus* and *Dryopithecus*) appear to have matured at rates and durations like those of living great apes, adding support to the conclusion that maturational delay was a feature of the biology of the LCA of living hominids (Kelley 1997, Chapter 15, this volume). Many have made the link between maturational delay and intelligence, whether it is related to brain growth or the duration of the infant and juvenile learning periods (Kelley, Chapter 15 this volume; Martin 1990 and references therein; Ross, Chapter 8, this volume)

Frugivorous diet

Most hominoids are described as frugivorous in the sense that fruit comprises a major portion of their diets. Frugivorous hominid diets are dominated by fruits during most of the year, but generally also include significant percentages of nonfruits (leaves, stems, shoots, buds, bark, gums, invertebrates, small vertebrates, etc.). Only siamangs and mountain gorillas rely heavily on more fibrous (folivorous) resources (Chivers 1975; Watts 1984). Other hylobatids and subspecies of *Gorilla* are known to consume large amounts of fruit (Chivers 1984; Tutin *et al.* 1997; Yamagiwa *et al.* 1992), and it is likely that folivory, though an important aspect of the adaptation of these taxa, is a relatively recent development, postdating their divergence from their closest living relatives (Smith 1999). Gorillas and *Oreopithecus* are the only hominids to have folivorous dentitions, with a predominance of shearing crests and tall, pointed cusps designed to cut herbaceous fibers. All other hominids, living and extinct, are known or thought to have been frugivorous, based on behavioral observations and jaw and molar occlusal morphology (Kay & Ungar 1997; Singleton, Chapter 16, this volume; Smith 1999; Ungar 1996; Ungar & Kay 1995).

Suspensory positional behavior

All hominoids other than humans spend significant amounts of time in the trees and all, including humans, have morphological features of the forelimbs widely interpreted as functionally related to suspensory arboreality (Larson 1988, 1996, 1998; Larson & Stern 1986; Rose 1973, 1988, 1994, 1996, 1997; Schultz 1930, 1969, 1973; Stern et al. 1976; Stern & Larson 2001). By suspensory, I mean forelimb dominated suspensory positional behavior, including locomotor behavior, as opposed to the suspensory positional behavior of some monkeys (e.g., *Alouatta*) that position themselves below branches but generally move on top of them. Other than humans, all living hominoids have elongated forelimbs compared with lower limb or trunk length, broad thoraxes, highly mobile shoulder joints, fully extendible elbows with a unique combination of stability with a wide range of motion in flexion/extension, wrist joints designed to resist torque in suspended postures, large hands and long and powerful digits (see references above). A number fossil hominids are also known to share many of these morphological features (*Dryopithecus, Oreopithecus*), though some have a curious mixture of suspensory and other characters suggestive of overall patterns of positional behavior without obvious modern analogues (*Ouranopithecus, Sivapithecus, Morotopithecus*) (MacLatchy et al. 2000; Pilbeam et al. 1990; and author's personal observations). While humans have lost a number of these characters, particularly forelimb elongation, it has been recognized for many years that we retain numerous features of our trunk anatomy, forelimb joints, and forelimb musculature that reveal our suspensory arboreal heritage (Schultz 1930, 1936, 1961). Thus, the preponderance of evidence from the fossil and neontological records indicates that the hominid LCA engaged in a suspensory form of positional behavior. Suspension is an essential part of the positional behavior characterized as "clambering" (Povinelli & Cant 1995), which has been linked by these authors to the evolution of great ape intelligence.

Forest ecology

All living hominoids other than *Homo* live exclusively in the tropics, and our genus is widely believed to have originated in the tropics. Among living hominids, *Pongo* is the only genus found almost exclusively in forest settings. African apes are mostly restricted to forests and certainly excluded from completely open environments, while hylobatids are, like *Pongo*, confined to forests (Bourliere 1985; Chivers 1980; Fleagle & Reed 1996; Fleagle, Janson & Reed 1999; Fossey 1983; Ghiglieri 1984; Goodall 1986; Kano 1992; Schaller 1963; Susman 1984; Terborgh 1992; Tuttle 1987). Many fossil hominids are associated with deposits indicative of forested and often closed forested conditions (*Dryopithecus, Lufengpithecus*), while others may have had more mixed ecological preferences, but with evidence of some dependence on forest ecology (*Sivapithecus, Ouranopithecus*) (Andrews 1992, 1996; Andrews, Begun & Zylstra 1997; Badgley et al. 1988; Bonis, Bouvrain & Koufos 1999; Guoqin 1993; Kordos 1982; Kordos & Begun 2001b; Potts, Chapter 13, this volume). Even fossil humans, traditionally placed in more open ecological settings in an attempt to explain their peculiar adaptations (e.g. bipedalism), are in fact also first found in forested settings (Reed 1997; Ward, Leakey & Walker 1999, 2001; White, Suwa & Asfaw 1994; WoldeGabriel et al. 1994; Wynn 2000). Finally, most extant primates live in forests. Thus, it is very likely that the ancestor of living hominids was also a forest dwelling taxon. A forest ecology is consistent with other aspects of the LCA as reconstructed here. Suspensory positional behavior and frugivory as described here require forests, although other known adaptations are obviously possible for hominids in forests (e.g., the more terrestrial and folivorous mountain gorilla.) Arboreality, frugivory, and the adaptability that goes along with those strategies have been related by many researchers to the evolution of enhanced intelligence in the great apes (discussed in this volume by Potts, Chapter 13, Hunt, Chapter 10, and Yamagiwa, Chapter 12).

Social organization and communication

The remaining characters listed in Figure 2.1 for the node representing the hominid LCA are less certain because they cannot be detected in the fossil record. This is because the fossil record cannot preserve them, not because they were absent. However, as these attributes tend to distinguish hominids from hylobatids and other primates, and as they are often associated with some characters discussed earlier that can be detected in the fossil record, it is legitimate to suggest that they may have characterized the hominid LCA. Strong sexual dimorphism in the hominid LCA may be correlated to intra-group male social strategies more like those of orangutans and gorillas than *Pan* and *Homo* (more direct male–male competition, less covert competition (e.g., sperm), and less apparent coalitionary behavior) (Nishida & Hiraiwa-Hasegawa 1986; Nishida & Hosaka 1996; van Schaik & van Hooff 1996; Watts 1996; Wrangham 1999; Wrangham *et al.* 1996). Fission–fusion tends to characterize all nonhuman hominids (and possibly many human hominid populations as well), although this is strongly affected by specific ecological and social factors (van Schaik *et al.*, Chapter 11, Yamagiwa, Chapter 12, this volume). Complex and flexible social roles and communication also distinguish hominids (Blake, Chapter 5, Parker, Chapter 4, van Schaik *et al.*, Chapter 11, this volume). It is not clear, however, to what extent these attributes and those involving social organization distinguish hominids from hylobatids because hylobatid social organization and communication are unusual and perhaps autapomorphic for that group. These same attributes none the less distinguish hominids from other primates, suggesting that they may well have been present in the hominid LCA.

Tool use and self awareness

Other capabilities requiring the sophisticated intelligence distinctive to the great apes were probably also present in the hominid LCA. These include tool using and some degree of self-awareness. Both have been argued to have adaptive value. Theoretical considerations and experimental evidence both suggest that self-awareness characterizes all great apes (Parker, Mitchell & Miles 1999; Povinelli & Cant 1995; Russon, Bard & Parker 1996). All great apes are clearly capable of making and using tools under experimental or semi-natural conditions although tool use is common in the wild only in chimpanzees, and in humans (Yamakoshi, Chapter 9, this volume, and references therein). Although neither is evenly distributed across living hominids, both capabilities appear in all hominids and are parsimoniously interpreted to have characterized the hominid LCA. Both are probably inevitable effects of elevated levels of intelligence overall, as suggested by their likely presence in at least one other mammal renowned for its intelligence (the dolphin) and evidence for other sophisticated intellectual abilities in all living great apes (Janik 2000; Leatherwood & Reeves 1990; Marino 1998; Parker & McKinney 1999; see various contributors to this volume).

SUBSEQUENT DEVELOPMENTS

Figure 2.1 also illustrates some of the changes that may have characterized later phases of hominid evolution. These are peripherally related to the theme of this book, but often contribute to resolving debates about ancestral conditions.

The LCA of the African apes and humans was probably a knuckle-walker, exploiting a broader range of foods and environments than its ancestor with *Pongo* or other hominines (*e.g. Dryopithecus* and *Ouranopithecus*) (Begun 1993, 1994; Richmond & Strait 2000; Richmond *et al.* 2001). Gorillas appear to have specialized in increased folivory along with associated anatomical changes, large body mass (possibly related to folivory as well), and certain aspects of their social organization. Even gorillas that are characterized as relatively frugivorous have molars that reveal their essentially folivorous adaptations (Smith 1999; Tutin & Fernandez 1993). Chimpanzees and humans are the only primates that engage in tool making and using in high frequencies in most populations, and in frequent hunting of prey of substantial body mass. Unusual patterns of cooperation, coalition, and reconciliation in chimpanzee societies distinguish them from all other hominoids except humans (de Waal 1989, 1993, 1996; van Schaik *et al.*, Chapter 11, this volume). Violence also characterizes chimpanzee and human societies and could well have existed in the societies of their LCA (Wrangham 1999). Humans are highly unusual primates in our positional behavior, environmental ubiquity, generalized diet, and skeletal anatomy. In many ways, the extent to which humans are flexible in responding to ecological challenges is an extreme expression of the capacity for higher-level intelligence already present in the hominid LCA. Finally, the genus *Pan* is the most conservative living hominid anatomically in the sense that is still closely resembles its LCA with its sister taxon (humans). *Pan* is distinguished by reduced levels of sexual dimorphism in body mass compared with other hominids except *Homo*, but this is probably homoplastic in *Pan* and *Homo* given the evidence of elevated levels of sexual dimorphism in fossil humans. *Pan paniscus* is unique in a number of anatomical and behavioral characters, some of which may also be homoplasies with similar aspects of the biology of *Homo* (Susman 1984; Zihlman 1984; Zihlman & Cramer 1978). Of the two, *Pan troglodytes* may be closer to the ancestral morphotype of the *Pan–Homo* LCA (Begun 1994).

CONCLUSIONS

The fossil record of hominid evolution and the argument from phylogenetic parsimony both suggest that most if not all characters that have been related to great ape levels of intelligence already existed in the hominid LCA. There is no way to know if this LCA was indeed as intelligent as living hominids. I suspect that it was close but not equal. Many features necessary (but perhaps not sufficient) for higher levels of intelligence are metabolically expensive (e.g., big brains) or ecologically risky (e.g., delayed maturation), so they are unlikely to exist in the absence of strong selection resulting from a significant reproductive benefit. These characters are known to have been present in fossil hominids (*Dryopithecus* and *Sivapithecus*) and it seems unlikely that they evolved numerous times independently in different hominid lineages because of the stringent conditions associated with their existence. The capacity for higher levels of intelligence was thus probably present in the hominid LCA. Aside from *Homo*, is there evidence that this capacity or its expression has been modified in the descendants of the hominid LCA? Probably.

Enhanced cognitive capacity is a contingent fact of hominid phylogeny, in that hominids are intelligent by virtue of a number of shared, primitive characters, the existence of which can only be explained by an adaptation that leads to significant increases in fitness in all hominids regardless of their specific adaptations. Hominids are in a sense obliged to be smart, to maintain the infrastructure of their intelligence. This suggests that differences in measures of intelligence in living great apes are more perceived than real. In the more than 13 million year history of the hominids it is not surprising to find that a capacity for superior intelligence would be expressed differently in different lineages, one of the facts that makes characterizing intelligence in hominids difficult. Furthermore, intelligence is probably an autocatalytic phenomenon; the more you have, the more you accumulate, and the more you need. Hominids set up social and ecological relations that require, or select for, high levels of intelligence, and it is likely that these levels have increased over the course of hominid evolutionary history in each lineage. The apparently more primitive brains of fossil hominines, in cerebral proportions but not in size, are possible evidence for changes in cerebral morphology (and intellectual capabilities) occurring independently in living hominines and pongines (*Pongo* vs. African apes) (Begun & Kordos, Chapter 14 this volume). These lines of evidence combine to suggest that while hominids evolved from a last common ancestor that was more intelligent than most living monkeys, each lineage of living hominid has since evolved a unique form of that intelligence. Current evidence on how this may have unfolded is offered in the chapters that follow.

ENDNOTE

1 It is frequently stated that parsimony is unacceptable as a justification for selecting one evolutionary hypothesis over another because evolution is not parsimonious. This is an unfortunate but common error of logic. Phylogenetic systematics (cladistics) uses parsimony as a means of choosing among alternative phylogenies, not because of an assumption that evolution is parsimonious but from the understanding that, all other things being equal, the more straightforward explanation for an observed pattern is preferable to any more complex competing explanation, the rule of logic known as Occam's Razor. Ironically, phylogenetic parsimony reveals the degree to which evolution is un-parsimonious. There is no judgment about the process of evolution. Regardless of its inherent complexity, the simplest explanation that conforms to the rules of the system is logically better than a more complex one, at least as a null hypothesis. For example, it is more parsimonious to claim that God created the heavens and the earth than to explain the existence and properties of the Universe by evoking numerous and complex principles of natural science. Scientists prefer the more complex alternative because it is consistent with the rules of the system as they have observed them. Hypotheses proposing previously unknown processes or phenomena are often needed,

and are in fact the stuff of scientific progress (gravity, evolution, quantum mechanics, the big bang, etc.), but they can only follow an exhaustive survey of competing, existing explanations (uniformitarianism).

REFERENCES

Andrews, P. (1992). Evolution and environment in the Hominoidea. *Nature*, **360**, 641–6.

(1996). Palaeoecology and hominoid palaeoenvironments. *Biological Reviews*, **71**, 257–300.

Andrews, P., Begun, D. R. & Zylstra, M. (1997). Interrelationships between functional morphology and paleoenvironments in Miocene hominoids. In *Function, Phylogeny, and Fossils: Miocene Hominoid Evolution and Adaptations*, ed. D. R. Begun, C. V. Ward & M. D. Rose, pp. 29–58. New York: Plenum Press.

Badgley, C., Guoqin, Q., Wanyong, C. & Defen, H. (1988). Paleoecology of a Miocene, tropical, upland fauna: Lufeng, China. *National Geographic Research*, **2**, 178–95.

Bean, A. (1999). Ecology of sex differences in great ape foraging. In *Comparative Primate Socioecology*, ed. P. C. Lee, pp. 339–62. Cambridge, UK: Cambridge University Press.

Begun, D. R. (1993). Knuckle walking ancestors. *Science*, **259**, 294.

(1994). Relations among the great apes and humans: new interpretations based on the fossil great ape *Dryopithecus*. *Yearbook of Physical Anthropology*, **7**, 11–63.

(1999). Hominid family values: morphological and molecular data on relations among the great apes and humans. In *The Mentalities of Gorillas and Orangutans: Comparative Perspectives*, ed. S. T. Parker, R. W. Mitchell & H. L. Miles, pp. 3–42. Cambridge, UK: Cambridge University Press.

(2002). European hominoids. In *The Primate Fossil Record*, ed. W. Hartwig, pp. 339–68. Cambridge, UK: Cambridge University Press.

(in press). How to identify (as opposed to define) a homoplasy: examples from fossil and living great apes. In *Homoplasy in Primate and Human Evolution*, ed. C. Lockwood & F. G. Fleagle. Cambridge, UK: Cambridge University Press.

Begun, D. R., Ward, C. V. & Rose, M. D. (1997). Events in hominoid evolution. In *Function, Phylogeny, and Fossils: Miocene Hominoid Origins and Adaptations*, ed.

D. R. Begun, C. V. Ward & M. D. Rose, pp. 389–415. New York: Plenum Press.

Bonis, L. de & Koufos, G. D. (1994). Our ancestor's ancestor: *Ouranopithecus* is a Greek link in human ancestry. *Evolutionary Anthropology*, **3**, 75–83.

(1997). The phylogenetic and functional implications of *Ouranopithecus macedoniensis*. In *Function, Phylogeny, and Fossils: Miocene Hominoid Origins and Adaptations*, ed. D. R. Begun, C. V. Ward & M. D. Rose, pp. 317–26. New York: Plenum Press.

Bonis, L. de, Bouvrain, G. & Koufos, G. (1999). Palaeoenvironments of late Miocene primate localities in Macedonia, Greece. In *The Evolution of Neogene Terrestrial Ecosystems in Europe*, ed. J. Agusti, L. Rook & P. Andrews, pp. 413–35. Cambridge, UK: Cambridge University Press.

Bourliere, F. (1985). Primate communities: their structure and role in tropical ecosystems. *International Journal of Primatology*, **6**, 1–26.

Chivers, D. J. (1975). Long term observations of siamang behavior. *Folia Primatologica*, **23**, 1–49.

(ed.) (1980). *Malayan Forest Primates*. New York: Plenum Press.

(1984). Feeding and ranging in gibbons: a summary. In *The Lesser Apes: Evolutionary and Behavioural Biology*, ed. H. Preuschoft, D. J. Chivers, W. Y. Brockelman & N. Creel, pp. 267–81. Edinburgh: Edinburgh University Press.

de Waal, F. B. M. (1989). *Peacemaking Among Primates*. Cambridge, MA: Harvard University Press.

(1993). Reconciliation among primates: a review of empirical evidence and unresolved issues. In *Primate Social Conflict*, ed. W. A. Mason & S. P. Mendoza, pp. 111–44. Albany, NY: State University of New York Press.

(1996). Conflict as negotiation. In *Great Ape Societies*, ed. W. C. McGrew, L. F. Marchant & T. Nishida, pp. 159–72. Cambridge, UK: Cambridge University Press.

Elton, S., Bishop, L. C. & Wood, B. (2001). Comparative context of Plio-Pleistocene hominin brain evolution. *Journal of Human Evolution*, **41**, 1–27.

Falk, D. (1980). Hominid brain evolution: the approach from paleoneurology. *Yearbook of Physical Anthropology*, **23**, 93–107.

(1983). A reconsideration of the endocast of *Proconsul africanus*: Implications for primate brain evolution. In *New Interpretations of Ape and Human Ancestry*, ed.

R. L. Ciochon & R. S. Corruccini, pp. 239–48. New York: Plenum Press.

(1987). Hominid paleoneurology. *Annual Review of Anthropology*, **16**, 13–30.

Fleagle, J. G. & Reed, K. E. (1996). Comparing primate communities: a multivariate approach. *Journal of Human Evolution*, **30**, 489–510.

Fleagle, J. G., Janson, C. H. & Reed, K. E. (ed.) (1999). *Primate Communities*. Cambridge, UK: Cambridge University Press.

Fossey, D. (1983). *Gorillas in the Mist*. Boston, MA: Houghton Mifflin.

Gebo, D. L., MacLatchy, L., Kityo, R., Deino, A., Kingston, J. & Pilbeam, D. (1997). A hominoid genus from the early Miocene of Uganda. *Science*, **276**, 401–4.

Ghiglieri, M. P. (1984). *The Chimpanzees of Kibale Forest*. New York: Columbia University Press.

Goodall, J. (1986). *The Chimpanzees of Gombe*. Cambridge, MA: Harvard University Press.

Guoqin, Q. (1993). The environment and ecology of the Lufeng hominoids. *Journal of Human Evolution*, **24**, 3–11.

Hartwig-Scherer, S. (1993). Body weight prediction in early fossil hominids: towards a taxon "independent" approach. *American Journal of Physical Anthropology*, **92**, 17–36.

Harvey, P. H. (1988). Allometric analysis and brain size. In *Intelligence and Evolutionary Biology*, ed. H. J. Jerison & I. Jerison, pp. 199–210. Berlin: Springer-Verlag.

Holloway, R. L. (1983). Human paleontological evidence relevant to language behaviour. *Human Neurobiology*, **2**, 105–14.

(1995). Toward a synthetic theory of human brain evolution. In *Origins of the Human Brain*, ed. J.-P. Changeux & J. Chavaillon, pp. 42–60. Oxford: Clarendon Press.

Janik, V. M. (2000). Whistle matching in wild bottlenose dolphins (*Tursiops truncatus*). *Science*, **289**, 1355–7.

Jungers, W. L. (1987). Body size and morphometric affinities of the appendicular skeleton in *Oreopithecus bambolii* (IGF 11778). *Journal of Human Evolution*, **16**, 445–56.

(1988). New estimates of body size in Australopithecines. In *Evolutionary History of the "Robust" Australopithecines*, ed. F. E. Grine, pp. 115–25. New York: Aldine de Gruyter.

Kano, T. (1992). *The Last Ape: Pygmy Chimpanzee Behavior and Ecology*. Stanford: Stanford University Press.

Kappelman, J. (1996). The evolution of body mass and relative brain size in fossil hominids. *Journal of Human Evolution*, **30**, 243–76.

Kay, R. F. & Ungar, P. S. (1997). Dental evidence for diet in some Miocene catarrhines with comments on the effects of phylogeny on the interpretation of adaptation. In *Function, Phylogeny and Fossils: Miocene Hominoid Evolution and Adaptations*, ed. D. R. Begun, C. V. Ward & M. D. Rose, pp. 131–51. New York: Plenum Publishing Co.

Kelley, J. (1995a). Sex determination in Miocene catarrhine primates. *American Journal of Physical Anthropology*, **96**, 391–417.

(1995b). Sexual dimorphism in canine shape among extant great apes. *American Journal of Physical Anthropology*, **96**, 365–89.

(1997). Paleobiological and phylogenetic significance of life history in Miocene hominoids. In *Function, Phylogeny, and Fossils: Miocene Hominoid Evolution and Adaptations*, ed. D. R. Begun, C. V. Ward & M. D. Rose, pp. 173–208. New York: Plenum Press.

(2002). The hominoid radiation in Asia. In *The Primate Fossil Record*, ed. W. Hartwig, pp. 369–84. Cambridge, UK: Cambridge University Press.

Kelley, J. & Etler, D. (1989). Hominoid dental variability and species number at the late Miocene site of Lufeng, China. *American Journal of Physical Anthropology*, **18**, 15–34.

Kelley, J. & Pilbeam, D. R. (1986). The Dryopithecines: taxonomy, comparative anatomy, and phylogeny of Miocene large hominoid. In *Comparative Primate Biology. Volume 1. Systematics, Evolution and Anatomy*, ed. D. R. Swindler & J. Erwin, pp. 361–411. New York: Alan R. Liss.

Kelley, J. & Qinghua, X. (1991). Extreme sexual dimorphism in a Miocene hominoid. *Nature*, **352**, 151–3.

Kordos, L. (1982). The prehominid locality of Rudabánya (NE Hungary) and its neighbourhood: a paleogeographic reconstruction. *A Magyar Àllami Földtani Intézet Évi Jelentése az 1980, évröl*, 395–406.

Kordos, L. & Begun, D. R. (1998). Encephalization and endocranial morphology in *Dryopithecus brancoi*: implications for brain evolution in early hominids. *American Journal of Physical Anthropology*, **Suppl. 26**, 141–2.

(2001a). A new cranium of *Dryopithecus* from Rudabánya, Hungary. *Journal of Human Evolution*, **41**, 689–700.

(2001b). Fossil catarrhines from the late Miocene of Rudabánya. *Journal of Human Evolution*, **40**, 17–39.

Larson, S. G. (1988). Subscapularis function in gibbons and chimpanzees: Implications for interpretation of humeral head torsion in hominoids. *American Journal of Physical Anthropology*, **76**, 449–62.

(1996). Estimating humeral head torsion on incomplete fossil anthropoid humeri. *Journal of Human Evolution*, **31**, 239–57.

(1998). Unique aspects of quadrupedal locomotion in nonhuman primates. In *Primate Locomotion: Recent Advances*, ed. E. Strasser, J. Fleagle, A. Rosenberger & H. McHenry, pp. 157–73. New York: Plenum Press.

Larson, S. G. & Stern, J. T. (1986). EMG of scapulohumeral muscles in the chimpanzee during reaching and "arboreal" locomotion. *American Journal of Physical Anthropology*, **176**, 171–90.

Leatherwood, S. & Reeves, R. R. (ed.) (1990). *The Bottlenose Dolphin*. San Diego: Academic Press.

Lockwood, C. A., Kimbel, W. H. & Johanson, D. C. (2000). Temporal trends and metric variation in the mandibles and dentition of *Australopithecus afarensis*. *Journal of Human Evolution*, **39**, 23–55.

Lockwood, C. A., Richmond, B. G., Jungers, W. J. & Kimbel, W. H. (1996). Randomization procedures and sexual dimorphism in *Australopithecus afarensis*. *Journal of Human Evolution*, **31**, 537–48.

MacLarnon, A. (1999). The comparative method: principles and illustrations from primate socioecology. In *Comparative Primate Socioecology*, ed. P. C. Lee, pp. 5–22. Cambridge, UK: Cambridge University Press.

MacLatchy, L., Gebo, D., Kityo, R. & Pilbeam, D. (2000). Postcranial functional morphology of *Morotopithecus bishopi*, with implications for the evolution of modern ape locomotion. *Journal of Human Evolution*, **39**, 159–83.

Marino, L. (1998). A comparison of encephalization between odontocete cetaceans and anthropoid primates. *Brain, Behavior & Evolution*, **51**, 230–38.

Martin, R. D. (1981). Relative brain size and basal metabolic rate in terrestrial vertebrates. *Nature*, **293**, 57–60.

(1983). *Human Brain Evolution in an Ecological Context: Fifty-Second James Arthur Lecture on the Evolution of the Human Brain*. New York: American Museum of Natural History.

(1990). *Primate Origins and Evolution*. Princeton, NJ: Princeton University Press.

(1993). Allometric aspects of skull morphology in *Theropithecus*. In *Theropithecus: The Rise and Fall of a Primate Genus*, ed. N. G. Jablonski, pp. 273–98. Cambridge, UK: Cambridge University Press.

Martin, R. D. & Harvey, P. H. (1985). Brain size allometry: ontogeny and phylogeny. In *Size and Scaling in Primate Biology*, ed. W. L. Jungers, pp. 147–74. New York: Plenum Press.

McHenry, H. M. (1988). New estimates of body weight in early hominids and their significance to encephalization and megadontia in robust australopithecines. In *The Evolution of the "Robust" Australopithecines*, ed. F. E. Grine, pp. 133–48. New York: Aldine de Gruyter.

(1992). Body size and proportions in early hominids. *American Journal of Physical Anthropology*, **87**, 407–30.

Morbeck, M. E. & Zihlman, A. L. (1989). Body size and proportions in chimpanzees, with special reference to *Pan troglodytes schweinfurthii* from Gombe National Park, Tanzania. *Primates*, **30**, 369–82.

Nishida, T. & Hiraiwa-Hasegawa, M. (1986). Chimpanzees and bonobos: cooperative relationships among males. In *Primate Societies*, ed. B. B. Smuts, D. L. Cheney, R. M. Seyfarth, R. W. Wrangham & T. T. Struhsaker, pp. 165–77. Chicago, IL: University of Chicago Press.

Nishida, T. & Hosaka, K. (1996). Coalition strategies among adult male chimpanzees of the Mahale Mountains, Tanzania. In *Great Ape Societies*, ed. W. C. McGrew, L. F. Marchant & T. Nishida, pp. 114–34. Cambridge, UK: Cambridge University Press.

Page, S. L. & Goodman, M. (2001). Catarrhine phylogeny: noncoding DNA evidence for a diphyletic origin of the mangabeys and for a human-chimpanzee clade. *Molecular Phylogenetics and Evolution*, **18**, 14–25.

Parker, S. T. & McKinney, M. L. (1999). *Origins of Intelligence: The Evolution of Cognitive Development in Monkeys, Apes, and Humans*. Baltimore, MD: Johns Hopkins.

Parker, S. T., Mitchell, R. W. & Miles, H. L. (ed.) (1999). *The Mentalities of Gorillas and Orangutans: Comparative Perspectives*. Cambridge, UK: Cambridge University Press.

Pilbeam, D. R. (1969). Tertiary Pongidae of east Africa: Evolutionary relationships and taxonomy. *Bulletin of the Peabody Museum of Natural History*, **31**, 1–185.

Pilbeam, D. R., Rose, M. D., Barry, J. C. & Shah, S. M. I. (1990). New *Sivapithecus* humeri from Pakistan and the relationship of *Sivapithecus* and *Pongo*. *Nature*, **384**, 237–9.

Plavcan, J. M. (1999). Mating systems, intrasexual competition and sexual dimorphism in primates. In

Comparative Primate Socioecology, ed. P. C. Lee, pp. 241–70. Cambridge, UK: Cambridge University Press.

Povinelli, D. J. & Cant, J. G. H. (1995). Arboreal clambering and the evolution of self-cognition. *Quarterly Review of Biology*, **70**, 393–421.

Reed, K. (1997). Early hominid evolution and ecological change through the African Plio-Pleistocene. *Journal of Human Evolution*, **32**, 289–322.

Richmond, B. G. & Strait, D. S. (2000). Evidence that humans evolved from a knuckle-walking ancestor. *Nature*, **404**, 382–5.

Richmond, B. G., Begun D. R. & Strait D. S. (2001). Origin of human bipedalism: the knuckle-walking hypothesis reconsidered. *Yearbook of Physical Anthropology*, **44**, 70–105.

Rodman, P. S. (1984). Foraging and social systems of orangutans and chimpanzees. In *Adaptations for Foraging in Nonhuman Primates*, ed. P. S. Rodman & J. G. H. Cant, pp. 134–160. New York: Columbia University Press.

Rose, M. D. (1973). Quadrupedalism in primates. *Primates*, **14**, 337–58.

(1988). Another look at the anthropoid elbow. *Journal of Human Evolution*, **17**, 193–224.

(1994). Quadrupedalism in some Miocene catarrhines. *Journal of Human Evolution*, **26**, 387–411.

(1996). Functional morphological similarities in the locomotor skeleton of Miocene catarrhines and platyrrhine monkeys. *Folia Primatologica*, **66**, 7–14.

(1997). Functional and phylogenetic features of the forelimb in Miocene hominoids. In *Function, Phylogeny, and Fossils: Miocene Hominoid Evolution and Adaptations*, ed. D. R. Begun, C. V. Ward & M. D. Rose, pp. 79–100. New York: Plenum Press.

Russon, A. E., Bard, K. A. & Parker S. T. (ed.) (1996). *Reaching into Thought: The Minds of the Great Apes*. Cambridge, UK: Cambridge University Press.

Satta, Y., Klein, J. & Takahata, N. (2000). DNA archives and our nearest relative: The trichotomy revisited. *Molecular Phylogenetics and Evolution*, **14**, 259–75.

Schaller, G. B. (1963). *The Mountain Gorilla: Ecology and Behavior*. Chicago, IL: University of Chicago Press.

Schultz, A. H. (1930). The skeleton of the trunk and limbs of higher primates. *Human Biology*, **2**, 303–438.

(1936). Characters common to higher primates and characters specific to man. *Quarterly Review of Biology*, **11**, 425–55.

(1941). The relative size of the cranial capacity in primates. *American Journal of Physical Anthropology*, **28**, 273–87.

(1961). Vertebral column and thorax. *Primatologica*, **4**, 1–66.

(1969). The skeleton of the chimpanzee. In *The Chimpanzee*, ed. G. H. Bourne, pp. 50–103. Basel: Karger.

(1973). The skeleton of the Hylobatidae and other observations on their morphology. *Gibbon and Siamang*, **2**, 1–54.

Schwartz, J. H. (1990). *Lufengpithecus* and its potential relationship to an orang-utan clade. *Journal of Human Evolution*, **19**, 591–605.

(1997). *Lufengpithecus* and hominoid phylogeny. Problems in delineating and evaluating phylogenetically relevant characters. In *Function, Phylogeny, and Fossils. Miocene Hominoid Evolution and Adaptations*, ed. D. R. Begun, C. V. Ward & M. D. Rose, pp. 363–88. New York: Plenum Press.

Shoshani, J., Groves, C. P., Simons, E. L. & Gunnel, G. F. (1996). Primate phylogeny: morphological vs. molecular results. *Molecular Phylogenetics and Evolution*, **5**, 101–53.

Smith, E. J. (1999). A functional analysis of molar morphometrics in living and fossil hominoids using 2-D digitized images. Doctoral Dissertation, University of Toronto.

Smith, R. J. & Jungers, W. L. (1997). Body mass in comparative primatology. *Journal of Human Evolution*, **32**, 523–59.

Stern, J. T. J. & Larson, S. G. (2001). Telemetered electromyography of the supinators and pronators of the forearm in gibbons and chimpanzees: Implications for the fundamental positional adaptation of hominoids. *American Journal of Physical Anthropology*, **115**, 253–68.

Stern, J. T., Wells J. P., Vangor, A. K. & Fleagle, J. G. (1976). Elecromyography of some muscles of the upper limb in ateles and lagothrix. *Yearbook of Physical Anthropology*, **20**, 498–507.

Susman, R. (ed.) (1984). *The Pygmy chimpanzee*. New York: Plenum Press.

Temerin, B., Wheatley, P. & Rodman, P. S. (1984). Body size and foraging in primates. In *Adaptations for Foraging in Primates*, ed. P. S. Rodman & J. G. H. Cant, pp. 217–48. New York: Columbia University Press.

Terborgh, J. (1992). *Diversity and the Tropical Rain Forest*. New York: Freeman.

Tobias, P. V. (1971). *The Brain in Hominid Evolution*. New York: Columbia University Press.

(1975). Brain evolution in Hominoidea. In *Primate Functional Morphology and Evolution*, ed. R. H. Turtle, pp. 353–92. The Hague: Mouton.

(1983). Recent advances in the evolution of the hominids with special reference to brain and speech. In *Recent Advances in the Evolution of Primates*, ed. C. Chagas, pp. 85–140. Citta del Vaticano: Pontificia Academia Scientiarum.

(1995). The brain of the first hominids. In *Origins of the Human Brain*, ed. J.-P. Changeux & J. Chavaillon, pp. 61–83. Oxford: Clarendon Press.

Tutin, C. E. G. & Fernandez, M. (1993). Composition of the diet of chimpanzees and comparisons with that of sympatric lowland gorillas in the Lope Reserve, Gabon. *American Journal of Primatology*, **30**, 195–211.

Tutin, C. E. G., Ham R., White, L. J. T. & Harrison, M. J. S. (1997). The primate community of the Lopé Reserve in Gabon: Diets, responses to fruit scarcity, and effects on biomass. *American Journal of Primatology*, **42**, 1–24.

Tuttle, R. H. (1987). *Apes of the World*. Park Ridge, NJ: Noyes.

Ungar, P. S. (1996). Dental microwear of European Miocene catarrhines: evidence for diets and tooth use. *Journal of Human Evolution*, **31**, 335–66.

Ungar, P. S. & Kay, R. F. (1995). The dietary adaptations of European Miocene catarrhines. *Proceeding of the National Academy of Sciences USA*, **92**, 5479–81.

van Schaik, C. P. & van Hooff, J. A. R. A. M. (1996). Toward an understanding of the orangutan's social system. In *Great Ape Societies*, ed. W. C. McGrew, L. F. Marchant & T. Nishida, pp. 3–15. Cambridge, UK: Cambridge University Press.

Walker, A. C., Falk, D., Smith, R. & Pickford, M. F. (1983). The skull of *Proconsul africanus*: Reconstruction and cranial capacity. *Nature*, **305**, 525–7.

Ward, C., Leakey, M. G. & Walker, A. (1999). The new hominid species *Australopithecus anamensis*. *Evolutionary Anthropology*, **7**, 197–205.

(2001). Morphology of *Australopithecus anamensis* from Kanapoi and Allia Bay, Kenya. *Journal of Human Evolution*, **41**, 255–368.

Watts, D. P. (1984). Composition and variability of mountain gorilla diets in the Central Virungas. *American Journal of Primatology*, **7**, 323–56.

(1996). Comparative socio-ecology of gorillas. In *Great Ape Societies*, ed. W. C. McGrew, L. F. Marchant & T. Nishida, pp. 15–28. Cambridge, UK: Cambridge University Press.

White, T., Suwa, G. & Asfaw, B. (1994). *Australopithecus ramidus:* a new species of early hominid from Aramis, Ethiopia. *Nature*, **371**, 306–12.

WoldeGabriel, G., White, T. D., Suwa, G., Renne, P., de Heinzelin, J., Hart, W. K. & Heiken, G. (1994). Ecological and temporal placement of early Pliocene hominids at Aramis, Ethiopia. *Nature*, **371**, 330–3.

Wrangham, R. W. (1999). Evolution of coalitionary killing. *Yearbook of Physical Anthropology*, **42**, 1–30.

Wrangham, R. W., Chapman, C. A., Clark-Arcadi, A. P. & Isabirye-Basuta, G. (1996). Social ecology of Kanyawara chimpanzees: implications for understanding the costs of great ape groups. In *Great Ape Societies*, ed. W. C. McGrew, L. F. Marchant & T. Nishida, pp. 45–57. Cambridge, UK: Cambridge University Press.

Wynn, J. G. (2000). Paleosols, stable carbon isotopes, and paleoenvironmental interpretation of Kanapoi. *Journal of Human Evolution*, **39**, 411–32.

Yamagiwa, J., Mwanza, N., Yumoto, T. & Maruhashi, T. (1992). Travel distances and food habits of eastern lowland gorillas: a comparative analysis. In *Topics in Primatology: Behaviour, Ecology and Conservation*, ed. N. Itoigawa, Y. Sugiyama, G. P. Sackett & R. Thompson, pp. 267–81. Tokyo: University of Tokyo Press.

Zihlman, A. L. (1984). Body build and tissue composition in *Pan pansicus* and *Pan troglodytes*, with comparisons to other hominoids. In *The Pygmy Chimpanzee: Evolutionary Biology and Behavior*, ed. R. L. Susman, pp. 179–200. New York: Plenum Press.

Zihlman, A. L. & Cramer D. L. (1978). Sexual differences between pygmy (*Pan paniscus*) and common chimpanzees (*Pan troglodytes*). *Folia Primatologica*, **29**, 86–94.

Part I
Cognition in living great apes

ANNE E. RUSSON

Psychology Department, Glendon College of York University, Toronto

INTRODUCTION

This first section offers a compact overview of great ape cognition. We did not attempt to review this material comprehensively because others have done so recently (e.g., Byrne 1995; Matsuzawa 2001; Parker & McKinney 1999; Parker, Mitchell & Miles 1999; Russon, Bard & Parker 1996; Suddendorf & Whiten 2001; Thompson & Oden 2000; Tomasello & Call 1997). Our primary aim was to revisit cognitive phenomena in living great apes considered to need evolutionary explanations beyond those applicable to other anthropoid primates. We then favored discussions of cognition as it develops in species-typical rearing conditions and applies to species-typical problems, and we emphasized the social and ecological cognition that have been the focus of discussions on primate cognitive evolution. We also revisited this topic to bring newer findings on great ape cognition to the broader community of scholars interested in cognitive evolution. Great apes are regularly taken as the best living models of the ancestral cognitive platform from which human cognition evolved (e.g., Donald 1991; Mithen 1996), so accurate portrayals of their cognition are essential to reconstructing human cognitive evolution accurately.

Byrne, Chapter 3, discusses "technical" skills, which have been major candidates for the defining force in great ape cognition. He argues that research focus on great apes' tool-based foraging skills, while important, has distracted attention from other impressive achievements equally likely to represent cognitive adaptations (Yamakoshi, Chapter 9, this volume takes a similar view) and that great apes' technical skills may be as cognitively complex as those of some pre-modern hominins. Great apes' manual foraging skills probably employ the same cause–effect cognitive processes as their tool-based ones, at comparable levels of complexity. Archaeological evidence on tools suggests human-like cognitive sophistication evolved by degrees, successively building on pre-modern hominins' more primitive capacities. Analyzing great apes' technical skills for features of ancestral hominin tool remains that have been used to infer their makers' cognitive abilities shows great apes to be cognitively comparable to early hominin stone toolmakers. Cognitive capacities credited to the human lineage then have a much longer evolutionary history, reaching back to the common great ape–human ancestor.

Parker, Chapter 4, considers social complexity and social cognition in great apes, extending her efforts to consider both social and ecological pressures in great ape cognitive evolution. Standard measures of social complexity based on group size have not shown differences between great apes and other anthropoids, leaving it unclear whether social problems challenge their cognition and why great apes show more complex social cognition than other anthropoids. Parker argues that differences in social complexity are evident, in fission–fusion patterns – particularly, in ephemeral activity groups that assemble for specific activities. Comparing fission–fusion patterns in chimpanzees and Hamadryas baboons, especially apprenticeship activities that support immatures' acquisition of expertise, shows a larger number of roles and more complex routines, scripts, and event representations in chimpanzees. All these increase the range, flexibility, and unpredictability of social activities and so require more complex cognition. Parker's analysis rests on two particular species and Hamadryas sociality is considered a unique case, but her exploration of a fission–fusion basis for greater social complexity is consistent with other views on great ape sociality in this volume (see van Schaik, Preuschoft & Watts, Chapter 11, Yamagiwa, Chapter 12).

In Chapter 5, Blake, a specialist in early human language development, analyzes great apes' gestural communication relative to language and cognitive development in humans to explore communication–cognition links. Gesture has been suggested as an important stepping-stone in the evolution of language and may offer a valuable avenue for exploring cognitively complex communication in wild great apes (e.g., Hewes 1976; and see MacLeod, Chapter 7, Potts, Chapter 13, this volume). In fact great apes' linguistic communication has been most prominent in cognitive discussions, probably because language has been seen as closely linked to cognition in human ontogeny and evolution. Equally if not more important to questions of great ape cognitive evolution, however, is great apes' species-typical communication. If its links with cognition are strong and if its complexity is similar to that of their language achievements, communication may have been significant in great ape cognitive evolution. Blake's assessment is that great apes' gestural and linguistic communication are similar in cognitive complexity, comparable to levels seen in two-year-old human children. This does not consider all communicative phenomena, so more extensive analyses may show greater complexity.

Russon, Chapter 6, examines great ape cognition as an integrated system distinct from the cognitive systems of other anthropoid primates. Discussions of great ape cognition have typically focused on specific problem types or domains (e.g., tool use, social cognition). Many, however, ultimately distinguish great ape cognition in terms of features that appear system-wide (e.g., hierarchization, symbolism, corrective guidance by schema, event representation) rather than problem or domain specific ones. Empirical studies showing roughly comparable achievement levels in the primitive symbolic range in all domains along with limited capacities for interconnecting cognitive abilities across domains also point to system-wide cognitive enhancements. If great apes' cognitive advantages span their cognitive system as a whole, they probably derive from centralized processes that generate problem-specific achievements.

These chapters all stress great apes' high level cognitive achievements and situate them as intermediate between other anthropoid primates and modern humans. It is this great ape system of intermediate level cognition, including the processes that generate it, that is in need of its own evolutionary explanation – not a narrow suite of problem- or domain-specific abilities. One important implication is that the evolutionary origins

of symbolic cognition need reconsideration. Evidence on great ape cognition indicates it originated with their common hominid ancestor in the Miocene, not within the human lineage.

REFERENCES

Byrne, R. W. (1995). *The Thinking Ape: Evolutionary Origins of Intellect in Monkeys, Apes and Humans*. Oxford: Oxford University Press.

Donald, M. W. (1991). *The Origins of the Modern Mind: Three Stages in the Evolution of Culture and Cognition*. Cambridge, MA: Harvard University Press.

Hewes, G. (1976). The current status of the gestural theory of language origins. In *Origins and Evolution of Language and Speech, Volume 280*, ed. S. Harnald, H. Steklis & J. Lancaster, pp. 482–504. New York: New York Academy of Sciences.

Matsuzawa, T. (ed.) (2001). *Primate Origins of Human Cognition and Behavior*. Tokyo: Springer-Verlag

Mithen, S. (1996). *The Prehistory of the Mind*. London: Thames & Hudson.

Parker, S. T. & McKinney, M. (1999). *Origins of Intelligence: The Evolution of Cognitive Development in Monkeys, Apes, and Humans*. Baltimore, MD: The Johns Hopkins Press.

Parker, S. T., Mitchell, R. W. & Miles, H. L. (eds.) (1999). *Mentalities of Gorillas and Orangutans: Comparative Perspectives*. Cambridge, UK: Cambridge University Press.

Russon, A. E., Bard K. A. & Parker, S. T. (eds.) (1996). *Reaching into Thought: The Minds of the Great Apes*. Cambridge, UK: Cambridge University Press.

Suddendorf, T. & Whiten, A. (2001). Mental evolution and development: evidence for secondary representation in children, great apes, and other animals. *Psychological Bulletin*, **127**(5), 629–50.

Thompson, R. K. R. & Oden, D. L. (2000). Categorical perception and conceptual judgments by nonhuman primates: the paleological monkey and the analogical ape. *Cognitive Science*, **24**(3), 363–96.

Tomasello, M. & Call, J. (1997). *Primate Cognition*. New York: Oxford University Press.

3 · The manual skills and cognition that lie behind hominid tool use

RICHARD W. BYRNE

School of Psychology, University of St. Andrews, St. Andrews

Tool use is an important aspect of being human that has assumed a central place in accounts of the evolutionary origins of human intelligence. This has inevitably focused a spotlight on any signs of tool use or manufacture in great apes and other nonhuman animals, to the relative neglect of skills that do not involve tools. The aim of this chapter is to explore whether this emphasis is appropriate. Suppose we take a broader view, accepting evidence from *all* manifestations of manual skill, what can we learn of the mental capacities of the great apes and the origins of human intelligence? My own ultimate purpose is to use comparative evidence from living species to reconstruct the evolutionary history of the many cognitive traits that came together to make human psychology. The cognition of great apes is the obvious starting point, to trace the more primitive (i.e., ancient) cognitive aptitudes that are still important to us today. In this chapter, I focus on great ape cognition as it is expressed in manual skills, based on cognitive aspects of tool use and manufacture considered significant in the human evolutionary lineage.

WHY IS TOOL USE IMPORTANT IN THE STUDY OF HUMAN EVOLUTION?

Consider first what aspects of tool use have recommended it as "special" to physical anthropologists and archaeologists. Most obviously, tools are *convenient* things for investigators. As physical objects, they can be collected, measured, and compared with ease. Often durable, they can be investigated long after the tool user or maker is dead. For archaeologists, this characteristic alone adds enormous value to tools in the study of human origins. Convenience alone would be little recommendation if tool skills were trivial. But of course, quite the reverse is believed to be the case. The significance of

tools is what they imply about the *cognitive abilities* of their users. From examining the products of tool making and using, researchers hope to discern the thinking that governed these activities: everyday physics, means–end analysis, coordination of dextrous manipulations towards a predefined goal, recognizing and coping with local difficulties in a complex process, and so on. I contend that these cognitive abilities are equally required by many tasks that do not involve tool use, especially complex manual skills, and that neglect of the study of manual skills has been an impediment to understanding great apes' technological abilities and therefore to understanding the evolutionary origins of human technological ability (Byrne 1996, 1999b; see Yamakoshi, Chapter 9, this volume, for a related view).

Moreover, psychology has offered little help in understanding the origins of complex manual skills. The mechanisms controlling reaching and grasping and their development have been carefully analyzed (Connolly 1998; Fitts & Posner 1967) but until recently little attention has been paid to what is done with an object once grasped (Bril, Roux & Dietrich 2000; Roux 2000). This chapter, then, also constitutes a plea for broader-based research on the psychology of skill acquisition and the relationship between complex manual tasks and mental abilities.

WHAT IS SPECIAL ABOUT HUMAN TOOL USE?

Identification of qualitative differences between human and nonhuman tool making may enable us to identify which facets of cognition and which particular aspects of technical skills were likely to have been crucial in human evolution. Some clear differences have been suggested. All known human populations fabricate composite tools out of *many component parts* (Gosselain 2000; Reynolds 1982): nets, hafted spears, and even boats. Also, human tools are often used to *make or assist other tools*, in a range of ways, from slings and throwing sticks, to carving wood with adze and knife or smelting metal. And human tool *materials* range from wood and stone to hair, bone, horn, plant seeds, and metal, according to the task in hand. Tools of nonhumans are poor things in comparison (unless bee and wasp nests, and the bowers and nests of birds, are treated as tools, in which case the contest is a closer one: but these feats are species-typical and presumably innately coded).

However, those highly distinctive characteristics of human tool making – composite tools, tool use to make tools, the rich array of raw material – may be relatively new even in human evolution (Mellars & Stringer 1989). In hominin deposits that date from before the arrival of anatomically modern humans, archaeologists are hard pressed to find uncontroversial evidence of tools going beyond single items, made by removing parts rather than combining items, and only of stone or sometimes wood. The animals that made these simpler tools were anatomically much closer to modern humans than they were to living great apes. We may fairly ask, then, whether there exist critical aspects of human tool making, and the cognitive abilities that they imply, that predate anatomically modern humans yet represent derived cognitive features of the human *lineage* – less ostentatious but "special" none the less because they are shared by some of our extinct relatives but not by any other living animal.[1] Alternatively, the tools of the earliest members of the human lineage and their relatives may have tapped only primitive cognitive capacities shared with living nonhuman primates. It is of course possible that all these accounts are true for different aspects of tool-related cognition: some uniquely human, some unique to the human line, some shared with living great apes, and some shared with a much wider range of species. Modern human competence with tools may have a long evolutionary history.

To find out, the most crucial species to examine are our closest living relatives, the great apes. All great apes sometimes use tools (McGrew 1989). The lesser apes do not use tools and little is known of their manual skills, so they will not be considered further and "apes," hereafter, refers to great apes. Are there unique features of the tool skills of the great ape clade (including humans)? To find out, it is necessary first to establish a baseline, asking: what tools do *non*-ape animal species use?

Tool using in animals other than apes

Although most species use none, tool use is quite widely distributed across animal taxa (Beck 1980). In many cases, the available evidence constitutes only a single reported instance of a particular individual making a particular tool, and provides little detail on method (e.g., Chevalier-Skolnikoff & Liska 1993). This picture is as true of monkeys as it is of most other animals. Some species of animal, but not monkeys, do use tools habitually, in consistent ways. In these cases, individuals are often found to use one sort of tool for one purpose, but members of closely related species show no tool use or any other exceptional behavior. Famous examples are the woodpecker finches of the Galapagos (where no true woodpeckers exist), which use cactus spines as probes; Egyptian vultures, which use stones to break ostrich eggs; and Californian sea otters, which use stones to break the shells of molluscs. Other species of Darwin's finch and other species of otter do not use tools.

Since habitual tool use of non-apes is generally limited to one species in a large clade, and the tool use functions in a highly specific way, the distinction might be that ape tool use is learnt from experience, while tool use in other animals is innate, coded on the species' genes. Unfortunately for this tidy partitioning, some non-apes *do* learn their tool-using habits. Only certain populations of Egyptian vultures have discovered how to use stones to break into ostrich eggs, others cannot exploit this valuable resource, and learning is likely involved (Goodall & van Lawick 1966). The most telling signs of learning tool use come from the sea otter, where traditional transmission is involved. Alaskan sea otters do not use tools but Californian otters, the same species, show either of two tool techniques, with different sized stones, to break either abalones and crabs. Abalones and crabs occur in the range of all otters, but individuals specialize on one or the other, and daughters acquire the same tool technique as their mothers (Estes *et al.* 2003).

If incidental/habitual and learned/innate distinctions do not hold up, will some other split serve to characterize the uniqueness of ape tool use? Otters, vultures, and finches may *use* tools but do not manufacture them (unless detaching cactus spines is accepted as manufacture). Orangutans (van Schaik 1994) and chimpanzees

(Goodall 1964) do sometimes manufacture tools. Until recently, the possibility that the crucial distinction is that *manufacture* of tools is unique to apes worked well. Now, however, some local populations of New Caledonia crows, *Corvus moneduloides*, have been found to manufacture tools, modifying stems and leaves to create hooks and barbs for extracting insects from crevices (Hunt 1996, 2000).

A chimpanzee's view of the archaeological record of tool use

Although no such simple dichotomy marks out ape tool use from that of all other animals, for an evolutionist the behavior of apes nevertheless has special relevance to the understanding of human origins. When humans and apes are alike in some trait, not shared by other anthropoid primates, the trait is likely to exist by virtue of common descent – whereas tool-related similarities with crows and otters doubtless result from convergent evolution. What ape tool use amounts to and whether it differs in any crucial way from that of the early representatives of the human lineage, then, potentially informs us about the *evolutionary history* of distinctively human mental abilities.

One living ape, the chimpanzee, is famous for its range of habitual tool use and tool manufacture in the wild. Found stones are employed to hammer open nuts on wood or stone anvils, sticks are used to pick up ants, stems and vines to "fish" for termites in their mounds or ants within arboreal nest-holes, leaves as rain shelters or sponges or cleaning tissues, etc. (Boesch & Boesch 1990; Goodall 1986; McGrew 1992; Nishida 1986; Sugiyama 1994; Sugiyama & Koman 1979; Yamakoshi, Chapter 9, this volume). Their methods of making tools from plants alone include detaching plant material, cutting it to length, removing leaves or bark, and sharpening the end – or all of these methods, for a single tool. Wild orangutans also make tools of plant material (Fox, Sitompul & van Schaik 1999), for probing and scraping, using similar methods.

These discoveries make it impossible to disregard the potential importance of plant material and found objects in ancestral humans' tool use, although archaeologists will seldom detect them. Archaeology none the less reveals one characteristic not shown by any living great ape under natural conditions, percussion flaking of stone. One line of argument would have us stop there.

Humans make stone tools, (some of) their extinct ancestors and relatives did too, but apes (and other animals) do not.

However, the significance of stone tool use and manufacture is not simply the kind of physical material used and modified, but what we can glean about the cognitive capacities of the tool user. In exploring what stone flaking tells us about the mind of the tool maker, we are very dependent on the quality of the evidence and the level of analysis it can sustain. At one level, percussion flaking is simply the modification of a found object by detachment of parts. As such, it falls in the same category as a chimpanzee stripping off the leaves and sharpening the tip of a stem. At another level, the particular *method* of stone modification may imply greater cognitive abilities in our extinct ancestors, but in the case of stone tools traditionally described as Oldowan, this implication has been disputed (Wynn & McGrew 1989).

Without necessarily accepting the conclusion, the process of point-by-point comparison in this argument is informative (see Joulian 1996, for detailed analysis of chimpanzee behavior from an archaeological perspective). Percussion flaking shows the ability to aim blows with care and precision. So, however, does a chimpanzee's hammering open a *Panda* nut (Boesch & Boesch 1990). Young chimpanzees' blows are usually ineffective, only knocking the nut away, and it takes years for a chimpanzee to become proficient, so comparable levels of precision may arguably be required. Oldowan tools are clearly made to a simple design concept: small enough to hold and with a sharp edge (Roche 1989; Toth 1985b). But so are a chimpanzee's termiting probes: sufficient flexibility, length, and smoothness to penetrate deep into a termite mound (Goodall 1964). Because the probes are sometimes made in advance of reaching the mound, the chimpanzee must possess some concept of an adequate tool (Goodall 1986). Percussion flaking also normally implies bimanual handling. But Oldowan stones may sometimes have been flaked by throwing them against a hard surface, a technique that one bonobo discovered and used successfully (Toth 1985b; Toth *et al.* 1993).

For the more advanced products of the middle and upper Paleolithic, and even some lower Paleolithic stone cultures (e.g., Acheulean), a very different picture emerges. Precision handling (Napier 1961), bimanual role differentiation in which the two hands perform in different but complementary tasks (Elliott & Connolly 1984), very precise aiming of powerful blows, and a sequential plan of flake removal are all essential to make tools characteristic of these cultures. Also evident within the process are hierarchical organization and exquisite guidance by an anticipatory schema of the finished product – in Acheulean hand axes, for instance, iterative detachment of large flakes followed by corrective detachments towards a straight edge (Wynn 1988), and in Levallois "tortoiseshell flakes," elaborate preparations before the final blow (Oakley 1949). Until recently, it was possible to argue for a relatively late origin for these distinctively human skills in tool manufacture and to portray the earlier, Oldowan skills as ape-like, differing only in the material used.

Remarkable evidence now shows that this picture is incorrect and illustrates the fragility of deduction from the patchy archaeological record, especially concerning ascriptions of incompetence. Roche *et al.* (1999) excavated stone material from Lokalalei, Kenya, dating from 2.34 Ma (well before typical Oldowan dates), which allows reconstruction of the process of detachment of up to 30 flakes from a single artifact by a "refit" of the debris. A large series of stone cores show the same principles applied to each, and knappers were clearly able to maintain the precise strike angle for successful flaking throughout these long manufacturing sequences. The tools and their debris are not associated with skeletal material, so which species made them is not known. What is clear is that these human-like capacities of tool manufacture, well beyond the cognitive capacities shown in ape tool use, are more ancient than ever suspected.

Assessing the cognition of ape manual skills

We can then use the distinctive, cognitively governed features known to be associated with tool use in *Homo sapiens*, whether extinct hominins or modern people who still make their own tools, as a guide to what evolutionary precursors of modern human skills might in principle be found in living apes. In the following list, I have omitted features that seem characteristic only of anatomically modern humans: use of tools to make tools, constructing new objects out of multiple components, and using a wide range of raw material for tools. I also largely omit discussion of an important archaeological feature, material transport, in which raw material is carried to a suitable working site or working takes place where material occurs and finished tools are then transported.

Living apes are primarily forest animals with relatively small home ranges: they have little need of systematic transport of materials, so lack of it tells us little of their cognitive capacities. The aim of the following list is to concentrate on features likely to offer helpful guidance to manual skills in apes, skills that therefore predate the divergence of the human lineage.

- **Precision handling**: e.g., tip-to-tip precision grips, rather than whole-hand power grips useful only in rough and ready manipulation (Christel 1993; Marzke & Wullstein 1995).
- **Accurate aiming of powerful blows**: e.g., to detach a useful flake, blows must be highly accurate in placement, yet still forceful (Inizian *et al.* 1999).
- **Bimanual role differentiation**: e.g., holding a stone securely while aiming a blow at it with a hammer – the two hands perform different actions but in a complementary way, so that they work together to achieve a single purpose (Connolly 1998; Elliott & Connolly 1984).
- **Regular and sequential plan**: e.g., in percussion flaking, the order of detachments is normally crucial to the final result (Inizian *et al.* 1999; Pelegrin 2000).
- **Hierarchical organization** with use of subroutines (Bruner 1970; Elliott & Connolly 1974; Lashley 1951; Miller, Galanter & Pribram 1960): e.g., flake detachments may be grouped into several series, of variable length according to the particular properties of the stone being worked, and each series achieves a distinct purpose – the overall plan of manufacture therefore consists of several subroutines, each performed to a local criterion of completeness (Inizian *et al.* 1999; Pelegrin 2000).
- **Corrective guidance by anticipatory schema**, i.e., actions corrected to attain a goal specified in advance (de Groot 1965): e.g., if there is no appropriate angle for flaking, then first construct a platform; if the main flakes do not produce a straight cutting edge, then make additional small detachments until straightness is achieved (Pelegrin 2000; Wynn 1988).
- **High individual manual laterality**: e.g., in a community of tool makers, each individual always uses the same hand for the same task (Marchant, McGrew & Eibl-Eibesfeldt 1995).
- **Population right-handedness**: e.g., for tool making, a significant majority of individuals are similarly lateralized, such that the left hand provides support

while the right applies precision actions (Marchant *et al.* 1995; McManus 1984).

With this perspective, we can now examine evidence from the living great apes. As emphasized earlier, this evidence need not be restricted to tools. In particular, manipulation of plant material may tap the same cognitive processes as tool use: both deal with operations that change the physical world, often by applying force, so both rely heavily on understanding cause–effect and organizing simple movements into complex programs (and see Yamakoshi, Chapter 9, this volume). I therefore draw also on the gathering and processing of plant foods by great apes as manifestations of manual skill.

Two provisos should be mentioned, to avoid misunderstanding: both concern the meaning of *skill*. First, to ethnographers and social anthropologists, skilled manual activity is seen as necessarily "situated": not the actions of single, clever individuals but within a social network of knowledge and support. Almost all the great ape skills described here are sometimes performed in social circumstances, and social transmission of this expertise is considered crucial. However, to date, most study has focused on individual apes rather than a social nexus. The aim of this chapter is therefore a more modest one, to characterize the cognitive processes of great apes relating to their ability to master complex manual skills. Second, in some branches of psychology (e.g., sport psychology), "skill" refers to the degree of perfection of muscle control in a movement, such as throwing a ball or pushing a cursor. Primatologists appreciate that differences in muscle control are sometimes important to apes engaged in the sort of activities discussed here; however, we have no way of studying this in wild animals. Rather, the focus of all work discussed here will usually be higher-level analysis, the organization of individual elements of action (such as throwing or pushing) into complexes that serve to accomplish tasks. *Skill*, therefore, is here taken to be the sum of psychological processes that enable the development and perfection of complex, goal-directed techniques by individuals (hereafter, skills).

COGNITION IN THE MANUAL SKILLS OF LIVING GREAT APES

Evidence on living apes comes from very different sources, for good reason when it reflects ecological specialization in different species, and also because of coincidences of what aspects have attracted the most research. This means comparison is not straightforward. I first examine the different genera before attempting an overall summary of cognitive capacities common to the clade.

Pan: the chimpanzee species

The two species of chimpanzee, *Pan troglodytes* and *Pan paniscus*, were recognized as distinct from each other in the 1920s but scientific study has always focused on the more common species, *Pan troglodytes*. The discovery that common chimpanzees not only use but make tools (Goodall 1964) has accentuated this research emphasis. Few field data were available on *Pan paniscus*, the bonobo or pygmy chimpanzee, until the 1970s (Badrian & Badrian 1984; Badrian & Malenky 1984; Kano 1982, 1983). Long-term study at two sites has subsequently failed to find any tool use involving skilled manipulation or tool manufacture in bonobos (but see Ingmanson, 1989, 1996) and their foraging and food processing skills have yet to be studied. One captive bonobo, however, readily learned to make stone flakes to cut rope securing a food box (Toth *et al.* 1993). He largely worked by throwing his stone core at a hard substrate, with no need of careful aim, so there may be no real difference in mental capacities between *Pan paniscus* and *Pan troglodytes* (see McGrew 1989). The manual skills of *Pan*, then, are better gauged from behavior recorded in *Pan troglodytes*.

Most chimpanzee tool use shows no particular sign of mental capacities beyond the association of tool and task and it is hard to establish how deeply tool users understand the cause-and-effect relations of what they are doing – the main cognitive ability that is relevant (but see Limongelli, Boysen & Visalberghi 1995). However, there is evidence that chimpanzee tool use is pre-planned, not simply evoked by stimuli in the situation of use. Stone hammers are sometimes selected in advance of use, according to criteria of fitness for purpose, then transported up to 0.5 km to the nut-cracking site (Boesch & Boesch 1983, 1984); insect fishing probes are sometimes made in advance to a simple pattern, then carried to the termite mound (Goodall, 1973, 1986). Tool selection or manufacture prior to transport to the site of use clearly indicates mental specification of the goal in advance of need (anticipatory schema). Their tool making has not shown corrective guidance, on-line, with detailed comparison with the goal specification. However, this may be a function of the difficulty of detection. Error correction in tool use has been seen, a stone wedge used to straighten an anvil stone for more efficient use (Matsuzawa 1996). This may have been a response to practical failure, so we cannot be sure that the modification reflected recognizing a mismatch with a preconceived plan. Chimpanzees at Mt. Niéniokoué, Ivory Coast, apparently evaluate nuts they intend to crack, giving a single blow and then sometimes abandoning the task: they appear to observers to be testing nuts' weight, maturity, and density, all factors affecting the task of cracking (F. Joulian pers. commun.). These impressions, if confirmed, may reflect on-line corrective guidance of the nut-cracking task.

Evidence from throwing detached objects suggests chimpanzees have poor aim (Goodall 1964; Kortlandt 1967), although humans who have been hit by chimpanzee-thrown rocks have claimed otherwise. When using hammer-stones to break hard nuts on stone anvils, however, their blows are both powerful and precise (Boesch & Boesch 1983, 1990), so it seems that chimpanzees can develop precise aim with long practice.

Despite the awkwardness of the chimpanzee hand, with relatively long fingers and short thumb (Napier 1960), chimpanzees show precision handling and bimanual role differentiation in numerous ways when they make and use tools. They make probe tools used to "dip" for ants by holding a stick in one hand and stripping protruding leaves or bark with the other, using a precision grip; they steady wobbly anvils with one hand while the other wields the hammer-stone, etc. (Boesch & Boesch 1983, 1990). In addition, bimanual role differentiation and precise, visually guided handling are shown in manual body grooming (Goodall 1986) and in manual preparation of woolly surfaced leaves of the sugar mulberry, *Broussonettia*, which is difficult to eat without rolling (Stokes & Byrne 2001).

One task that nicely shows the advantage of precision handling and bimanual control is eating aggressive *Dorylus* ants. To capture these ants, chimpanzees insert a stick into a mass of ants and agitate it, provoking the ants to attack and climb the stick. At Taï, chimpanzees wait until about 10 cm of a relatively short stick is covered with ants and then pick off the ants with the lips (Boesch & Boesch 1990). At Gombe, chimpanzees wait until around 30 cm of a long wand is covered with ants. Then, with a sweeping movement of the other hand,

half-closed in a precision grip, they accumulate a mass of ants, which they eat rapidly (McGrew 1974). This difference in technique probably reflects local adaptation to the various species of *Dorylus* ants, which vary in aggressiveness, since at Bossou, Guinea, where several species occur, individual chimpanzees use both techniques differentially, according to the species of ant and its current activity (Humle & Matsuzawa 2002).

The manual technique for eating *Broussonettia* leaves gives some evidence of hierarchical organization (Stokes & Byrne 2001). Hierarchical structure has not been explicitly shown for any chimpanzee tool-using task, although probably only because cognitive organization has seldom been examined. Matsuzawa (2001) suggests that a wide range of chimpanzee tool using shows hierarchical organization, in a somewhat different way.

Certainly, sequential task organization towards an eventual goal, sometimes in several stages, is shown in many chimpanzee tool-using tasks. Examples are insect fishing and hammer-and-anvil use, in each of which the sequence may start with preparing or selecting a tool and transporting it to the site of use. Iteration of a regular string of actions until a criterion is reached gives evidence that the iterated string constitutes a subroutine of the main process (Byrne 1999a): repeatedly agitating a dipping stick until ants reach a predetermined point, or repeatedly moving a stone anvil about until it is level, suggest this form of organization. Moreover, the manufacture and use of several different tools in series to obtain a single goal has occasionally been noted (Brewer & McGrew 1990; Sugiyama 1997; Suzuki, Kuroda & Nishihara 1995), again suggesting the ability to apply a regular and systematic sequence of actions, although each tool might have been made in response only to the outcomes of the previous tool's use.

Manual laterality in chimpanzees has been studied for many years, with controversy in both methods and conclusions (see Finch 1941; Hopkins & Morris 1993; Marchant & McGrew 1991; McGrew & Marchant 1991). Now a clearer picture is beginning to emerge. In most spontaneous manual actions in the wild, individuals are generally ambidextrous (Marchant & McGrew 1996). However, individual lateralization has been found to be high in termite fishing (McGrew & Marchant 1992), stone tool use (Boesch 1991; Sugiyama et al. 1993), and manually cracking large *Strychnos* fruits (fruits must be pounded, accurately, against a hard object to break them: McGrew *et al.* 1999). This collection of tasks suggests that lateralization is a strategy for enhancing manual precision, particularly if accurately aimed blows are necessary: presumably, one hand can specialize in perfecting a particularly difficult skill. In termite fishing, exclusively lateralized chimpanzees worked more quickly than weakly lateralized or ambidextrous individuals (McGrew & Marchant 1999), although they must suffer corresponding disadvantages, compared with ambidextrous individuals, from having to adjust their posture to each termite mound. However, manual laterality is found in a plethora of other contexts and animal species (see review by Bradshaw & Rogers 1993), suggesting there may be reasons for lateralized function beyond simple efficiency. Bimanual role differentiation, which also serves task complexity, may be one of these other reasons: in captive chimpanzees, strong laterality was evoked by a task requiring bimanual solution, prising out food from a hollow object held in the other hand (Hopkins 1995).

Assessment of whether population right-handedness occurs depends on which statistical methods are considered adequate (Hopkins 1999; McGrew & Marchant 1991), but on current evidence there is no sign of this distinctively human trait in wild chimpanzees (McGrew & Marchant 1996).

Gorilla: the gorilla species

Gorillas exist in two widely separated populations whose obvious morphological differences are at last leading towards their recognition as species (Groves 2000), the western *Gorilla gorilla* and the eastern *Gorilla beringei*. All captive studies concern the former, whereas all field studies on manual behaviour concern one subspecies of the latter, the mountain gorilla *G. b. beringei*. The gorilla hand is more human proportioned than the chimpanzee hand (Christel 1993; Napier 1960, 1961) and in captivity, western gorillas readily make tools (McGrew 1989; Parker *et al.* 1999), but no tool use of any sort has been reported from the wild. Instead, mountain gorillas use complex and skilful manual techniques for gathering and processing plant material (Byrne & Byrne 1991, 1993; Byrne, Corp & Byrne 2001a; Schaller 1963), and it is in these tasks that the clearest evidence of cognitive skill is shown.

Mountain gorillas need considerable manual skill to obtain adequate nutrition, because their four major foods are all plants that are "defended" physically in ways that impede consumption (Byrne 1999b). Thus, *Laportea* nettles are covered with stinging hairs, *Carduus* thistles with spines, *Peucedanum* celery with hard outer casing, and *Galium* bedstraw with tiny hooks. Gorillas' techniques for dealing with these problems are complex, with several different actions organized into a regular sequence that is effective in removing or rendering harmless the various defenses while efficient in rapidly amassing plant matter ready for eating (Byrne, Corp & Byrne 2001b).

Individual actions show considerable precision handling and bimanual role differentiation (Byrne *et al.* 2001a), for instance the deft folding and re-grasping of a whorl of sting-covered nettle leaves, which wraps the worst stings safely inside a single leaf. The organization of the tasks is not simply a chain, but a flexible hierarchy of control (Byrne & Byrne 1993; Byrne & Russon 1998). Hierarchical control means that processing stages which are occasionally unnecessary may be omitted (e.g., cleaning off debris before eating), local difficulties during the execution of a sequence can be handled by several alternative processes (e.g., substituting unimanual for normal bimanual accumulation of leaves, when one hand is needed for support in a tree), and series of processes may be treated as a single subroutine (such as iteration of the processes of procuring a nettle, stripping the stem of leaves, and removing petioles from those leaves, to the criterion of an adequately sized handful).

Finally, mountain gorilla manual lateralization is very strong in plant preparation. Techniques for consuming these four main foods are all bimanual, with left- and right-handed forms differing in which particular actions are done by each hand. For all four foods, almost every individual in the study population of 38 showed very strong preference for either right- or left-handed methods; almost none were ambidextrous (Byrne & Byrne 1991). Moreover, for processing both celery pith and thistle leaves, individuals with the strongest lateralization were the quickest to prepare handfuls for eating; as in chimpanzees, lateralization makes for more efficient performance (Byrne & Byrne 1991). Intriguingly, a mountain gorilla seems to have two "hand preferences," for leaf and stem processing respectively. For the three very different techniques by which leaves are processed, gorillas that were (say) right-handed on one task were also right-handed on the other two, just as in most skilled manual tasks in humans. However, knowing an individual's laterality of leaf processing does not in the slightest predict its (equally strong) hand preference for stem processing. At the population level, no handedness was found for stem processing but hand preferences for each of the three leaf-processing tasks were significantly skewed. More individuals preferred performing the most delicate actions right-handed, with the left hand giving grip support (Byrne & Byrne 1991). The strongest bias was found for *Carduus* thistle, which was 64% right-handed and individuals with an exclusive hand preference (i.e., 100% for left or right) were significantly right biased (McGrew & Marchant 1996). This pattern seems to be the closest to human handedness yet found in any animal: several different tasks each evoke strong behavioral laterality, such that individuals have the same preference for each, and over the population there is a significant bias towards right-handed fine manipulation in all of them.

Comparing the gorilla data with the cognitive skills evidenced in human tool making, conspicuous absences are the lack of aimed blows (which may simply reflect lack of need) and of corrective guidance based on anticipatory schema (which, as already noted, is difficult to detect). What is striking is that gorilla plant feeding without tools provides remarkably similar evidence of complex cognition to that provided by chimpanzee tool use and manufacture. Indeed, but for west African chimpanzees' stone hammer and anvil use, gorillas would furnish *better* evidence of cognitive sophistication in manual skill, because hierarchical organization is more firmly established and hand preferences at individual and population level are stronger. This *Pan/Gorilla* comparison supports a picture of the two genera as cognitively rather similar (Byrne 1996), rather than of the gorilla as having lost many cognitive capacities still present in the chimpanzee (Povinelli 1993).

Pongo: the orangutan subspecies

Although Bornean *Pongo pygmaeus pygmaeus* and Sumatran *Pongo pygmaeus abelii* orangutans differ as much genetically as the two chimpanzee species (Begun 1999), they are usually treated as one species. Comparing the two is complicated by the fact that their forest ecology differs markedly, at least at well-studied sites. Bornean forests are impoverished in fruit production compared with Sumatran forests. In Sumatra, tigers still range and large fruiting trees occur that act as magnets to

orangutans. Therefore, Sumatrans more often congregate in groups and Borneans may rely more heavily on difficult fallback foods like bark (van Schaik, Deaner & Merrill 1999).

Wild orangutan tool use is rare. Individuals probe into arboreal bees' nests with a stick to obtain grubs and honey; they use a stick to scrape out irritating hairs within *Neesia* fruit, then prise the edible seeds from the husk so they can be safely eaten (Fox *et al.* 1999). These tool habits reveal a basic difference from African apes. Whereas chimpanzees or gorillas would use their hands for the fine motor control needed in comparable tasks, orangutans often transfer the stick to the mouth (Fox *et al.* 1999; see also O'Malley & McGrew 2000; Russon 2002). Orangutan tool using may involve precision "mouthing" more than precision "handling." The much greater mobility of chimpanzee lips from those of gorillas has often been remarked, but the difference with orangutans is apparently even more marked.

Like gorillas, orangutans also confront many challenging plant foods, which often present multiple rather than single defenses (Fox *et al.* 1999; Russon 1998, 1999a, 2003), and in addition their efforts to copy various complex human activities have been studied closely (Russon 1997, 1999b; Russon & Galdikas 1993, 1995). These behaviors provide a rich source of data on manual skill. Delicate care in visually guided precision handling is evident: for instance, when a rehabilitant poured kerosene onto smoldering embers of a fire, poured coffee from one narrow necked bottle into another, or threaded a rope through a metal ring. Hierarchical organization of plans has also been described: for instance, attaining the (prohibited) goal of "washing" laundry with stolen soap entailed a whole series of actions – untying a canoe, rocking it side to side to remove the bilge water, punting it to the otherwise inaccessible raft where laundry and soap could be had, etc. (Byrne & Russon 1998). In eating meristematic tissue from the base of new *Borassodendron borneensis* palm leaves, free-ranging rehabilitants show a systematic and hierarchically organized approach (Russon 1998). The long action sequence in this process is evident, from constructing a clear working zone in the palm's crown, to complex subdivision and extraction of the leaf, to final departure while still eating carefully cached leftovers. Orangutans often begin by lightly fingering the leaf; they seem to be evaluating it because then they either abandon it or proceed with extracting it (Russon pers. commun.). Like Mt. Niéniokoué chimpanzees' nut testing, this and other adjustments to the sequence suggest to observers the use of on-line corrective guidance. Like gorillas, orangutans may have little natural need for percussive tool use, but rehabilitants spontaneously bang termite nest chunks together to crack them open and once hammered a hole through the concrete floor of their cage with scavenged chunks of cement (Russon 2000, pers. commun.), and one captive readily learned to flake stone to produce sharp flakes (but with human tuition and assistance: Wright 1972). Most of these examples concern individuals with some degree of human experience, so better confirmation from wild orangutans is desirable, but present evidence indicates that orangutans show most of the cognitive attributes evident in African great apes.

Evidence on manual lateralization in orangutans is relatively sparse, but Rogers and Kaplan (1996) found no population trend in hand preferences when food processing, and even individual lateral preferences varied widely. Considerable use of bimanual role differentiation in feeding was found in some individuals (Rogers & Kaplan 1996, figure 5), suggesting that such motor control is quite possible in orangutans.

CONCLUSIONS

Table 3.1 offers an attempt to summarize current evidence on the cognitively driven manual skills of the living great apes. Living great apes, to summarize briefly, can use their considerable abilities of precise handling of objects and bimanual role differentiation to construct motor skills that involve a regular, sequential plan of many actions, some of which are hierarchically organized – with resulting flexibility of tool and manual problem solving in the physical domain. Characteristically, these complex skills involve lateralized processing in individuals.

Inevitably, these judgments have an element of subjectivity but in general the *lack* of difference across species is clear, especially among the African apes, as are the cognitive similarities underlying manual and tool skills (see Yamakoshi, Chapter 9, this volume, for consistent findings using a different approach). Partly, this conclusion may reflect the lack of descriptive work sensitive enough to characterize fine details of motor control and planning, and real differences may yet appear. But on the aspects analyzed here, it seems more likely that as evidence accumulates, especially from the less-studied

Table 3.1. *Current evidence on the cognitively driven manual skills of living great apes*

	Pan	Gorilla	Pongo
Precision handling	√	√	√
Accurately aimed, powerful blows	√	?	(√)
Bimanual role differentiation	√	√	(√)
Regular, sequential plan	√	√	(√)
Hierarchical organization	(√)	√	(√)
Corrective guidance by schema	(X)	(X)	?
Strong individual lateralities	√	√	(√)
Population right-handedness	(X)	√	X

Note: The symbol √ indicates substantial positive evidence; X indicates lack of such evidence despite extensive study; brackets indicate that evidence is inconclusive, usually because it came from only one or two individuals; and ? implies that the topic has apparently not been studied. Sources on which these subjective judgments were based are included in the text.

Pongo and *Gorilla*, apparent differences are more likely to disappear except insofar as they reflect responses to ecological need.

In many ways, this suite of capacities in living apes closely resembles that inferred for extinct, bipedal apes on the human line (*Paranthropus, Australopithecus, Homo habilis*, etc.). Other aspects of the behavior of living apes also suggest similarities in cognition to those extinct species. In the elegantly flaked tools from Lokalalei, conclusive evidence of on-line guidance from comparison with a mental anticipation (schema) is apparently lacking (Roche *et al.* 1999). Only with the visible traces of corrections during the manufacture of much later hand axes, by *Homo erectus* and subsequent species, does this become incontrovertible. Alternatively, many archaeologists would argue that guidance by mental schema is strongly suggested in more ancient stone tool repertoires; but equally, many primatologists would argue the same from the skilled behavior of living apes. Nothing like the 90% right-handedness typical of modern human populations is known in any living ape population. However, claims of right-handed manufacture of early stone tools are also controversial (Toth 1985a).

Further, until the reasons for laterality in living apes are better understood, no useful comparison can be made. Strong individual laterality is associated with increased efficiency (*Gorilla*: Byrne & Byrne 1991; *Pan*: McGrew & Marchant 1999), but population right-handedness is at best relatively weak, and shown convincingly in the wild only in gorilla leaf-processing tasks (Byrne & Byrne 1991; McGrew & Marchant 1996).

The question then becomes, where do the real *differences* in cognitive capacities lie and are these differences likely to be critical ones for human evolution? In Table 3.1, chimpanzees (and, less conclusively, orangutans) are noted as able to aim relatively accurate and powerful blows, most clearly shown by chimpanzees using stone hammers and anvils. Hammer and anvil use is much slower to acquire than any other manual skill in any ape species (Boesch & Boesch 1983) and experimental induction of stone flaking in one orangutan and one bonobo did not result in either individual learning to strike flakes off a hand-held core. Evidently, accurately aimed hitting does not come easily to living apes. Moreover, apes' level of accuracy is unlikely to be anywhere near that shown in the stone tools of Lokalalei, where there is evidence of very precise control of force and blows to detach flakes in a regular, planned sequence. The ability to control blows this precisely aimed but still powerful seems to be a crucial adaptation of the human lineage. Incorporating these refined actions into an organized, planned sequential program is something that apes already do. (It seems a suspicious coincidence that the cognitive capacity that emerges as crucial happens to be almost the only one that current archaeological methods are capable of showing before 2 Ma. Perhaps the Lokalalei tool makers had other skills we can only guess at.)

These conclusions suggest that human manual skill has a relatively *long* evolutionary history, which can usefully be studied in living apes as well as archaeologically. The extinct bipedal apes of 4 to 2 Ma that made stone tools were very different animals to any other living species, but the cognitive capacities of chimpanzees, gorillas and orangutans are appropriately compared to those of these first stone tool makers.

In beginning to make such comparisons and developing an integrated understanding of the cognition that lies behind manual skills in both human and nonhuman apes, it is important that all evidence of advanced manual skill be utilized. My frequent and (I hope) telling uses of data from plant processing show that evidence should not be sought only from skills involving tools. Tool use *per se* has a mystique that risks distorting our perspective away from recognizing other manifestations of complex manual ability. Focusing attention on the cognitive capacities that skilled behavior can indicate, whether or not tools are involved, should allow a better understanding of great ape as well as human intellectual origins.

ENDNOTE

1 A single name for this group of species would be convenient, and traditionally the term was "hominids," defined as extinct relatives of modern humans that were bipedal, more closely related to ourselves than to any living animal, but not quite human. Usage varied slightly as to whether *Homo* species other than *sapiens* were described as "hominids" or simply humans, but genera like *Australopithecus*, *Paranthropus*, and *Ardipithecus* were always referred to as (early) hominids. Unfortunately, "hominid" now has at least two meanings. Modern taxonomy recognizes the remarkably close relationship between humans and the living great apes (Begun 1999) and now includes some or all of them among the hominids. Sometimes only the African great apes, including *Homo*, are included, with the Asian orangutans remaining in the family Pongidae; sometimes all the great apes, including *Homo* and also *Pongo*, are treated as hominids. Meanwhile, many paleontologists keep to the original usage. For clarity, the term is avoided in this chapter.

REFERENCES

Badrian, A. & Badrian, N. (1984). Social organisation of *Pan paniscus* in the Lomako forest, Zaire. In *The Pygmy Chimpanzee: Evolutionary Biology and Behavior*, ed. R. L. Susman, pp. 325–36. New York: Plenum Press.

Badrian, N. L. & Malenky, R. K. (1984). Feeding ecology of *Pan paniscus* in the Lomako Forest, Zaire. In *The Pygmy Chimpanzee: Evolutionary Biology and Behavior*, ed. R. L. Susman, pp. 275–99. New York: Plenum.

Beck, B. B. (1980). *Animal Tool Behavior*. New York: Garland Press.

Begun, D. R. (1999). Hominid family values: morphological and molecular data on relations among the great apes and humans. In *The Mentalities of Gorillas and Orangutans: Comparative Perspectives*, ed. S. T. Parker, R. W. Mitchell & H. L. Miles, pp. 3–42. Cambridge, UK: Cambridge University Press.

Boesch, C. (1991). Handedness in wild chimpanzees. *International Journal of Primatology*, **12**, 541–58.

Boesch, C. & Boesch, H. (1983). Optimisation of nut-cracking with natural hammers by wild chimpanzees. *Behaviour*, **83**, 265–86.

(1984). Mental map in wild chimpanzees: an analysis of hammer transports for nut cracking. *Primates*, **25**, 160–70.

(1990). Tool use and tool making in wild chimpanzees. *Folia Primatologica*, **54**, 86–99.

Bradshaw, J. & Rogers, L. (1993). *The Evolution of Lateral Asymmetries, Language, Tool Use, and Intellect*. San Diego, CA: Academic Press.

Brewer, S. & McGrew, W. C. (1990). Chimpanzee use of a tool-set to get honey. *Folia Primatologica*, **54**, 100–4.

Bril, B., Roux, V. & Dietrich, G. (2000). Skills involved in the knapping of chalcedony beads: motor and cognitive characteristics of a complex situated action. In *Cornaline de L'Inde: Des Pratiques Techniques de Cambay aux Techno-systèmes de L'Indus*, ed. V. Roux, pp. 207–325. Paris: Éditions de la Maison des Sciences de l'Homme.

Bruner, J. (1970). The growth and structure of skill. In *Mechanisms of Motor Skill Development*, ed. K. J. Connolly, pp. 63–94. New York: Academic Press.

Byrne, R. W. (1996). The misunderstood ape: cognitive skills of the gorilla. In *Reaching Into Thought: The Minds of the Great Apes*, ed. A. E. Russon, K. A. Bard, & S. T. Parker, pp. 111–30. Cambridge, UK: Cambridge University Press.

(1999a). Imitation without intentionality. Using string parsing to copy the organization of behaviour. *Animal Cognition*, **2**, 63–72.

(1999b). Cognition in great ape ecology. Skill-learning ability opens up foraging opportunities. *Symposia of the Zoological Society of London*, **72**, 333–50.

Byrne, R. W. & Byrne, J. M. E. (1991). Hand preferences in the skilled gathering tasks of mountain gorillas (*Gorilla g. beringei*). *Cortex*, **27**, 521–46.

(1993). Complex leaf-gathering skills of mountain gorillas (*Gorilla g. beringei*): variability and standardization. *American Journal of Primatology*, **31**, 241–61.

Byrne, R. W. & Russon, A. E. (1998). Learning by imitation: a hierarchical approach. *Behavioural and Brain Sciences*, **21**, 667–721.

Byrne, R. W., Corp, N. & Byrne, J. M. E. (2001a). Manual dexterity in the gorilla: bimanual and digit role differentiation in a natural task. *Animal Cognition*, **4**, 347–61.

(2001b). Estimating the complexity of animal behaviour: how mountain gorillas eat thistles. *Behaviour*, **138**, 525–57.

Chevalier-Skolnikoff, S. & Liska, J. (1993). Tool use by wild and captive elephants. *Animal Behaviour*, **46**(2), 209–19.

Christel, M. (1993). Grasping techniques and hand preferences in Hominoidea. In *Hands of Primates*, ed. H. Preuschoft & D. J. Chivers, pp. 91–108. New York: Springer Verlag.

Connolly, K. J. (1998). *The Psychobiology of the Hand*. Cambridge, UK: MacKeith Press.

de Groot, A. D. (1965). *Thought and Choice in Chess*. The Hague, NL: Mouton.

Elliott, J. M. & Connolly, K. J. (1974). Hierarchical structure in skill development. In *The Growth of Competence*, ed. K. Connolly & K. Bruner, pp. 135–68. London: Academic Press.

(1984). A classification of manipulative hand movements. *Developmental Medicine & Child Neurology*, **26**, 283–96.

Estes, J. A., Riedman, M. L., Staedler, M. M., Tinker, M. T. & Lyon, B. E. (2003). Individual variation in prey selection by sea otters: patterns, causes and implications. *Journal of Animal Ecology*, **72**, 144–55.

Finch, G. (1941). Chimpanzee handedness. *Science*, **94**, 117–8.

Fitts, P. M. & Posner, M. I. (1967). *Human Performance*. Belmont, CA: Brooks/Cole Publishing Company.

Fox, E., Sitompul, A. & van Schaik, C. P. (1999). Intelligent tool use in wild Sumatran orangutans. In *The Mentality of Gorillas and Orangutans: Comparative Perspectives*, ed. S. T. Parker, R. W. Mitchell & H. L. Miles, pp. 99–116. Cambridge, UK: Cambridge University Press.

Goodall, J. (1964). Tool-using and aimed throwing in a community of free-living chimpanzees. *Nature*, **201**, 1264–6.

(1986). *The Chimpanzees of Gombe: Patterns of Behavior*. Cambridge, MA: Harvard University Press.

Goodall, J. van Lawick (1973). Cultural elements in a chimpanzee community. In *Precultural Primate Behaviour*, ed. E. W. Menzel, pp. 144–84. Basel: Karger.

Goodall, J. van Lawick & van Lawick, H. (1966). On the use of tools by the Egyptian vulture, *Neophron percnopterus*. *Nature*, **212**, 1468–9.

Gosselain, O. P. (2000). Materializing identities: an African perspective. *Journal of Archaeological Method and Theory*, **7**(3), 187–217.

Groves, C. (2000). What, if anything, is taxonomy? *Gorilla Journal*, **21**, 12–5.

Hopkins, W. D. (1995). Hand preferences for a coordinated bimanual task in 110 chimpanzees (*Pan troglodytes*): cross-sectional analysis. *Journal of Comparative Psychology*, **109**, 291–7.

(1999). On the other hand: statistical issues in the assessment and interpretation of hand preference data in nonhuman primates. *International Journal of Primatology*, **20**, 851–66.

Hopkins, W. D. & Morris, R. D. (1993). Handedness in great apes: a review of findings. *International Journal of Primatology*, **14**, 1–25.

Humle, T. & Matsuzawa, T. (2002). Ant-dipping among the chimpanzees of Bossou, Guinea, and some comparisons with other sites. *American Journal of Primatology*, **58**, 133–48.

Hunt, G. R. (1996). Manufacture and use of hook-tools by New Caledonian crows. *Nature*, **379**, 249–51.

(2000). Tool use by the New Caledonian crow *Corvus moneduloides* to obtain Cerambycidae from dead wood. *Emu*, **100**, 109–14.

Ingmanson, E. J. (1989). Branch dragging by pygmy chimpanzees at Wamba, Zaire – the use of objects to facilitate social communication in the wild. *American Journal of Physical Anthropology*, **78**, 244.

(1996). Tool-using behavior in wild *Pan paniscus*: social and ecological considerations. In *Reaching into Thought: The Minds of the Great Apes*, ed. A. E. Russon, K. A. Bard & S. T. Parker, pp. 190–210. Cambridge, UK: Cambridge University Press.

Inizian, M.-L., Reduron-Ballinger, M., Roche, H. & Tixier, J. (1999). *Technology and Terminology of Knapped Stone*. Nanterre: CREP.

Joulian, F. (1996). Comparing chimpanzee and early hominid techniques: some contributions to cultural and cognitive questions. In *Modelling the Early Human Mind*, ed. P. Mellars & K. R. Gibson, pp. 173–89. Cambridge, UK: McDonald Institute for Archaeological Research.

Kano, T. (1982). The social group of pygmy chimpanzees of Wamba. *Primates*, **23**, 171–88.

(1983). An ecological study of the pygmy chimpanzees (*Pan paniscus*) of Yalosidi, Republic of Zaire. *International Journal of Primatology*, **4**, 1–31.

Kortlandt, A. (1967). Experimentation with chimpanzees in the wild. In *Neue Ergebnisse der Primatologie*, ed. D. Starck, R. Schneider, & H. J. Kuhn, pp. 208–24. Stuttgart: Gustau Fisher Verlag.

Lashley, K. S. (1951). The problem of serial order in behaviour. In *Cerebral Mechanisms in Behaviour: The*

Hixon Symposium, ed. L. A. Jeffress, pp. 112–36. New York: Wiley.

Limongelli, L., Boysen, S. T. & Visalberghi, E. (1995). Comprehension of cause–effect relations in a tool-using task by chimpanzees (*Pan troglodytes*). *Journal of Comparative Psychology*, **109**, 18–26.

Marchant, L. F. & McGrew, W. C. (1991). Laterality of function in apes: a meta-analysis of function. *Journal of Human Evolution*, **21**, 425–38.

(1996). Laterality of limb function in wild chimpanzees of Gombe National Park: comprehensive study of spontaneous activities. *Journal of Human Evolution*, **30**, 427–43.

Marchant, L. F., McGrew, W. C. & Eibl-Eibesfeldt, I. (1995). Is human handedness universal? Ethological analyses from three traditional cultures. *Ethology*, **101**, 239–58.

Marzke, M. W. & Wullstein, K. L. (1995). Chimpanzee and human grips: a new classification with a focus on evolutionary morphology. *International Journal of Primatology*, **17**(1), 117–39.

Matsuzawa, T. (1996). Chimpanzee intelligence in nature and captivity: isomorphism of symbol use and tool use. In *Great Ape Societies*, ed. W. C. McGrew, L. F. Marchant, & T. Nishida, pp. 196–209. Cambridge, UK: Cambridge University Press.

(2001). Primate foundations of human intelligence: a view of tool use in nonhuman primates and fossil hominids. In *Primate Origins of Human Cognition and Behavior*, ed. T. Matsuzawa, pp. 3–25. Tokyo: Springer-Verlag.

McGrew, W. C. (1974). Tool use by wild chimpanzees feeding on driver ants. *Journal of Human Evolution*, **3**, 501–8.

(1989). Why is ape tool use so confusing? In *Comparative Socioecology: The Behavioural Ecology of Humans and Other Mammals*, ed. V. Standen & R. A. Foley, pp. 457–72. Oxford: Blackwell Scientific Publications.

(1992). *Chimpanzee Material Culture: Implications for Human Evolution*. Cambridge, UK: Cambridge University Press.

McGrew, W. C. & Marchant, L. F. (1991). Laterality of function in apes: a critical review. *American Journal of Physical Anthropology*, **84**(Supplement 12), 129–30.

(1992). Chimpanzees, tools, and termites: hand preference or handedness? *Current Anthropology*, **33**, 114–19.

(1996). On which side of the apes? Ethological study of laterality of hand use. In *Great Ape Societies*, ed. W. C. McGrew, L. F. Marchant, & T. Nishida, pp. 255–72. Cambridge, UK: Cambridge University Press.

(1999). Laterality of hand use pays off in foraging success for wild chimpanzees. *Primates*, **40**(3), 509–13.

McGrew, W. C., Marchant, L. F., Wrangham, R. W. & Klein, H. (1999). Manual laterality in anvil use: wild chimpanzees cracking *Strychnos* fruits. *Laterality*, **4**, 79–87.

McManus, I. C. (1984). Genetics of handedness in relation to language disorder. *Advances in Neurology*, **42**, 125–38.

Mellars, P. A. & Stringer, C. (ed.) (1989). *The Human Revolution: Behavioural and Biological Perspectives on the Origins of Modern Humans*. Princeton, NJ: Princeton University Press.

Miller, G. A., Galanter, E. & Pribram, K. (1960). *Plans and the Structure of Behavior*. New York: Holt, Rinehart and Winston.

Napier, J. R. (1960). Studies of the hands of living primates. *Proceedings of the Zoological Society of London*, **134**, 647–57.

(1961). Prehensility and opposability in the hands of primates. *Symposia of the Zoological Society of London*, **5**, 115–32.

Nishida, T. (1986). Local traditions and cultural transmission. In *Primate Societies*, ed. B. B. Smuts, D. L. Cheney, R. M. Seyfarth, R. W. Wrangham & T. T. Struhsaker, pp. 462–74. Chicago: University of Chicago Press.

Oakley, K. P. (1949). *Man the Tool Maker*. London: Trustees of the British Museum.

O'Malley, R. C. & McGrew, W. C. (2000). Oral tool use by captive orangutans (*Pongo pygmaeus*). *Folia Primatologica*, **71** (5), 334–41.

Parker, S. T., Kerr, M., Markowitz, H. & Gould, J. (1999). A survey of tool use in zoo gorillas. In *The Mentalities of Gorillas and Orangutans: Comparative Perspectives*, ed. S. T. Parker, R. W. Mitchell & H. L. Miles, pp. 188–93. Cambridge, UK: Cambridge University Press.

Pelegrin, J. (2000). The knapping methods and techniques practiced at Cambay. In *Cornaline de l'Inde: Des Pratiques Techniques de Cambay aux Techno-systèmes de l'Indus*, ed. V. Roux, pp. 55–93. Paris: Éditions de la Maison des Sciences de l'Homme.

Povinelli, D. J. (1993). Reconstructing the evolution of mind. *American Psychologist*, **48**, 493–509.

Reynolds, P. C. (1982). The primate constructional system: the theory and description of instrumental tool use in humans and chimpanzees. In *The Analysis of Action*, ed. M. Van Cranach & R. Haas, pp. 243–385. Cambridge, UK: Cambridge University Press.

Roche, H. (1989). Technological evolution in early hominids. *Ossa*, **14**, 97–8.

Roche, H., Delegnes, A., Brugal, J.-P., Feibel, C., Kibunjia, M., Mourre, V. & Texier, P.-J. (1999). Early hominid stone production and technical skill 2.34 Myr ago in West Turkana, Kenya. *Nature*, **399**, 57–60.

Rogers, L. J. & Kaplan, G. (1996). Hand preferences and other lateral biases in rehabilitated orang-utans, *Pongo pygmaeus pygmaeus*. *Animal Behaviour*, **51**(1), 13–25.

Roux, V. (ed.) (2000). *Cornaline de l'Inde: Des Pratiques Techniques de Cambay aux Techno-systèmes de l'Indus*. Paris: Éditions de la Maison des Sciences de l'Homme.

Russon, A. E. (1997). Exploiting the expertise of others. In *Machiavellian Intelligence II: Extensions and Evaluations*, ed. A. Whiten & R. W. Byrne, pp. 174–206. Cambridge, UK: Cambridge University Press.

(1998). The nature and evolution of intelligence in orangutans (*Pongo pygmaeus*). *Primates*, **39**, 485–503.

(1999a). Naturalistic approaches to orangutan intelligence and the question of enculturation. *International Journal of Comparative Psychology*, **12**, 1–22.

(1999b). Orangutans' imitation of tool use: a cognitive interpretation. In *The Mentalities of Gorillas and Orangutans: Comparative Perspectives*, ed. S. T. Parker, R. M. Mitchell & H. L. Miles, pp. 117–46. Cambridge, UK: Cambridge University Press.

(2000). *Orangutans: Wizards of the Rainforest*. Toronto: Key Porter Publications.

(2002). Return of the native: cognition and site-specific expertise in orangutan rehabilitation. *International Journal of Primatology*, **23**, 461–78.

(2003). Comparative developmental perspectives on culture: the great apes. In *Between Biology and Culture: Perspectives on Ontogenetic Development*, ed. H. Keller, Y. H. Poortinga & A. Schoelmerich, pp. 30–56. Cambridge, UK: Cambridge University Press.

Russon, A. E. & Galdikas, B. M. F. (1993). Imitation in free-ranging rehabilitant orangutans. *Journal of Comparative Psychology*, **107**, 147–61.

(1995). Constraints on great ape imitation: model and action selectivity in rehabilitant orangutan (*Pongo pygmaeus*) imitation. *Journal of Comparative Psychology*, **109**, 5–17.

Schaller, G. B. (1963). *The Mountain Gorilla*. Chicago, IL: Chicago University Press.

Stokes, E. J. & Byrne, R. W. (2001). Cognitive capacities for behavioural flexibility in wild chimpanzees (*Pan troglodytes*): the effect of snare injury on complex manual food processing. *Animal Cognition*, **4**, 11–28.

Sugiyama, Y. (1994). Tool use by wild chimpanzees. *Nature*, **367**, 327.

(1997). Social tradition and the use of tool-composites by wild chimpanzees. *Evolutionary Anthropology*, **6**, 23–7.

Sugiyama, Y. & Koman, J. (1979). Tool-using and tool-making behaviour in wild chimpanzees at Bossou, Guinea. *Primates*, **20**, 513–24.

Sugiyama, Y., Fushimi, T., Sakura, O. & Matsuzawa, T. (1993). Hand preference and tool use in wild chimpanzees. *Primates*, **34**, 151–9.

Suzuki, S., Kuroda, S. & Nishihara, T. (1995). Tool-set for termite-fishing by chimpanzees in the Ndoki Forest, Congo. *Behaviour*, **132**, 219–35.

Toth, N. (1985a). Archaeological evidence for preferential right-handedness in the lower and middle Pleistocene, and its possible implications. *Journal of Human Evolution*, **14**, 607–14.

(1985b). The Oldowan reassessed: a close look at early stone artifacts. *Journal of Archaeological Science*, **12**, 101–20.

Toth, N., Schick, K. D., Savage-Rumbaugh, E. S., Sevcik, R. A. & Rumbaugh, D. M. (1993). *Pan* the tool-maker: investigations into the stone-tool-making and tool-using capabilities of a bonobo (*Pan paniscus*). *Journal of Archaeological Science*, **20**, 81–91.

van Schaik, C. P. (1994). Tool-use in wild Sumatran orangutans (*Pongo pygmaeus*). Presentation at XVth Congress of International Primatological Society, August 3–8, Kuta, Bali.

van Schaik, C. P., Deaner, R. O. & Merrill, M. Y. (1999). The conditions for tool-use in primates: implications for the evolution of material culture. *Journal of Human Evolution*, **36**, 719–41.

Wright, R. V. S. (1972). Imitative learning of a flaked-tool technology – the case of an orang-utan. *Mankind*, **8**, 296–306.

Wynn, T. (1988). Tools and the evolution of human intelligence. In *Machiavellian Intelligence: Social Expertise and the Evolution of Intellect in Monkeys, Apes and Humans*, ed. R. W. Byrne & A. Whiten, pp. 271–84. Oxford: Clarendon Press.

Wynn, T. & McGrew, W. C. (1989). An ape's view of the Oldowan. *Man*, **24**, 383–98.

4 · The cognitive complexity of social organization and socialization in wild baboons and chimpanzees: guided participation, socializing interactions, and event representation

SUE TAYLOR PARKER

Anthropology Dept., Sonoma State University, Rohnert Park

INTRODUCTION

Is the intelligence of monkeys and apes primarily a "Machiavellian" adaptation for social life, an ecological adaptation for resource exploitation, or both a social *and* an ecological adaptation? How can we evaluate these competing models of intelligent adaptation? Are great apes really smarter than monkeys? How can we assess the cognitive complexity of social life? Are the social organizations and/or social roles of great apes more complex than those of monkeys? If they are, do different developmental and acquisition processes underlie that greater complexity, and which models are useful for studying those processes? Similarly, how is socioecological knowledge distributed and transmitted from one generation to the next, and what models are useful for studying this process? All these questions plague efforts to characterize great ape cognition and to reconstruct its evolution.

Systematic evaluation of social and/or ecological hypotheses about primate cognition calls for an integrated framework for comparing the sophistication of social organization and socialization in wild primates. Despite keen interest in specific topics in social cognition (e.g., social learning, social communication), a comprehensive ecologically based framework for comparing primate social cognition is currently lacking.

In this chapter, I propose such a framework. Specifically, this comparison can be made in terms of (1) the number, function, and composition of *activity subgroups* within primate *range groups*; (2) the *social roles* the members of these activity subgroups play; and (3) the typical routines or *scripts* in which these roles are played. I further propose that individuals acquire their roles through age-, sex-, species-, and context-typical *socializing interactions*, which occur during *guided participation* in routine activities with other group members. Finally, I propose that the nature and cognitive complexity of these socializing interactions depend on species-typical forms of *event representation*.

In an attempt to characterize unique features of great ape social and ecological adaptations, I focus primarily on comparing chimpanzees with Hamadryas baboons (though both are also compared with their close relatives). As our closest living relatives, chimpanzees are an obvious choice, but the choice of Hamadryas is less obvious. It is based on several factors. First, baboons probably are the most intelligent species in their Old World monkey clade. They display the longest development and the largest brains of any Old World monkeys (Harvey, Martin & Clutton-Brock 1987), characteristics that correlate with intelligence. Although there are few systematic comparative studies of baboon intelligence, field workers credit baboons with great behavioral adaptability (Hamilton, Buskirk & Buskirk 1978; Hamilton & Tilson 1985; Strum 1987). Therefore, they are unlikely to under-represent the cognitive complexity of monkey social life and can stand for other Old World monkey species. Second, like chimpanzees (and possibly all great apes: see van Schaik, Preuschoft & Watts, Chapter 11, Yamagiwa, Chapter 12, this volume), Hamadryas display fission–fusion social organization, that is, larger groups that break into spatially distinct subgroups (Kummer 1968). Therefore, the comparison is between like social forms. Third, the *socialization* of both Hamadryas and chimpanzees has been studied in the wild. Similar data from other wild populations are unavailable. Before comparing these two species, a brief

Table 4.1. *Socioecological model for primate social organization*

Units	Descriptions	Functions
Populations	Demes or breeding populations containing range groups	Reproductive gene pool from which mates are drawn
Aggregations	Two or more range groups in daily or seasonal association	Sharing a scarce resource such as water or sleeping cliffs
Range groups	Groups sharing (and sometimes defending) an annual home range; members of at least one sex disperse at puberty	Socioecological and reproductive units sharing distributed knowledge of home range
Activity subgroups	Subgroups of range groups engaging in particular daily and seasonal activities	Subunits participating in socioecological and reproductive roles and scripts, locus of socialization

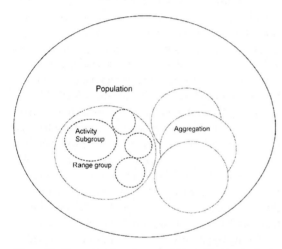

Figure 4.1. Representation of proposed hierarchical levels of primate social organization: *Population* or demes include *range group*, which may sometimes *aggregate* at scarce resources or during breeding seasons, and which are composed of *activity subgroups* of various degrees of spatial and/or temporal discreteness; relative degrees of permeability of various units are denoted by dotted lines.

analysis of primate social organization is offered in Table 4.1 and Figure 4.1, which summarize the model of social organization and socialization and its hierarchical structure respectively.

PRIMATE SOCIAL GROUPS IN THE WILD

Range groups and aggregations

Breeding populations or demes are the basic units of evolution. Populations of nonhuman primates are composed of interbreeding groups that inhabit contiguous or overlapping annual home ranges containing most of the resources they use throughout the year. Groups that share a home range, which I call *range groups*, differ among species in their size, composition, stability, cohesiveness, activity subgroupings, mating systems, and dispersal patterns. I use the term "range group" rather than the familiar "group" to distinguish this from other levels of social organization, and to highlight the level associated with use of a shared home range. Range groups and/or their included females are usually defended by their resident male(s). Many primate species display more than one range-grouping pattern across their species range. Likewise, primate range groups may display more than one pattern over time. Finally, in some cases, range groups aggregate at sites of such scarce resources as water or safe sleeping sites. These *aggregations* can be daily (e.g., Kummer 1968) or seasonal (e.g., Cords 1986).

Like the larger breeding populations of which they are constituents, range groups undergo *demographic cycles* as individuals are born, mature, emigrate, immigrate, and die. Individuals emigrate and immigrate among range groups within a breeding population. In some species, for example, one-male harem groups are periodically overtaken by all-male groups whose members mate with resident females before competing among themselves to take over the group as harem master (Cords 1988). The core of range groups may be composed of philopatric matrilines (e.g., most Old World monkeys) or patrilines (e.g., chimpanzees, bonobos, red and olive colobus), with members of the opposite sex dispersing at sexual maturity (Pusey & Packer 1987).

Alternatively, individuals of both sexes may disperse at puberty (e.g., Hamadryas baboons (Kummer 1997), gibbons (Fuentes 2000), several New World monkeys (Strier 1994, 1996)). According to range group size and composition, and male and female mating strategies, mating systems may be monogamous, polygynous, polyandrous, promiscuous (polygynous/polyandrous), or a combination of these.

Range groups differ in their internal structure and spatial cohesiveness as well as their size, composition, and mating systems. Some large multimale–multifemale range groups (as in many Old World monkeys) move together as monounits throughout their daily activity cycle, maintaining visual and/or auditory contact in accord with the visual and acoustic properties of their habitats. Others, like chimpanzees and bonobos (Fruth, Hohmann & McGrew 1999), orangutans (van Schaik 1999), Hamadryas baboons (Kummer 1971), and spider monkeys (Chapman, Wrangham & Chapman 1995), display a fission–fusion pattern, dispersing and regrouping according to resource distribution and densities.

Like other animals, wild primates display *daily activity cycles* that include sleeping, socializing, mating, progressing to feeding, drinking, and sleeping sites, and, often, interacting with other range groups in the population. These patterns are often seasonal (e.g., Altmann & Altmann 1970; Goodall 1986).

I envision differences in primate social organization in terms of the composition and spatial distribution of *activity subgroups* within range groups, that is, the ephemeral groups involved in specific activities within the daily or annual activity cycle.

Activity subgroups

Each activity performed by range group members can be characterized by the number and composition of participants (i.e., the activity subgroup), the subgroup's spatial relationship with other activity subgroups, and the location and duration of the activity. Common activity subgroups in wild primates are those for sleeping, progressing, foraging/feeding, socializing/grooming, playing, caretaking, mating, range boundary defense, anti-predator defense, and hunting (primarily in chimpanzees). Species differences in social organization arise from differences in the kinds of range groups and activity subgroups they display as well as these subgroup qualities. In some species, for example, all range group members forage and feed together while in other

species, they break into separate subgroups to forage and feed. Subgroups in some species may rejoin in the evening, in others they may travel separately for days.

Range groups are persistent organizations that contain the relatively stable pool of interactants from which activity subgroups are formed. They accommodate the formation and dissolution of various daily, seasonal, and annual activity subgroups that promote the survival, reproductive, and/or genealogical interests of their participants. The size, composition, cohesiveness, and even the existence of activity subgroups constitute flexible responses to the density and dispersion of various resources, the reproductive status of females, and the demography of other range groups in the same population.

In some terrestrial Old World monkeys' range groups, the composition of activity subgroups may be cryptic because many daily activities (e.g., progressing from place to place, foraging) involve the entire range group. Other activities, like playing, caretaking, or mating, are performed by a subgroup while the whole range group remains in close physical proximity and moves as a stable monounit. This pattern is characteristic of vervets (Struhsaker 1967), olive baboons (Smuts 1985), yellow baboons (Altmann & Altmann 1970), and Barbary macaques (Mehlman 1989), which live in large multimale–multifemale groups and mate promiscuously.

Fission–fusion social organization in Hamadryas baboons and chimpanzees

Activity subgroups within range groups are spatially distinct and obvious in the fission–fusion social organization of Hamadryas baboons, which exhibits a multilevel social structure that accommodates to differing distributions of sleeping sites, water, and food. The smallest unit, the *one-male harem* or family unit, is the minimum foraging unit feeding on dispersed acacia trees in the dry scrubland of Ethiopia. The next largest unit, the *clan*, is composed of a few families of related males that gather to drink at water holes at midday. The *band*, composed of several clans, uses the same home range. It sets out traveling in the same direction each day and members of its harems socialize (mate, groom, play) at night and in the morning near their sleeping cliffs. Bands may move independently to different sleeping sites, where their males sometimes fight with stranger bands and abduct their females. Finally, the *troop* is an *aggregation* of several familiar bands that assembles at night at sleeping cliffs. Band size varies from 30 to 60 individuals; aggregates can include up to 750 (Kummer 1968, 1997). In the terminology used here, *the band is the range group*, the harem is a foraging and mating subunit, and the clan is a drinking activity subunit. This multi-level structure accommodates different activity subgroups and supergroups appropriate to the daily use both of scarce clumped sleeping cliffs (the troop) and of dispersed food and water resources (harem and clan subgroups). The harem subgroup optimizes reproduction, caretaking, foraging in single trees in desert habitat, and defense against desert predators. The band unit optimizes range use and socialization. The troop optimizes defense against nocturnal predators (Kummer 1968, 1997).

Hamadryas' multilevel harem-based social organization contrasts with the more cohesive monounit social organization of the closely related olive baboon. The two subspecies (Jolly 1993) produce fertile hybrids at the boundaries of their parapatric (overlapping) ranges in Ethiopia. They share many of the same morphological and social signals, but diverge in those crucial to differences in their social organizations.

Olive baboons live in large multimale–multifemale range groups characterized by female *philopatry* and male *dispersal* at puberty. In contrast to Hamadryas, they mate promiscuously, forming short-term consortships in which males follow and groom females for several hours or days during the breeding season. Outside the breeding season, females move freely through the range group; males spend most of their social time interacting with other males and females with other females and their young. Their diet is omnivorous but they depend heavily on grass rhizomes and acacia trees, which are abundant and fairly evenly distributed in their savanna woodland to forest habitats. Their home ranges overlap with those of other range groups but most groups sleep in trees in nonoverlapping core areas (Rowell 1972; Smuts 1985; Strum 1987). In appearance, adult olive baboons are larger than Hamadryas with olive colored fur and black faces, hands, and genitals.

Notable among the *uniquely derived characteristics* of Hamadryas are: (1) bright red coloration of the face and genitalia and long silvery manes of adult males, which are attractive to Hamadryas females, (2) jealousy-motivated possessive herding and neck biting behavior

used by harem leaders to herd female harem members, and (3) juvenile and subadult Hamadryas males' nurturing of female infants, who form their initial harem.[1] All these characteristics apparently arose as adaptations to harem formation and maintenance in a desert environment (Kummer 1997).[2]

Hamadryas' rigidly patterned fission–fusion organization also contrasts sharply with the flexible fission-fusion social organization of chimpanzees (see also van Schaik, Preuschoft & Watts, Chapter 11, this volume). Chimpanzees live in large multimale–multifemale "communities" (range groups) that break into "parties" (activity subgroups) of variable composition and duration. Unlike Hamadryas, chimpanzee species' subgroups only occasionally coalesce into a single group. Unlike Hamadryas subgroups, which are always composed of both sexes, chimpanzee subgroup composition varies temporally and regionally. Chimpanzee subgroups are most often composed of both adult males and females, but they can range in size from large parties to consortships and vary from community to community (see table 5.5 in Boesch & Boesch-Achermann 2000). Females and their offspring often travel alone in all chimpanzee populations, for instance, but they travel together more often in Taï than in Gombe.

Boesch and Boesch-Achermann (2000) argue that party (i.e., activity subgroup) size and composition in Taï chimpanzees is influenced directly by fruit availability, sexual opportunities, and hunting rates. They believe that interpopulation variation in party size and sex ratio is explained by community (i.e., range group) size and adult sex ratio. Since fission–fusion social organization in common chimpanzees and closely related bonobos allows for flexible adaptation to factors that vary across time and space, they argue ". . . we should expect to find a gradient from male bonded to bi-sexually bonded societies" (p. 108).

Other great apes show similar patterns. Bonobos resemble chimpanzees in that their large multimale–multifemale communities break into parties of variable composition, only occasionally coalescing into a single group. Bonobos display greater bi-sexual bonding than chimpanzees, however. Their subgroups also differ: mixed subgroups of adult males and females and subgroups of females are more common, and subgroups of males are less common (Fruth *et al.* 1999).

Orangutans, often described as solitary, are now viewed as displaying a type of dispersed sociality characterized by fission–fusion (van Schaik 1999; van Schaik *et al.*, Chapter 11, this volume). Although mountain gorillas live in cohesive groups of one or two adult males and several adult females and young (Fossey 1983), some lowland gorillas live in multimale–multifemale groups that break into small feeding subgroups when food is scarce (Doran & McNeilage 1998; Goldsmith 1999; Yamagiwa, Chapter 12, this volume). Fuentes (2000) argues that all African great apes are best characterized by "some variant of multimale–multifemale group that exhibits variable cohesion and group/subgroup size along a continuum" (p. 53) and that ancestral Miocene great apes may have lived in multimale–multifemale groups with fission–fusion tendencies.

ROLES AND SCRIPTS IN WILD PRIMATES

Just as the concept of activity subgroups is useful because it focuses attention on the activities, subgroupings, and ecological settings in which wild primates interact, so the concept of social *roles* is useful because it focuses on the *situated behaviors* of participants in activity subgroups. This framework emphasizes that social role performances are situated in particular social, temporal, and ecological settings.

Kummer (1971) and Crook (1970) extended the sociological concept of *social roles* (Mead 1970) to nonhuman primates, to describe typical recurrent behaviors of individuals of a given age, sex, and status. Hinde (1974) also noted that nonhuman primate roles are biologically constrained, usually following from individuals' characteristics of sex, age, and parental and competitive status (see Fedigan 1982 on primatologists' use of role concepts).

Likewise, some primatologists (Mitchell 1999) adopted the term "*script*," first used by cognitive anthropologists and psychologists (Schank & Abelson 1977; Suchman 1987), to describe context-specific sequences of actions or activities that characterize routine events in nonhuman primates. Primatological usage follows that of developmental psychologists, describing the activity-specific frames in which youngsters develop social knowledge: "scripts are temporally organized structures composed of sequences of actions leading to a goal within a specified temporal and spatial locations . . ." (Nelson 1996: 188).

The *script* concept is useful because it focuses attention on the overall *structure and sequence of role performances of activity subgroups* situated in particular settings. Both concepts of roles and scripts are useful for comparing the cognitive complexity of social organization and social behavior in nonhuman primates. I illustrate their utility for contrasting the social behaviors of Hamadyras and chimpanzees.

Kummer's (1997) description of male roles in Hamadryas' daily progression away from the sleeping cliffs provides a clear-cut application of the role concept to nonhuman primates. This activity also involves one of the most complex sets of roles described in wild Old World monkeys. Harem leaders play the role of "initiators" or "determiners" of the band's travel direction each morning, before setting off to forage in the desert. Younger harem leaders play the *initiator role*, "proposing" a direction of travel by setting out in that direction, stopping, and looking back to see how older harem leader males respond. This continues until one of the older males, in the *determiner role*, gets up and follows (Kummer 1997). The choice is critical because it determines which water hole will be used that day.

This daily routine of Hamadryas baboons is easily translated into script language. The daily identifier–determiner (I–D) role interaction among adult males, for example, is part of a *decision-making script* for determining the band's daily travel direction/choice of water hole. This script allows the band to benefit from the distributed knowledge of its members about the state of various food and water resources. Other Hamadryas routines that can be cast into the script language include: (1) daily march script for the family unit going out to forage, drink, and return to sleeping cliffs; (2) male follower script for assessing the vitality of the harem leader and the strength of his females' attachments to him, preparatory to taking over the harem leader role; (3) band males' fighting script in interband disputes; and (4) harem take-over script in the interband context.

An even more complex example of roles is in the collaborative hunting of Taï forest chimpanzees (Boesch & Boesch 1989; Boesch & Boesch-Achermann 2000). Unlike chimpanzees in other regions, forest-living Taï chimpanzees apparently hunt intentionally rather than opportunistically, in groups rather than individually (92% of hunts versus 36% and 23% in Gombe and Mahale), collaboratively more often than just synchronously (68% of hunts versus 19% in Gombe), and

Table 4.2. *Definitions of cooperation in group hunts: four levels of growing complexity of organization between hunters (adapted from Boesch & Boesch 1989: 550)*

Category	Definition	Variation
Similarity	All hunters concentrate similar actions on the same prey, but without any spatial or time relation between them	Similar actions are varying elements of pursuing a prey, i.e., stalk, chase . . .
Synchrony	Each hunter concentrates similar actions on the same prey and tries to relate in time to each other's action	Hunters may begin at the same time or adjust their speed to remain in time
Coordination	Each hunter concentrates similar actions on the same prey and tries to relate in time and space to each other's actions	Hunters may begin from different directions or adjust their position and speed to remain coordinated
Collaboration	Hunters perform different complementary actions, all directed toward the same prey	Examples are driving, blocking escape way, and encirclement

with a specific prey image (red colobus monkeys) rather than adventitiously.

Taï hunting subgroups range in size from one to more than six individuals, but parties of three to five were most frequent and most successful. Most significant is their collaboration, the highest form of cooperation in hunting, defined as performance of "different complementary actions, all directed toward the same prey" (Boesch & Boesch 1989: 550). The Boeschs contrast collaboration with coordination, synchrony, and similarity (defined in Table 4.2), noting that hunting baboons display similarity and synchrony but not collaboration.

Hunting roles include driver, blocker, chaser, encircler or ambusher, and capturer, which the Boeschs describe as complementary, *interchangeable, and varying in difficulty*. Driver and chaser roles are reactions to prey movement. The blocker role involves some coordination and anticipation of prey movements. The *ambusher role* is the most complex because it requires anticipating the movements of drivers and their prey through three dimensions, without visual access to their movements. "The ambusher (or encircler) is the hunter who *anticipates* the escape route of the quarry long enough in advance to be able to force it to turn backwards towards its pursuers or to move downwards into the lower canopy, where chimpanzees have a very good chance of catching it . . ." (Boesch & Boesch-Achermann 2000: 172–3) (my italics). Males typically progress sequentially through these roles during their socialization.

The Boeschs argue that these unequal roles are maintained by a sophisticated reciprocity system that rewards hunters for their *type of contribution* during the hunt (Boesch & Boesch-Achermann 2000). At Gombe, in contrast, meat is distributed according to hunters' dominance status (Stanford 1998).[3]

Like the Hamadryas initiator–determiner routine, the chimpanzee collaborative hunting routine is easily cast in the language of scripts. The entire *collaborative hunting script* involves the following steps: (1) search (silently stopping, clumping, looking up and listening, and changing direction without vocalizing); (2) sighting prey (red colobus monkeys); (3) responding to the prey's reactions on being sighted (freezing, fleeing, attacking, or mobbing); (4) collaborative capture by driving, chasing, blocking, ambush, and capture; (5) killing adult prey; and (6) dividing the prey (through respect, theft, transfer, or division) (Boesch & Boesch 1989; Boesch & Boesch-Achermann 2000).

In addition to the collaborative hunting script, Boeschs' work and other studies reveal other complex routines in wild chimpanzees that can be cast into script language. These include: (1) adult males' boundary patrolling script; (2) males' and estrus females' consort-forming script; (3) nut-cracking script; and (4) nut-cracking-school script.

The *nut-cracking script*, for example, can entail anticipatory (1) searching for appropriate hammerstones; (2) transporting them to distant anvils; and (3) transporting nuts to the anvils; before (4) cracking the nuts with the hammers on the anvils; and (5) extracting and consuming nuts. Like the hunting script, this script involves considerable anticipation and planning (Boesch & Boesch 1984; Boesch & Boesch-Achermann 2000).

The related *nut-cracking-school script* facilitates learning how to crack nuts. It entails infants and juveniles (1) watching their mothers crack nuts; (2) playing with hammerstones, nuts, and anvils in progressively more complex and functional combinations; and (3) consuming nuts their mothers cracked. The mothers' role involves encouraging their offsprings' learning by (1) leaving nuts and hammerstones lying around to be played with; (2) sharing nuts with their offspring; and, in rare cases, (3) demonstrating correct positioning of the hammer for opening nuts (Boesch 1991; Boesch & Boesch-Achermann 2000; Boesch *et al.* 1994). Acquisition of nut-cracking skills occurs over several years, through progressively more complete participation in nut-cracking school, from about 3 to 8 years of age (Inoue-Nakamura & Matsuzawa 1997), as discussed in the socialization section.

Assessing the complexity of roles and scripts

Kummer (1967) pioneered using roles as a basis for assessing the complexity of primate social interactions. He focused on "multipartite," especially tripartite, role interactions as the most complex social interactions described in Old World monkeys. In his classic paper, he described tripartite relations in Hamadryas as ". . . sequences in which *three individuals simultaneously* interact in *three essentially different roles* and *each of them aims its behavior at both* of its partners" (Kummer 1967: 64; author's emphasis). For example, *protected threat* typically occurred when two females competed for position near the harem leader: "the female closer to the male now tries to threaten her opponent away from the male, staying as much as possible between the two, while presenting to the male . . ." (Kummer 1967: 65). All these interactions occurred in agonistic contexts and entailed highly ritualized rather than cognitively complex behaviors. Hamadryas infants also learned to establish triangles, with themselves in the role of protégé, when they ran to their mothers or subadult males for comfort when frightened in play.

Drawing on Kummer's tripartite concept, Tomasello and Call (1997) proposed a model of primate social cognition that promotes the capacity for understanding third party relationships (TPRs) as the cognitive ability that characterizes anthropoid primates and sets them apart from other mammals. Their so-called TPRs involve understanding the interactions of others, which, they argue ". . . require[s] special observational skills to learn, as the observer gains understanding by watching social interactions in which it is not directly participating" (Tomasello & Call 1997: 199). TPRs they attribute to all anthropoid primates include protected threat, grooming competition, recruitment screams, redirected aggression, separating interventions, mediating reconciliations, and respect for "ownership" (Tomasello 1998).[4] They argue that TPRs in all anthropoid species have the same cognitive complexity.

The material presented here suggests, on the contrary, that two taxa of anthropoid primates – Hamadryas and chimpanzees – differ in the cognitive complexity of their social interactions. As described above, chimpanzees' complex and flexible scripts and roles contrast with the more stereotyped scripts and roles of Hamadryas and other baboons, in all the proposed dimensions. This contradicts Tomasello and Call's proposition that all anthropoid primates display the same level of cognitive complexity in their tripartite interactions. It further suggests that the number of participants in an interaction is only one measure of social complexity (Whiten 2000).

My scheme contrasts to Tomasello and Call's in emphasizing the more comprehensive and situated concepts of scripts and roles for describing and assessing the complexity of two forms of primate social organization. According to this framework, script complexity can be measured in terms of the flexibility of response and the degree of anticipation and planning involved, i.e., by (1) the number of interdependent sequential actions preceding the goal action or outcome; (2) the number of alternative actions that can serve the same function at each stage in the sequence; and (3) the number and complexity of interacting roles of the participants. The number of scripts a range group displays is also a measure of cognitive complexity. Finally, cultural variation in scripts among the 14 populations of wild chimpanzees, including hunting and tool-using scripts, is another indication of the cognitive complexity of this species (McGrew 1992; Whiten *et al.* 1999).

Similarly, my socioecological framework suggests that the complexity of roles can be measured by (1) the number of alternative actions that can subserve a given role (which is inversely related to the degree of canalization by innate displays); and/or (2) the degree and kind of contingency and complementarity between one actor's actions and another's reactions; and/or (3) the number of complementary and interdependent roles involved in a script. Finally, the number of roles in a range group is another measure of the cognitive complexity of social organization.

SOCIALIZATION AND APPRENTICESHIP IN HAMADRYAS AND CHIMPANZEES

I suggest that the study of cognitive mechanisms involved in *socialization* is the key to understanding differences in social organization, scripts, and social roles of wild Hamadryas and chimpanzees. Apprenticeship and activity theory from developmental psychology provide useful frameworks for investigating socialization in nonhuman primates.

The concept of *apprenticeship* focuses on the developing individual's *repeated participation* in various activity groups through the regular activity cycle. This guided participation leads to increasingly complex appropriation of constituent behaviors. The concept of apprenticeship inspired by Vygotsky (1978) refers to collaborative problem solving with guidance by a more skilled partner, supporting or scaffolding the child's performance in accord with an intuitive understanding of the child's "proximal zone of development" or current potential for development (Rogoff 1990). *Guided participation* refers to the process of communication in guidance and collaboration. *Appropriation or participatory appropriation* refers to individual change through involvement in activities, i.e., "the process by which individuals transform their skills and understanding through their participation" (Rogoff 1993: 138).

Much of this participatory appropriation occurs during role development. While participating in regular activities, youngsters are learning their roles through socializing interactions and simultaneously learning about progression routes, the spatial and temporal distribution of food, water, and sleeping sites, parts of plants that can be eaten, food preparation techniques, group versus nongroup members, predators and antipredator tactics, etc. (Hall 1968; Hall & DeVore 1965).

Application of these concepts to nonhuman primate socialization is useful because it focuses attention on the means by which immatures come to play their roles in species-typical scripts. According to this model, young nonhuman primates serve an apprenticeship in species-typical behaviors, skills, and roles through repeated guided participation in the scripts of various activity subgroups during their daily and seasonal activity cycles. Although only a few studies of wild nonhuman primates have focused on socialization, available evidence suggests significant differences in the apprenticeships of monkeys, great apes, and humans.

Abegglen's (1984) study of Hamadryas socialization suggests that their apprenticeship is achieved through simple social learning mechanisms and canalized by innate stage-specific affinities, dispositions and attachments, constrained by interference from third parties. Boesch and Boesch-Achermann's (2000) studies suggest that chimpanzee apprenticeship is achieved not only through simple social learning mechanisms but also through imitation and teaching. Humans, unlike Hamadryas or chimpanzees, receive symbolically mediated instruction.

Abegglen's (1984) study demonstrates how interactions based on stage- and species-typical affinities, constrained by triadic interference, socialize young baboons into certain roles. For Hamadryas males, he identifies the following age classes: infant (to 12 months old); juvenile (to 5 years old); subadult (to 11 years old), adult (11 through 14 years old) and old adult (over 14 years old). Infants play in male playgroups and are periodically kidnapped and groomed by younger juvenile males. Older juvenile males periodically kidnap and groom female infants; some of them form "initial units" with these females. Subadult males attach themselves to and follow a harem leader. Adult males, with fully developed mantles and hair at the sides of the head, become harem leaders aggressively herding and protecting their females. Old adults are deposed but continue as troop leaders.

Specifically, Abegglen shows how the peripheral tendencies of young Hamadryas males combine with their attraction to and maternal behavior toward infants, to produce a male and a female socialization script. For female infants the process is continuous, and for male infants, discontinuous. The growing attachment of a young female to an attentive (juvenile) bachelor male, fostered by repeated experiences of being carried and groomed by him, culminates in separation from her mother. By repeatedly participating in these reinforcing interactions, the two become increasingly bonded, the female being rewarded for remaining within a few feet of her male, grooming him, and rushing to him when he threatens her. In contrast, male infants experience only a brief period of interaction with bachelors that ends when bachelors begin to gain access to female infants, who attract them more. The harem leader's greater protection of female infants encourages young bachelors to kidnap male rather than female infants. It also delays and prolongs the separation phase of formation of the bachelor's initial unit. The attachments younger juvenile males form to older juvenile (bachelor) males that kidnap and groom them re-emerge several years later, when they attach themselves to a harem leader as a subadult follower (Abegglen 1984; Kummer 1997).

This study demonstrates that young Hamadryas baboons socialize one another into gender-specific roles through dyadic and triadic interactions based on stage- and population-specific motivations and behaviors. These *socializing interactions* include highly stereotyped "displays" (e.g., male notifying, grooming, neck biting) by individuals of one age/sex category toward others of different age/sex categories. By shaping, rewarding, and reinforcing particular role performances in their senders and receivers, these socializing interactions constitute population-typical developmental scaffolds.

Through repeated guided participation in selective use of local resources (scattered acacia trees, water holes, sleeping sites), subadult and younger adult males gradually acquire ecological knowledge regarding seasonal and longer-term resource variations from older, more experienced males (Kummer 1997). This is a good example of the distributed nature of local knowledge. Acquisition of foraging and feeding behaviors by immatures through participatory appropriation and simple forms of social learning are widely described examples of what I call *socializing interactions* in anthropoid primates (Chevalier-Skolnikoff & Poirier 1977; Hall & DeVore 1965; Poirier 1972).

In contrast, descriptions of socialization of young chimpanzees in the wild suggest that participatory appropriation and social learning occur over a longer time scale and are supplemented by imitation and demonstration teaching, which are unknown among Old World monkeys (Custance, Whiten & Fredman 2002; Visalberghi & Fragaszy 2002). Chimpanzees are classified as infant (0 to 5 years old), juvenile (to 10 years old), adolescent (to 13 (female), and 15 (male) years old), and prime adult (to 40 years old in both sexes). They spend the first 10 years of life primarily with their mothers and siblings. By 10 years of age, males are spending considerable time with adult males (Boesch & Boesch-Achermann 2000).

Apprenticeship in nut cracking illustrates the process. In Taï, infants go through four phases: (1) at about 2 years of age, they begin to try to hit the nuts without a hammer; (2) at about 3 years, they understand that they need an anvil and a hammer to open the nut but are unable to hit the nut hard enough; (3) at about 4 years, they succeed in opening Coula nuts; (4) after 6 years of age, they succeed in opening the harder Panda nuts; and (5) from this time until they become adult, they slowly gain in efficiency (Boesch & Boesch-Achermann 2000). A similarly long apprenticeship in nut cracking has been reported in other groups of West African chimpanzees (Inoue-Nakamura & Matsuzawa 1997; Matsuzawa 1994).

Mother chimpanzees at Taï use at least three forms of pedagogy to help their offspring acquire nut-cracking skills: (1) stimulation, by setting up the three elements, i.e., leaving the hammer and nut on the anvil; (2) facilitation, by giving a better hammer to their offspring once they have begun to open nuts; and (3) (rarely) teaching, by demonstrating the technique of positioning and holding the hammer for infants experiencing continued difficulty. Throughout this prolonged period, mothers continue to share nuts they have opened with their offspring (Boesch & Boesch-Achermann 2000). As predicted (Caro & Hauser 1992), these pedagogical activities cost mothers considerable effort that otherwise could be used to open more nuts for themselves. Also, as predicted, they allow offspring to learn skills that otherwise would be beyond their reach (Boesch & Boesch-Achermann 2000).

The script for the *nut-cracking school*, i.e., apprenticeship, involves progressively and flexibly changing roles as mothers tailor their nut sharing to their offsprings' developing abilities. In some cases, offspring also change their begging behaviors in response to these changes. Whereas most young chimpanzees decrease their begging as their mothers reduce sharing, at least two young males responded by collecting nuts, bringing them to or placing them on their mother's anvil, and waiting. Their mothers in turn changed their own roles:

both responded by cracking and sharing most of these nuts with them (Boesch & Boesch-Achermann 2000).

Apprenticeship in hunting skills in Taï chimpanzees begins later and continues longer than acquisition of nut cracking skills. Although juveniles from 6 to 8 years of age approach colobus monkeys, they are easily frightened by adult colobus who chase them. Active hunting apprenticeship usually begins by age nine or ten, when young chimpanzee males spend more time with adult males participating in hunts (Boesch & Boesch-Achermann 2000). The distributed nature of chimpanzee knowledge is indicated by the shift from maternal to adult male models. As they age, young chimpanzees become less afraid and begin to hunt more efficiently, first as drivers, then in other roles as they acquire new skills. It is also likely that they continue to develop cognitively for several years. In any case, they continue their apprenticeship for up to *20 years* (Boesch & Boesch-Achermann 2000).

This late start in hunting apprenticeship may occur because mothers are not the models and hunting apprenticeship depends upon participating in hunts with adult males. An orphaned male's apprenticeship shows the importance of joining adult males. He was adopted by "Brutus," the best hunter in Taï, and followed Brutus everywhere from 5 years of age when his mother died. He learned more quickly than other males, successfully anticipating the colobus' movements and acting as a blocker at age 12 (Boesch & Boesch-Achermann 2000).

Because orangutans were the earliest extant lineage to branch off the great ape common ancestor, recent reports that wild orangutans use tools and tool sets in extractive foraging (Fox, Sitompul & van Schaik 1999; van Schaik, Fox & Sitompul 1996) are highly significant phylogenetically, in reconstructing the evolution of great ape cognition. Likewise, the suggestion that orangutans experience intensive apprenticeship in foraging techniques (Galdikas 1981, 1995) is supported by reports of tool use traditions in one community, social learning of foraging techniques in juvenile rehabilitant orangutans returned to forest life (Russon 2003), and the prolonged association (approximately 7 years) between wild orangutan mothers and their immature offspring. These discoveries support the suggestion that the capacity for intelligent tool use arose in the common ancestor of great apes (Parker & Gibson 1977).

These examples reveal that socialization in great apes differs from that of Hamadryas in several respects: it involves (1) much longer apprenticeships (till age 30 in Taï chimpanzees versus 14 in Hamadryas); (2) greater diversity and flexibility of role interactions; (3) dependence on learning through imitation and greater dependence on pedagogy (Boesch 1991; Boesch & Boesch 1984; Boesch & Boesch-Achermann 2000; Boesch *et al.* 1994); and (4) less dependence on and constraints from innate displays. In contrast, Hamadryas socialization, though prolonged, depends significantly more on simpler social learning mechanisms canalized by innate grouping tendencies and stereotyped communicative displays. In other words, Hamadryas socialization is limited by simple learning mechanisms and heavily scaffolded by fairly stereotyped species-typical behaviors.

If the patterns of Hamadryas and chimpanzee socialization described here can be taken as representative of each taxon's abilities, it seems likely that socializing interactions in monkeys, apes, and humans fall along a continuum from a limited number of highly stereotyped, context-bound displays to a large variety of flexible interactions. As young individuals participate in taxon-specific scripts during daily activities, they are socialized into specific roles. These scripts and roles, in turn, guide them in the acquisition of ecological and social expertise. The kinds of social learning involved in these acquisitions reflect taxon-specific cognitive mechanisms, including different mechanisms for encoding and representing events.

Cognitive processes underpinning socialization

According to developmental psychologists, human children's participatory appropriation of new schemes, roles, and scripts is accompanied by gradual development of *event representations* (Nelson 1983; Nelson & Gruendel 1986; Nelson & Seidman 1984), i.e., ". . . representations of objects, persons, and person roles, and sequences of actions appropriate to a specific scene . . . include[ing] specific social and cultural components essential for carrying through a particular activity" (Nelson 1983: 135). Early language, for instance, takes the form of symbolic event representations of simple familiar scripts in which children have participated. These simple verbal formulas provide a frame in which words for new objects, persons, and roles can be substituted.

Application of the concept of *event representation* to comparative studies is useful because it focuses attention on taxon-specific forms of representation involved in participatory appropriation of roles and scripts by young nonhuman primates during their routine daily activities. Although they do not represent events in symbolic grammars as humans do (see Blake this volume), they probably generate other forms of representation. Great apes apparently have the capacity to represent events through imitative rehearsals and reviews (Donald 1998) of their schemes, roles, and scripts (Russon 1996). They also supplement imitation of object manipulation with individual learning of local physical constraints (Russon 1999). Their functional amalgam of imitation, focused individual learning, and imitative rehearsal and review might be called an *imitation complex*. In contrast, it seems likely that Old World monkeys represent events primarily in the form of conditioned habits and associations generated through repeated participation in daily routines.

Together, these complementary developmental approaches – apprenticeship and event representation – suggest that repeated experiences of guided participation in activities of the daily routine shape apprentices' minds in taxon-typical manners. Evidence suggests, for example, that Hamadryas represent the routine event of determination of the daily march, and the script of the daily march itself, its members, its direction, and resource distribution and utilization patterns (Kummer 1997). At the simpler cognitive level characteristic of Old World monkeys, guided participation engenders non-symbolic event representation based on motor habits and simple social learning (e.g., social reinforcement, priming of stimuli, responses, or goals) (Byrne 1995). At a more sophisticated cognitive level characteristic of great apes, imitative rehearsal and review engender primitive symbolic event representations. In some cases, these may be accelerated and elaborated through teaching by creating opportunity, coaching, or demonstration (Caro & Hauser 1992). Finally, humans acquire grammars that can encode information about the relations among agents, actions, objects, instruments, etc. of events, through participatory appropriation of verbal formulas during daily routines symbolically marked by such caretaker utterances as "now

we are going to drink our milk . . . go to bed . . . wash our face" (Nelson & Gruendel 1986).

In other words, in anthropoid primates, scripts and roles are emergent phenomena co-constructed by interaction based on embodied knowledge *distributed* among participants (Strum, Forster & Hutchins 1997). Participatory appropriation in each species occurs according to stage-specific developmental readiness (zones of proximal development) and species-typical cognitive abilities. In the case of imitative learning in great apes and humans, for example, imitation is specific to respected models performing activities in the novices' zones of proximal development (Russon & Galdikas 1995).

These participatory approaches to the acquisition and representation of scripts, roles, and associated skills provide tools for comparing social cognition among nonhuman and human primates. They suggest the following pattern: chimpanzees, and perhaps all great apes, participate in scripts characterized by more flexible intermediate steps to a common goal and playing collaborative roles entailing the capacity for role reversal, both of which entail planning several steps ahead. This can be seen, for example, in the ambusher role in the hunting script, which entails reconstructing the actions of collaborating hunters that the ambusher has played in the past. They acquire these abilities through participatory appropriation based on rehearsal and review through imitation and play, which provide a primitive symbolic form of event representation.

Old World monkeys participate in simpler scripts characterized by fewer and less flexible intermediate steps to a goal, as seen for example in the stereotyped roles of males in the initiator–determiner script of Hamadryas. Like chimpanzees, they acquire these abilities through participatory appropriation but acquisition mechanisms are apparently limited to such associationist mechanisms as classical and operant conditioning and simpler forms of socially mediated learning. Their event representations, for example in simple planning of the daily march and rendezvous, may arise through operant conditioning and habit formation rather than through the imitation complex and symbol formation.

DISCUSSION

This socioecological–participatory appropriation framework is offered as a means for establishing comparative measures of cognitive complexity of social behavior in wild primates. It has the advantage of situating primate cognition in social and ecological context. It reveals the highly canalized nature of social roles, scripts, and socialization in Old World monkeys as compared with the more flexible, complex, and cooperative nature of these features in chimpanzees and other great apes. My assessment of differences in cognitive complexity of social organization, scripts, roles, and socialization in wild Hamadryas and chimpanzees, based on an extension of the concept of event representation, is consistent with earlier conclusions regarding differences in intelligence between great apes and Old World monkeys.

Numerous studies have revealed that all the great apes share a constellation of cognitive abilities that are absent in monkeys (e.g., for summaries, see Byrne 1997; Parker & McKinney 1999; Russon, Chapter 6, this volume): intelligent tool use, (manual, gestural, and facial) imitation, simple pretend play, mirror self recognition, and the capacity to understand and use symbols. Chimpanzees and orangutans are known to display the ability for demonstration teaching (Miles, Mitchell & Harper 1996). All these cognitive abilities of adult great apes have been shown, repeatedly, to resemble those of 3- to 4-year-old human children in Piaget's preoperations period of intellectual development (Parker & Gibson 1990; Parker, Mitchell & Miles 1999; Parker, Mitchell & Boccia 1994; Premack 1976; Russon, Bard & Parker 1996). The monkey species that have been studied from this perspective display cognitive abilities similar to those of human children less than one year of age in Piaget's sensorimotor period (Antinucci 1989, 1990; Antinucci *et al.* 1982; Parker 1977). In addition to lacking intelligent tool use, they display neither imitation of novel schemes (Parker 1977) nor the more elaborated pretend play (Visalberghi & Fragaszy 1990, 2002).

CONCLUSIONS

Tentative answers to the questions that open this chapter follow. Great apes are smarter than monkeys. It is possible to demonstrate that great ape social organizations are more complex than those of monkeys, using the concepts of activity subgroups and their associated scripts and roles. Chimpanzees' activity subgroups are demonstrably more flexible and variable than those of Hamadryas in their composition, scripts, and roles. Their scripts are longer and more complex than those

of Hamadryas, and the associated roles more numerous and more complexly coordinated.

As to how socioecological knowledge is transmitted from generation to generation, young anthropoid primates acquire socioecological roles and scripts through repeated episodes of *guided participation in activity subgroups* with various of their conspecifics. Concerning how taxonomic differences in the complexity of roles and scripts arise, I propose that they reflect different mechanisms of event representation associated with different terminal levels of cognitive ability. Whereas young monkeys form event representations primarily through simpler mechanisms of associative learning, young great apes supplement this kind of learning with an imitation complex and rudimentary symbolic event representation.

Field data on wild chimpanzees support the hypothesis that great ape cognition is an adaptation for *prolonged cooperative learning, i.e., prolonged apprenticeship*, in *both* social and ecological skills. In chimpanzees it is, and in the common ancestor of great apes it probably was, associated with tool-aided exploitation of new embedded food sources (Boesch & Boesch-Achermann 2000; Byrne 1995; Parker 1996; Parker & McKinney 1999; see Byrne, Chapter 3, Yamagiwa, Chapter 12, this volume for related views).

Both Hamadryas and chimpanzees display both ecological and social intelligence. Hamadryas coordinate their daily march to resource (water and food) distribution, and their initiator–determiner script is instrumental in adapting to sparse desert resources (Kummer 1997). Taï chimpanzees coordinate their nut and tool collection and their hunting relative to resource distribution. However, chimpanzees have more cognitively complex adaptations than Hamadryas in *both* social and ecological domains. If these adaptations are typical of their respective taxa, we can say that the intelligence of monkeys and apes is neither primarily an adaptation for Machiavellian social manipulation, nor is it primarily an adaptation for resource location and exploitation. Rather, intelligence is a co-adaptation to both social and ecological selection pressures.

ACKNOWLEDGMENTS

I thank the following people for their helpful comments on earlier drafts on this paper: David Begun, Marina Cords, Diana Divecha, Karin Enstam, Hans Kummer,

Constance Milbrath, Anne Russon, Larissa Swedell, and two anonymous reviewers. Their generous help does not imply endorsement of the ideas contained herein.

ENDNOTES

1 Juvenile and subadult Hamadryas males display a more extreme and ritualized interest in infants than anubis baboons do; attraction to and interaction with infants is a common theme among adult male baboons and macaques (Taub & Redican 1984).

2 Moreover, the nature of these differences suggest that selection favored retention of infantile skin coloration in adult males, and transfer of maternal behaviors to juvenile and subadult males. It also suggests a transfer of neck-biting from the context of male–male aggression to male–female aggression. Two lines of evidence support the hypothesis that differences in social organization between the two subspecies are primarily the result of innate differences in male characteristics (Sugawara 1988). First, male hybrids between Hamadryas and olive baboons display behavioral characteristics that correlate with their morphological characteristics rather than with behaviors of other group members, for example males that resembled Hamadryas engaged in more herding behavior and had more females than males who resembled olive baboons (Sugawara 1988). Second, this conclusion is supported by data from transplantation experiments in the wild demonstrating that female olive are more flexible in their behaviors than are male olive and Hamadryas; specifically, naïve adult female olive can be herded into harems (Kummer, Goetz, & Angst 1970).

3 Although the adult male chimpanzees at Taï formed a linear dominance hierarchy, the eldest male, Brutus, the best hunter and "war leader" frequently formed coalitions against the dominant male and his attacks were tolerated, perhaps because he shared meat with the dominant male. Likewise, perhaps for the same reason, Brutus received more grooming from more dominant males than would be expected from his fourth rank (Boesch & Boesch-Achermann 2000).

4 Although Tomasello & Call define TPRs as those in which the observer is not participating, their examples actually involve the third party as participant. This is also true of Kummer's (1967) tripartite relationships.

REFERENCES

Abegglen, J. (1984). *On Socialization in the Hamadryas Baboons*. London and Toronto: Associated University Presses.

Altmann, S. & Altmann, J. (1970). *Baboon Ecology, African Field Research*. Chicago, IL: University of Chicago Press.

Antinucci, F. (1989). *Cognitive Structure and Development in Nonhuman Primates*. Hillsdale, NJ: Erlbaum.

(1990). The comparative study of cognitive ontogeny in four primate species. In *"Language" and Intelligence in Monkeys and Apes*, ed. S. T. Parker & K. R. Gibson, pp. 157–71. New York: Cambridge University Press.

Antinucci, F., Spinozzi, G., Visalberghi, E. & Volterra, V. (1982). Cognitive development in a Japanese macaque (*Macaca fuscata*). *Annali dell Istituto Superiore Sanita*, 18(2), 177–83.

Boesch, C. (1991). Teaching among wild chimpanzees. *Animal Behaviour*, 41, 530–2.

Boesch, C. & Boesch, H. (1984). Mental map in wild chimpanzees: an analysis of hammer transports for nut cracking. *Primates*, 25(2), 160–70.

(1989). Hunting behavior of wild chimpanzees in the Taï National Park. *American Journal of Physical Anthropology*, 78, 547–73.

Boesch, C. & Boesch-Achermann, H. (2000). *The Chimpanzees of the Taï Forest*. Oxford: Oxford University Press.

Boesch, C., Marchesi, P., Marchesi, N., Fruth, B. & Joulian, F. (1994). Is nut cracking in wild chimpanzees a cultural behaviour? *Journal of Human Evolution*, 26, 325–38.

Byrne, R. W. (1995). *The Thinking Ape: Evolutionary Origins of Intelligence*. Oxford, UK: Oxford University Press.

(1997). The technical intelligence hypothesis: an alternative evolutionary stimulus to intelligence? In *Machiavellian Intelligence II: Extensions and Evaluations*, ed. R. W. Byrne & A. Whiten, pp. 289–311. Cambridge, UK: Cambridge University Press.

Caro, T. M. & Hauser, M. D. (1992). Is there teaching in nonhuman animals. *The Quarterly Review of Biology*, 67(2), 151–74.

Chapman, C. A., Wrangham, R. & Chapman, L. (1995). Ecological constraints on group size: an analysis of spider monkey and chimpanzee subgroups. *Behavioural Ecology and Sociobiology*, 36, 59–70.

Chevalier-Skolnikoff, S. & Poirier, F. (ed.) (1977). *Primate Biosocial Development*. New York: Garland.

Cords, M. (1986). Forest guenons and patas monkeys: male–male competition in one-male groups. In *Primate Societies*, ed. B. Smuts, D. Cheney, R. Seyfarth, R. Wrangham & T. Struhsaker, pp. 98–111. Chicago, IL: University of Chicago Press.

(1988). Mating systems of forest guenons: a preliminary review. In *A Primate Radiation: Evolutionary Biology of the African Guenons*, ed. A. Gautier-Hion, F. Bourliere, J. P. Gautier & J. Kingdon, pp. 323–39. Cambridge, UK: Cambridge University Press.

Crook, J. (1970). Social organization and the environment: aspects of contemporary ethology. In *Primates on Primates*, ed. D. Quiatt, pp. 75–93. Minneapolis, MN: Burgess.

Custance, D., Whiten, A. & Fredman, T. (2002). Social learning and primate reintroduction. *International Journal of Primatology*, **23**(3), 479–99.

Donald, M. (1998). Commentary on Michael Tomasello's "Uniquely primate, uniquely human." *Developmental Science*, **1**(1), 17–18.

Doran, D. M. & McNeilage, A. (1998). Gorilla ecology and behavior. *Evolutionary Anthropology*, **6**, 120–31.

Fedigan, L. (1982). *Primate Paradigms: Sex Roles and Social Bonds*. Montreal, Quebec: Eden Press.

Fossey, D. (1983). *Gorillas in the Mist*. New York: Houghton-Mifflin.

Fox, E., Sitompul, A. F. & van Schaik, C. P. (1999). Intelligent tool use in wild Sumatran orangutans. In *The Mentality of Gorillas and Orangutans*, ed. S. T. Parker, R. W. Mitchell & H. L. Miles, pp. 99–116. Cambridge, UK: Cambridge University Press.

Fruth, B., Hohmann, G. & McGrew, W. (1999). The *Pan* species. In *The Nonhuman Primates*, ed. P. Dolhinow & A. Fuentes, pp. 64–72. Mountain View, CA: Mayfield.

Fuentes, A. (2000). Hylobatid communities: changing views of pair bonding and social organization in hominoids. *Yearbook of Physical Anthropology*, **43** (Supplement **31** to *American Journal of Physical Anthropology*), 33–60.

Galdikas, B. (1981). Orangutan reproduction in the wild. In *Reproductive Biology of the Great Apes*, ed. C. E. Graham, pp. 281–99. New York: Academic Press.

(1995). *Reflections of Eden: My Years with the Orangutans of Borneo*. New York: Little Brown.

Goldsmith, M. (1999). Gorilla socioecology. In *The Nonhuman Primates*, ed. P. Dolhinow & A. Fuentes, pp. 58–63. Mountain View, CA: Mayfield.

Goodall, J. (1986). *The Chimpanzees of Gombe: Patterns of Behavior*. Cambridge, MA: The Belnap Press of Harvard University Press.

Hall, K. R. L. (1968). Social learning in monkeys. In *Primates: Studies in Adaptation and Variability*, ed. P. C. Dolhinow, pp. 383–97. New York: Holt, Rinehart and Winston.

Hall, K. R. L. & DeVore, I. (1965). Baboon social behavior. In *Primates: Field Studies of Monkeys and Apes*, ed. I. DeVore, pp. 53–110. New York: Holt, Rinehart and Winston.

Hamilton, W. J. I. & Tilson, R. L. (1985). Fishing baboons at desert waterholes. *American Journal of Primatology*, **8**, 255–7.

Hamilton, W. J. I., Buskirk, R. & Buskirk, W. (1978). Environmental developmental determinants of object manipulation by chacma baboons (*Papio ursinus*) in two southern African environments. *Journal of Human Evolution*, **7**, 205–16.

Harvey, P., Martin, R. D. & Clutton-Brock, T. (1987). Life histories in comparative perspective. In *Primate Societies*, ed. B. Smuts, D. Cheney, R. Seyfarth, R. Wrangham & T. Struhsaker pp. 181–96. Chicago, IL: University of Chicago Press.

Hinde, R. (1974). *Biological Bases of Human Social Behavior*. New York: McGraw Hill.

Inoue-Nakamura, N. & Matsuzawa, T. (1997). Development of stone tool use by wild chimpanzees (*Pan troglodytes*). *Journal of Comparative Psychology*, **111**(2), 159–73.

Jolly, C. (1993). Species, subspecies and baboon systematics. In *Species, Species-concepts and Primate Evolution*, ed. W. Kimbel & L. B. Martin, pp. 67–107. New York: Plenum.

Kummer, H. (1967). Tripartite relations in Hamadryas baboons. In *Social Communication among Primates*, ed. S. A. Altmann, pp. 63–71. Chicago, IL: University of Chicago Press.

(1968). *Social Organization of Hamadryas Baboons*. Chicago, IL: University of Chicago Press.

(1971). *Primate Societies*. New York: Aldine.

(1997). *In Search of the Sacred Baboon*. Princeton, NJ: Princeton University Press.

Kummer, H., Goetz, W. & Angst, W. (1970). Cross-species modifications of social behavior in baboons. In *Old World Monkeys: Evolution, Systematics, and Behavior*, ed. J. R. Napier & P. H. Napier, pp. 351–63. New York: Academic Press.

Matsuzawa, T. (1994). Field experiments on use of stone tools by chimpanzees in the wild. In *Chimpanzee Cultures*, ed. R. W. Wrangham, W. C. McGrew, F. B. M. de Waal & P. G. Heltne, pp. 351–70. Cambridge, MA: Harvard University Press.

McGrew, W. (1992). *Chimpanzee Material Culture*. New York: Cambridge University Press.

Mead, G. H. (1970). *Mind, Self and Society*. Chicago: University of Chicago Press.

Mehlman, P. T. (1989). Comparative density, demography, and ranging behavior of Barbary macaques (*Macaca sylvanus*) in marginal and prime conifer habitats. *International Journal of Primatology*, **10**, 269–92.

Miles, H. L., Mitchell, R. & Harper, S. (1996). Simon says: the development of imitation in an enculturated orangutan. In *Reaching into Thought: The Minds of the Great Apes*, ed. A. E. Russon, K. A. Bard & S. T. Parker, pp. 278–99. Cambridge, UK: Cambridge University Press.

Mitchell, R. W. (1999). Deception and concealment as strategic script violations in great apes and humans. In *The Mentalities of Gorillas and Orangutans: Comparative Perspectives*, ed. S. T. Parker, R. W. Mitchell & H. L. Miles, pp. 295–315. Cambridge, UK: Cambridge University Press.

Nelson, K. (1983). The derivation of concepts and categories from event representations. In *New Trends in Conceptual Representation: Challenges to Piagetian Theory*, ed. E. Scholnick, pp. 129–49. Hillsdale, NJ: Erlbaum.

(1996). *Language in Cognitive Development: The Emergence of the Mediated Mind*. New York: Cambridge University Press.

Nelson, K. & Gruendel, J. (1986). Children's scripts. In *Event Knowledge*, ed. K. Nelson, pp. 21–45. Hillsdale, NJ: Erlbaum.

Nelson, K. & Seidman, S. (1984). Playing with scripts. In *Symbolic Play*, ed. I. Bretherton, pp. 45–72. New York: Academic Press.

Parker, S. T. (1977). Piaget's sensorimotor series in an infant macaque: a model for comparing unstereotyped behavior and intelligence in human and nonhuman primates. In *Primate Biosocial Development*, ed. S. Chevalier-Skolnikoff & F. Poirier, pp. 43–112. New York: Garland.

(1996). Apprenticeship in tool-mediated extractive foraging: The origins of imitation, teaching, and self-awareness in great apes. In *Reaching into Thought: The Minds of the Great Apes*, ed. A. E. Russon, K. A. Bard & S. T. Parker, pp. 348–70. Cambridge, UK: Cambridge University Press.

Parker, S. T. & Gibson, K. R. (ed.) (1990). *"Language" and Intelligence in Monkeys and Apes*. New York: Cambridge University Press.

(1977). Object manipulation, tool use, and sensorimotor intelligence as feeding adaptations in cebus monkeys and great apes. *Journal of Human Evolution*, **6**, 623–41.

Parker, S. T. & McKinney, M. L. (1999). *Origins of Intelligence: The Evolution of Cognitive Development in Monkeys, Apes, and Humans*. Baltimore, MD: Johns Hopkins University Press.

Parker, S. T., Mitchell, R. W. & Boccia, M. L. (ed.) (1994). *Self-awareness in Animals and Humans*. New York: Cambridge University Press.

Parker, S. T., Mitchell, R. W. & Miles, H. L. (ed.) (1999). *The Mentalities of Gorillas and Orangutans: Comparative Perspectives*. Cambridge, UK: Cambridge University Press.

Poirier, F. (ed.) (1972). *Primate Socialization*. New York: Random House.

Premack, D. (1976). *Intelligence in Ape and Man*. Hillsdale, NJ: Erlbaum.

Pusey, A. E. & Packer, C. (1987). Dispersal and philopatry. In *Primate Societies*, ed. B. Smuts, D. Cheney, R. Seyfarth, R. Wrangham & T. Struhsaker, pp. 250–65. Chicago: University of Chicago Press.

Rogoff, B. (1990). *Apprenticeship in Thinking*. New York: Oxford University Press.

(1993). Children's guided participation and participatory appropriation in sociocultural activity. In *Development in Context*, ed. R. Wozniak & K. Fischer, pp. 121–53. Hillsdale, NJ: Erlbaum.

Rowell, T. (1972). *Social Behaviour of Monkeys*. London: Penguin Books.

Russon, A. E. (1996). Imitation in everyday use: matching and rehearsal in the spontaneous imitation of rehabilitant orangutans (*Pongo pygmaeus*). In *Reaching into Thought: The Minds of the Great Apes*, ed. A. E. Russon, K. A. Bard & S. T. Parker, pp. 152–76. Cambridge, UK: Cambridge University Press.

(1999). Imitation of tool use in orangutans. In *The Mentalities of Gorillas and Orangutans*, ed. S. T. Parker, R. W. Mitchell & H. L. Miles, pp. 117–46. Cambridge, UK: Cambridge University Press.

(2003). Comparative developmental perspectives on culture: The great apes. In *Between Biology and Culture: Perspectives on Ontogenetic Development*, ed. H. Keller, Y. H. Poortinga & A. Schoelmerich, pp. 30–56. Cambridge, UK: Cambridge University Press.

Russon, A. E. & Galdikas, B. M. F. (1995). Constraints on great ape imitation: model and action selectivity in rehabilitant orangutans (*Pongo pygmaeus*) imitation. *Journal of Comparative Psychology*, **109**, 5–17.

Russon, A. E., Bard, K. A. & Parker, S. T. (ed.) (1996). *Reaching into Thought: The Minds of the Great Apes*. Cambridge, UK: Cambridge University Press.

Schank, R. & Abelson, R. (1977). Scripts, plans and knowledge. In *Thinking: Readings in Cognitive Science*, ed. P. Johnson-Laird & P. Wason, pp. 421–32. Cambridge, UK: Cambridge University Press.

Smuts, B. (1985). *Sex and Friendship in Baboons*. New York: Aldine de Gruyter.

Stanford, C. (1998). *Chimpanzees and Red Colobus*. Cambridge, MA: Harvard University Press.

Strier, K. B. (1994). Myth of the typical primate. *Yearbook of Physical Anthropology*, **37**, 233–71.

(1996). Male reproductive strategies in New World monkeys. *Human Nature*, **7**(2), 105–23.

Struhsaker, T. T. (1967). Behavior of vervet monkeys (*Cercopithecus aethiops*). *University of California Publications in Zoology*, **82**, 1–74.

Strum, S. (1987). *Almost Human: A Journey into the World of Baboons*. New York: Random House.

Strum, S. C., Forster, D. & Hutchins, E. (1997). Why Machiavellian intelligence may not be Machiavellian. In *Machiavellian Intelligence II: Extensions and Evaluations*, ed. R. W. Byrne & A. Whiten, pp. 50–85. Cambridge, UK: Cambridge University Press.

Suchman, L. (1987). *Plans and Situated Actions: The Problem of Human Machine Communication*. New York: Cambridge University Press.

Sugawara, K. (1988). Ethological study of the social behavior of hybrid baboons between *Papio anubis* and *P. Hamadryas* in free-ranging groups. *Primates*, **22**(4), 429–48.

Taub, D. M. & Redican, W. K. (1984). Adult male–infant interactions in Old World monkeys and apes. In *Primate Paternalism*, ed. D. M. Taub, pp. 377–406. New York: van Nostrand Reinhold Publishing Co.

Tomasello, M. (1998). Uniquely primate, uniquely human. *Developmental Science*, **1**(1), 1–16.

Tomasello, M. & Call, J. (1997). *Primate Cognition*. New York: Oxford University Press.

van Schaik, C. P. (1999). The socioecology of fission–fusion sociality in orangutans. *Primates*, **40**, 69–86.

van Schaik, C. P., Fox, E. A. & Sitompul, A. F. (1996). Manufacture and use of tools in wild Sumatran orangutans: implications for human evolution. *Naturwissenschaften*, **83**, 186–8.

Visalberghi, E. & Fragaszy, D. M. (1990). Do monkeys ape? In *"Language" and Intelligence in Monkeys and Apes*, ed. S. T. Parker & K. R. Gibson, pp. 247–73. New York: Cambridge University Press.

(2002). Do monkeys ape? Ten years after. In *Imitation in Animals and Artifacts*, ed. K. Dautenhahn & C. Nehaniv, pp. 477–99. Cambridge, MA: MIT Press.

Vygotsky, L. S. (1978). *Mind in Society: The Development of Higher Psychological Processes*. Cambridge, MA: Harvard University Press.

Whiten, A. (2000). Primate culture and social learning. *Cognitive Science*, **24**(3), 477–508.

Whiten, A., Goodall, J., C., M. W., Nishida, T., Reynolds, V., Sugiyama, Y., Tutin, C. E. G., Wrangham, R. W. & Boesch, C. (1999). Cultures in chimpanzees. *Nature*, **399**, 682–5.

5 · Gestural communication in the great apes

JOANNA BLAKE

Department of Psychology, York University, Toronto

INTRODUCTION

This chapter will focus on aspects of great ape gestural communication that have implications for cognition and human language. From an evolutionary perspective, gestures may provide an important link across primate species in communicative systems. As a researcher interested in the development of language in human infants, and its precursors, I will stress similar developments, or their absence, in great apes. These developments will include communicative gestures, symbolic gestures, sign acquisition, and sign combinations.

Gesture has a very broad usage; here it will be restricted to movements of the hand, arm, head, and body with communicative functions. Postural and tactile gestures, as well as facial expressions, although often communicative, are not a focus because their implications for language are unclear. It must be kept in mind, however, that visual communication is a complex, redundant system in which these various components are usually combined (Marler 1965).

I will begin with a brief review of communicative gestures according to their functions and then focus on specific gestures involved in exchange, requests involving cognizance of agency, and pointing, as well as on the degree to which great ape gestures are intentional and inventive. These are the kinds of gestures and their characteristics that are related to the development of language in human infants. I then discuss the extent to which apes display symbolic gestures, both in captivity and in the wild, because such gestures in human infants emerge during the transition to language. I distinguish between communicative and symbolic gestures because symbolic gestures, while they can be communicative, do not have communication as their primary function. Finally, I discuss the ape language studies with regard to apes' ability to learn a human sign system and to combine signs, as well as the degree to which this mimics human language acquisition. My coverage of the literature aims to be illustrative of great apes' communicative capacities, rather than comprehensive.

COMMUNICATIVE GESTURES

The typical functions of communicative gestures across the great apes are aggression/threat, display, submission, reconciliation, courtship, maternal care, food begging, request to be groomed or carried, refusal, and readiness to play. These gestures are summarized in Table 5.1 according to function, kind of gesture, and species displaying the gesture. The table also includes gestures that are discussed more fully below: exchange, requests involving cognizance of agency, and pointing.

Some of these gestures are used by several great ape species and some are species specific. Two vary across communities in chimpanzees. In the grooming–hand-clasp gesture, each of the participants, usually of different sexes, extends an arm overhead and clasps the other's hand (McGrew & Tutin 1978). This stylized gesture was seen in four chimpanzee communities (Taï, Mahale M group, Mahale K group, Kibale) and not in three others (Bossou, Gombe, Budongo), suggesting cultural transmission (McGrew 1998; McGrew & Tutin 1978; Whiten *et al.* 1999). Similarly, leaf clipping has been observed in only three chimpanzee communities, with different functions for each: in play at Bossou, as part of a tree drumming display at Taï (see below), and in courtship, food begging, and other frustrating contexts at Mahale (Boesch 1996; Boesch & Boesch-Achermann 2000; Nishida 1980). In leaf clipping, the blade is removed by the incisors and the mid-rib is left. As a courtship signal, at Mahale, it may have been

Table 5.1. *Communicative gestures in the great apes*

Function	Gesture	Pan troglodytes	Pan paniscus	Gorilla	Pongo
			Species		
Aggression/Threat	Arm flap/wave	X	X		X
	Branch wave		X		
	Hit	X			X
	Head tip	X		MG	
Display	Stamp feet	X	X		
	Rock	X			X
	Branch wave				X
	Branch drag		X		
	Leaf clip	X			
	Chest beat			X	
Frustration	Leaf clip	X			
Submission	Present	X	X		X
	Crouch	X	X	MG	X
	Bob/bow	X			
	Extend arm/hand	X	X		
	Bend away	X			
	Duck, limbs under body				X
Reconciliation	Beg	C	C		
	Extend hand	C			X
Courtship	Beckon	X			
	Knuckle-knock	X			
	Leaf clip	X			
	Extend hand/sway		X		
	Stretch over		C		
Maternal care	Beckon	X			
	Gather				X
	Extend hand		X		
Play readiness	Leaf clip	X		MG	
	Slap ground	C		X	
	Beg		C		
	Hit, shake arm, chestbeat			X	
	Branch wave, run		X		
	Dive				X
Request: food	Beg with cupped hand under mouth	X	X		X
	Leaf clip	X			
Request: to be carried/ groomed	Raise arm	X			
	Beg		C		
	Clap hands or feet	X	C		
	Clasp other's hand	X	C		

Table 5.1. (*cont.*)

Function	Gesture	Species			
		Pan troglodytes	*Pan paniscus*	*Gorilla*	*Pongo*
Request: other's action	Place other's hand on specific place	X	X	C	C
	Place other's hand + gesture		X		X
	Give object	X	C		X
	Give object + gesture	C	C		
	Trace desired act			C	X
Object share	Offer/give food	X	X		X
	Offer/give tool/objects	X			C
Refusal	Push/bat away/down	X	X	MG	
	Turn away/flap arm		X		
	Head shake	C			
Information share	Point		X		C

Notes: X, found in both wild and captive individuals; MG, mountain gorillas; C, reported only in captive individuals; blank, no reports.

Sources: Wild chimpanzees: Boesch 1996; Boesch & Boesch-Achermann 2000; Goodall 1986; McGrew & Tutin 1978; Nishida 1980; Plooij 1984; van Lawick-Goodall 1968. *Captive chimpanzees*: de Waal & Aureli 1996; Gardner & Gardner 1969; Hayes & Nissen 1971; Russon, 1990; Tomasello 1990; Tomasello *et al.* 1985; van Hooff 1973. *Wild bonobos*: Hohmann & Fruth 1993; Ingmanson 1996; Kano 1992; King, 1994; Kuroda 1980; Mori 1984; Veà & Sabater-Pi 1998. *Captive bonobos*: de Waal 1988; Savage-Rumbaugh *et al.* 1986; Savage-Rumbaugh *et al.* 1998. *Wild mountain gorillas*: Fossey 1979; Schaller 1965; Yamagiwa 1992. *Captive lowland gorillas*: Gomez 1990; Tanner & Byrne 1996. *Wild orangutans*: King 1994; MacKinnon 1974; Rijksen 1978; van Schaik *et al.*, Chapter 11, this volume. *Ex-captive orangutans*: Bard 1990, 1992; Russon 1995. *Captive orangutans*: Miles 1990.

devised by low-ranking males so as not to attract the attention of high-ranking males (Nishida 1980). It was usually performed by adolescent or juvenile males to estrous females or the reverse (Nishida 1980). The ripping sound is noisy and attracts the attention of the potential mate but provokes adult males less than other signals.

The gestures in Table 5.1 have been documented in the wild and/or captivity as having clear communicative functions, although conspecifics' responses have not always been systematically recorded. Most appear to be intentionally goal oriented (see below). Several are common to humans (e.g., arm wave, arm flap, head tip, beckon) but often serve very different functions. I will now review evidence in great apes for those gestures linked to language acquisition in humans.

Exchange

The frequency with which human infants engage in give and take exchanges with objects (object sharing) has been related to early vocabulary acquisition (Blake 2000). In wild great apes, sharing focuses on food (Goodall 1968, 1986; Reynolds 2002) but can include tools (Boesch & Boesch-Achermann 2000). In chimpanzees, females give their infants food in response to begging and adults share meat (Goodall 1986) and other foods (e.g., Bethell *et al.* 2000). Orangutan and bonobo mothers offer food to their young without it being solicited (King 1994). In wild orangutans, an adolescent female offered a fruit-laden twig to an approaching male (Rijksen 1978), and food sharing occurs between adult females and between consorting females and males (van Schaik, Preuschoft & Watts, Chapter 11, this volume). Two captive chimpanzees under one year of age engaged in give and take of objects (Russon 1990), as did ex-captive orangutans (Russon 1995) and a captive language-trained orangutan, Chantek (Miles 1990). Female mountain gorillas were not seen to give their infants solid food (Fossey 1979). In wild bonobos, females as well as a dominant male have given infants food (Kuroda 1980); adults also share food, including meat, particularly among females (Hohmann & Fruth 1993). Thus, food sharing is widespread in the wild; sharing of other objects appears to be frequent in captivity and has been observed infrequently thus far in the wild.

Requests involving cognizance of agency

These request gestures develop in the second year in human infants and have also been related to vocabulary acquisition (Blake 2000; Camaioni et al. 1991). They are more sophisticated than food begging or simple reach–request gestures in that they involve taking another's hand and placing it in a specific place or giving another an object, in both cases with the clear demand that the other do something. Thus, they involve the understanding that the other can accomplish something that the individual making the gesture often cannot do. Food begging or reaching, in contrast, is focused on the desired food or object rather than on the agent who can deliver it, at least in the early stages.

Plooij (1984) observed wild infant chimpanzees taking the mother's hand in the context of a tickling game. Another good example is Nina, a juvenile female, giving her mother the hammer she has been using unsuccessfully to crack a nut, as a clear request that her mother crack it for her (Boesch & Boesch-Achermann 2000). The wild-born infant son of a free-ranging rehabilitant orangutan handed an extremely hard nut to his mother and then waited while she cracked it for him (Russon personal observation). Bonobo females grasp or touch another female and shake her, to request genito-genital rubbing (Kano 1992).

Like the wild infant chimpanzees, a language-trained chimpanzee, Washoe, took her human trainers' hands and placed them where she wanted to be tickled (Gardner & Gardner 1969), as did the orangutan Chantek (Miles 1990). Viki, another language-trained chimpanzee, placed the fingers of humans on injuries that she wanted them to fix (Hayes & Nissen 1971). She also brought them objects that she wanted them to do

something with and prodded their hands, while nodding her head, if they did not respond. Finally, she put their hands on one knob to get a key if she wanted to go for a ride and on another if she wanted to go to bed. The captive infant gorilla Muni took her caretaker's hand in the context of a play chase game. She also led him by the hand to a forbidden door and guided his hand to the latch to request that he open it (Gomez 1990). Like Muni, the captive adult female bonobo Matata would often take a human's hand and lead this person to a door that she wanted opened. She also would hand a person an empty bowl and gesture to it, indicating that she wanted it filled. When some chimpanzees displayed at her, she would give people a hose so that they would spray the offending chimpanzees. Her adoptive son, Kanzi, as an infant, put people's hands on a tree to request that they climb with him and also gave them a flashlight to use to tickle him (Savage-Rumbaugh, Shanker & Taylor 1998).

Thus, it is clear that captives in all great apes demonstrate a concept of agency in that they clearly communicate requests that others do something for them. This type of communication has less often been documented in the wild and deserves greater attention. It would be useful to know how often great apes make requests of this nature to conspecifics rather than to humans.

Pointing

Pointing is a referential gesture that is considered to be critically important in children's language acquisition (e.g., Bates 1979). In examining the evidence for pointing in the great apes, there are two aspects to consider: morphology and function. The morphological definition of pointing in human infants is index finger extension, typically with the other fingers curled, combined with arm extension, at least to a degree (Blake 2000; Blake et al. 1992; Butterworth 1998). Index finger extension with object contact is a poke, rather than a point, unless the contact is to a book (Blake, O'Rourke & Borzellino 1994). This morphological distinction is important because poking is exploratory rather than referential, but pointing to pictures in a book with contact is typically referential.

Povinelli and Davis (1994) demonstrated that young children and adults show significantly more protrusion of the index finger than do chimpanzees when the hand is relaxed. Index finger extensions have none the less been observed in chimpanzees and other great apes. Juveniles at Yerkes pointed with the index finger to the spot where they wanted to be tickled or groomed (Tomasello *et al.* 1985). So did the orangutan Chantek (Miles 1990). Since pointing in American Sign Language is a gesture indicating pronouns, e.g., "I" and "you," language- versus non-language-trained chimpanzees should be distinguished in terms of pointing. Two language-trained adult chimpanzees exhibited index finger extension (Krause & Fouts 1997), but so did three non-language-trained adult chimpanzees (Leavens, Hopkins & Bard 1996). Whole-hand extension, with no index finger extension, is none the less much more common (Leavens & Hopkins 1998).

The function of pointing is referential or declarative and *not* imperative. It is considered important in human language acquisition *because* it involves a simple sharing of information, a step beyond object sharing. Traditionally, such a "rule of sharing" (Trevarthen & Hubley 1978), descriptive function (Thorpe 1978), or informative function (Halliday 1975) has been viewed as absent in nonhuman primates (e.g., Gomez 1996; Tomasello & Camaioni 1997). Since the whole-hand extensions observed in captive chimpanzees were directed mostly at food or at a computer apparatus that delivered food (Leavens *et al.* 1996), they appear to have had an imperative function. Both Chantek and Kanzi extended their arms in the direction that they wanted to take (Miles 1990; Savage-Rumbaugh *et al.* 1998), but these gestures also had an imperative function. Chantek progressed after 26 months of age, however, to pointing with index finger extension with a declarative function, for example in answer to questions about where things were (Miles 1990). In addition, spontaneous pointing with a declarative function has now been observed in wild bonobos (Veà & Sabater-Pi 1998). One individual called and then extended his arm with his hand half closed except for the index and ring finger towards observers in the undergrowth. The behavior was repeated until other members of the group approached and looked at the observers.

Intentionality

Gestures may inform the observer without a conscious intention to communicate (Ekman & Friesen 1969); gestures that inform intentionally are considered important because they indicate awareness of what is required for communication. Two common criteria for inferring

communicative intentionality in human infant communication are eye contact with the person to whom the gesture is directed (Bates 1979) and execution of the gesture only when an observer is present (Franco & Butterworth 1996). Eye contact in both human and nonhuman primates is not always evident in habitual communication, however, and may occur only when a communication needs to be checked (Gomez 1990). For example, even with show gestures, when a human infant was facing the parent, no eye contact occurred half of the time (Blake *et al.* 1992).

In wild infant chimpanzees, food begging was accompanied with intermittent looks at the mother's face (Plooij 1984). Captive adult chimpanzees almost always established eye contact before hand extension (Krause & Fouts 1997). Gaze alternation between the experimenter and an out-of-reach banana occurred frequently combined with food begs and whole-hand or index finger extensions (Leavens & Hopkins 1998). It was seen more often in adult chimpanzees than in juveniles and more often in those that had been nursery-reared than those mother-reared.

The captive gorilla Muni, in her request behavior described above, looked alternately at the latch and at her caretaker's eyes from 20 months, though she began to coordinate simpler request gestures with eye contact at 1 year (Gomez 1996). Captive adult Western lowland gorillas in Gabon, however, avoided direct eye contact with the experimenter and failed to use the experimenter's gaze alone (without head orientation) as a cue to a baited container (Peignot & Anderson 1999). Similarly, six of seven adolescent captive chimpanzees failed to respond above chance level to gaze alone (and four of seven to point alone) to a distant box, whereas 3-year-old children had little difficulty (Povinelli *et al.* 1997). However, in this study, the experimenter pointed to a location while looking down at the floor, a combination that lacks ecological validity.

In contrast to captive lowland gorillas, wild mountain male gorillas in the Virunga volcanoes engaged in prolonged "social" staring (non-threatening) at group members from a short distance without a gesture, to solicit non-agonistic interaction (e.g., play), reduce social tension, or supplant older males at feeding spots (Yamagiwa 1992). Prolonged peering behavior (staring) from a close distance also occurred frequently in wild adolescent bonobos to solicit non-agonistic interactions (Kano 1992). In great apes, in general, social staring is used in initiating play and copulation, inviting reconciliation, greeting, and intervening in conflict (see Yamagiwa this volume).

Whole-hand extensions by captive chimpanzees were exhibited almost exclusively in the presence of a human observer (Leavens *et al.* 1996). Both adult and subadult chimpanzees gestured more frequently when the experimenter holding a banana was facing them versus facing away with the banana behind his back (Hostetter, Cantero & Hopkins 2001). Again, the vast majority of chimpanzees did not gesture to an out-of-reach banana when no observer was present. Captive juvenile chimpanzees called to an inattentive observer to gain her attention and then engaged in eye contact before making a food request (Gomez 1996). Chantek inhibited pointing when the experimenter could not see him, but another orangutan was less able to inhibit the gesture in the absence of an audience (Call & Tomasello 1994).

In a baited-box experiment, half of the chimpanzees tested gestured to the correct box as soon as a naïve human observer arrived (i.e., within 30 s), on the first trial (Leavens & Hopkins, unpublished). By comparison, rhesus macaques needed an average of 428 trials to learn to gesture to communicate the location of a baited box to a naïve human observer (Blaschke & Ettlinger 1987). Some of the monkeys did engage in eye contact while gesturing. Thus, there appear to be species differences in the speed with which nonhuman primates learn to signal communicatively to a baited box, as well as in their sensitivity to an audience.

Inventiveness

It has been said that nonhuman primates, unlike humans, do not invent gestures (Donald 1991). Observations in the wild belie this statement, however. Goodall (1986) reported that a female juvenile chimpanzee, Fifi, briefly used wrist shaking as a threat. Fossey (1983) observed for a brief period a female juvenile gorilla twirling her head as a greeting before she groomed her mother's wounds. Both of these gestures were engaged in briefly but were clearly idiosyncratic. Taï chimpanzees suddenly began to leaf clip while resting on the ground; this was not a novel gesture but a new context for it, and in this context leaf clipping was no longer restricted to adult males (Boesch 1996). In addition, males had their own idiosyncratic ways of warming up for drumming displays (Boesch 1996). Since Donald (1991) was more concerned with the inventiveness of symbolic gestures, I will return to this topic below.

SYMBOLIC GESTURES AND SYMBOLIC PLAY

Symbolic gestures represent an object or action in a similar way to the representational function of language. The term iconic has been used synonymously (McNeill 1992; Tanner & Byrne 1996). In Tanner and Byrne (1996), gestures characterized as iconic involved tracing the path of the movement or action desired from another individual, for example, beckoning, arm extension or knocking a surface to indicate the desired direction of joint movement, and moving a hand on a recipient's body to request a desired direction for her movement. These are communicative gestures and, furthermore, deictic, but not symbolic as the term is typically defined. Deictic gestures are referential, but not symbolic. There is no symbol.

For human infants, symbolic gestures have been defined as having some resemblance to that which they signify, typically in form, such as a plate used as a steering wheel. However, they must also be differentiated from the significate, that is, distant or decontextualized (Piaget 1945/1962; Werner & Kaplan 1963). Miming a desired action directly on another, such as moving the hand down another's body, does not meet this last criterion because there is no distance between the desired action and the person who should make it. It thus fits into the same category as taking a person's hand and putting it where an animal wants to be tickled. In contrast, making a twisting action as a request for someone to open a jar (from Goldin-Meadow & Mylander 1984) is symbolic because it represents the action even though cued by the jar. If the child did this action with hands on the jar while looking at the adult, then this gesture would also belong in the category of request involving cognizance of agency, like the tracing on the body example. The distance between the jar and the action makes it a symbolic action gesture. Playfighting in nonhuman primates (e.g., Liska 1994) does not quality because it is not decontextualized despite the play face accompanying it. It is a game in itself, and the symbol is difficult to define unless it is simply "the projection of a supposed situation onto an actual one" (Lillard 1993, p. 349), a definition that would seem to cover all of play.

Symbolic play is an important context for early symbolic gestures. Russon, Vasey and Gauthier (2001) analyzed eye-covering play in captive orangutans and Japanese macaques; they noted that neither the orangutans nor the macaques tried to distort reality constraints, a feature of symbolic play for Piaget (1945/1962). While the orangutans' eye-covering play involved planning of travel routes and imagery, it is a different type of planning from that involved in symbolic play, for example looking for an object needed to complete the symbolic play sequence (McCune 1995). They resorted to imagery to guide their movements while traveling blind; little groping or peeking occurred, typically at discontinuous points or before a difficult transfer. Therefore, while their behavior reflects Piagetian stage 6 representation in terms of planning and representation, it does not qualify as symbolic play. Lower levels of symbolic play can be found in captive language-trained great apes. For example, Chantek "cooked" his cereal by putting it in a pot on top of the stove (Miles, Mitchell & Harper 1996). Washoe bathed a doll in a tub, soaping it and then drying it with a towel (Gardner & Gardner 1969).

Ritualized eating by a dominant individual in the presence of a subordinate in orangutans has been interpreted as symbolic (Russon 2002). The goal of the eating was not nourishment but calming the subordinate and perhaps encouraging its approach. It was clearly simulation, as in cases of deception (see Mitchell 1994); but whereas simulation does involve pretense, pretense is not always symbolic. In the orangutan case of ritualized eating, what is symbolized has to be inferred; there is no clear symbol. It could be compared to human infants' (Blake 2000) feeding of imaginary food to human adults. However, in the latter case, that the empty plates symbolize food is clear, and the behavior is classified as level 3, the first level of symbolic play in McCune's (1995) model. The observed orangutan case of ritualized eating does nevertheless communicate a message of friendly intention to a conspecific.

Boesch and Boesch-Achermann (2000; Boesch 1991) observed an alpha male chimpanzee at Taï, Brutus, who appeared to convey three different messages by his drumming on tree buttresses. Drumming on two different trees signaled a change in travel direction, the new direction being between the trees. Drumming twice on the same tree signaled a resting period. Drumming once on one tree and twice on another tree, or the reverse, signaled both messages, a rest followed by a change of direction. The messages were inferred from the responses of group members who were out of Brutus' sight when he drummed. Brutus' tree drumming qualifies as symbolic because it seems to have been decontextualized and to have symbolized the travel direction. Tree drumming also occurs in bonobo males (Ingmanson 1996), where it appears to serve as an added emphasis to charging displays. In this species, then, it qualifies as a communicative rather than symbolic gesture.

"True," high level, symbolic gestures have been observed. Chantek held his thumb and finger together and blew through them to represent a balloon (Miles et al. 1996). Kanzi and Mulika made twisting motions towards containers they wanted opened (Savage-Rumbaugh et al. 1986), as in the human example above. They also made hitting motions towards nuts they wanted cracked. One ex-captive orangutan did likewise (Russon personal observation); another moved sticks across her hair with a cutting motion to represent scissors (Russon 1996). These appear to be spontaneous, invented gestures, criteria for symbolic gestures often invoked by Piaget (1945/1962). Thus, in contrast to some traditional beliefs (e.g., Donald 1991), the great apes, like humans, do have the ability to invent gestures, even symbolic ones.

SIGNS

Projects teaching sign language have been conducted with chimpanzees, orangutans, and a gorilla using similar methodology. Training generally involved modeling a sign, molding the ape's hands, and non-food reinforcement (e.g., tickling). Training began by one year of age, or two years in the case of Chantek, a captive male orangutan. In the case of Rinnie, a free-ranging female rehabilitant orangutan, it began near adulthood (at 11 years) and included food reinforcement (Shapiro & Galdikas 1999). Across projects, routine activities were often ritualized in the teaching of signs, as is also done in human language acquisition, and non-native signers were used as trainers, sometimes combined with native signers. Non-native signers typically use English expressed in manual signs, which is not the same as American Sign Language (ASL) (Stokoe 1983). Speech was not used for the chimpanzee Washoe (Gardner & Gardner 1971) or for Chantek during the first several years of training (Miles 1990) but was for the gorilla Koko, simultaneously with sign (Patterson 1978).

Criteria across projects for acquisition of a sign were similar: a sign had to be reported by more than one observer and used on at least 15 days of the month, sometimes 15 consecutive days. By about 4 years of age, by these criteria, apes across projects (except Rinnie) had acquired about 130 signs. Loulis, trained by his adoptive mother, Washoe, acquired only 51 signs by 6 years (Fouts, Fouts & Van Cantfort 1989). For most apes, the signs represented a broad semantic range, with only 10 to 25% food items (Fouts et al. 1989; Gardner & Gardner 1975; Miles 1990; Patterson 1978). This is important because it means that their signing went beyond simple requests for food, presumably in part because food reinforcement was not used in most of the training. Most of Koko's signs in her 50-sign lexicon are also found in the 50-sign lexicons of deaf children learning ASL (Bonvillian & Patterson 1999).

Ape signing has been criticized for being nonspontaneous, i.e., cued by the trainer (Terrace et al. 1979). In the above projects, however, high levels of spontaneity have been reported, typically 40% (Miles 1990; Patterson 1981); and in the 50th month of Project Washoe, Washoe still used her 130 credited signs spontaneously (Gardner & Gardner 1975). These apes also generalized their signs appropriately: Washoe signed "open" as a request that the water faucet be turned on; Chantek signed "cat" to a bird, dog, and opossum and "nut" to small round pistol caps; Koko signed "straw" for an antenna and plastic tubing. They also invented signs, for example, Chantek signed "eye drink" for contact lens solution. Washoe also signed with an informative function, for example, signing "toothbrush" not as a request to brush her teeth but simply to name it. Chantek's signs were often decontextualized, for example, "Brock-hall" when the building was not in view and names of former caretakers after his move back to Yerkes. Some of Washoe's ASL signs were very similar, however, to the communicative gestures reported above in wild chimpanzees: beckoning for "come-gimme," wrist shaking for "hurry," arm extension upward for "up" (see also Petitto & Seidenberg 1979). This may indicate that these signs, at least, were already in her repertoire of communicative gestures and did not have to be learned as "signs."

In a test of Washoe's ability to understand syntax at 5 years, she was presented with signed questions introduced by different wh words (who, whose, where, and what). She responded with the appropriate grammatical category 84% of the time (Gardner & Gardner 1975). Some argued that Washoe was simply learning to reply to category questions with signs from the appropriate category, chosen from a limited set of response alternatives (Terrace et al. 1979). Recently, however, the adult Washoe and three other chimpanzees that had been taught sign language since birth were tested on four types of wh-questions that required openended answers: questions indicating failure to understand the chimpanzee's signs, wh-questions including a sign made previously by the chimpanzee, wh-questions unrelated to the chimpanzee's previous utterance, and negative statements (e.g., "can't") (Jensvold & Gardner 2000). Their responses were appropriately contingent on the human's questions in that they maintained or altered the signs in their previous utterances appropriately. For example, in response to the first type of question, all four chimpanzees responded using signs that differed from the experimenter's signs, but in response to the second type of question, their responses incorporated or expanded the experimenter's signs. Therefore, these chimpanzees appeared to understand conversational contingency in responding to wh questions.

Sign acquisition is consistently much slower in great apes than human language acquisition, including deaf children of deaf parents, and their sign repertoires are ultimately much smaller. These limits may owe in part to methodological confounds, such as the use of non-native signers and simultaneous input from speech and sign, as well as to their slower rate of cognitive development and lower cognitive ceiling. These great apes did, none the less, use their acquired signs as symbols in that most used them spontaneously, generalized them appropriately, represented absent referents, and invented their own signs. These characteristics may or may not be found in great apes learning sign systems that involve lexigrams.

Another method of language training has been to use lexigrams, a visual symbol system consisting of geometric figures on a keyboard. Lexigrams stand for words and brighten when touched. The system was originally used with food reinforcement with two chimpanzees, Sherman and Austin. The experimenter showed an object; the chimpanzee was to depress the key that named that object and received a food reward if correct (Savage-Rumbaugh 1979). As with children beginning to read, the chimpanzees had difficulty understanding that the geometric figure stood for the object. They needed training before they could name foods without being allowed to eat them; initially, punching keys was what they did when they were hungry, i.e., the lexigrams acted as conditioned stimuli. Thus, "naming as a skill divorced from consuming had to be acquired" (Savage-Rumbaugh 1979: 9). Sherman and Austin eventually mastered using symbols to name hidden (decontextualized) food items.

Two captive bonobos learned lexigrams without training, one male from the age of 2.5 years (Kanzi) and one female from the age of 11 months (Mulika) (Savage-Rumbaugh *et al.* 1986). For these bonobos, symbol use was neither regimented nor defined as correct by food rewards. Rather, it was spontaneous and integrated into normal activities, where it was considered correct when it occurred with appropriate behavior, for example pushing the lexigram for "banana" followed by selecting a banana out of a group of fruits. Such behavioral demonstration was used to confirm bonobos' knowledge of the correspondence between a symbol and its referent. By 46 months, Kanzi met this symbol use criterion for 44 lexigrams. Kanzi did not use symbols only to request, as Sherman and Austin had done originally; for example, he signed ball to himself and then looked for a ball. Mulika used "milk" as a general request symbol. As in the ASL projects, both bonobos often learned symbol use within routines; for example, Kanzi pushed the lexigram for "strawberries" only in the specific context in which this symbol was first demonstrated to him.

To explore whether there are species differences between chimpanzees and bonobos in their ability to acquire a symbol system, one female bonobo (Panbanisha) and one female chimpanzee (Panpanzee) were reared together using the lexigram system almost from birth (Brakke & Savage-Rumbaugh 1996). Panbanisha began signing around 11 months; Panpanzee did not use symbols much until she was 2 years old, relying more on gestures. Once she began, however, her acquisition curve of frequently used symbols was identical to Panbanisha's, though Panpanzee used more of the infrequent symbols not included in this curve. Differences between the two in symbol production appeared to be marginal at 4 years.

One major issue is the degree to which lexigrams are being used associatively rather than referentially. Kanzi's symbols, even when decontextualized, nevertheless seemed to be requests to be taken to a place habitually associated with the symbol. Food symbols also formed a higher proportion of the vocabulary of lexigram-learning than of sign-learning great apes (24/44 of Kanzi's signs; 18/50 of Panbanisha's and Panpanzee's signs). Thus, requests for food predominate more among these great apes than among the ASL-learning great apes. Lexigram studies do overcome the ASL disadvantage of gestural ambiguity, and the more recent lexigram studies have the advantage of providing a learning situation more similar to that of human infants, i.e., observational and conversational. The system is also more successful at providing a measure of comprehension. While comprehension is beyond the scope of this review, it is worth mentioning that, as in studies of human children, sign production is easier to assess than sign comprehension in great apes. Studies using ASL seem, however, to have an advantage over lexigram studies when it comes to combinations of symbols in production, the area perhaps of greatest controversy.

Sign combinations

There has been great interest in the potential ability of these great apes to combine signs, that is, in the degree to which they can acquire syntax. Whereas syntax involves more than simple combinations, this would be a first step. ASL projects that have assessed mean length of utterance (MLU) report a ceiling of two signs on average length (Miles 1990; Patterson 1978). This average, of course, does not preclude longer utterances.

Washoe began producing sign combinations 10 months after the onset of training, at about 20 months of age (Gardner & Gardner 1971). Nine signs occurred in most of her combinations, four as emphasizers ("please," "come-gimme," "hurry," and "more") and five to amplify meaning ("go," "out," "in," "open," and "hear-listen") (Gardner & Gardner 1969). Later, the pronouns "you" and "me" were added, as well as "up" and "food" (Gardner & Gardner 1971). Washoe tended to repeat an introductory sign, which is similar to the constant-plus-variable form of early combinations produced by many children (Braine 1976). Most of Washoe's combinations were her own inventions, not copies of trainers' combinations. The components of her combinations were appropriate to her contextual referents, and tests in two situations elicited restricted combinations but with the elements in a variable order. Washoe's two-sign combinations could be classified into semantic relations similar to those typical of children's early two-word combinations (Brown 1973), for example, object–attribute ("drink red"), action–location ("look out"), agent–action ("you drink"), and action–object ("open blanket") (Gardner & Gardner 1971). This implies direct links with cognition, because these combinations require understanding of these relations.

Chantek began combining signs in his second month of training at about 24 months of age. Many of his combinations included the gesture "point." This gesture can have a pronominal meaning in ASL, but, in the examples provided, "point" appears to be a referential point to a location, for example, "carrot point" (at refrigerator). Whereas some investigators of child language treat a gesture plus a word as a combination (Greenfield & Smith 1976), most would not. Also, several of his combinations involved two food items, for example, "cookie cracker point." Others, especially those produced in play contexts, seem more like early child combinations, for example "pull beard," "Jeannie Chantek chase."

Rinnie began combining signs after one month of training. Her combinations were usually requests for food or contact activities, two- to three-signs in length (Shapiro & Galdikas 1999). These combinations began with "you," "more," "give," or "food/feed." After 4 months, her combinations increased in length but with duplication, for example "rice you rice." However, this type of duplication is not unusual for children.

During her third month of training, at about 15 months, Koko began combining the sign for "more" with signs for food and drink (Patterson 1978). She also produced several of the semantic relations found in children's early word combinations: nomination ("that bird"), attributive ("hot potato"), genitive ("Koko purse"), agent–action ("me listen"), action–object ("open bottle"). "Open" was a general request form (Patterson & Cohn 1990.) Patterson (1978) calculated Koko's MLU as somewhat less than two at 41 months, having excluded immediate repetition of signs. Koko's longest utterance without repetition was seven signs: "come sorry out me please key open." This is not an utterance that would be typical for children. One apparent problem in deciphering Koko's sign combinations is that she has sometimes been reported to have produced three or four signs simultaneously, and it is not clear how they were discriminated. It is apparently physically impossible to produce these signs simultaneously in ASL (Petitto & Seidenberg 1979). Perhaps what was meant was that Koko produced three or four signs sequentially, not simultaneously, or that she produced one sign that stood for multiple words.

Nim's longer sign combinations are said to show no evidence of elaborating or qualifying his two-sign combinations (Terrace et al. 1979). Many of his four-sign combinations were repetitions of two-sign combinations, for example "eat drink eat drink." Based on Nim's performances, these authors argued that great apes are learning linear sequences of symbols without understanding any grammatical relationships, perhaps not even semantic relations. In contrast, Washoe's longer combinations added new information and new relations at least half of the time, for example, specification of both the subject and object of an action ("you peekaboo me") (Gardner & Gardner 1971).

Kanzi also began combining signs soon after the onset of training. At 5.5 years, 10.4% were combinations, of which half were spontaneous two-symbol combinations (Greenfield & Savage-Rumbaugh 1990). Many of these seem to be double-item requests (e.g., "hotdog coke") (Savage-Rumbaugh et al. 1986) like Chantek's. Also, many combinations included a gesture that was not a sign (e.g., pushing one person's hand towards another). The problem in this case appears to be that people's names were not always represented by a lexigram, so Kanzi had to indicate them by physically touching them. This was also true for demonstratives (Greenfield & Savage-Rumbaugh 1990). The consequence is that he evidenced fewer combinations that were strictly lexigrams. However, like Washoe and Koko, Kanzi's two-lexigram combinations replicated young children's early semantic relations, for example, agent–action, action–object, agent–object, location–entity, but Kanzi also conjoined actions (e.g., "chase hide"), which is rare in child language (Greenfield & Savage-Rumbaugh 1990). For Panbanisha and Panpanzee, the number of strictly lexigram combinations increased steadily but was never more than 12% of the total number of utterances (Brakke & Savage-Rumbaugh 1996).

COGNITIVE IMPLICATIONS OF GREAT APE GESTURAL COMMUNICATION

Great apes clearly use their gestures intentionally, in a goal-directed fashion, and communicatively, though conspecifics' responses have not always been recorded. Intentionality is seen in their deliberate use of eye contact, in the wild and in captivity. Sensitivity to an audience, or to the attentional orientation of an observer, has been documented most thoroughly in captive chimpanzees, which thus meet both criteria for intentional communication in human infants. Request gestures involving cognizance of agency have also been observed more often in captivity, though there are some observations from the wild. Greater attention to this type of gesture is likely to yield more instances of its occurrence between conspecifics.

All of these aspects of great ape gestural communication – eye contact, sensitivity to audience, and cognizance of agency – have important cognitive implications for awareness of others and their capabilities. These features of communication, in themselves, do not go so far as demonstrating that great apes understand the mental states of others. It is clear, however, that great apes understand that others can accomplish things that they, themselves, cannot and that communication is enhanced by eye contact and requires an audience.

Sharing of objects appears to be limited in the wild, but objects are much used in communication (see Table 5.1). Branches and leaves are used in threat gestures, in display, in courtship, and in play. Tree buttresses are used to communicate movement signals. Thus, while give-and-take of objects may be more evident in captive great apes with many human-supplied objects available, wild great apes utilize objects in their communications. If object sharing is, indeed, a precursor of language acquisition in human infants (Blake 2000), then great apes' use of objects in communication may be a similar expression of this precursor.

Sharing of information about entities in the environment through pointing and showing also appears to be rare in the wild, though observations are beginning to appear. In captive great apes, it seems clear that the request function predominates over the declarative function, perhaps because they do not control their environment. Great apes are clearly able to use pointing and showing gestures with a declarative function with human scaffolding (Miles 1999). The one clear case from the wild (Veà & Sabater-Pi 1998) suggests that this is within great apes' independent capacity.

Captive and ex-captive apes use gestures that are symbolic according to criteria applied to human infants. This means that such gestures are idiosyncratic and decontextualized. There is still little evidence that such gestures occur in the wild. Since symbolic gestures represent the object or action they stand for, they have a cognitive link through mental representation. Great apes can also learn a human symbol system and use it spontaneously, creatively, and referentially, Terrace *et al.* (1979) notwithstanding. Both Chantek and Koko signed spontaneously 40% of the time. The semantic relations expressed in sign combinations across great ape species are similar to those first used by children at the two-word stage. These reflect cognitive abilities for understanding simple relations between agents, actions, objects, and attributes that develop at about the same time. Although it is often difficult to assign a semantic interpretation to great apes' utterances (Petitto & Seidenberg 1979), this is also true for children. It is nevertheless interesting that the number of vocabulary items acquired by age 4 years is similar across great ape species in the ASL studies (about 130) and that the mean length of utterance achieved appears to be limited to two signs. This may indicate a general production limit, but given their performance in other cognitive areas (memory, problem solving), it seems unlikely to reflect a capacity limit. More likely, it reflects the slow speed with which they add vocabulary items, compared with children, although all species began to combine signs very soon after the onset of training.

In any event, all great apes demonstrate many of the gestural precursors that have been highlighted as important for human language acquisition. These gestures have cognitive underpinnings and constitute the developmental foundations for language, so their presence in great apes underlines the degree to which hominid cognition formed the platform for human advances in communication. It seems evident that continuity in the evolution of language is cognitively based and that great apes share not only the necessary underlying cognitive abilities but also language-like communicative capacities. From the evidence presented to this point, they can master a human symbolic communication system to the level of a 2-year-old child.

REFERENCES

Bard, K. A. (1990). "Social tool use" by free-ranging orangutans: a Piagetian and developmental perspective on the manipulation of an animate object. In *"Language" and Intelligence in Monkeys and Apes: Comparative Developmental Perspectives*, ed. S. T. Parker & K. R. Gibson, pp. 356–78. New York: Cambridge University Press.

(1992). Intentional behavior and intentional communication in young free-ranging orangutans. *Child Development*, **63**, 1186–97.

Bates, E. (1979). *The Emergence of Symbols: Cognition and Communication in Infancy*. New York: Academic Press.

Bethell, E., Whiten, A., Muhumuza, G. & Kakura, J. (2000). Active plant food division and sharing by wild chimpanzees. *Primate Report*, **56**, 67–71.

Blake, J. (2000). *Routes to Child Language: Evolutionary and Developmental Precursors*. New York: Cambridge University Press.

Blake, J., McConnell, S., Horton, G. & Benson, N. (1992). The gestural repertoire and its evolution over the second year. *Early Development and Parenting*, **1**, 127–36.

Blake, J., O'Rourke, P. & Borzellino, G. (1994). Form and function in the development of pointing and reaching gestures. *Infant Behavior and Development*, **17**, 195–203.

Blaschke, M. & Ettlinger, G. (1987). Pointing as an act of social communication by monkeys. *Animal Behaviour*, **35**, 1520–3.

Boesch, C. (1991). Symbolic communication in wild chimpanzees. *Human Evolution*, **6**, 81–90.

(1996). Three approaches for assessing chimpanzee culture. In *Reaching into Thought: The Minds of the Great Apes*, ed. A. E. Russon, K. A. Bard, & S. T. Parker, pp. 404–29. Cambridge, UK: Cambridge University Press.

Boesch, C. & Boesch-Achermann, H. (2000). *The Chimpanzees of the Taï Forest: Behavioural Ecology and Evolution*. Oxford: Oxford University Press.

Bonvillian, J. D. & Patterson, F. G. P. (1999). Early sign-language acquisition: comparisons between children and gorillas. In *The Mentalities of Gorillas and Orangutans: Comparative Perspectives*, ed. S. T. Parker, R. W. Mitchell & H. L. Miles, pp. 240–64. Cambridge, UK: Cambridge University Press.

Braine, M. D. S. (1976). Children's first word combinations. *Monographs of the Society for Research in Child Development*, **41**, Serial No. 164.

Brakke, K. E. & Savage-Rumbaugh, E. S. (1996). The development of language skills in *Pan*. II. Production. *Language and Communication*, **16**, 361–80.

Brown, R. (1973). *A First Language: The Early Stage*. Cambridge, MA: Harvard University Press.

Butterworth, G. (1998). What is special about pointing in babies? In *The Development of Sensory, Motor and Cognitive Capacities in Early Infancy: From Perception to Cognition*, ed. F. Simion & G. Butterworth, pp. 171–90. East Sussex, UK: Psychology Press.

Call, J. & Tomasello, M. (1994). Production and comprehension of referential pointing by orangutans. *Journal of Comparative Psychology*, **108**, 307–17.

Camaioni, L., Caselli, M. C., Longobardi, E. & Volterra, V. (1991). A parent report instrument for early language assessment. *First Language*, **11**, 345–59.

de Waal, F. B. M. (1988). The communicative repertoire of captive bonobos (*Pan paniscus*) compared to that of chimpanzees. *Behaviour*, **106**, 183–251.

de Waal, F. B. M. & Aureli, F. (1996). Consolation, reconciliation, and a possible difference between macaques and chimpanzees. In *Reaching into Thought: The Minds of the Great Apes*, ed. A. E. Russon, K. A. Bard & S. T. Parker, pp. 80–110. Cambridge, UK: Cambridge University Press.

Donald, M. (1991). *Origins of the Modern Mind: Three Stages in the Evolution of Culture and Cognition*. Cambridge, MA: Harvard University Press.

Ekman, P. & Friesen, W. (1969). The repertoire of nonverbal behavior: categories, origins, usage, and coding. *Semiotica*, **1**, 49–97.

Fossey, D. (1979). Development of the mountain gorilla (*Gorilla gorilla beringei*): the first thirty-six months. In *The Great Apes*, ed. D. A. Hamburg & E. R. McCown, pp. 139–84. Menlo Park, CA: Benjamin/Cummings.

Fossey, D. (1983). *Gorillas in the Mist*. Boston: Houghton Mifflin.

Fouts, R. S., Fouts, D. H. & Van Cantfort, T. E. (1989). The infant Loulis learns signs from cross-fostered chimpanzees. In *Teaching Sign Language to Chimpanzees*, ed. R. A. Gardner, B. T. Gardner & T. E. Van Cantfort, pp. 280–92. Albany, NY: SUNY Press.

Franco, F. & Butterworth, G. (1996). Pointing and social awareness: declaring and requesting in the second year. *Journal of Child Language*, **23**, 307–36.

Gardner, B. T. & Gardner, R. A. (1971). Two-way communication with an infant chimpanzee. In *Behavior of Nonhuman Primates: Modern Research Trends*, ed. A. M. Schrier & F. Stollnitz, pp. 117–84. New York: Academic Press.

(1975). Evidence for sentence constituents in the early utterances of child and chimpanzee. *Journal of Experimental Psychology: General*, **104**, 244–67.

Gardner, R. A. & Gardner, B. T. (1969). Teaching sign language to a chimpanzee. *Science*, **165**, 664–72.

Goldin-Meadow, S. & Mylander, C. (1984). Gestural communication in deaf children: the effects and noneffects of parental input on early language development. *Monographs of the Society for Research in Child Development*, **49**, Serial No. 207.

Gomez, J. C. (1990). The emergence of intentional communication as a problem-solving strategy in the gorilla. In *"Language" and Intelligence in Monkeys and Apes: Comparative Developmental Perspectives*, ed. S. T. Parker & K. R. Gibson, pp. 333–55. New York: Cambridge University Press.

(1996). Ostensive behavior in great apes: the role of eye contact. In *Reaching into Thought: The Minds of the Great Apes*, ed. A. E. Russon, K. A. Bard & S. T. Parker, pp. 131–51. Cambridge, UK: Cambridge University Press.

Goodall, J. van Lawick (1968). A preliminary report on expressive movements and communication in the Gombe Stream Chimpanzees. In *Primates: Studies in Adaptation and Variability*, ed. P. C. Jay, pp. 313–74. New York: Holt, Rinehart, & Winston.

Goodall, J. (1986). *The Chimpanzees of Gombe: Patterns of Behavior*. Cambridge, MA: The Belnap Press of Harvard University Press.

Greenfield, P. M. & Savage-Rumbaugh, E. S. (1990). Grammatical combination in *Pan paniscus*: processes of learning and invention in the evolution and development of language. In *"Language" and Intelligence in Monkeys and Apes: Comparative Developmental Perspectives*, ed. S. T. Parker & K. R. Gibson, pp. 540–78. New York: Cambridge University Press.

Greenfield, P. M. & Smith, J. H. (1976). *The Structure of Communication in Early Language Development*. New York: Academic Press.

Halliday, M. A. K. (1975). *Learning How to Mean: Explorations in the Development of Language*. London: Edward Arnold.

Hayes, K. J. & Nissen, C. H. (1971). Higher mental functions of a home-raised chimpanzee. In *Behavior of Nonhuman Primates*, Vol. 4, ed. A. M. Schrier & F. Stollnitz, pp. 60–115. New York: Academic Press.

Hohmann, G. & Fruth, B. (1993). Field observations on meat sharing among bonobos (*Pan paniscus*). *Folia Primatologica*, **60**, 225–9.

Hostetter, A. B., Cantero, M. & Hopkins, W. D. (2001). Differential use of vocal and gestural communication by chimpanzees (*Pan troglodytes*) in response to the attentional status of a human (*Homo sapiens*). *Journal of Comparative Psychology*, **115**, 337–43.

Ingmanson, E. J. (1996). Tool-using behavior in wild *Pan paniscus*: social and ecological considerations. In *Reaching into Thought: The Minds of the Great Apes*, ed. A. E. Russon, K. A. Bard & S. T. Parker, pp. 190–210. Cambridge, UK: Cambridge University Press.

Jensvold, M. L. A. & Gardner, R. A. (2000). Interactive use of sign language by cross-fostered chimpanzees (*Pan troglodytes*). *Journal of Comparative Psychology*, **114**, 335–46.

Kano, T. (1992). *The Last Ape: Pygmy Chimpanzee Behavior and Ecology*. Transl. by E. O. Vineberg. Stanford, CA: Stanford University Press.

King, B. J. (1994). *The Information Continuum: Evolution of Social Information Transfer in Monkeys, Apes, and Hominids*. Santa Fe, NM: Sar Press.

Krause, M. A. & Fouts, R. S. (1997). Chimpanzee (*Pan troglodytes*) pointing: hand shapes, accuracy, and the role of eye gaze. *Journal of Comparative Psychology*, **111**, 330–36.

Kuroda, S. (1980). Social behavior of the pygmy chimpanzees. *Primates*, **21**, 181–97.

Leavens, D. A. & Hopkins, W. D. (1998). Intentional communication by chimpanzees: a cross-sectional study of the use of referential gestures. *Developmental Psychology*, **34**, 813–22.

Leavens, D. A., Hopkins, W. D. & Bard, K. A. (1996). Indexical and referential pointing in chimpanzees (*Pan troglodytes*). *Journal of Comparative Psychology*, **110**, 346–53.

Lillard, A. S. (1993). Pretend play skills and the child's theory of mind. *Child Development*, **64**, 348–71.

Liska, J. (1994). The foundation of symbolic communication. In *Hominid Culture in Primate Perspective*, ed. D. Quiatt & J. Itani, pp. 233–51. Niwot, CO: University Press of Colorado.

MacKinnon, J. (1974). The behaviour and ecology of wild orang-utans. *Animal Behaviour*, **22**, 3–74.

Marler, P. (1965). Communication in monkeys and apes. In *Primate Behavior: Field Studies of Monkeys and Apes*, ed. I. DeVore, pp. 544–84. New York: Holt, Rinehart, & Winston.

McCune, L. (1995). A normative study of representational play at the transition to language. *Developmental Psychology*, **31**, 198–206.

McGrew, W. C. (1998). Culture in nonhuman primates? *Annual Review of Anthropology*, **27**, 301–28.

McGrew, W. C. & Tutin, C. E. G. (1978). Evidence for a social custom in wild chimpanzees? *Man*, **13**, 234–51.

McNeill, D. (1992). *Hand and Mind: What Gestures Reveal about Thought*. Chicago, IL: University of Chicago Press.

Miles, H. L. (1990). The cognitive foundations for reference in a signing orangutan. In *"Language" and Intelligence in Monkeys and Apes: Comparative Developmental Perspectives*, ed. S. T. Parker & K. R. Gibson, pp. 511–39. New York: Cambridge University Press.

(1999). Symbolic communication with and by great apes. In *The Mentalities of Gorillas and Orangutan: Comparative Perspectives*, ed. S. T. Parker, R. W. Mitchell & H. L. Miles, pp. 197–210. Cambridge, UK: Cambridge University Press.

Miles, H. L., Mitchell, R. W. & Harper, S. E. (1996). Simon says: The development of imitation in an enculturated orangutan. In *Reaching into Thought: The Minds of the Great Apes*, ed. A. E. Russon, K. A. Bard & S. T. Parker, pp. 278–99. Cambridge, UK: Cambridge University Press.

Mitchell, R. W. (1994). The evolution of primate cognition: simulation, self-knowledge, and knowledge of other minds. In *Hominid Culture in Primate Perspective*, ed. D. Quiatt & J. Itani, pp. 177–232. Niwot, CO: University Press of Colorado.

Mori, A. (1984). An ethological study of pygmy chimpanzees in Wamba, Zaire: a comparison with chimpanzees. *Primates*, **25**, 255–78.

Nishida, T. (1980). The leaf-clipping display: a newly-discovered expressive gesture in wild chimpanzees. *Journal of Human Evolution*, **9**, 117–28.

Patterson, F. G. (1978). The gestures of a gorilla: language acquisition in another Pongid. *Brain and Language*, **5**, 72–97.

(1981). Ape language. Response to Terrace *et al.* (1979). *Science*, **211**, 86–7.

Patterson, F. G. P. & Cohn, R. H. (1990). Language acquisition by a lowland gorilla: Koko's first ten years of vocabulary development. *Word*, **42**, 97–121.

Peignot, P. & Anderson, J. R. (1999). Use of experimenter-given manual and facial cues by gorillas (*Gorilla gorilla*) in an object-choice task. *Journal of Comparative Psychology*, **113**, 253–60.

Petitto, L. A. & Seidenberg, M. S. (1979). On the evidence for linguistic abilities in signing apes. *Brain and Language*, **8**, 162–83.

Piaget, J. (1962). *Play, Dreams, and Imitation in Childhood*. Transl. by C. Gattegno & F. M. Hodgson. New York: Norton. (Original work published 1945.)

Plooij, F. X. (1984). *The Behavioral Development of Free-living Chimpanzee Babies and Infants*. Norwood, NJ: Ablex.

Povinelli, D. J. & Davis, D. R. (1994). Differences between chimpanzees (*Pan troglodytes*) and humans (*Homo sapiens*) in the resting state of the index finger: implications for pointing. *Journal of Comparative Psychology*, **108**, 134–9.

Povinelli, D. J., Reaux, J. E., Bierschwale, D. T., Allain, A. D. & Simon, B. B. (1997). Exploitation of pointing as a referential gesture in young children, but not adolescent chimpanzees. *Cognitive Development*, **12**, 327–65.

Reynolds, P. C. (2002). Pretending primates: play and simulation in the evolution of primate societies. In *Pretending and Imagination in Animals and Children*, ed. R. W. Mitchell, pp. 196–209. Cambridge, UK: Cambridge University Press.

Rijksen, H. D. (1978). *A Field Study on Sumatran Orang Utans (Pongo pygmaeus abelii Lesson 1927): Ecology, Behavior and Conservation*. Wageningen: H. Veenman & Zonen.

Russon, A. E. (1990). The development of peer social interaction in infant chimpanzees: comparative social, Piagetian, and brain perspectives. In *"Language" and Intelligence in Monkeys and Apes: Comparative Developmental Perspectives*, ed. S. T. Parker & K. R. Gibson, pp. 379–419. New York: Cambridge University Press.

(1995, November). Reaching into thought: the minds of the great apes. Invited Address, York University, Toronto.

(1996). Imitation in everyday use: matching and rehearsal in the spontaneous imitation of rehabilitant orangutans (*Pongo pygmaeus*). In *Reaching into Thought: The Minds of the Great Apes*, ed. A. E. Russon, K. A. Bard & S. T. Parker, pp. 152–76. Cambridge, UK: Cambridge University Press.

(2002). Pretending in free-ranging rehabilitant orangutans. In *Pretending and Imagination in Animals and Children*, ed. R. W. Mitchell, pp. 229–40. Cambridge, UK: Cambridge University Press.

Russon, A. E., Vasey, P. L. & Gauthier, C. (2001). Seeing with the mind's eye: eye-covering play in orangutans and Japanese macaques. In *Pretending and Imagination in Animals and Children*, ed. R. W. Mitchell, pp. 241–54. Cambridge, UK: Cambridge University Press.

Savage-Rumbaugh, E. S. (1979). Symbolic communication – its origins and early development in the chimpanzee. *New Directions for Child Development*, **3**, 1–15.

Savage-Rumbaugh, S., McDonald, K., Sevcik, R. A., Hopkins, W. D. & Rubert, E. (1986). Spontaneous symbol acquisition and communicative use by pygmy chimpanzees (*Pan paniscus*). *Journal of Experimental Psychology: General*, **115**, 211–35.

Savage-Rumbaugh, S., Shanker, S. G. & Taylor, T. J. (1998). *Apes, Language, and the Human Mind*. New York: Oxford University Press.

Schaller, G. B. (1965). The behavior of the mountain gorilla. In *Primate Behavior: Field Studies of Monkeys and Apes*, ed. I. DeVore, pp. 324–67. New York: Holt, Rinehart & Winston.

Shapiro, G. L. & Galdikas, B. M. F. (1999). Early sign performance in a free-ranging, adult orangutan. In *The Mentalities of Gorillas and Orangutans: Comparative Perspectives*, ed. S. T. Parker, R. W. Mitchell & H. L. Miles, pp. 266–79. Cambridge, UK: Cambridge University Press.

Stokoe, W. C. (1983). Apes who sign and critics who don't. In *Language in Primates: Perspectives and Implications*, ed. J. de Luce & H. T. Wilder, pp. 147–58. New York: Springer-Verlag.

Tanner, J. E. & Byrne, R. W. (1996). Representation of action through iconic gesture in a captive lowland gorilla. *Current Anthropology*, **37**, 162–73.

Terrace, H. S., Petitto, L. A., Sanders, R. J. & Bever, T. G. (1979). Can an ape create a sentence? *Science*, **206**, 891–9.

Thorpe, W. H. (1978). *Purpose in a World of Chance*. London: Oxford University Press.

Tomasello, M. (1990). Cultural transmission in the tool use and communicatory signaling of chimpanzees. In *"Language" and Intelligence in Monkeys and Apes: Comparative Developmental Perspectives*, ed. S. T. Parker & K. R. Gibson, pp. 274–311. New York: Cambridge University Press.

Tomasello, M. & Camaioni, L. (1997). A comparison of the gestural communication of apes and human infants. *Human Development*, **40**, 7–24.

Tomasello, M., George, B. L., Kruger, A. C., Farrar, M. J. & Evans, A. (1985). The development of gestural communication in young chimpanzees. *Journal of Human Evolution*, **14**, 175–86.

Trevarthen, C. & Hubley, P. (1978). Secondary intersubjectivity: confidence, confiding and acts of meaning in the first year. In *Action, Gesture and Symbol: The Emergence of Language*, ed. A. Lock, pp. 183–229. New York: Academic Press.

van Hooff, J. A. R. A. M. (1973). A structural analysis of the social behaviour of a semi-captive group of chimpanzees. In *Social Communication and Movement: Studies of Interaction and Expression in Man and Chimpanzee*, ed. M. von Cranach & I. Vine, pp. 75–162. London: Academic Press.

Veà, J. J. & Sabater-Pi, J. (1998). Spontaneous pointing behaviour in the wild pygmy chimpanzee (*Pan paniscus*). *Folia Primatologica*, **69**, 289–90.

Werner, H. & Kaplan, B. (1963). *Symbol Formation*. New York: Wiley.

Whiten, A., Goodall, J., McGrew, W. C., Nishida, T., Reynolds, V., Sugiyama, Y., Tutin, C. E. G., Wrangham, R. W. & Boesch, C. (1999). Cultures in chimpanzees. *Nature*, **399**, 682–5.

Yamagiwa, J. (1992). Functional analysis of social staring behavior in an all-male group of mountain gorillas. *Primates*, **33**, 523–44.

6 · Great ape cognitive systems

ANNE E. RUSSON

Psychology Department, Glendon College of York University, Toronto

INTRODUCTION

This chapter considers cognition in great apes as integrated systems that orchestrate the many abilities that great apes express, systems for which satisfactory characterizations remain elusive. In part, difficulties owe to research trends. Empirical studies have been guided by diverse and sometimes contradictory models, questions, measures, tasks, and living conditions. Performance levels have proven inconsistent across individuals, rearing conditions, and testing conditions, and evidence is patchy across species for virtually any facet of cognition. Evidence on wild great apes, the most important from an evolutionary perspective, is especially patchy because research has favored captives; much of what is available was collected for other purposes, so it was neither described nor analyzed with cognition in mind. The issues at stake are also hard-felt ones that touch on the human–nonhuman boundary, so entrenched beliefs infect how the literature is interpreted and even what of it is read.

Attempts have none the less been made to develop an integrated model of great ape cognition using available evidence. They include both edited survey volumes (Matsuzawa 2001a; Parker, Mitchell & Miles 1999; Russon, Bard & Parker 1996) and integrative reviews, three of the latter as major books (Byrne 1995 (RWB), Parker & McKinney 1999 (P&M); Tomasello & Call 1997 (T&C)) and others as articles (e.g., Byrne 1997; Suddendorf & Whiten 2001; Thompson & Oden 2000; Whiten & Byrne 1991). My aim is not to analyze this terrain, yet again, in detail, but to offer a compact *mise à date* to ground evolutionary reconstruction. Guiding questions are "what, if anything, about great ape cognition requires evolutionary explanations beyond those developed for other nonhuman primates?", and "how is great ape cognition best characterized with respect to evolutionary questions?"

CONCEPTS AND MODELS OF COGNITION

Situating great ape cognition comparatively hinges on mental processes that support symbolism, notably representation, metarepresentation, and hierarchization. Weaker and stronger conceptualizations exist for each and which is used affects assessments of great apes' capabilities.

Weak meanings of symbolism include reference by arbitrary convention (Peirce 1932/1960), using internal signs like mental images to stand for referents rather than using direct sensations or motor actions, and solving problems mentally versus experientially. In the strong sense, symbolism refers to self-referring systems wherein phenomena owe their significance and even existence to other symbols in the system rather than to sensorimotor entities (e.g., Deacon 1997; Donald 2000; Langer 2000). Representation can refer to any form of mental coding that stands for entities, perceptual included (Perner 1991; Whiten 2000) or, more strongly, to recalling to mind or "re-presenting" mental codes for entities and simple object relations in the absence of normal sensorimotor cues (P&M; Whiten 2000). Meanings of metarepresentation range from representing other representations (e.g., Leslie 1987; Matsuzawa 1991; Whiten & Byrne 1991) to representing a representation *as a representation*, i.e., an interpretation of a situation (Perner 1991). Meanings of hierarchization span creating new, higher-order cognitive structures from lower-level ones (i.e., structures with superordinate-subordinate features: Byrne & Byrne 1991; Case 1985; Langer 1998) to generating cognitive structures that show embedding (e.g., classification showing nesting of classes: Langer 1998).

Developmentalists commonly consider weak and strong forms to be related in humans, as basic and advanced ontogenetic achievements of early and later childhood respectively (Table 6.1). Comparative primate cognition often shares this view (P&M; Whiten 2000). Insofar as symbols must be grounded in real world referents at some point (Donald 1991) and weak symbolism is the more likely in great apes, I consider great ape cognition relative to weak symbolism and its associated processes (strong representation, weak hierarchization, weak metarepresentation). The terms symbolic, representation, hierarchization, and metarepresentation henceforth refer to these meanings.

The models guiding empirical studies of great ape cognition also contribute to disparities because of the ways in which they shape the generation of evidence and the interpretive frameworks they impose. Several important models are sketched below to suggest their strengths and limitations for understanding great ape cognition.

Animal models designed for nonhuman mentality have been frequent frameworks for studies of great ape cognition. They concentrate on the non-symbolic, associative processes presumed to govern nonhuman cognition, for example trial-and-error experiential learning or behavior chains. This leaves them conceptually and methodologically impoverished concerning symbolic cognition (Anderson 1996; RWB; Rumbaugh 1970; T&C), quantification and logic being important exceptions (e.g., Boysen & Hallberg 2000; Thompson & Oden 2000), so relatively little of the evidence they have generated helps determine whether great apes, or any species, are capable of symbolic cognition.

Generality–modularity models are potentially important because they concern cognitive architecture. In this view, favored by evolutionary psychologists and neo-nativists, there exist cognitive "modules," problem-specific cognitive structures that represent innately specified neurological systems and operate with relative autonomy, as well as general purpose or central processes that apply across problem types and affect system-wide properties (e.g., representation, executive control structures, working memory). These models have influenced understandings of great ape cognition with their assumption that modular architecture characterizes nonhuman cognition and generality evolved uniquely in humans (e.g., Mithen 1996; T&C; Tooby & Cosmides 1992). Little if any empirical study has examined cognitive architecture in great apes, however. Studies of great ape cognition have typically assumed modularity and have aimed for clean tests of individual

problem-specific structures – effectively eliminating chances for detecting use of multiple or general purpose processes. Given the lack of relevant empirical evidence, these models remain speculative concerning great apes.

Cognitive science models portray the mind as a device for processing, storing, integrating, and transforming information. Some of their concepts have been incorporated into models of cognitive development (e.g., Case 1985; Leslie 1987; Pascual-Leone 1987), others have aided in detecting hierarchization in great ape cognition (Byrne & Byrne 1991; Byrne, Corp & Byrne 2001; Byrne & Russon 1998; P&M; Russon 1998). Limitations concern portraying cognition in static, mechanistic terms that may not apply to living beings.

Models of human cognitive development have proven valuable for assessing primate cognition comparatively because they provide conceptual and methodological tools for assessing non-symbolic and symbolic cognition within one unified framework and the generation and structure of cognition. Piaget's model supported the first developmental studies of great ape cognition; among its greatest contributions is its portrayal of cognition as constructed progressively during ontogeny and directly affected by interaction with the environment. Early piagetian studies focused on sensorimotor (human infant) cognition, which relies on pre-symbolic processes similar to those portrayed in animal models, so similar limits apply. Recently, neo-piagetian models have supported studies of the rudimentary symbolic range (for an overview, see P&M). Models inspired by Vygotsky, which portray socio-cultural forces like apprenticeship or enculturation as fundamental to cognitive development (e.g., Donald 2000; T&C), have spawned many studies on social cognition and cognitive development in great apes. Given how richly primate lives are socially embedded, their merits are obvious. Among these models may be included models of understanding others' mental states, or theory of mind, which some propose to underpin much cognitive progress in early human childhood (e.g., Carruthers & Smith 1996). Two such models have been applied to great apes, both proposing that general-purpose cognitive processes in the rudimentary symbolic range underwrite this progress (second-order representation – Leslie 1987; secondary representation – Perner 1991). While both offer useful tools for assessing rudimentary symbolic processes, their focus on one ability series in the social domain risks underrepresenting the breadth of great apes' achievements.

Table 6.1. *Processes and structures posited in rudimentary sybolic level cognition, in human developmental perspective*

Age (yr)	Piaget (general)	Case[1] (causality)	Langer[2] (logic–math)	Leslie[3] (Theory of Mind)	Perner[4] (Theory of Mind)
0	Sensorimotor stage Schemata	Sensorimotor stage	1st-order cognition	1st-order representation	Primary representation
		Operational consolidation			
1.5		(Inter-) relational cognition unifocal relational operations	2nd-order cognition	2nd-order representation metarepresentation	Secondary representation re-represent primary representations multiple representations
2	stage 6, symbols, **representation**				
3.5	Pre-operational stage Symbolic subperiod	Bifocal relational operations (1st-order symbolic)			
5	Intuitive subperiod	Elaborated relational operations (2nd-order symbolic)	3rd-order cognition	Reason across metarepresentational structures	**Metarepresentation symbols**

Notation: Symbol, representation, metarepresentation defined in text; normal/**bold** type indicates weak/**strong** meanings respectively; major cognitive periods are underlined; significant processes or structures within periods are indented.

[1] Case (1985, 1996) models causal cognition at the level of operating on object–object relations, i.e., relational (properly, inter–relational) cognition. Children develop structures first for single, simple relations, then relations-between-relations, then coordinating increasing numbers of relational structures: 12–20 mo – represent one relationship between two items (operational consolidation); 20–27 mo – represent one inter–relational structure (unifocal operations); 27–40 mo – represent two inter–relational structures (bifocal relational operations, first order symbolic); 40–60 mo – inter-relate more inter-relational structures (elaborated relational operations, second order symbolic).

[2] Langer (1998, 2000) models logical operations on subjects' spontaneous object groupings. 0–12 mo – make one set of objects with one class property, map 1st-order operations onto it (1st-order cognition); 18–36 mo – make two contemporaneous sets, map 2nd-order operations onto the sets (2nd-order cognition); > 36 mo – make three contemporaneous sets, map 3rd-order operations onto them (e.g., construct correspondences) (3rd-order cognition).

[3] Leslie (1987). First-order (primary) representations encode entities in an accurate, literal way; they are perceptually based and defined in sensorimotor codes by direct semantic relation with the world; multiple primary representations of a situation can exist. Second-order representation creates a decoupled copy of a primary representation then reconstructs or redescribes it; making a decoupled copy entails metarepresentation; second order refers to being derived from a primary representation; second-order representations typically remain anchored to parts of the primary representation.

[4] Perner (1991). 0–1 yr – representation(s) of the situation exists; 1–4 yr – other representation(s) of the situation (past, future) (secondary representations) are entertained simultaneously with the primary representation; > 4 yr – representations of other representations are created and understood *as* representations (i.e., as interpretations) (strong metarepresentation).

More broadly, reservations are that Vygotsky-based models tend to emphasize socio-cultural factors to the neglect of individual and biological ones, while Piaget-based models suffer the opposite bias. Together, these models offer rich portrayals of cognitive development and have spawned comparative models situating primate cognition in developmental and evolutionary perspective (e.g., P&M).

I favor development frameworks because they allow assessment of symbolic processes, their constitution, and continuities as well as discontinuities between human and nonhuman primates. I adopt them here as the basis for interpreting evidence.

EVIDENCE

For evidence, I relied on recent integrative reviews (RWB, P&M, T&C) more than edited volumes, to privilege syntheses over the breadth of current views, plus findings appearing since their publication (post 1998). I concentrated on achievements linked with symbolism as the critical cognitive threshold and feral great apes[1] as most relevant to evolutionary questions. Table 6.2 summarizes this evidence, arranged by the cognitive structures inferred in terms of cognitive domain (broad areas of knowledge, typically physical, logico-mathematical, social, linguistic), problem-specific structures (ability series), and complexity (level). This arrangement derives from models of human cognitive development near the rudimentary symbolic range (Table 6.1).

My coverage of the evidence is inevitably incomplete but sufficient to establish broad patterns. Evidence for complex achievements is substantial, for instance, and the relevance of complex skills to feral life is clear in all cognitive domains even though little evidence derives from feral subjects. Equally clear and needing explanation are the repertoire's impressive breadth and "openness" (i.e., including apparently "atypical" language and mathematical abilities). Disputes in any case lie less with what great apes achieve than with cognitive inferences, so more important cautions are that inferences are controversial, numerous factors complicate interpretation, and I inevitably glossed over subtleties and debates in working towards an overall picture.

A long-standing concern is variability in achievements across problem types, individuals, species, and contexts. Some report great apes outperforming 5- to 6-year-old humans (e.g., Call & Rochat 1996),

others report them failing at simple levels of understanding (e.g., Povinelli 2000). While this variability may be meaningful (e.g., cognitive differences between species, significant features of cognitive development, module-like cognitive architecture), it also reflects confounding factors extensive enough to undermine interpretation.[2] Because quantitative breakdowns remain un-interpretable, I have not provided them. Most experts in any case consider that all great apes share roughly equivalent cognitive capacities (RWB, P&M, T&C) and it is these similarities that are of primary interest here.

COGNITIVE LEVELS: THE HIGH-MINDED

An important consideration in analyzing the cognition governing great apes' complex achievements is that it may involve higher *levels* of cognitive abstraction, not just very rapid processing, extended working memory, or new types of abilities (Roberts & Mazmanian 1988). Humans, great apes, and some monkeys can master making and using tools, for instance, so all share the means–end *type* of cognition; great apes and humans differ in achieving higher *levels* of means–end cognition that support more complex tool use (e.g., Visalberghi & Limongelli 1996). What levels great apes attain is a major focus of current debate. Three levels recognized in human development beyond pre-symbolic, sensorimotor cognition (with its schemata, i.e., first-order or primary representations) are probably important to resolving this debate (see Table 6.2). These are:

(1) **Emergence of rudimentary symbols.** Around 1.5 years of age, humans begin creating and using simple symbols, like mental images, to stand for referents instead of having to use direct sensorimotor information. A classic example is inferring where an item is hidden after watching it be displaced "invisibly," along a trajectory that passes behind barriers; this shows that the actor can mentally reconstruct events it did not directly perceive (de Blois, Novak & Bond 1998). Early symbols have been attributed to strong representation (Piaget 1952, 1954; P&M), understanding relational categories between entities external to the actor (Herrnstein 1990; Rumbaugh & Pate 1984; Spearman 1927; Thompson & Oden 2000; T&C), or representing single object–object relations (Case 1985). This level is usefully viewed as a transition, i.e., a phase

Table 6.2. *Great apes' cognitive achievements and cognitive abilities*

Physical Domain

Cognition Series/level	Achievement	Comments – examples	Sources[a]
Object concept: *developing the concept of "object" in the environment (extends only to sensorimotor stage 6)*			
Transitional (1.5–2 yrs)	Track invisible displacements		1 (2) 3 4 16
Causality: *dynamic relations between objects when external forces affect them*			
Transitional (1.5–2 yrs)	Inconsistent but insightful make & use rake tool	(Inconsistent success)	1 2 3 12
	Single object–object relations	Tertiary relations between objects	1 2 3
Rudimentary symbolic (2–3.5 yrs)	Consistent raking	Rake with consistent success	1 (2) 8
	Advance tool preparation	Emerges in children > 2 yrs old (T&C)	1 2 3 11 12
	Hierarchical techniques	Manual and tool (sets, series, meta–)	1 (2) 3 5 7 8 10 17 18 20
	Composite tools	i.e, multi-tool assemblages	1 (2) 3 6 15 17
	Inter-relational object use	i.e, relations-between-relations	1 (2) 3 9 17 20
	Cooperative hunting	Arboreality–prey–hunter relations	1 3 12
Space: *spatial understanding (knowledge, relations) and reasoning*			
Transitional	Detour	re barrier, check food in advance	(1) 2 3
	Navigate 2-dimensional maze		2 21
	Arboreal "clamber" travel		3
Rudimentary symbolic	Stack blocks	Put objects in containers, stack	1
	Block assembly	Two or more blocks, variously related	1
	Draw circle or cross		1
	Tie simple knot	Winding and inserting	1 13
	"Map" read	Use scale models, TV, photos	1 (2) 14 19
	Euclidean mental maps	Minimize site-site travel distance	1 (2) 3 12 22[b]
	Plan travel routes	Least distance, arboreal routes	1 (2) 3

Notes: [a] Cognitive attributions based on criteria discussed in text; source bracketed when my attribution differs from that of the authors cited. [b] Symbol-trained great apes tested.

Sources: 1, Parker & McKinney 1999; 2, Tomasello & Call 1997; 3, Byrne 1995; 4, de Blois, Novak & Bond 1998; 5, Corp & Byrne 2002; 6, Sugiyama 1997; 7, Russon 1998; 8, Byrne & Russon 1998; 9, Russon & Galdikas 1993; 10, Russon 1999a; 11, Fox, Sitompul & van Schaik 1999; 12, Boesch & Boesch–Achermann 2000; 13, Maple 1980; 14, Kuhlmeier *et al.* 1999; 15, Bermejo & Illera 1999; 16, Call 2001a; 17, Yamakoshi, Chapter 9, this volume; 18, Stokes & Byrne 2001; 19, Kuhlmeier & Boysen 2001; 20, Matsuzawa 2001b; 21, Iverson & Matsuzawa 2001; 22, Menzel, Savage-Rumbaugh & Menzel 2002.

Table 6.2. (cont.)

Logical–Mathematical Domain

Cognition Series/level	Achievement	Comments and examples	Sources[a]
***Classification:** organize objects by features and categories*			
Transitional	Double sets + exchange	2nd-order classification	1 6^b
	Concept formation	Predators, foods, other species	2 4^b
	Simple relational category	TPR, e.g., identity, odd, same–different	2 4^b 8
Rudimentary	2nd-order classify, operations	To levels like humans 24–30 mo	1 (2) 5 6^b 7^b 15
symbolic	Analogical reasoning	Abstract relations between relations	1 (2) 4^b 7^b
	Use abstract codes		1 4^b
	Multiplicative classification	Simultaneous multi-feature sort	1 (2)
	Classify by function	"Tool" class, sort bottles with caps	(2) 4^b 10
	Minimal 3rd-order classify	Prerequisite for hierarchical classification	5 6^b 7^b
	Hierarchical part–whole relations		16^b
	Routine structure	For object–object relations; hierarchical	16^b
***Seriation:** organize object sets with respect to ordinality and transitivity*			
Transitional	Seriate nesting cups ("pot")	By "pot" strategy (one cup into another)	(11)
rudimentary	Spontaneous seriation	Order sticks, order tools in tool set	1 (2) 12
symbolic	Seriate nesting cups (nesting)^b	By subassembly (so hierarchical)	1 (2) (11)
	Seriate: transitivity based	Operational logic, 2-way relations	1 (2) 4^b 11 19
	Transitivity	In social rank; in a serial learning task	1 (2) 3 21

Table 6.2. (*cont.*)

Logical–Mathematical Domain

Cognition *Series/level*	Achievement	Comments and examples	Sources[a]
Number/Quantity: assess object sets with respect to number or quantity			
Transitional	Sequentially tag several items		1 2
	Sequentially tag + label number		1 2
Rudimentary symbolic[b]	Count	Exact number of items in arrays	1 (2) 14
	Compare proportions	Fraction, quantity–based analogy	1 (2)
	Conserve number (1:1)	Understand 1:1 correspondence	1 (2)
	Summation	Adding items increases quantity	(2) 9 13 14
	Quantified (social) reciprocity	Meat share rules, exchange groom/favors	1 (2) 3 9 20 22
	Planned numerical ordering	Sequence all items before acting	15
	Reverse contingency task	Choose smaller of 2 arrays to get more	17 18
	Symbolic quantity judgment	Select array for quantity using symbols	17
Conservation: conserve properties of objects that undergo transformations			
Rudimentary symbolic[b]	Conserve quantity (conceptual)	Physically transformed (solid & liquid)	1 (2)

Notes: [a] cognitive attributions based on criteria discussed in text; source bracketed when my attribution differs from that of the authors cited. [b] symbol-trained great apes tested.

Sources: 1, Parker & McKinney 1999; 2, Tomasello & Call 1997; 3, Byrne 1995; 4, Thompson & Oden 2000; 5, Langer 2000; 6, Potí *et al.* 1999; 7, Spinozzi *et al.* 1999; 8, Tanaka 2001; 9, Sousa & Matsuzawa 2001; 10, Russon 1999a; 11, Johnson-Pynn *et al.* 1999; Johnson-Pynn & Fragaszy 2001; 12, Bermejo & Illera 1999; 13, Call 2000; 14, Beran 2001; 15, Biro & Matsuzawa 1999; 16, Spinozzi & Langer 1999; 17, Boysen & Berntson 1995; Boysen *et al.* 1996; 18, Shumaker *et al.* 2001; 19, Kawai & Matsuzawa 2000; 20, Boesch & Boesch-Achermann 2000; 21, Tomonaga & Matsuzawa 2000; 22, Mitani & Watts 2001; Mitani, Watts & Muller 2002. (Note: related studies are grouped.)

Table 6.2. (cont.)

Social domain

Cognition Series/level	Achievement	Comments and examples	Sources[a]
Social learning & imitation: socially influenced learning; imitation is learning to do new acts by seeing them done			
Transitional	Deferred imitation	Delayed imitation of novel actions	1 24[b]
	Action-level imitation	"Impersonation," to some	1 3
	• gestures	Spontaneous gestures, gesture signs	1 2 3 15 22
	• actions on objects	Simple tool use, object manipulation	1 2 3 4 14 18 24[b] 29 34 37
	Imitate action sequence	2-action sequence or longer	2 3 6 24[b]
Rudimentary	Program-level imitation	Routine structure, relations–between–relations	1 3 4 5 14 23 24[b]
Symbolic	Mime intent, request, teach	Act out (for other), express intent	1 (2) 3 7 33[b]
Pretense: re-enact actions outside their usual context and without their usual objectives			
Transitional	Re-enact events (scripts)	"Feed" doll, "take" photo with camera	1 2 7 8
	Basic symbolic play		1
Rudimentary	Symbolic object use		1 (2)
Symbolic	Advanced symbolic play	With substitute object (e.g., log baby)	1 (2) 3
	Role play	Play mother's or another's role	1
	Demonstration teach		1 3
Social knowledge & theory of mind: understanding others: behaviors, roles, and mental states			
Transitional	Mirror self-recognition		1 2 3
	Gain other's attention	Wait for, vocalize/gesture to gain attention	1 2 3 30 32 33[b] 39
	Interpret visual perspective	Track other's gaze (e.g., to get food)	1 2 3 11 27 28 35 36
	Third-party relations (T&C)		2
	Pre-select allies	Curry favor with potential helpers	2
	Conversational contingency	Context-appropriate responses	31[b]
	Impute intentions	Unfinished, deliberate (vs. accidental) acts	1(2) 3 11 (17) 34 39

Table 6.2. (cont.)

Social domain

Cognition *Series*/level	Achievement	Comments and examples	Sources[a]
Rudimentary	Impute knowledge, competence	Knowledge-sensitive social activity	1 (2) 3 11 19 21 28 29 39
Symbolic	Empathy	Console (nb. post death), mediate reconciliation	1 10 11
	Take complementary role	Cooperative hunt, role-based teamwork	1 (2) 3
	Role reversal		1 (2) 3 28
	Cooperation (enact, plan)		1 (2) 3 11
	Quantified reciprocity	Balanced revenge, share, loser help	1 (2) 3 12 38
	2nd-order intentionality	2nd-order tactical deception (withhold item, mislead, counter-deceive), teach	1 (2) 3 28
	Transitivity in social rank	Social seriation	1 (2) 3
	Complex coalitions/alliances		3 9

***Sense of self**: self* (*awareness and self understanding* (*cognitions about the self*)

Cognition *Series*/level	Achievement	Comments and examples	Sources[a]
Transitional	Mirror self recognition		1 2 3
	Self label	Personal pronouns	1 3
	Self conscious behavior		1
	Self concept	As a causal agent	1 2 3
	Understand see-know in self	Know if you know, based on what you saw	26
Rudimentary	Indirect self recognition	Picture, shadow	1 12
Symbolic	Sense of possession		13
	Self evaluative emotions	Shame, guilt, pride	1

Notes: [a] cognitive attributions based on criteria discussed in text; source bracketed when my attribution differs from that of the authors cited. [b] symbol-trained great apes tested.

Sources: 1, Parker & McKinney 1999; 2, Tomasello & Call 1997; 3, Byrne 1995; 4, Myowa–Yamakoshi & Matsuzawa 1999; 5, Byrne & Russon 1998; 6, Whiten 1998b; 7, Russon 2002b; 8, Suddendorf & Whiten 2001; 9, Parker, Chapter 4, this volume; 10, de Waal & Aureli 1996; 11, Boesch & Boesch–Achermann 2000; 12, Patterson & Linden 1981; 13, Noe, de Waal & van Hooff 1980; 14, Russon 1999a; 15, Tanner & Byrne unpublished; 16, Hare *et al.* 2000; 17, Call & Tomasello 1998; 18, Stoinski *et al.* 2001; 19, Whiten 2000; 20, Whiten 1998a; 21, Boysen 1998; 22, Call 2001b; 23, Stokes & Byrne 2001; 24, Bering, Bjorkland & Ragan 2000; Bjorkland *et al.* 2002; 25, Call, Agnetta & Tomasello 2000; 26, Call & Carpenter 2001; 27, Call, Hare & Tomasello 1998, 2000; Hare *et al.* 2000, 2001; Tomasello, Call & Hare 1998; Tomasello, Hare & Agnetta 1999; Tomasello, Hare & Fogleman 2001; 28, Hirata & Matsuzawa 2001; 29, Hirata & Matsuzawa 2000; 30, Hostetter, Cantera & Hopkins 2001; 31, Jensvold & Gardner 2000; 32, Leavens & Hopkins 1999; 33, Menzel 1999; 34, Myowa–Yamakoshi & Matsuzawa 2000; 35, Itakura & Tanaka 1998; 36, Peignot & Anderson 1999; 37, Custance *et al.* 2001; 38, Mitani & Watts 2001; Mitani, Watts & Muller 2002; 39, Blake, Chapter 5, this volume. (Note: related studies are grouped.)

Table 6.2. (cont.)

Linguistic (Symbolic Communication) Domain

Cognition Series/level	Achievement	Comments and examples	Sources[a]
Lexical			
Transitional	Word production		1 3
	Create new sign	Create "Dave missing finger" name	6
Semantic			
Transitional	Simple referential symbols	Tree drum, leaf groom, linguistic symbol	1 2 3 9
	Create new meaning		2 8
Rudimentary	Teach signs		1 3
Symbolic	Mould or mime request		(2) 5 8
	Symbol-based solution	Solve reverse contingency task with symbols	(2) 7
Grammar			
Transitional	Novel 2-word combinations		1 2 3 8
	Wild symbol combinations	drums, drum 2 trees in sequence	2 4 8
	Reliable use of word order	as a contrastive symbolic device	2
	Nested combinations		1 3
Rudimentary	Three-word utterances	(Plus comprehension beyond)	1 (2) 3 8
Symbolic	Grammatical morphemes		1 8
	Functional sign classes	Call attention, request, declare	(2) 3

Note: [a] cognitive attributions based on criteria discussed in text; source bracketed when my attribution differs from that of the cited authors.
Sources: 1, Parker & McKinney 1999; 2, Tomasello & Call 1997; 3, Byrne 1995; 4, Boesch & Boesch-Achermann 2000; 5, Russon 2002b; 6, Miles, Mitchell & Harper 1996; 7, Boysen & Berntson 1995; Boysen *et al.* 1996; 8, Blake, Chapter 5, this volume; 9, Menzel, Savage-Rumbaugh & Menzel 2002.

of reorganizing or transforming lower-level structures into new, higher-level ones (Case 1985, 1992). Eating with a spoon, for instance, can be achieved either using a complex action strategy governed by a combination of sensorimotor-level motor action schemata or using a simple higher-level strategy that consolidates this combination of schemata into one operation on a relationship. Importantly, behavior in transitional periods may owe to cognitive structures at either the lower or higher level – here, sensorimotor schemata or simple symbols.

(2) **Rudimentary (first-order) symbolic-level structures**. From about 1.5–2 to 3.5–4 years of age, human children create cognitive structures that represent simple events and relationships among them (Case 1985; P&M). Behavioral examples are simple word combinations, using two tools in interrelated fashion, and symbolic pretend play.

Several models portray cognitive development in this phase in terms of creating higher-level cognitive structures derived from sensorimotor ones, i.e., they represent, in the sense of recoding or redescribing, existing representations. Case (1985) construes this as operating on relations-between-relations, where one relationship is subordinated to another or used as a way to effect change in another. Included are coordinating two different relationships into one "inter-relational" cognitive structure (e.g., hammer-hit-nut with nut-on-anvil) and coordinating two inter-relational structures. Other models are *second-order cognition* (Langer 2000) and *second-order representations* (Leslie 1987). Second-order cognition is exemplified by creating two sets concurrently, so that items are similar within each set and different between sets (e.g., red balls, blue balls); this involves simultaneously managing the relationship within each set (same item class) and a higher-order relationship between two sets (different classes). Second-order representations are derivatives of realistic (first-order) representations, for example using a banana as a telephone. To avoid confusion, Leslie argued, "banana-as-telephone," must remain linked to its first-order representation, "banana-as-banana," yet decoupled from it (i.e., marked as an imagined copy). Making a decoupled copy requires re-representing an existing representation, so second-order representations are higher-level structures.

A competing model of cognition in this range is *secondary representations* (Perner 1991), where re-presentations are *subsequent* presentations of something

previously present in the mind. Examples are entertaining past or future representations of a situation or bringing schemata to mind without their normal sensorimotor cues. Secondary representation may be what allows coordinating multiple models of a situation, which may enable tracking where an object went after it moved along an invisible trajectory, pretending that an empty cup is full, or interpreting external representations of a situation. Children in the secondary representation phase can represent how things *might* be as well as how they actually are; previously, they could only represent the latter (Whiten 1996). Secondary representations, like second-order ones, are representations of a situation entertained concurrently with the situation's realistic or current representation and they represent something about the relations among multiple representations of a situation; differently, secondary representations are not higher-level structures. They remain pre-symbolic in Perner's view; strong metarepresentation, which follows them, is the simplest symbolic process.

(3) **Strong (second-order) symbolic-level structures**. Strong symbolism emerges around 4 years of age. Understanding that people can hold false beliefs about the world is the accepted benchmark (Whiten 1996). To Perner, this requires appreciating that others may have different thoughts about reality than oneself, i.e., understanding re-presentations *as* re-presentations (interpretations) or strong metarepresentation. An alternative model is third-order cognition, where third-order structures are structures that encompass multiple second-order ones in superordinate–subordinate fashion (Langer 1998, 2000). An example is composing three matching sets of items, which creates hierarchical correspondences between the sets, i.e., a superordinate category subsuming two subordinate, second-order ones. Three sets is the minimum needed for hierarchical classification, which enables truly hierarchical cognition (Langer 1998).

Levels in great ape cognition

I attributed cognitive levels to great apes' complex achievements, per Table 6.2, using recognized indicators of early symbolic processes in humans. Indicators of rudimentary symbolic-level cognition included weak hierarchization (e.g., routines that subsume subroutines), tasks first solved by children between 2 and 3.5 years of age, tests of abilities accepted as higher-order ones (e.g., analogies), and manipulating relations-between-relations. I interpreted achievements emerging in 1.5- to 2-year-old humans and taken to mark the threshold of weak symbolism as transitional. I considered levels that original authors attributed but privileged the indicators noted above. T&C did not analyze great ape achievements individually, for example, so some of their cognitive attributions lack substance. On this basis, I consider the four positions currently entertained on the cognitive levels great apes attain.

(1) Great ape cognition operates with same low-level associative processes attributed to all nonhuman species (e.g., Balda, Pepperberg & Kamil 1998). All three reviews reject this position because of substantial evidence for higher-level cognitive processes in great apes. Recent informed opinion concurs (e.g., Matsuzawa 2001b; Suddendorf & Whiten 2001; Thompson & Oden 2000; Table 6.2). Low-level associative processes like trial-and-error learning and sequential chaining are necessary but not sufficient to account for great apes' achievements.

(2) Great apes share with all anthropoid primates a cognitive level beyond other mammals, understanding third-party relations (TPRs) (T&C). T&C define TPRs as interactions among third parties in which the actor does not participate, for example separating interventions and mediating reconciliations. Tomasello's group advocates this position but most other experts disagree (Matsuzawa 2001b, P&M, Russon 1999b, RWB, Suddendorf & Whiten 2001, Thompson & Oden 2000). T&C consistently interpret great apes' achievements with undue skepticism and monkeys' with undue generosity; for instance, no evidence supports their claim that monkeys can perceive, let alone judge, relations-between-relations (Parker 1998, Chapter 4, this volume; Rumbaugh 2000; Russon 1999b; Thompson & Oden 2000).

T&C's relational cognition model is itself problematic (Russon 1999b), although many agree with them that understanding relational categories and relations-between-relations is among great apes' crowning achievements. T&C characterize great apes' relational achievements as understanding TPRs, construed as a generalized ability governed by advanced sensorimotor cognition (stages 5 and 6). This cognition reaches into a transitional range where either sensorimotor or symbolic structures can generate achievements. Stage 6 also supports understanding *single* relational

categories but not relations-between-relations; the latter requires rudimentary symbolic cognition because it concerns relations between abstract entities (Case 1985). T&C rely exclusively on sensorimotor measures, so they fail to assess whether early symbolic or sensorimotor processes generate achievements and they underrate achievements involving relations-between-relations, such as great apes' meta-tool and tool set use. Their TPR model also conflates transitional with rudimentary symbolic achievements, confounding two levels of probable significance in distinguishing great ape from monkey cognition.

(3) Great apes surpass other nonhuman primates in attaining secondary representation, which may characterize the 1.5- to 3.5-year phase in human cognitive development, but fall short of strong symbolic levels (e.g., Suddendorf & Whiten 2001; Whiten 1996). Suddendorf and Whiten's (2001) review of great apes' achievements on invisible displacements, means–end reasoning, pretense, mirror self-recognition, mental state attribution, and understanding imitation supports their conclusion that great apes achieve secondary representation up to the level of 2-year-old humans. This is consistent with the common characterization that great apes acquire language abilities up to the level of human 2 year olds (e.g., Blake, Chapter 5, this volume).

This review neglects to consider great apes' highest level achievements in pretense and means–ends reasoning, however, or any of their achievements in logico-mathematical or spatial reasoning (e.g., Langer 1996; Mitchell 2002; Table 6.2) so it does not provide a thorough test of position 3. It also emphasizes human achievements in the second year and underplays the third, situating it closer to position 2 than position 3. Scale model use and minimal third-order classifying, which humans master in their third year, have been shown in great apes (Kuhlmeier, Boysen & Mukobi 1999; Poti *et al.* 1999; Spinozzi *et al.* 1999). Scale model use in particular may involve using models *as* representations, putting great apes on the brink of strong metarepresentation.

Secondary representation also fails to account for the higher level structures that can enrich cognition beyond sensorimotor levels, especially those concerning relations-between-relations. Great apes' complex feeding techniques and their logical and quantitative achievements offer prime evidence of such higher-level cognitive structures (Byrne & Byrne 1991; Byrne *et al.*

2001; Langer 2000; Matsuzawa 1996, 2001b; P&M; Russon 1998; Spinozzi & Langer 1999). The secondary representation concept fails to address the structure that individuals may add to a representation by re-presenting it or precisely how multiple representations of a situation are related; models proposing higher-order structures fill this gap (Case 1985; Leslie 1994; Whiten 2000). Actors may not only recall alternative realistic representations of a situation (e.g., past, future), for instance, they may also re-represent the situation differently from any reality they have experienced (e.g., a banana as a telephone) and/or at a higher level (one relationship vs. multiple schemata). While the secondary representation concept is valuable in suggesting where higher-level cognitive structures are *not* used to entertain multiple representations of a situation, it fails to consider circumstances in which they are.

(4) Great apes surpass other nonhuman primates in attaining rudimentary symbolic-level cognition (e.g., RWB; Langer 1996; Matsuzawa, 2001b; P&M; Russon 1998, 1999a). P&M, RWB, and many recent studies (Table 6.2) support this position for all great ape species, in all cognitive domains, based on recognized indices of weak symbolism (weak hierarchization, abilities recognized to involve higher-order processes, relations-between-relations). Comparable achievements claimed for monkeys have been shown to involve performances based on response rules generated by simpler processes, probably associative ones (Parker, Chapter 4, this volume; Thompson & Oden 2000).

Many current disagreements stem from what assessment tools are used and what meanings of symbolism, metarepresentation, and hierarchization are applied (Whiten 1996). With the meanings and assessments used here, the best interpretation of current evidence is that great apes attain rudimentary symbolic-level cognition and in this, they surpass other nonhuman primates.

The levels that great apes achieve within the rudimentary symbolic range are relatively uncharted. Assessment remains an impediment because many current tests for symbolism use threshold criteria (e.g., metarepresentation, hierarchization). Indices of early symbolic levels have been used in a few cases, e.g., number of relational operators, complexity of classification, depth of hierarchies, or human age norms (Byrne & Russon 1998; Kuhlmeier *et al.* 1999; Matsuzawa 2001b; P&M; Poti' *et al.* 1999; Russon & Galdikas unpublished;

Spinozzi *et al.* 1999; Thompson & Oden 2000). These suggest great apes' cognitive ceiling at a hierarchical depth of about three levels (e.g., use a hammer stone to hit (a nut placed on (an anvil stone placed on a wedge, to level it)) – Matsuzawa 2001b; and see Yamakoshi, Chapter 9, this volume), coordinating three object–object relations in one inter-relational structure (e.g., coordinate *anvil-on-wedge*, *nut-on-anvil*, and *hammer-hit-nut* – P&M; Russon & Galdikas unpublished), or minimal third-order classification (e.g., create three contemporaneous sets with similar items within sets and differences between sets – Langer 2000; Poti' *et al.* 1999; Spinozzi *et al.* 1999). All remain consistent with Premack's (1988) rule of thumb, that great apes reach levels like 3.5-year-old children but not beyond.

COGNITIVE INTERCONNECTION: THE ORCHESTRALLY MINDED

Cognitive facilitation refers to achievements made through interplay among different types of cognition. It is an important source of an actor's cognitive power: tasks that require interconnecting several abilities can be solved, and individual abilities can advance by exploiting other abilities (Langer 1996). Cracking a nut, for example, might require using a stone hammer–anvil tool set (means–end reasoning), identifying a substitute when the best hammer is unavailable (logical reasoning), and obtaining the substitute from a companion (social cognition), or classification abilities might be extended by categorizing according to causal utility. Cognitive facilitation almost certainly occurs in great apes. Chimpanzees skilled in symbol use solved analogy problems better than chimpanzees without symbol skills, for example (Premack 1983; Thompson, Oden & Boysen 1997).

Facilitation has received little attention in great apes despite its implications for cognitive architecture. If it occurs, especially across domains, then qualitatively different cognitive structures can operate and interact beyond the bounds of the problem types for which they were designed: that is, the cognitive system cannot simply comprise a collection of independent, special-purpose modules. Facilitation is also important comparatively because it has been claimed to be uniquely or at least characteristically human, for whom it has been likened to fluidity of thought, multiple intelligences functioning seamlessly together, a passion for the analogical, and mapping across knowledge systems (e.g., Gardner 1983; Mithen 1996). What enables facilitation is unresolved. Hypotheses include analogical reasoning, which transfers knowledge from one problem type to another (Thompson & Oden 2000), or synchronizing developmental progress across distinct types of cognition so that their structures build upon common experiences, which promote interplay by serving as bridges between cognitive structures (Langer 1996, 2000). Possibilities typically require hierarchization; analogical reasoning, for instance, involves judging if one relationship is the same as another, i.e., logical equivalences at abstract levels, which is founded on the ability to judge relations-between-relations.

In part, systematic evidence on facilitation in great apes is meager because studies of nonhuman cognition have tended to control *against* using multiple abilities in aiming for clean tests of single abilities. Among the few sources of systematic evidence are studies of logic, which show that analogical reasoning is within the normal reach of great apes but not other nonhuman primates (Oden, Thompson & Premack 1990; Thompson *et al.* 1997; Thompson & Oden 2000). For feral great apes, P&M is the only review to have systematically considered achievements that may involve facilitation. I consider evidence for facilitation across physical, logical, and social domains as the most important in comparative perspective.

Logical–Physical

Great apes interconnect logical with physical cognition when they classify items by function or functional relations, for example sort items into sets of toys and tools or sort bottles with caps (Savage-Rumbaugh *et al.* 1980; Tanaka 1995; Thompson & Oden 2000), use substitute tools (Figure 6.1), or classify foods on the basis of the technique for obtaining them (Russon 1996, 1999a, 2002a). A rehabilitant orangutan stored termite nest fragments on specific parts of his body, in the order in which he planned to open them, to streamline his termite foraging (Russon 2002a) and a rehabilitant chimpanzee made and used a seriated set of stick tools (ordered from smallest to largest) to extract honey from a bee's nest (Brewer & McGrew 1990).

Social–physical

Great apes use socially mediated learning in acquiring food processing skills (Boesch 1991; Byrne & Byrne

Figure 6.1. Princess, an adult female rehabilitant orangutan, blows on the burning tip of a mosquito coil. A paper marked with two dots is at her feet. She had drawn the dots by touching the coil's burning tip to the paper, i.e., substituting the coil for a pen. She often scribbled in notebooks with pens, so she used a functional similarity between pens and the coil, that both have tips that can mark paper. She did not simply confuse the two tools. She drew differently with the coil (touch vs. scribble) and she fixed it differently (if a pen did not mark when she scribbled, she fixed it by biting at its tip or by clicking the pen's switch to advance the tip; to fix her coil, she blew on its tip).

1993; Inoue-Nakamura & Matsuzawa 1997; Matsuzawa & Yamakoshi 1996; Russon 1999a, 2003a,b). When they use imitation or demonstration to advance complex food processing skills, social cognition contributes to physical cognition at rudimentary symbolic levels. The most complex cases known concern stone nut-cracking in west African chimpanzees: mothers demonstrate to their offspring how to use stone hammers, and offspring replicate the techniques they were shown (Boesch 1991, 1993). Mithen (1996) argued that food sharing, used as a medium for social interaction with formalized sharing rules, uses "natural history" cognition to enhance social problem solving. If so, chimpanzees show this capability: they share meat in rule-governed fashion to serve social functions and social relationships are important in distribution (Boesch & Boesch-Achermann 2000; Goodall, 1986; Mitani & Watts 2001; Mitani, Watts & Muller 2002).

Logical–Social–Symbolic

Boysen et al. (1996) used a reverse contingency task to test if chimpanzees could select the smaller of two arrays to gain greater rewards against a social competitor. Boysen showed two dishes of candies to a dyad of symbol-trained chimpanzees, had one choose a dish by pointing, and then gave the chosen dish to the other chimpanzee and the leftover dish to the chooser. Shown real candies, choosers consistently picked the dish with more – to their disadvantage. When number symbols replaced candies, choosers consistently picked the dish with fewer – to their advantage. Symbols improved these chimpanzees' ability to solve a quantification (logical) problem. Orangutans also solve this task, without symbol skills and using real candies (Shumaker et al. 2001). If subjects interpreted this as a competitive social task, as intended, their quantification (logical) abilities assisted their social problem solving.

Complex facilitations

Some expertise taps all three domains interactively. The most complex is chimpanzee cooperative hunting in the Taï forest (Boesch & Boesch-Achermann 2000). Once a hunting group detects a red colobus group, the ideal hunt has four phases involving four roles (driver, chaser, ambusher, captor). Participants must be able to alter their actions flexibly and rapidly to track colobus' attempts to escape; they also take different roles and accommodate their actions to chimpanzees in other roles. If successful, they share the meat formally according to each participant's role in the hunt, age, and dominance. Successful cooperative hunting in the forest, a three-dimensional space with low visibility, requires hunters to "perceive other hunters as independent agents with their own intentions and competence, attribute abilities to the prey that differ from those of conspecifics, and understand the causality of the external relation between prey and other hunters" (Boesch & Boesch-Achermann 2000: 242). It requires cognitive abilities in the physical domain (space – arboreal locomotion and routes; causality – predicting how chasing, blocking, or driving will affect colobus' flight path and the canopy), the social domain (self-manipulating the presentation of oneself to the colobus; figuring one's weight into arboreal travel; enacting complementary roles), and the logical domain (quantifying how to distribute meat sharing). Hunters can change roles repeatedly over the course of a hunt, so some must have all or most of these cognitive capabilities and use them in interconnected fashion.

Evidence for cognitive facilitation jibes with the complex, varying, and multifaceted challenges facing great apes in their natural habitat (Boesch & Boesch-Achermann 2000; Russon in 2003b). Their foraging offers a prime example: it calls for a wide spectrum of abilities to organize biological knowledge, construct foraging techniques, acquire alternative strategies, and negotiate cooperative and competitive social foraging situations (Russon 2002a; Stokes 1999). The multifaceted nature of complex foraging tasks calls for combining high-level abilities, and interactions among task components call for interconnecting them. Evidence for cognitive facilitation also jibes with evidence that great apes spontaneously transfer expertise from one domain to another (Thompson & Oden 2000), with Parker's (1996) apprenticeship model of interconnected physical and social abilities, and with arguments that interconnecting mechanisms of some sort are essential to cognitive systems that handle different types of information in parallel using distinct modules (Mithen 1996). It clearly refutes strictly modular models of cognition in great apes.

GENERATING GREAT APE COGNITION

The variability and flexibility of great apes' cognitive abilities, including the capacity to generate unusual abilities as needed and the roughly consistent cognitive ceiling across abilities, domains, and species, suggest that their cognitive systems may be better characterized by the processes that generate them than by specific abilities such as tool use or self-concept. Generative processes are considered below.

Development and culture

Developmental models of human cognition have probably been fruitful in studying great apes because their cognitive structures develop in similar fashion (Langer 1996, 2000; P&M). Like humans, great apes experience extensive and lengthy sensory, motor, socio-sexual, brain, and cognitive development that is affected by age and experience and is concentrated in immaturity (Boesch & Boesch-Achermann 2000; Inoue-Nakamura & Matsuzawa 1997; Langer 1996; Matsuzawa 2001b; P&M; Poti *et al.* 1999; Russon in 2003b; Spinozzi *et al.* 1999). Their complex structures develop on the basis of simpler ones and emerge late in immaturity (Langer 1996, 2000; Matsuzawa 2001b; P&M). Their complex foraging techniques, for example, develop piecemeal over many years with youngsters first acquiring basic elements, next assembling them into a basic strategy, then gradually elaborating it (Fox, Sitompul & van Schaik 1999; Inoue-Nakamura & Matsuzawa 1997; Russon 2002a; 2003a).

In life history perspective, developmental models are also consistent with evidence that: (1) cognitive capacity peaks in juveniles and levels off after adolescence; (2) parents contribute to acquiring advanced juvenile as well as basic infant skills; (3) rudimentary symbolic level abilities emerge post-infancy, around the move to semi-independent life; (4) most adult-level expertise is mastered by adolescence, around the move to fully independent life; and (5) post-adolescent learning seems less flexible (Boesch 1991; Boesch & Boesch-Achermann 2000; Ingmanson 1996; Inoue-Nakamura & Matsuzawa 1997; King 1994; Parker 1996; P&M). All correlate with the slower pace and disproportionately prolonged immaturity that distinguish great ape development from that of other nonhuman primates (P&M; Kelley, Chapter 15, Ross, Chapter 8, this volume). Compared with humans, great apes' cognitive development is faster in the first year of life but subsequently slower (P&M; Poti *et al.* 1999; Spinozzi *et al.* 1999), which explains why some of the distinctive abilities they share with humans develop later and persist longer.

Social–cultural influences, interwoven with individual experience, also contribute to cognitively governed achievements in great apes, as they do in humans (e.g., P&M; Tomasello 1999; T&C; van Schaik *et al.* 2003; Whiten *et al.* 1999). The distribution of "atypical" abilities and some complex skills in the wild, for instance, shows that great apes may not realize some complex achievements without appropriate socio-cultural support despite appropriate individual opportunities (van Schaik *et al.* 2003; Whiten *et al.* 1999). If their achievements are products of combining socio-cultural with individual experience during development, then enculturation should be primarily responsible. In great apes enculturation probably resembles apprenticeship (guided participation in shared activities of a routine nature; Rogoff 1992) and supports and perhaps extends their natural behavioral repertoires (Boesch & Boesch-Achermann 2000; Matsuzawa *et al.* 2001; Parker 1996; P&M; Russon 1999b, 2003b; Suddendorf & Whiten 2001). It has been assigned responsibility for achievement variability across wild, captive-reared, and human-enculturated great apes (e.g., Donald 2000; T&C).

Great apes' cultural and cognitive processes are more tightly interwoven than this scenario suggests. Cultural processes depend on what information can be shared and how, which depend on information processing capabilities, i.e., cognition. Great apes' cultural processes may be exceptionally powerful among nonhuman primates because they access high-level cognitive capabilities unique to great apes and humans (e.g., imitation, self-awareness, demonstration; Parker 1996). Conversely, great apes' cognitive achievements are probably boosted by cultural processes. Chimpanzee cultures show ratcheting, for instance, the accumulation of cultural variants over time, in the form of cumulative modifications to complex techniques (McGrew 1998; Yamakoshi & Sugiyama 1995). This probably allows learners to acquire more complex techniques than they would have constructed independently. If enculturation has a special role to play in cognitive development, it may primarily affect high levels, as it typically does in humans (P&M; Tomasello 1999). No convincing evidence exists, however, for claims that human enculturation induces higher-level cognitive structures in great apes than species-normal enculturation (Boesch & Boesch-Achermann 2000; Langer 2000; P&M; Russon 1999b; Spinozzi *et al.* 1999; Suddendorf & Whiten 2001).

Generating cognitive structures

A final issue is what mental processes generate great ape cognitive development and how, especially their distinctive cognitive structures. Great apes consistently attain the same cognitive level across cognitive domains, rudimentary symbolism, which suggests that centralized generative processes that operate across the whole cognitive system govern their cognitive development, rather than processes specific to a single cognitive domain or problem type. That the level achieved supports simple symbols suggests hierarchization as a good candidate for that centralized generative process: It is considered essential to the cognitive abilities and achievements that distinguish the great apes among nonhuman primates (e.g., simple language, abstract level problem solving, complex tool and manual foraging techniques) (Case 1985; Piaget 1954) and, among nonhuman primates, only great apes show evidence of hierarchization (e.g., Byrne 1997; Langer 1996; Matsuzawa 2001b; Russon *et al.* 1998).

Great apes' flexible range of high-level cognitive abilities could be generated by hierarchization used in conjunction with combinatorial mechanisms, in the form of hierarchical mental construction (e.g., Byrne 1997; Gibson 1990, 1993; Langer 1994, 1996). Combinatorial mechanisms are centralized generative processes that combine, decompose, and recombine multiple mental units at a time, as in combining actions or objects in sequence; they probably generate cognitive structures in all primates (Langer 2000). The pattern in which great apes acquire food processing techniques is consistent with a hierarchical mental construction model of cognitive development (e.g., Inoue-Nakamura & Matsuzawa 1997; Russon 2002a; Stokes & Byrne 2001). Infant chimpanzees acquiring stone nut-cracking skills, for instance, first learn the individual basic actions needed to crack nuts and apply single actions to single objects (only stone, only nut); next, they apply multiple actions to multiple objects (some stones, some nuts, stones and nuts) combined in sequence (some are ineffective, e.g., put a nut on a stone but hit the nut with a hand then pick up a piece of kernel from a broken shell on the ground and eat that); finally, they integrate appropriate combinations into more complex, hierarchically organized techniques showing understanding of action–object relationships (Inoue-Nakamura & Matsuzawa 1997). To date, other nonhuman primates have not shown hierarchically organized techniques (Harrison 1996). Great apes reach only rudimentary symbolic levels, however, and their achievements are rougher-grained than humans', i.e., focused primarily on general problem features and less able to incorporate fine ones (Langer 1996; P&M; Spinozzi & Langer 1999). Their low symbolic ceiling may reflect limited hierarchization relative to humans, described as shallow (Byrne 1997; Matsuzawa 2001b) or protohierarchical (Langer 2000). The rougher grain may reflect lower limits on the number of units they can combine at once.

Cognitive facilitation may take great ape cognition beyond modularity, and it may hinge on hierarchization (RWB; Case 1985; Karmiloff-Smith 1992; Thompson & Oden 2000; see Langer 1996, 2000 for alternatives).[3] This link is supported by evidence that cognitive facilitation in great apes is limited, because this is consistent with shallow hierarchization. Shallow hierarchization generates only rudimentary hierarchical cognitive structures, which remain more isolated than the higher-level cognitive structures that human hierarchization generates (Case 1985).

This sort of model, which characterizes great ape cognition in terms of central generative processes, may help explain several features that have puzzled scholars. The "atypical" abilities that emerge in great apes under highly nurturing human rearing conditions (e.g., linguistic and mathematical abilities: Gardner, Gardner & van Cantfort 1989; Tomasello 1999) may simply be customized abilities of the sort expected from generative cognitive systems that build structures to suit the specific challenges encountered during development (Boesch & Boesch-Achermann 2000; P&M; Rumbaugh 2000; Swartz, Sarauw & Evans 1999). Marked individual differences in achievements may similarly be normal features of generative cognitive systems. "Atypical" abilities may also have feral counterparts, making them less unusual than suggested. Feral communication suggestive of symbolism has been reported, for example tree drumming, leaf clipping, knuckle-knocking, demonstration teaching (Boesch 1991, 1993, 1996), symbolic eating (Schaller 1963, Russon 2002b), miming requests (Russon 2002b), and placing leaves to indicate travel direction (Savage-Rumbaugh *et al.* 1996) (see also Blake, Chapter 5, this volume), as have complex quantitative abilities such as seriation (arranging items in a graded series: Brewer & McGrew 1990) and body-part counting (using body parts to order items: Russon 2002a).

Generative models also suggest how modularity–generality may play out in great apes. In humans, module-like structures may be products of generative processes operating in the context of problem-specific constraints and innately founded structures (e.g., Elman *et al.* 1996; Greenfield 1991; Karmiloff-Smith 1992; Langer 2000; P&M). Human cognitive structures change with development: they have been characterized as relatively undifferentiated at their earliest (i.e., applicable generally, across problem types), subsequently differentiated into domain- and problem-specific structures with module-like features (applicable to specific problem types), and finally interconnected (applicable across problem types when used in combination) (e.g., Case 1996; Greenfield 1991; Langer 1996, 2000; P&M). Generalized capabilities may then exist in undifferentiated and interconnected forms, module-like structures may be developmental products, and development may affect the qualities of modularity and generality and the balance between them. Great apes' cognitive development shows similar patterns although comparatively, differentiation and interconnection proceed more slowly after the first year and are ultimately less powerful (e.g., P&M). Great ape cognition then involves both modularity and generality, and characterizations in terms of generality–modularity are likely flawed if they fail to consider developmental change or to distinguish undifferentiated from interconnected forms of generality (e.g., Mithen 1996).

DISCUSSION

Evidence consistently supports conclusions that great apes differ cognitively from other nonhuman primates. Virtually all experts agree, there is no longer any justification for reducing great ape cognition to associative processes or lumping great apes with other nonhuman primates. What sets great ape cognitive achievements apart is not specific problem-specific abilities such as tool use, imitation, or self-concept. It is rather the broader and more open repertoire of abilities, rudimentary symbolic levels achieved across domains, and limited interconnectedness among them. What underpins this suite of cognitive structures may be centralized generative processes that operate ontogenetically, limited hierarchization and perhaps facilitation being the best current candidates. This characterization is not new. Revisiting it, however, helps articulate what needs evolutionary explanation: more powerful generative processes that produce rudimentary symbolism and limited fluidity of thought.

This characterization helps explain why it has been difficult to get a handle on great ape mentality. First, if variable achievement is intrinsic to great ape cognition, then studies that have tested great apes as immatures or reared in non-stimulating environments have failed to tap their full potential. Second, achievements during the transition from sensorimotor to rudimentary symbolic cognition may be governed by either advanced sensorimotor or primitive symbolic-level structures (Case 1985). It is possible to distinguish the two behaviorally, and studies that failed to do probably underestimated subjects' level of cognitive functioning (e.g., Byrne & Russon 1998; Russon 1998). Third, if entertaining multiple representations of a situation underpins rudimentary symbolic-level cognition then great apes, like 2- to 3.5-year-old children, should be able to entertain symbolic and perceptual representations concurrently. In such children, when the two representations conflict, perceptual representations tend to override symbolic ones for control of behavior; they have been described as perception bound because they are easily swayed by perceptual cues (Case 1996; P&M). Chimpanzees have shown similar tendencies. They solved a reverse contingency task (what you pick goes to your partner) when it was presented with symbols but failed when it was presented with real candies, so they can function symbolically but not when perceptual cues are salient (Boysen et al. 1996, 1999). This suggests that their symbol use is unstable and they, like young children, may fail symbolic tasks not because they lack the capability but because perceptual cues activate this bias. Orangutans without symbol skills solved the reverse contingency task with real candies (Shumaker et al. 2001), so even great apes can sometimes privilege symbolic over perceptual solutions. These difficulties do not render it futile to study rudimentary symbolic cognition in great apes: many difficulties are assessment related and have been resolved for humans. What is needed is greater attention to the qualities of rudimentary symbolic cognition and factors that contribute to variability in its development and application.

The characteristics of great ape cognition that require evolutionary explanation are among those currently treated as evolutionary achievements of the human lineage. That these qualities appeared earlier in primate evolutionary history does not alter their significance but it does change their role, from innovations of the human lineage to foundations for its elaborations. This affects evolutionary reconstructions of human cognition that use great apes to represent the ancestral cognition from which human cognition evolved because they typically assume great apes to be incapable of symbolic cognition (see Russon, Chapter 1, this volume). We now know that this assumption is incorrect, in at least one form. Accurate models of great ape cognition are then important next steps towards better understanding of great ape and human cognitive evolution.

ACKNOWLEDGMENTS

My thanks to David Begun and Sue T. Parker for valuable comments on earlier drafts of this paper. The research on which this chapter was based was supported by funding from the Natural Sciences and Engineering Council of Canada, Glendon College, and York University.

ENDNOTES

1 Feral, here, includes wild and reintroduced individuals living free in natural habitat. I grouped reintroduced with wild great apes because both face species-typical rather than human-devised problems. Their achievements may differ in their specific nature (e.g., reintroduced orangutans often show complex tool use but wild ones rarely do) but not in cognitive complexity, which is the major concern here (Russon 1999b).

2 (1) This body of evidence is expected to be small because complex achievements should be rare relative to average ones. (2) If evidence on great ape cognition is notoriously patchy, evidence on complex cognition should be even more so. (3) On tasks tapping an actor's highest-level capabilities, high performance variability is expected (Spinozzi et al. 1999; Swartz et al. 1999). (4) Methodological confounds can cause performance variation, especially misleading cues that undermine performance and scaffolding that boosts it. The number of items that must be held concurrently in working memory to solve a task affects success for example, and how a task is presented can increase or decrease that number (Pascual-Leone & Johnson 1999). If threshold tests are used, such confounds can affect assessments of cognitive levels. (5) Few studies have verified that their tests for great apes are commensurate with human benchmarks; close scrutiny often shows they are not (e.g., P&M). (6) In children at rudimentary symbolic levels, perceptual processes readily dominate symbolic ones and unstable achievement is common. If great apes function at this level, comparable instabilities are probable (e.g., Boysen et al. 1996; Case 1985; Boysen, Mukobi & Berntson 1999).

3 Similar suggestions use terms like representational redescription (Karmiloff-Smith 1992), abstract level generalization (RWB), higher levels of abstraction (Case 1985), and analogical reasoning (Thompson & Oden 2000). Langer's (1996, 2000) alternative is that facilitation may owe to developmental synchronization, i.e., yoking developmental progress across distinct types of cognition so they develop together rather than independently; this offers the best timing pattern possible for interconnecting them.

REFERENCES

Anderson, J. R. (1996). Chimpanzees and capuchin monkeys: comparative cognition. In *Reaching into Thought: The Minds of the Great Apes*, ed. A. E. Russon, K. A. Bard & S. T. Parker, pp. 23–56. Cambridge, UK: Cambridge University Press.

Balda, R. P., Pepperberg, I. M. & Kamil, A. C. (1998). Preface. In *Animal Cognition in Nature: The Convergence of Psychology and Biology in Laboratory and Field*, ed. R. Balda, I. Pepperberg & A. Kamil, pp. vii–ix. New York: Academic Press.

Beran, M. J. (2001). Summation and numerousness judgments of sequentially presented sets of items by chimpanzees (*Pan troglodytes*). *Journal of Comparative Psychology*, 115(2), 181–91.

Bering, J. M., Bjorkland, D. F. & Ragan, P. (2000). Deferred imitation of object-related actions in human-reared juvenile chimpanzees and orangutans. *Developmental Psychobiology*, 36, 218–32.

Bermejo, M. & Illera, G. (1999). Tool-set for termite-fishing and honey extraction by wild chimpanzees in the Lossi Forest, Congo. *Primates*, 40(4), 619–27.

Biro, D. & Matsuzawa, T. (1999). Numerical ordering in a chimpanzee (*Pan troglodytes*): planning, executing, and monitoring. *Journal of Comparative Psychology*, 113(2), 178–85.

Bjorkland, D. F., Yunger, J. L., Bering, J. M. & Ragan, P. (2002). The generalization of deferred imitation in enculturated chimpanzees (*Pan troglodytes*). *Animal Cognition*, 5, 49–58.

Boesch, C. (1991). Symbolic communication in wild chimpanzees. *Human Evolution*, 6(1), 81–90.

(1993). Aspects of transmission of tool-use in wild chimpanzees. In *Tools, Language, and Cognition in Human Evolution*, ed. K. R. Gibson & T. Ingold, pp. 171–83. Cambridge, UK: Cambridge University Press.

(1996). Three approaches for understanding chimpanzee culture. In *Reaching into Thought: The Minds of the Great Apes*, ed. A. E. Russon, K. A. Bard, & S. T. Parker, pp. 404–29. Cambridge, UK: Cambridge University Press.

Boesch, C. & Boesch-Achermann, H. (2000). *The Chimpanzees of the Taï Forest: Behavioural Ecology and Evolution*. Oxford: Oxford University Press.

Boysen, S. T. (1998). Attribution processes in chimpanzees: Heresy, hearsay, or heuristic? Paper presented at the 17th Congress of the International Primatological Society, Antananarivo, Madagascar.

Boysen, S. T. & Berntson, G. G. (1995). Responses to quantity: Perceptual versus cognitive mechanisms in chimpanzees (*Pan troglodytes*). *Journal of Experimental Psychology, Animal Behavior Processes*, A21 (1), 82–6.

Boysen, S. T. & Hallberg, K. I. (2000). Primate numerical competence: contributions toward understanding nonhuman cognition. *Cognitive Science*, 24(3), 423–43.

Boysen, S. T., Berntson, G. G., Hannan, M. B. & Cacioppo, J. T. (1996). Quantity-based inference and symbolic representations in chimpanzees (*Pan troglodytes*). *Journal of Experimental Psychology: Animal Behavior Processes*, 22, 76–86.

Boysen, S. T., Berntson, G. G. & Mukobi, K. L. (2001). Size matters: impact of size and quantity on array choice by chimpanzees (*Pan troglodytes*). *Journal of Comparative Psychology*, 115, 106–10.

Boysen, S. T., Mukobi, K. L. & Berntson, G. G. (1999). Overcoming response bias using symbol representation of number by chimpanzees (*Pan troglodytes*). *Animal Learning and Behavior*, 27, 229–35.

Brewer, S. & McGrew, W. C. (1990). Chimpanzee use of a tool-set to get honey. *Folia Primatologica*, 54, 100–4.

Byrne, R. W. (1995). *The Thinking Ape*. Oxford: Oxford University Press.

(1997). The technical intelligence hypothesis: an alternative evolutionary stimulus to intelligence? In *Machiavellian Intelligence II: Extensions and Evaluations*, ed. R. W. Byrne & A. Whiten, pp. 289–311. Cambridge, UK: Cambridge University Press.

Byrne, R. W. & Byrne, J. M. E. (1991). Hand preferences in the skilled gathering tasks of mountain gorillas (*Gorilla gorilla berengei*). *Cortex*, 27, 521–46.

(1993). Complex leaf-gathering skills of mountain gorillas (*Gorilla g. beringei*): variability and standardization. *American Journal of Primatology*, 31, 241–61.

Byrne, R. W. & Russon, A. E. (1998). Learning by imitation: a hierarchical approach. *Behavioural and Brain Sciences*, 21, 667–721.

Byrne, R. W., Corp, N. & Byrne, J. M. E. (2001). Estimating the complexity of animal behaviour: how mountain gorillas eat thistles. *Behaviour*, 138, 525–57.

(2000). Estimating and operating on discrete quantities in orangutans (*Pongo pygmaeus*). *Journal of Comparative Psychology*, 114, 136–47.

(2001a). Object permanence in orangutans (*Pongo pygmaeus*), chimpanzees (*Pan troglodytes*), and children (*Homo sapiens*). *Journal of Comparative Psychology*, 115(2), 159–71.

(2001b). Body imitation in an enculturated orangutan (*Pongo pygmaeus*). *Cybernetics and Systems*, 32, 97–119.

Call, J. & Carpenter, M. (2001). Do apes and children know what they have seen? *Animal Cognition*, 4, 207–20.

Call, J. & Rochat, P. (1996). Liquid conservation in orangutans (*Pongo pygmaeus*) and humans (*Homo sapiens*): individual differences and perceptual strategies. *Journal of Comparative Psychology*, 110(3), 219–32.

Call, J. & Tomasello, M. (1998). Distinguishing intentional from accidental actions in orangutans (*Pongo pygmaeus*), chimpanzees (*Pan troglodytes*), and human children (*Homo sapiens*). *Journal of Comparative Psychology*, 112(2), 192–206.

Call, J., Agnetta, B. & Tomasello, M. (2000). Cues that chimpanzees do and do not use to find hidden objects. *Animal Cognition*, 3, 23–4.

Call, J., Hare, B. A. & Tomasello, M. (1998). Chimpanzee gaze following in an object-choice task. *Animal Cognition*, 1, 89–99.

Carruthers, P. & Smith, P. K. (ed.) (1996). *Theories of Theories of Mind*. Cambridge, UK: Cambridge University Press.

Case, R. (1985). *Intellectual Development from Birth to Adulthood*. Orlando, FL: Academic Press.

(1992). *The Mind's Staircase: Exploring the Conceptual Underpinnings of Children's Thought and Knowledge*. Hillsdale, NJ: Lawrence Erlbaum Associates.

(1996). Introduction: reconceptualizing the nature of children's conceptual structures and their development in middle childhood. In *The Role of Central Conceptual Structures in the Development of Children's Thought* (Series No. 246, Vol. 61), ed. R. Case & Y. Okamoto, pp. 1–26. Chicago, IL: University of Chicago Press.

Corp, N. & Byrne, R. W. (2002). The ontogeny of manual skill in wild chimpanzees: evidence from feeding on the fruit of *Saba florida*. *Behaviour*, 139(1), 137–68.

Custance D. M., Whiten, A., Sambrook, T. & Galdikas, B. (2001). Testing for social learning in the "artificial fruit" processing of wildborn orangutans (*Pongo pygmaeus*), Tanjung Puting, Indonesia. *Animal Cognition*, 4, 305–13.

Deacon, T. W. (1997). *The Symbolic Species: The Coevolution of Language and the Brain*. New York: W. W. Norton.

de Blois, S. T., Novak, M. A. & Bond, M. (1998). Object permanence in orangutans (*Pongo pygmaeus*) and squirrel monkeys (*Saimiri sciureus*). *Journal of Comparative Psychology*, **112**(2), 137–53.

de Waal, F. B. M. & Aureli, F. (1996). Consolation, reconciliation, and a possible cognitive difference between macaques and chimpanzees. In *Reaching into Thought: The Minds of the Great Apes*, ed. A. E. Russon, K. A. Bard & S. T. Parker, pp. 80–110. Cambridge, UK: Cambridge University Press.

Donald, M. (1991). *Origins of the Modern Mind: Three Stages in the Evolution of Cognition and Culture*. Cambridge, MA: Harvard University Press.

(2000). The central role of culture in cognitive evolution: a reflection on the myth of the "isolated mind." In *Culture, Thought, and Development*, ed. L. P. Nucci, G. B. Saxe & E. Turiel, pp. 19–38. Mahwah, NJ: Lawrence Erlbaum Associates.

Elman, J., Bates, E., Johnson, M., Karmiloff-Smith, A., Parisi, D. & Plunkett, K. (1996). *Rethinking Innateness*. Cambridge, MA: MIT Press.

Fox, E. A., Sitompul, A. F. & van Schaik, C. P. (1999). Intelligent tool use in wild Sumatran orangutans. In *The Mentalities of Gorillas and Orangutans: Comparative Perspectives*, ed. S. T. Parker, R. W. Mitchell & H. L. Miles, pp. 99–116. Cambridge, UK: Cambridge University Press.

Gardner, H. (1983). *Frames of Mind*. New York: Basic.

Gardner, R. A., Gardner B. T. & van Cantfort, T. E. (ed.) (1989). *Teaching Sign Language to Chimpanzees*. Albany, NY: SUNY Press.

Gibson, K. R. (1990). New perspectives on instincts and intelligence: brain size and the emergence of hierarchical mental construction skills. In *"Language" and Intelligence in Monkeys and Apes: Comparative Developmental Perspectives*, ed. S. T. Parker & K. R. Gibson, pp. 97–128. New York: Cambridge University Press.

(1993). Beyond neoteny and recapitulation: new approaches to the evolution of cognitive development. In *Tools, Language and Cognition in Human Evolution*, ed. K. R. Gibson & T. Ingold, pp. 273–8. Cambridge, UK: Cambridge University Press.

Goodall, J. (1986). *The Chimpanzees of Gombe: Patterns of Behavior*. Cambridge, MA: The Belknap Press of Harvard University Press.

Greenfield, P. M. (1991). Language, tools and brain: the ontogeny and phylogeny of hierarchically organized sequential behavior. *Behavioral and Brain Sciences*, **14**, 531–95.

Hare, B., Call, J., Agnetta, B. & Tomasello, M. (2000). Chimpanzees know what conspecifics do and do not see. *Animal Behaviour*, **59**, 771–85.

Hare, B., Call, J. & Tomasello, M. (2001). Do chimpanzees know what conspecifics know? *Animal Behaviour*, **61**, 139–51.

Harrison, K. E. (1996). Skills used in food processing by vervet monkeys, *Cercopithecus aethiops*. Unpublished doctoral dissertation, University of St. Andrews, St. Andrews, Scotland.

Herrnstein, R. J. (1990). Levels of stimulus control: a functional approach. *Cognition*, **37**(1–2), 133–66.

Hirata, S. & Matsuzawa, T. (2000). Naive chimpanzees' (*Pan troglodytes*) observation of experienced conspecifics in a tool task. *Journal of Comparative Psychology*, **114**(3), 291–6.

Hirata, S. & Matsuzawa, T. (2001). Tactics to obtain a hidden food item in chimpanzee pairs (*Pan troglodytes*). *Animal Cognition*, **4**, 285–95.

Hostetter, A. B., Cantera, M. & Hopkins, W. D. (2001). Differential use of vocal and gestural communication by chimpanzees (*Pan troglodytes*) in response to the attentional state of a human (*Homo sapiens*). *Journal of Comparative Psychology*, **115**(4), 337–43.

Ingmanson, E. (1996). Tool-using behavior in wild *Pan paniscus*: social and ecological considerations. In *Reaching into Thought: The Minds of the Great Apes*, ed. A. E. Russon, K. A. Bard & S. T. Parker, pp. 190–210. Cambridge, UK: Cambridge University Press.

Inoue-Nakamura, N. & Matsuzawa, T. (1997). Development of stone tool use by wild chimpanzees (*Pan troglodytes*). *Journal of Comparative Psychology*, **111**, 159–73

Itakura, S. & Tanaka, M. (1998). Use of experimenter-given cues during object-choice tasks by chimpanzees (*Pan troglodytes*), an orangutan (*Pongo pygmaeus*), and human infants (*Homo sapiens*). *Journal of Comparative Psychology*, **112**, 119–26.

Jensvold, M. L. A. & Gardner, R. A. (2000). Interactive use of sign language by cross-fostered chimpanzees (*Pan troglodytes*). *Journal of Comparative Psychology*, **114**(4), 335–46.

Johnson-Pynn, J. & Fragaszy, D. M. (2001). Do apes and monkeys rely upon conceptual reversibility? A review of studies using seriated nesting cups in children and nonhuman primates. *Animal Cognition*, **4**, 315–24.

Johnson-Pynn, J., Fragaszy, D. M., Hirsch, E. M., Brakke, K. E. & Greenfield, P. M. (1999). Strategies used to combine seriated cups by chimpanzees (*Pan troglodytes*), bonobos (*Pan paniscus*), and capuchins (*Cebus apella*). *Journal of Comparative Psychology*, 113(2), 137–48.

Karmiloff-Smith, A. (1992). *Beyond Modularity: A Developmental Perspective on Cognitive Science*. Cambridge, MA: MIT Press.

Kawai, N. & Matsuzawa, T. (2000). Numerical memory span in a chimpanzee. *Nature*, 403, 39–40.

King, B. J. (1994). *The Information Continuum*. Santa Fe, NM: School of American Research.

Kuhlmeier, V. A. & Boysen, S. T. (2001). The effect of response contingencies on scale model task performance by chimpanzees (*Pan troglodytes*). *Journal of Comparative Psychology*, 115(3), 300–6.

Kuhlmeier, V. A., Boysen, S. T. & Mukobi, K. L. (1999). Scale-model comprehension by chimpanzees (*Pan troglodytes*). *Journal of Comparative Psychology*, 113(4), 396–402.

Langer, J. (1994). From acting to understanding: the comparative development of meaning. In *The Nature and Ontogenesis of Meaning*, ed. W. F. Overton & D. Palermo, pp. 191–214. Hillsdale, NJ: Erlbaum.

(1996). Heterochrony and the evolution of primate cognitive development. In *Reaching into Thought: The Minds of the Great Apes*, ed. A. E. Russon, K. A. Bard & S. T. Parker, pp. 257–77. Cambridge, UK: Cambridge University Press.

(1998). Phylogenetic and ontogenetic origins of cognition: Classification. In *Piaget, Evolution, and Development*, ed. J. Langer & M. Killen, pp. 33–54. Mahwah, NJ: Lawrence Erlbaum Associates.

(2000). The descent of cognitive development (with peer commentaries). *Developmental Science*, 3(4), 361–88.

Leavens, D. & Hopkins, W. (1998). Intentional communication by chimpanzees: a cross-sectional study of the use of referential gestures. *Developmental Psychology*, 34(5), 813–22.

Leavens, D. A. & Hopkins, W. D. (1999). The whole-hand point: the structure and function of pointing from a comparative perspective. *Journal of Comparative Psychology*, 113(4), 417–25.

Leslie, A. M. (1987). Pretense and representation in infancy: the origins of "theory of mind." *Psychological Review*, 94, 412–26.

(1994). Pretending and believing: issues in the theory of ToMM. *Cognition*, 50, 211–38.

Maple, T. L. (1980). *Orang-utan Behavior*. New York: Van Nostrand Reinhold.

Matsuzawa, T. (1991). Nesting cups and metatools in chimpanzees. *Behavioral and Brain Sciences*, 14(4), 570–1.

(1996). Chimpanzee intelligence in nature and in captivity: isomorphism in symbol use and tool use. In *Great Ape Societies*, ed. W. C. McGrew, L. F. Marchant & T. Nishida, pp. 196–209. Cambridge, UK: Cambridge University Press.

(ed.) (2001a). *Primate Origins of Human Cognition and Behavior*. Tokyo: Springer-Verlag.

(2001b). Primate foundations of human intelligence: a view of tool use in nonhuman primates and fossil hominids. In *Primate Origins of Human Cognition and Behavior*, ed. T. Matsuzawa, pp. 3–25. Tokyo: Springer-Verlag.

Matsuzawa, T. & Yamakoshi, G. (1996). Comparison of chimpanzee material culture between Bossou and Nimba, West Africa. In *Reaching into Thought: The Minds of the Great Apes*, ed. A. E. Russon, K. A. Bard & S. T. Parker, pp. 211–32. Cambridge, UK: Cambridge University Press.

Matsuzawa, T., Biro, D., Humle, T., Inoue-Nakamura, N., Tonooka, R. & Yamakoshi, G. (2001). Emergence of culture in wild chimpanzees: education by master-apprenticeship. In *Primate Origins of Human Cognition and Behavior*, ed. T. Matsuzawa, pp. 557–74. Tokyo: Springer-Verlag.

McGrew, W. C. (1998). Culture in nonhuman primates? *Annual Review of Anthropology*, 27, 301–28.

Menzel, C. R. (1999). Unprompted recall and reporting of hidden objects by a chimpanzee (*Pan troglodytes*) after extended delays. *Journal of Comparative Psychology*, 113(4), 426–34.

Menzel, C. R., Savage-Rumbaugh, E. S. & Menzel, E. W. Jr. (2002). Bonobo (*Pan paniscus*) spatial memory and communication in a 20-hectare forest. *International Journal of Primatology*, 23(3), 601–19.

Miles, H. L., Mitchell, R. W. & Harper, S. (1996). Simon says: the development of imitation in an enculturated orangutan. In *Reaching into Thought: The Minds of the Great Apes*, ed. A. E. Russon, S. T. Parker & K. A. Bard, pp. 278–99. Cambridge, UK: Cambridge University Press.

Mitani, J. C. & Watts, D. P. (2001). Why do chimpanzees hunt and share meat? *Animal Behaviour*, 61, 1–10.

Mitani, J. C. Watts, D. P. & Muller, M. N. (2002). Recent developments in the study of wild chimpanzee behavior. *Evolutionary Anthropology*, **11**, 9–25.

Mitchell, R. W. (ed.) (2002). *Pretending and Imagination in Animals and Children*. Cambridge, UK: Cambridge University Press.

Mithen, S. (1996). *The Prehistory of the Mind: The Cognitive Origins of Art and Science*. London: Thames and Hudson.

Myowa-Yamakoshi, M. & Matsuzawa, T. (1999). Factors influencing imitation of manipulatory actions in chimpanzees (*Pan troglodytes*). *Journal of Comparative Psychology*, **113**(2), 128–36.

(2000). Imitation of intentional manipulatory actions by chimpanzees (*Pan troglodytes*). *Journal of Comparative Psychology*, **114**(4), 381–91.

Noe, R., de Waal, F. B. M. & van Hooff, J. A. R. A. M. (1980). Types of dominance in a chimpanzee colony. *Folia Primatologica*, **34**, 90–110.

Oden, D. L., Thompson, R. K. R. & Premack, D. (1990). Infant chimpanzees (*Pan troglodytes*) spontaneously perceive both concrete and abstract same/different relations. *Child Development*, **61**, 621–31.

Parker, S. T. (1996). Apprenticeship in tool-mediated extractive foraging: imitation, teaching and self-awareness in great apes. In *Reaching into Thought: The Minds of the Great Apes*, ed. A. E. Russon, K. A. Bard & S. T. Parker, pp. 348–70. Cambridge, UK: Cambridge University Press.

(1998). Review of *Primate Cognition*. *Quarterly Review of Biology*, **73**, 540–1.

Parker, S. T. & McKinney, M. (1999). *Origins of Intelligence: The Evolution of Cognitive Development in Monkeys, Apes, and Humans*. Baltimore, MD: The Johns Hopkins Press.

Parker, S. T., Mitchell, R. W. & Miles, H. L. (ed.) (1999). *The Mentalities of Gorillas and Orangutans: Comparative Perspectives*. Cambridge, UK: Cambridge University Press.

Pascual-Leone, J. (1987). Organismic processes for neo-Piagetian theories: a dialectical causal account of cognitive development. *International Journal of Psychology*, **22**, 531–70.

Pascual-Leone, J. & Johnson, J. (1999). A dialectical constructivist view of representation: role of mental attention, executives, and symbols. In *Development of Mental Representation: Theories and Applications*, ed. I. E. Sigel, pp. 169–200. Mahwah, NJ: Erlbaum.

Patterson, F. G. & Linden, H. (1981). *The Education of Koko*. New York: Holt, Rinehart and Winston.

Peirce, C. S. (1932/1960). The icon, index, and symbol. In *Collected Papers of Charles Sanders Peirce*, Vol. II, ed. C. Hartshorne & P. Weiss, pp. 156–73. Cambridge, MA: Harvard University Press.

Peignot, R. & Anderson, J. R. (1999). Use of experimenter-given manual and facial cues by gorillas (*Gorilla gorilla*) in an object-choice task. *Journal of Comparative Psychology*, **113**(3), 253–60.

Perner, J. (1991). *Understanding the Representational Mind*. Cambridge, MA: Bradford/MIT.

Piaget, J. (1952). *The Origins of Intelligence in Children*. New York: International Universities Press.

(1954). *The Construction of Reality in the Child*. New York: Basic Books.

Poti', P., Langer, J., Savage-Rumbaugh, S. & Brakke, K. E. (1999). Spontaneous logico-mathematical constructions by chimpanzees (*Pan troglodytes, P. paniscus*). *Animal Cognition*, **2**, 147–56.

Povinelli, D. J. (2000). *Folk Physics for Apes*. New York: Oxford University Press.

Premack, D. (1983). The codes of man and beast. *Behavioral and Brain Sciences*, **6**, 125–37.

(1988). "Does the chimpanzee have a theory of mind?" revisited. In *Machiavellian Intelligence: Social Expertise and the Evolution of Intellect in Monkeys, Apes and Humans*, ed. R. W. Byrne & A. Whiten, pp. 160–79. Oxford: Clarendon Press.

Roberts, W. A. & Mazmanian, D. S. (1988). Concept learning at different levels of abstraction by pigeons, monkeys, and people. *Animal Behavior Processes*, **14**, 247–60.

Rogoff, B. (1992). *Apprenticeship in Thinking*. New York: Oxford University Press.

Rumbaugh, D. M. (1970). Learning skills in anthropoids. In *Primate Behavior: Developments in Field and Laboratory*, ed. L. A. Rosenblum, pp. 1–70. New York: Academic Press.

(2000). Review of *The Mentalities of Gorillas and Orangutans*. *International Journal of Primatology*, **21**(4), 769–71.

Rumbaugh, D. M. & Pate, J. L. (1984). Primates' learning by levels. In *Behavioral Evolution and Integrative Levels*, ed. G. Greenberg & E. Tobach, pp. 221–40. Hillsdale, NJ: Erlbaum.

Russon, A. E. (1996). Imitation in everyday use: matching and rehearsal in the spontaneous imitation of rehabilitant orangutans (*Pongo pygmaeus*). In *Reaching* into Thought: The Minds of the Great Apes, ed. A. E. Russon, K. A. Bard & S. T. Parker, pp. 152–76. Cambridge, UK: Cambridge University Press.

(1998). The nature and evolution of orangutan intelligence. Primates, 39, 485–503.

(1999a). Orangutans' imitation of tool use: a cognitive interpretation. In The Mentalities of Gorillas and Orangutans: Comparative Perspectives, ed. S. T. Parker, R. W. Mitchell & H. L. Miles, pp. 117–46. Cambridge, UK: Cambridge University Press.

(1999b). Naturalistic approaches to orangutan intelligence and the question of enculturation. International Journal of Comparative Psychology, 12(4), 181–202.

(2002a). Return of the native: cognition and site specific expertise in orangutan rehabilitation. International Journal of Primatology, 23(3), 461–78.

(2002b). Pretending in free-ranging rehabilitant orangutans. In Pretending and Imagination in Animals and Children, ed. R. W. Mitchell, pp. 229–40. Cambridge, UK: Cambridge University Press.

(2003a). Comparative developmental perspectives on culture: the great apes. In Between Biology and Culture: Perspectives on Ontogenetic Development, ed. H. Keller, Y. Poortinga & A. Schoelmerich, pp. 30–56. Cambridge, UK: Cambridge University Press.

(2003b). Developmental perspectives on great ape traditions. In Towards a Biology of Traditions: Models and Evidence, ed. D. Fragaszy & S. Perry, pp. 187–212. Cambridge, UK: Cambridge University Press.

Russon, A. E., Bard, K. A. & Parker, S. T. (ed.) (1996). Reaching into Thought: The Minds of the Great Apes. Cambridge, UK: Cambridge University Press.

Russon, A.E., Mitchell, R.W., Lefebvre, L. & Abravanel, E. (1998). The comparative evolution of imitation. In Piaget, Evolution, and Development, ed. J. Langer & M. Killen, pp. 103–43. Hillsdale, NJ: Lawrence Erlbaum Associates.

Savage-Rumbaugh, E. S., Williams, S. L., Furuichi, T. & Kano, T. (1996). Language perceived: paniscus branches out. In Great Ape Societies, ed. W. C. McGrew, L. F. Marchant & T. Nishida, pp. 173–84. Cambridge, UK: Cambridge University Press.

Savage-Rumbaugh, E. S., Rumbaugh, D. M., Smith, S. T. & Lawson, J. (1980). Reference: the linguistic essential. Science, 210, 922–5.

Schaller, G. B. (1963). The Mountain Gorilla: Ecology and Behavior. Chicago, IL: University of Chicago Press.

Shumaker, R. W., Palkovich, A. M., Beck, B. B., Guagnano, G. A. & Morowitz, H. (2001). Spontaneous use of magnitude discrimination and ordination by the orangutan (Pongo pygmaeus). Journal of Comparative Psychology, 115(4), 385–91.

Sousa, C. & Matsuzawa, T. (2001). The use of tokens as rewards and tools by chimpanzees (Pan troglodytes). Animal Cognition, 4, 213–21.

Spearman, C. (1927). The Abilities of Man: Their Nature and Measurement. New York: Macmillan.

Spinozzi, G. & Langer, J. (1999). Spontaneous classification in action by a human-enculturated and language-reared bonobo (Pan paniscus) and common chimpanzees (Pan troglodytes). Journal of Comparative Psychology, 113(3), 286–96.

Spinozzi, G., Natale, F., Langer, J. & Brakke, K. E. (1999). Spontaneous class grouping behavior by bonobos (Pan paniscus) and common chimpanzees (P. troglodytes). Animal Cognition, 2, 157–70.

Stoinski, T. S., Wrate, J. L., Ure, N. & Whiten, A. (2001). Imitative learning by captive western lowland gorillas (Gorilla gorilla gorilla) in a simulated food processing task. Journal of Comparative Psychology, 115(3), 272–81.

Stokes, E. J. (1999). Feeding Skills and the Effect of Injury on Wild Chimpanzees. Unpublished PhD dissertation, University of St. Andrews, St. Andrews, Scotland.

Stokes, E. J. & Byrne, R. W. (2001). Cognitive abilities for behavioural flexibility in wild chimpanzees (Pan troglodytes): the effect of snare injury on complex manual food processing. Animal Cognition, 4, 11–28.

Suddendorf, T. & Whiten, A. (2001). Mental evolution and development: evidence for secondary representation in children, great apes, and other animals. Psychological Bulletin, 127(5), 629–50.

Sugiyama, Y. (1997). Social tradition and the use of tool-composites by wild chimpanzees. Evolutionary Anthropology, 6, 23–27.

Swartz, K. B., Sarauw, D. & Evans, S. (1999). Comparative aspects of mirror self-recognition in great apes. In The Mentalities of Gorillas and Orangutans: Comparative Perspectives, ed. S. T. Parker, R. W. Mitchell & H. L. Miles, pp. 283–294. Cambridge, UK: Cambridge University Press.

Tanaka, M. (1995). Object sorting in chimpanzees (Pan troglodytes): classification based on physical identity, complementarity, and familiarity. Journal of Comparative Psychology, 109, 151–61.

(2001). Discrimination and categorization of photographs of natural objects by chimpanzees (*Pan troglodytes*). *Animal Cognition*, **4**, 201–11.

Thompson, R. K. R. & Oden, D. L. (2000). Categorical perception and conceptual judgments by nonhuman primates: the paleological monkey and the analogical ape. *Cognitive Science*, **24**(3), 363–96.

Thompson, R. K. R., Oden, D. L. & Boysen, S. T. (1997). Language-naive chimpanzees (*Pan troglodytes*) judge relations between relations in a conceptual match-to-sample task. *Journal of Experimental Psychology: Animal Behavior Processes*, **23**, 31–43.

Tomasello, M. (1999). *The Cultural Origins of Human Cognition*. Cambridge, MA: Harvard University Press.

Tomasello, M. & Call, J. (1997). *Primate Cognition*. New York: Oxford University Press.

Tomasello, M., Call, J. & Hare, B. (1998). Five primate species follow the visual gaze of conspecifics. *Animal Behaviour*, **55**, 1063–9.

Tomasello, M., Hare, B. & Agnetta, B. (1999). Chimpanzees, *Pan troglodytes*, follow gaze geometrically. *Animal Behaviour*, **58**, 769–77.

Tomasello, M., Hare, B. & Fogleman, T. (2001). The ontogeny of gaze following in chimpanzees, *Pan troglodytes*, and rhesus macaques, *Macaca mulatta*. *Animal Behaviour*, **61**, 335–43.

Tomonaga, M. & Matsuzawa, T. (2000). Sequential responding to Arabic numerals with wild cards by the chimpanzee (*Pan troglodytes*). *Animal Cognition*, **3**, 1–11.

Tooby, J. & Cosmides, L. (1992). The psychological foundations of culture. In *The Adapted Mind*, ed. J. H. Barkow, L. Cosmides & J. Tooby, pp. 19–136. New York: Oxford University Press.

van Schaik, C. P., Ancrenaz, M., Borgen, G., Galdikas, B., Knott, C. D., Singleton, I., Suzuki, A., Utami, S. & Merrill, M. (2003). Orangutan cultures and the evolution of material culture. *Science*, **299**, 102–5.

van Schaik, C. P., Deaner, R. O. & Merrill, M. Y. (1999). The conditions for tool use in primates: implications for the evolution of material culture. *Journal of Human Evolution*, **36**, 719–41.

Visalberghi, E. & Limongelli, L. (1996). Acting and understanding: tool use revisited through the minds of capuchin monkeys. In *Reaching into Thought: The Minds of the Great Apes*, ed. A. E. Russon, K. A. Bard & S. T. Parker, pp. 57–79. Cambridge, UK: Cambridge University Press.

Whiten, A. (1996). Imitation, pretence, and mindreading: secondary representation in comparative primatology and developmental psychology? In *Reaching into Thought: The Minds of the Great Apes*, ed. A. E. Russon, K. A. Bard & S. T. Parker, pp. 300–24. Cambridge, UK: Cambridge University Press.

(1998a). Evolutionary and developmental origins of the mindreading system. In *Piaget, Evolution, and Development*, ed. J. Langer & M. Killen, pp. 73–99. Hillsdale, NJ: Lawrence Erlbaum Associates.

(1998b). Imitation of the sequential structure of actions by chimpanzees (*Pan troglodytes*). *Journal of Comparative Psychology*, **112**, 270–81.

(2000). Chimpanzee cognition and the question of mental re-representation. In *Metarepresentation: A Multidisciplinary Perspective*, Vancouver Studies in Cognitive Science, ed. D. Sperber, pp. 139–67. Oxford: Oxford University Press.

Whiten, A. & Byrne, R. W. (1991). The emergence of metarepresentation in human ontogeny and primate phylogeny. In *Natural Theories of Mind*, ed. A. Whiten, pp. 267–82. Oxford: Basil Blackwell.

Whiten, A., Goodall, J., McGrew, W. C., Nishida, T., Reynolds, V., Sugiyama, Y., Tutin, C. E. G., Wrangham, R. W. & Boesch, C. (1999). Culture in chimpanzees. *Nature*, **399**, 682–5.

Yamakoshi, G. & Sugiyama, Y. (1995). Pestle-pounding behaviour of wild chimpanzees at Bossou, Guinea: a newly observed tool-using behaviour. *Primates*, **36**(4), 489–500.

Part II
Modern great ape adaptation

ANNE E. RUSSON

Psychology Department, Glendon College of York University, Toronto

INTRODUCTION

This part explores shared adaptations and challenges acting upon living great apes in the wild that may be linked to their capacities and needs for high-level cognition. Its well-known premise is that their modern adaptations and pressures are valuable proxies for those of their common ancestor.

Efforts to assess the cognitive potential of great ape brains have turned up few distinctive features, most predictable from their large body sizes. Assessment remains hampered, however, by very small sample sizes, measurement problems, and extensive individual variability. Cognitive measures typically represent "encephalization," for instance, in the sense of relative brain size after body size effects have been removed (e.g., EQ (encephalization quotient), neocortical index), and these are problematic as proxies for cognitive potential. These measures also show no greater encephalization in great apes than other anthropoids, which is hard to reconcile with their distinctive cognitive capacities. Features potentially more germane to cognitive capacity have been suggested, such as exceptionally large absolute size, reorganization of information processing functions, or evolution of specific structures, but have received less attention.

Large brains are linked with slow life histories – specifically, in primates, with slow maturation concentrated in slow juvenile growth. This points to brain development as a pivotal factor, although how remains unclear. Hypotheses include energetically trading off body growth to support the brain, keeping energy needs low to improve chances of surviving to maturity, and extending time for learning foraging skills. All implicate the brain's high and inflexible energy demands, so pressures affecting juveniles' energy intake, such as food accessibility, competition, or nutritional support, may be involved. Great apes have larger brains and slower juvenile growth than other primates but both correlate with their larger body sizes, so the questions are whether anything differs about great ape life histories that either generates or affords their larger brains, and how their ecological and social challenges play into their brain–life history equation.

Studies of ecological challenges to great ape cognition have focused on two themes, diet and arboreality. Great apes' diets have shown the clearest links with enhanced cognition. They are characteristically fruit dominated, but it is equally if not more likely that cognitive challenges concern supplementary non-fruit foods richer in proteins and fats and/or fallback foods using during seasonal fruit scarcities. Spatial distribution, anti-predator defenses, and seasonal variation in food availability are all plausible cognitive challenges and all merit further study. Arboreal travel has also been suggested to pose distinctive cognitive challenges because great apes' exceptionally heavy bodies render problems associated with canopy compliance and discontinuity especially severe. So far, little evidence is available to determine if arboreal travel exacts exceptional cognitive capabilities across the great apes.

The search for social challenges to great ape cognition has been particularly vexing. Standard group-size measures of social complexity suggest great apes face no greater social challenges than other anthropoids, and social life differs dramatically within the great ape clade. Both patterns are hard to reconcile with the enhanced cognitive potential that all great apes share, and that they use to solve social problems that other anthropoids solve more simply. Efforts to identify other social problems that are especially cognitively challenging for all great apes have to date turned up little.

Our contributors offer insights in several directions. MacLeod, Chapter 7, reviews great ape neuroanatomy relative to cognitive potential. Currently, few distinguishing features have been identified. Large absolute brain size may be the most important among them because it predicts most of the others – although it is largely predicted by large body size. Important here are discussions of the role of subcortical structures in cognition, the possibility of a distinctive "ape" brain design, and the effects of brain size increase on structural components, interconnectedness, and localization of functions. An "ape" brain design combined with larger size might explain great apes' enhanced cognition, although clearly, conclusions await more substantial samples.

Ross, Chapter 8, explores life history–brain size links in anthropoid primates and potential environmental contributors. Analyses suggest the slow juvenile growth–large brain link concerns growth constraints, i.e., slowing body growth to support enlarged brains. None of Ross' indices of environmental complexity (diet – percentage folivory/frugivory, sociality – group size, habitat – forest/open) links with brain size once confounds are controlled. For great apes *per se*, analytical possibilities are limited because so few species survive. No special links between their brain size and life history are evident, although their juvenile growth and maturational rates are usually but not always slower than expected for anthropoids of their body size. The emerging picture is that life history correlates of large brain size likely represent lifting constraints against evolving large brains. One frustration is that analyses of environmental correlates are limited to measures currently available across a wide range of primates, and these may not be sophisticated enough to detect whether environmental complexity has affected great ape brains. Percentage frugivory, group size, and forest do not capture the complexities of great apes' diets, social life, or habitat.

Concerning ecological challenges, Yamakoshi, Chapter 9, compares foraging techniques and their ecological correlates across anthropoid primates as a basis for honing in on the unique features of great ape tool use, its ecology, and its cognitive demands. Using insertion tools to obtain social insects hidden in substrate-like nests emerges as unique to great apes and cognitively more complex than other anthropoid foraging techniques. This shows that there exist basic foraging techniques unique to the great apes that require their distinctively enhanced cognition. What remains to be explained is the additional cognitive complexity shown in their most advanced techniques, which coordinate multiple tools, manipulations, and operations.

Hunt, Chapter 10, compares locomotor and postural modes in great apes, lesser apes, and baboons to assess whether arboreal travel poses exceptional cognitive challenges to all great apes. Modes singled out as cognitively challenging are those involving flexible, figure-it-out-as-you-go arboreal positioning. His analysis indicates that arboreality does not impose special cognitive demands on all great apes: cognitively challenging arboreal modes are not prominent in African great apes even if they are in orangutans, but they also occur in some hylobatids, which do not show correspondingly enhanced cognition. There is little to support the hypothesis that arboreal positioning poses distinctive cognitive challenges to the great ape clade.

Van Schaik, Preuschoft, and Watts, Chapter 11, consider common features of great ape social life that could demand distinctively complex cognition, based on how their large body size and prolonged life histories affect their fission–fusion tendencies. Effects include greater individual independence from the group, notably in foraging, greater reliance on non-kin allies, greater conspecific competition, greater subordinate leverage, less rigid dominance structures, and greater social tolerance. The more flexible, less reliable, and more intermittent basis for interaction translates into more cognitively demanding social negotiations. Dominance and alliances, for instance, must be managed relative to a more complex set of factors (e.g., less constant reinforcement, less reliable support from non-kin than kin allies) and re-establishing contact after prolonged absence makes distance communication more important and demanding.

Yamagiwa, Chapter 12, considers how social foraging is affected by seasonal fruit scarcities. Sociality is recognized as linked with ecological pressures and these links probably differ in great apes compared with other anthropoids because of differences in their diets, especially seasonal reliance on fallback foods. Examining how great apes adjust their fission–fusion units during periods of fruit scarcity suggests how ecological pressures affect their sociality. One product is an indication of how great ape species differ socially: differences in fallback foods and sexual dimorphism interact to produce different social responses to food scarcities across great ape taxa. Another product, especially valuable in cognitive perspective, is highlighting a recurring and critical cognitive challenge to wild great apes wherein complex problems from two distinct domains, social and ecological, co-occur and interact.

These chapters consider facets of great ape adaptation already recognized as potentially significant to cognition. It is highly unlikely that these topics exhaust the possibilities. Understanding of great ape brains is still limited, for instance, and better understanding of brain development should enhance understanding of the energetic tradeoffs involved in ape life history profiles. More focused measures of ecological and social challenges are also probably needed, to dissect links between the brain and life history parameters in great apes, and further study of interactions between ecological and social challenges should advance our understanding of the roles and complexity of great ape cognition in the wild. The cognitive adaptations and challenges of living great apes offer great promise as avenues for developing a composite picture of the constellation of features that enabled and favored great apes' enhanced intelligence.

7 · What's in a brain? The question of a distinctive brain anatomy in great apes

CAROL E. MacLEOD

Department of Anthropology, Langara College, Vancouver

INTRODUCTION

Most scientists would not waste their time trying to teach sign language to a baboon or even a gibbon, but their success with the great apes is well known. The superior cognitive abilities of great apes are evident not only in their performance in such tasks as language learning, but also in their arithmetic, tool making and using, imitation, self-recognition, and feral skills indicative of a human-like intelligence. Although such skills do not represent the many dimensions of cognition and so cannot be generalized to all facets of intelligence, most researchers see a chasm with the great apes and humans on one side, and the lesser apes and monkeys on the other (see Tomasello & Call 1994 for another view). If this cognitive distinction is to be understood in terms of brain anatomy, then the neuroanatomy of the great apes should show more continuity with humans and less with the other anthropoids. Some headway has been made in discerning attributes of great ape anatomy that may parallel these cognitive patterns, but progress has been slow. This chapter will briefly discuss some of the more important findings in hominoid neuroanatomy that may have a bearing on our understanding of the great ape mind.

MEASURING THE BRAIN

Comparative studies are hindered by the rarity of ape brains and the time-consuming task of measuring the brain. Brains can be compared quantitatively by measuring volumes of the brain and its component structures, or qualitatitively by observing patterns in its gross morphology and cell organization, with some quantitative indices of these observations. The wide range of biological variability among brains of a given species limits the interpretive power of both studies of brain and brain structure volumes, and studies of cell architecture and organization, nuclei, cell columns, etc., because they are usually done with one or a handful of specimens. The most widely used data set for volumes of primate brains and their composite structures was compiled by Stephan and colleagues (1970, 1981), based on the Stephan collection of Nissl- and myelin-stained, serially sectioned brains representing 48 primate genera, but lacking *Pongo pygmaeus* and *Pan paniscus*. This lacuna has been rectified with a recently compiled data base of *in vivo* magnetic resonance (MR) scans of 47 primate specimens from the Yerkes Regional Primate Research Center (Rilling 1998; Semendeferi 1994), and with fixed-brain data from the collection of the Institut für Hirnforschung in Duesseldorf (MacLeod 2000). Both provide volumes from the full complement of extant ape genera. When volumes from several specimens are published, however, it is clear that the degree of biological variability within taxa is extensive, tempering ready interpretation of socio-ecological correlates of brain structure volumes at the species or genus level (MacLeod *et al.* 2003).

The most obvious and outstanding aspect of great ape neuroanatomy is absolute brain size (Table 7.1), and this may well prove to be the single most important aspect of brain anatomy that distinguishes great apes from the rest of the nonhuman primates. Large bodies come with large brains, and brain size (whether volume or weight) can be easily predicted from body size with linear regression, an example of allometry, or scaling (Jerison 1973). Residual values above or below this regression line are expressed as encephalization quotients (EQs), or actual brain size divided by expected brain size for animals of a given body weight. Any EQ above 1 is interpreted to mean that brain tissue is being used for nonsomatic functions, i.e., cognition (Jerison 1973). All apes do not have high EQs because some, notably gorillas, have enormous guts that inflate their

Table 7.1. *Whole brain volume from author's data acquired from the Yerkes and Hirnforschung samples*

Species	Number	cv HIRN & YERKES (%)	Trimmed combined means (cm³)	sem combined means (cm³)
Human	14	3.3	1265.1	40.1
Gorilla	5	1.2	401.3	24.1
Bonobo	6	7.4	322.7	13.1
Chimpanzee	14	8.1	361.7	13.6
Orangutan	8	14.0	391.6	19.9
Gibbon	9	11.7	91.4	4.9
Patas	1	—	89.0	—
Rhesus	7	14.8	90.6	5.2
Cercopithecus	2	—	64.0	7.1
Mangabey	5	8.8	101.5	2.9
Baboon	4	1.5	144.8	6.8
Spider	3	—	90.7	10.3
Howler	2	—	37.4	0.9
Cebus	6	12.2	69.0	4.8
Squirrel	6	10.7	22.0	0.9
Night	4	—	17.4	1.3
Grades				
Humans	14	3.3	1265.1	40.1
Great apes	33	0.4	369.3	17.7
Lesser apes	9	11.7	91.4	4.9
Old World monkeys	19	3.9	98.0	13.2
New World monkeys	21	6.1	47.3	14.1

body size. However, the apes as a whole have a significantly larger brain mass to body mass than monkeys for a given body weight, and monkeys in turn have larger EQs than prosimians (Figure 7.1) (MacLeod 2000; see also Begun and Kordos, Chapter 14, this volume). Apes do not show a significantly higher EQ under the statistical analysis of independent contrasts, however, thereby weakening the hypothesis of a higher degree of encephalization for apes (MacLeod *et al.*, 2003; see also Ross, Chapter 8, this volume).

An essential question in the evolution of the ape brain is whether size increase in brain structures has been uniform across the brain, or mosaic. Finlay and Darlington (1995) treated a large sample of brain structure volumes in insectivores, bats, and primates with linear regression to reveal allometric patterning in component structure size that was overwhelmingly predictable from brain volume alone. However, more

"progressive" structures such as the neocortex or cerebellum have higher exponents of increase, 0.445 and 0.341 respectively, than more "conservative" structures such as the medulla, 0.259, or the mesencephalon, 0.266 (Finlay & Darlington 1995). Consequently, those structures with the highest exponents will come to occupy more of the whole as brain size increases; this accounts for the different proportions of neocortex in the larger-brained macaque over the common marmoset, 72.2% and 60.4% respectively. In the Finlay and Darlington model, neither macaques nor humans have larger neocortices because of any specific evolutionary selection for that structure; only absolute brain size is subject to natural selection. Some studies have challenged this model with statistical methods other than pure linear regression (Barton & Harvey 2000), and de Winter and Oxnard's (2001) treatment of the complete Stephan data set of primates, insectivores, bats, tree shrews and

BRAIN TO BODY REGRESSION FROM COMBINED HIRNFORSCHUNG, YERKES AND STEPHAN SAMPLES

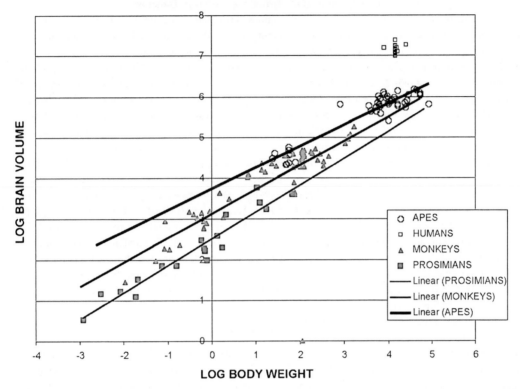

Figure 7.1. Regression of logged brain volume to logged body weight with three grades of prosimian, monkey, and ape. Regression analysis was done with a method that allowed the two *x* values of body weight and grade to interact, and was calculated using Systat 7. Results are slightly different than when calculated in graphic form using Excel v. X. Formulae for regressions are as follows: prosimian $y' = 2.799 + 0.761x$; monkey $y' = 3.314 + 0.542x$; ape $y' = 3.725 + 0.521x$. The y-intercepts are significantly different, r^2 value is 0.956, and standard error is 0.296.

elephant shrews in a multidimensional statistical model demonstrates definitively that brain structures can vary independently of absolute brain volume in response to adaptive demands. Some research to be discussed below demonstrates important breaks from allometry when comparing hominoids with monkeys.

GROSS ANATOMY

The external morphology of the great ape brain has long been a subject of study (Bolk 1902; Campbell 1905; Connolly 1950; Larsell & Jansen 1970; Retzius 1906; Semendeferi 1999; Tilney & Riley 1928; Zilles & Rehkämper 1988). For most of the sulci (furrows) and gyri (ridges) of the human brain, corresponding features are found on the surface of great ape brains. In fact, all primates share an overall homologous pattern in gross

anatomy and cyto- and myeloarchitecture (cell bodies and connecting fibres) (Zilles & Rehkämper 1988). With such commonality, it is meaningful to compare structures and their volumes within the primate order under the assumption that these structures have similar function. Differences in shape are more difficult to connect with function, but some differences in morphology in great apes may be related to aspects of great ape cognition.

Gyrification

Great ape brains are distinguished from those of other anthropoids and prosimians by the more complex pattern of convolution in their neocortex. This can be attributed largely to their size: there is a linear relation between absolute brain size and degree of convolution.

The gyrification index (GI) measures cortical folding in primates as the ratio of the total length of the cortical surface, including the hidden surface within the fissures and sulci, to the length of the contour of the superficial surface in brain sections (Armstrong *et al.* 1991; Zilles *et al.* 1988; 1989). The GI correlates well with brain weight in all primates, and even more with neopallial volume in anthropoids (the neopallium is the grey and white matter of the neocortex). Monkeys, apes, and humans follow the same regression line, i.e., show the same allometric relation between GI and brain or neopallial volume, whereas prosimians show a lower slope on a different regression line than anthropoids.

Zilles *et al.* (1989) suggest that the greater gyrification may be caused by higher growth in outer cortical layers (I–III) over inner cortical layers (IV–VI) (after Richman *et al.* 1975). This implies greater connectivity in the outer layers of the neocortex and a tendency towards more intracortical processing in anthropoids over prosimians. In this regard, nothing distinguishes great ape or human brains from other anthropoids. When the GI is tracked on a rostro-caudal trajectory, however, the great ape and human curves are closer to one another than to the other anthropoids in their absolute values and rostro-caudal patterning. Humans also show a higher GI in the prefrontal cortex and temporal region than the great apes (Zilles *et al.* 1988).

Although the relation of sulci to cytoarchitectural organization and function is still not understood, it is evident that they are connected with brain organization. Connolly (1950), who did the most complete examination of the external morphology of the cerebrum in great apes and other primates, noted an evolutionary progression in cerebral convolutions, even though it was obvious to him that smaller brains, with some exceptions such as in the slow loris, maintained sulcal complexity commensurate with their absolute size. In all three great ape genera, "the sulci are more tortuous, are provided with more branches, and with their greater size, have more tertiary furrows than the lower forms" (Connolly 1950: 118). Connolly viewed the complexity of fissuration in gibbon and siamang brains as intermediate between the great ape–human continuum and other anthropoids. Even though gibbons have brain volumes comparable to those of macaques, their sulcal patterning portends that of the larger hominoids. Notably, gibbons have a clear separation between the superior and inferior parietal lobules, and a well-defined, humanlike distinction between the supramarginal and angular gyri. These parts of the neocortex are primarily auditory and visuospatial association areas that become critical to language comprehension, reading, and writing in humans, so their emergence as discernable gyri becomes relevant to a distinctive hominoid anatomy.

Patterns close to those of humans could imply similar function, yet the gibbon brain is lacking so many of the specialized areas present in human brains that the linking of structure and function is problematic. This is also true for differences in overall shape, proportion, and sulcal patterning that are peculiar to each ape genus. For example, Zilles and Rehkämper (1988) noted the striking height of the frontal lobes, and the keel-like form of its orbital part in orangutans (cf. Connolly 1950; Retzius 1906). Yet, the higher frontal lobes do not correspond to *significantly* larger frontal lobe volumes (Semendeferi *et al.* 1997, 2002), and it is possible that some of these gross anatomical differences within the great ape genera may be functionally meaningless, merely consequences of skull shape and cranial base anatomy.

Asymmetries

The *planum temporale* (PT) is an auditory association area on the upper platform of the superior temporal gyrus, buried within the lateral sulcus, or Sylvian fissure. It is directly behind Heschl's gyrus, or primary auditory cortex, and is part of a major linguistic processing region in humans. As the left hemisphere is specialized for semantic speech processing, it is not surprising that the PT is larger on the left in humans (in 65% of adults – Geschwind & Levitsky 1968; in 86% of newborns and 81% of adults – Witelson & Pallie 1973).

Great apes share this PT asymmetry with humans. Yeni-Komishian and Benson (1976) measured the length of the Sylvian fissure as a proxy for the PT in 25 specimens each of humans, chimpanzees and macaques. According to the authors of the study, the width of the PT is equal in the two hemispheres, so any difference in length could be attributed to a larger *planum*. The left Sylvian fissure was significantly longer in humans and chimpanzees (84% and 80% longer respectively), but not macaques, although left over right Sylvian fissure lengths were found in Old World (OW) monkeys by Falk (1978) and Heilbroner and Holloway (1988), but not by Gannon (1995). Discrepancies may be due to measurement techniques and the small magnitude of left/right differences (Heilbroner & Holloway, 1988).

Three recent studies using different methodologies to measure the surface of the PT *directly*, have confirmed that chimpanzees, gorillas, and orangutans share a left-biased asymmetry with humans (Gannon *et al.* 1998, 2001; Gilissen 2001; Gilissen *et al.* 1998; Hopkins *et al.* 1998). In great apes, this bias is more robust in chimpanzees, possibly because of small sample sizes for gorillas and orangutans. Heschl's gyrus can be identified in the lesser apes, thereby demarcating the PT, but no hemispheric bias has been detected (Gannon *et al.* 2001). On this basis, Gannon *et al.* (2001) propose the emergence of a humanlike receptive language area in the common ancestor to all hominoids, while Hopkins *et al.* (1998) place this event later, as a homology shared among the great apes and humans.

If PT asymmetry is associated with human linguistic functions, why is it present at all in great apes? Gannon *et al.* (1998, 2001) are of the view that cognition in chimpanzees and other great apes is most developed in the realm of communication, including gestural and vocal, and that the PT is one anatomical marker of their communicative superiority over other primates. Cantalupo and Hopkins (2001) report that Brodmann's area 44, the region homologous to part of Broca's area, is larger in the left hemisphere than the right in chimpanzees, bonobos, and gorillas, as it is in humans. Broca's area is a classic speech region in humans, but also serves gestural functions. Cantalupo and Hopkins suggest that asymmetry in this region may be associated with the production of gestures accompanied by vocalizations in great apes, and this may have been a precursor in the common ancestor to the evolution of language in hominids. The asymmetries in great ape brains suggest a functional reorganization over monkey brains in the area of communication. Their linguistic capacity is certainly evident in the laboratory, but is less obvious in their natural gestural and vocal system (see Blake, Chapter 5, this volume).

The PT is not the only area of asymmetry in great ape brains. In human brains, for example, the posterior end of the Sylvian fissure usually angles more sharply upwards on the right than the left side, reflecting the expanded parietal operculum of the left hemisphere as part of a specialized language area (LeMay & Culebras 1972). This asymmetry is especially marked for right-handed individuals (Hochberg & LeMay 1975; LeMay & Culebras 1972). LeMay and Geschwind (1975) measured for this asymmetry in 30 monkeys, 11 lesser apes,

and 28 great apes. The only consistent pattern was found in the great apes, with 17 of 28 showing asymmetry of which 16 followed the human pattern. This high rate of asymmetry comes primarily from the orangutan sample, however; 10 of 12 orangutan brains showed right over left asymmetry, with less consistent asymmetries in chimpanzees and gorillas. If LeMay and Culebras (1972) were reluctant to interpret the function of this asymmetry in humans, it is even more difficult to interpret in apes.

A third area of cerebral asymmetry associated with handedness and lateralization in humans is petalias, asymmetries in hemisphere proportions. In humans, the posterior portion of the left hemisphere is wider and protrudes farther than the right, and the anterior portion of the right hemisphere is wider and protrudes farther than the left (LeMay 1976). LeMay (1985) found similar petalial patterns on some great ape brains. However, in Holloway and de la Coste-Lareymondie's (1982) study of a considerably larger sample size of bonobos, gorillas, chimpanzees, and orangutans, great ape patterns did not conform to the more consistent petalias found in fossil hominids and humans. The gorilla brain was more commonly left occipital petalial than the other great apes, but all lacked the human *combination* of right frontal and left occipital petalias.

The lack of evidence for clearly human petalial patterns in the great apes mirrors research on handedness. Although hand preference in wild and captive chimpanzees is well documented (summarized in McGrew & Marchant 1996), there is no overwhelming right-hand bias for manual activity as with humans (see Byrne, Chapter 3, this volume). The presence of brain asymmetries in great apes and humans suggest that both find advantages in systematic hand preferences for accomplishing varied and specialized tasks, but the inconsistency between human and great ape asymmetries could imply a shared propensity for cortical asymmetries rather than a shared complex inherited from the common ancestor. Instead, the asymmetries may be attributable to absolute brain volume, on which great apes and humans far surpass other anthropoids. The corpus callosum, connecting the two hemispheres of the brain, is smaller relative to brain volume and cortical surface in larger-brained primates, suggesting more localized processing of data in large brains. This could explain the great ape tendency towards anatomical asymmetry and handedness (Hopkins and Rilling 2000). As the number of neurons increases, there is an exponential increase in potential synapses, creating a strain on the computational system. One solution to this complexity of interconnectedness is a more localized processing of data, hence lateralization and areal specialization in the neocortex (Deacon 1990). The picture is clouded by the fact that brain lateralization occurs in birds, fish, reptiles, and amphibians in response to genetic, hormonal and environmental events (Rogers 1982, 2000; Vallortigara, Rogers & Brsazza 1999). Nonetheless, brain size may be a particularly marked influence in the lateralization of the primate brain because of our evolutionary tendency towards the corticalization of behavior.

CYTOARCHITECTURE

Increasingly, researchers are looking for more subtle differences in cerebral neuroanatomy with the realization that volume and shape of regions alone do not paint a complete picture of comparative primate brain anatomy (Preuss 2000, 2001). Cerebral cortex is organized in layers, with variations in cell shape, size, and density in each localized area. Each structural signature has its own pattern of intracortical and subcortical connections, although the actual links between structure and function are still not well known.

One measure of cortical differentiation is cell-packing density, indicated by the grey-level index (GLI) (Schleicher, Zilles & Kretschmann 1978). Comparisons of the GLI in each of the six layers of primary somatosensory cortex (area 3), primary motor cortex (area 4), primary visual cortex (area 17), and primary auditory cortex (areas 41 and 42) in chimpanzees, gorillas, orangutans, and humans show no significant differences in neocortical structure when layer thickness is expressed as a percentage of the cortex (Zilles & Rehkämper 1988). However, GLI values for the posterior cingulate cortex, an important component of the limbic system, do differ between anthropoids and prosimians but not between great apes and anthropoids (Armstrong et al. 1986). The authors suggest that anthropoids, with relatively larger outer and granular isocortical layers used in intracortical communication, might be capable of higher integration and differentiation of incoming information than prosimians.

Semendeferi (1994) and Semendeferi et al. (1998, 2001) measured areas 13 and 10 in comparative studies of the hominoid frontal cortex. Area 13 is part of the posterior orbitofrontal cortex. It is closely related to the limbic system, important in emotions and social behavior. Its cell shape, density, size of cortical layers, and space for connections are similar across hominoids, with slight variations that might indicate reliable species-specific differences (Semendeferi et al. 1998). Although qualitative observations were carried out on a sample of 22 primates, including all hominoids and the rhesus macaque, quantitative measures of cortical layer size were restricted to the right hemispheres from single specimens of orangutans, gorillas, chimpanzees, bonobos, and gibbons. Conclusions based on quantitative measures are limited by this small sample size. Compared with other hominoids, orangutans had a larger granular layer IV, with similarly sized infragranular and supragranular layers, whereas gorillas had a smaller layer IV and also larger infragranular to supragranular layers, a pattern more typical of limbic cortex. This suggests to Semendeferi that orangutans have a decreased representation of the limbic cortices in the frontal region compared with the rest of the hominoids (Semendeferi 1999). On the other hand, the orangutan specimen showed a lower density of neurons in area 13, in contrast to the high density of the gorilla (Semendeferi 1999). Low neuronal density is associated with greater neuronal connectivity, suggesting more complex organization for area 13 in orangutans.

Area 10, or the frontal pole, participates in working memory and attention, and is important in planning and taking of initiative (Semendeferi et al. 2001). In the same sample of five hominoids and one rhesus, area 10 showed a similar cytoarchitecture and cell density, but with some nuances. Cell density in area 10 was low in the gorilla and high in the orangutan relative to other hominoids. The density for area 10 was higher than for area 13 in all hominoids but the gorilla. The gorilla's frontal pole displayed a distinct appearance, with layers II and Va very prominent, raising a question as to the homology of the frontal pole cortex in gorillas within the hominoids (Semendeferi et al. 2001). No comparative statistical data on the range of variation in cytoarchitectural patterning in primate brains exists, and these observations require further exploration.

Spindle neurons of the anterior cingulate cortex are a rare case in which a qualitative distinction in great ape brain anatomy can be isolated (Nimchinsky et al. 1999). Spindle neurons participate in such mundane functions as control of heart rate, blood pressure, and digestion, but evidence on humans suggests their role in attention,

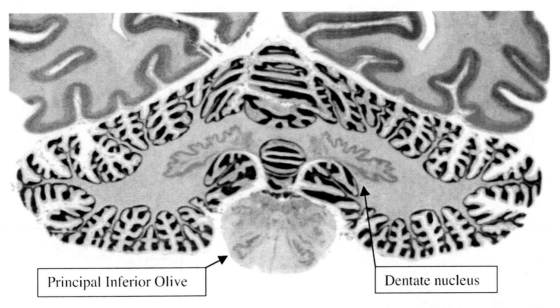

Principal Inferior Olive   Dentate nucleus

Figure 7.2. Cerebellum of *Pan troglodytes* demonstrating the hominoid pattern of convoluted dentate and principal infe-rior olivary nuclei, and extensive development of cerebellar hemispheres.

awareness of pain, and recognition of the emotional content of faces. These large projection neurons are found only in great apes and humans. Variations in spindle cell volume, distribution and density are apparent among the great apes, with bonobos closest to humans, followed by chimpanzees, gorillas, then orangutans. Spindle cell volume is strongly correlated with EQ in the sample. The spindle cells of the anterior cingulate cortex might constitute specialized neurons that integrate emotionally toned input and project to highly specific motor centers controlling vocalization, facial expression, or autonomic function (Nimchinsky *et al.* 1999). This view is consistent with the emergence of the planum temporale as a recognizable anatomical landmark in great apes, and would suggest selection for more specialized communication in great apes and humans (Nimchinsky *et al.* 1999).

SUBCORTICAL STRUCTURES

The neocortex is not the only arena of intelligence, but functions through its connections with subcortical structures such as the basal ganglia, the hippocampus and amygdala, the limbic system, the cerebellum, and brain stem structures. Tilney and Riley's (1928) early study compared subcortical structures of a number of primates. In hominoids but not monkeys, the principal inferior olivary nucleus and the dentate nucleus in the cerebellar nuclear complex are markedly convoluted, and the lateral cerebellum is more developed. This was later confirmed for chimpanzees (Figure 7.2), gorillas, and humans (Larsell & Jansen 1970). Tilney and Riley interpreted their findings of a differential increase in the size of the cuneatus nucleus, which relays information from the upper body, along with the expansion of the neocerebellar complex to mean that hominoids have greater coordination and dexterity in their upper limbs. Early anatomical studies tended to be impressionistic and devoid of statistical context, yet these observations are still basically true, that hominoids can be distinguished from monkeys in the cerebellar complex, and that gibbons group with hominoids and not monkeys, despite other aspects of their brain anatomy that appear intermediate.

BRAIN AND BRAIN STRUCTURE VOLUMES

Cerebral cortex

Of the many structures measured by Stephan and colleagues (1981), the neocortex is the most "progressive," i.e., increases with the highest exponent. Hominoids

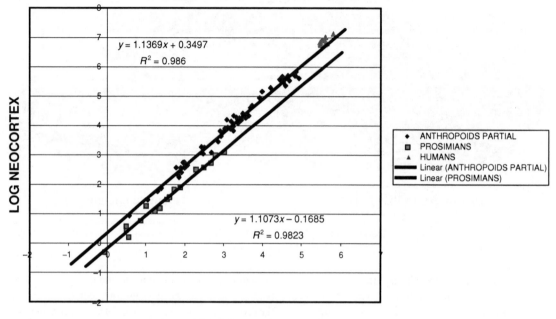

Figure 7.3. Logged neopallium to the rest of the brain in a combined sample of 95 specimens. Data are best explained by two regression lines for anthropoids in contrast to prosimians, with no distinction in neocortex regression line between apes and monkeys. Humans have a positive residual, which translates into significantly more neocortical brain tissue than an anthropoid of comparable brain volume.

do not show a differential expansion of the neocortex over monkeys, however, even in recent studies with more extensive data for ape brains (Rilling & Insel 1999; MacLeod unpublished data). The volume of neocortical grey and white matter, i.e., neopallium, was measured according to Frahm, Stephan and Stephan's (1982) protocols from the histological sections available from the Institut für Hirnforschung, including 18 apes, 8 humans and 21 monkeys. These data were combined with published data from the Stephan sample to produce a double logarithmic plot of neopallial volume regressed against the rest of the brain (Figure 7.3). There is nothing to distinguish hominoids from other anthropoids in neopallial volume under linear regression, but the regression line for anthropoids has a higher intercept than the line for prosimians, which is parallel, showing a grade shift (Martin 1980). Humans show a 22% increase in neopallial volume over that predicted by the anthropoid regression line, a substantial amount in absolute terms. Thus, the neocortex expanded differentially with the evolution of the anthropoids, and then again in humans but not great apes. If social intelligence is associated with the ratio of neocortex to the rest of the brain (Dunbar 1992), then the cognitive differences between monkeys and apes are not qualitative, but are differences of degree. The absolute mass of neocortex is greater in the large brains of the great apes, but no neural reorganization with regard to the neopallium is in evidence. The neocortex is not the exclusive seat of cognitive activity, however, nor is a measure of neocortical volume a fine enough instrument for discerning anatomical differences related to cognition.

Frontal lobes

Within the neocortex, the frontal lobes, specifically the prefrontal lobes, are associated with higher cortical functions because of their executive role in conscious control of behavior, planning and strategy, self-awareness, and abstract thinking. Several studies have concluded that humans have inordinately large frontal lobes compared with great apes (Blinkov & Glezer 1968; Brodmann 1912; Deacon 1997; Uylings & Van Eden 1990), but small sample sizes and inconsistent methodologies have undermined their conclusions. Semendeferi et al. (2002) have probably put to rest the notion that humans have greater *relative* frontal lobe expansion, although absolute values of human frontal lobes still remain impressive. Their study measured the grey matter of the frontal cortex and the entire cerebral cortex in the hemispheres (including also the hippocampus and amygdala) in 15 great apes, four gibbons, three macaques, two capuchin monkeys, and ten humans based on MR scans. Frontal cortex volume as a percentage of total hemisphere size was comparable in humans (37.7%) and great apes (35.9%), but significantly different from gibbons (29.4%), macaques (30.6%), and capuchins (30.6%). These results are consistent with two previous reports by Semendeferi et al. (1997, Semendeferi & Damasro 2000) of frontal lobe volumes in smaller samples using two different measurement protocols. Semendeferi et al. (2002, Semendeferi & Damasro 2000) interpret these results as an expansion of the frontal lobes in humans and great apes over lesser apes and monkeys in hominoid evolution.

The frontal cortex does occupy more of the hemispheric volume in humans and great apes than in gibbons, macaques, and capuchins, and could therefore play a more important part in cerebral processing. This does not imply differential expansion beyond expected allometry, however. When the logged frontal cortex is regressed against the cerebral hemispheres minus frontal cortex, the same regression line fits smaller-brained anthropoids and great apes if humans are omitted from the regression line (figure 2 in Semendeferi et al. 2002). The slope of this single line is 1.142, indicating that the frontal lobes have been expanding in the

anthropoids sampled at a higher rate than the rest of the cerebral hemispheres. Since a structure with a higher exponent comes to occupy a greater part of the whole as the brain expands in volume, the difference between the gibbon relative frontal lobe percentage of 29.4% and the chimpanzee percentage of 35.4%, for example, may be explained by the difference in absolute brain size alone (82.3 cm^3 and 320.9 cm^3, from Semendeferi et al. 2000). The logged values of frontal lobes to the rest of the cerebral hemispheres (figure 2 in Semendeferi et al. 2002) show humans slightly below expected values. A much larger sample size representative of the anthropoids would enable a more reliable regression analysis, to verify if there is any detectable difference in allometric proportions between humans and the great and lesser apes.

The frontal lobes can be divided into sectors that show variability in relative volume within the great ape clade. Semendeferi et al. (2002) calculated the volume of the frontal cortex rostral to the precentral gyrus to examine the frontal lobes minus primary motor cortex. Their results were comparable to those for the entire frontal lobes (contra Brodmann 1912, Deacon 1997, Preuss 2000). Semendeferi et al. (1997) also divided the frontal lobes into their dorsal, mesial, and orbital sectors, which did not show a discrepancy between gibbon and great ape proportions but did suggest specializations within the orangutan and gorilla frontal lobes. Area 13 of the orbitofrontal cortex, important in emotional reactions and social behavior, was smaller relative to the entire brain in gibbons, bonobos, and humans than in orangutans, gorillas, and chimpanzees (Semendeferi et al. 1998), but the small sample size precludes any interpretation based on differences in behavior. Area 10, a major cortical area of the prefrontal cortex, does show a larger relative value in humans compared with apes, and a smaller relative value in gibbons compared with other hominoids (Semendeferi et al. 2001). Nothing definitive can be concluded about the possibility of mosaic evolution within the frontal lobes of the great apes without a larger sample size.

Other forebrain structures

Semendeferi and Damasio (2000), found homogeneity in the relative size of the temporal lobes, parietal lobes, and insula in hominoids when expressed as a percentage of total hemisphere volume. Human values for temporal lobes appeared larger than expected for ape hemispheres scaled to human size, but differences were not significant. Rilling and Seligman (2002) measured temporal lobes using the same MR scans, but included a wider sample of monkey brains. They found human temporal lobes to be significantly above expected size, and detected a disparity between great ape and lesser ape temporal lobe proportions when subject to *regression* analysis, contrary to Semendeferi and Damasio. Only regression analysis, not ratios, will show whether a structure has increased differentially, but the regression line is affected by the number of species sampled (MacLeod 2000). Furthermore, Rilling and Seligman used only five sections per temporal lobe to obtain its volume, even using the first section to represent tissue both rostral and caudal to it. Less than ten sections per structure results in unreliable volumetric estimates, especially when the first section in the series is most vulnerable to the error of overestimation (Zilles, Schleicher & Pehlemann 1982). Hence, the issue of temporal lobe proportions remains unresolved.

The thalamic nuclei offer insight into the relative importance of incoming and outgoing information to particular regions of the brain, and constitute a kind of deep structure of the cerebral cortex. Within the hominoids, most of the thalamic nuclei scale allometrically, with no distinctions between apes and humans (summary in Armstrong & Frost 1988). One exception is the anterior nuclear complex, important in the limbic system and hence to emotions and their social expression (Armstrong 1986, 1991; Armstrong, Clarke & Hill 1987). The anterior thalamus has expanded in humans beyond expected allometry, perhaps because of the more elaborate social behavior of humans (Armstrong 1991). Armstrong *et al.* (1987) found a relation between social organization and the relative size of the anterior thalamic nucleus in 17 anthropoid species, but this correlation is not robust because of the overlap between categories of social organization, and the magnitude of anterior thalamic residuals. Armstrong and Frost (1988) suggest that differences in social behaviour will be found by looking at smaller neuroanatomical units or other limbic structures (cf. Semendeferi *et al.* 1998). Alternatively, social behavior and species-specific ecological adaptations may be functions of more generalized cognitive operations, with the plasticity of the brain allowing variations in behavior that have no observable neuroanatomical signatures.

The Cerebellum

As early anatomists observed, a distinctive ape brain anatomy can readily be seen in the cerebellum and its related nuclei. Matano conducted the first systematic study of cerebellar circuitry using the Stephan database, and concluded that the dentate nucleus, the pons and the principal inferior olive (PIO) were progressive structures that had increased significantly in humans and some other primates (Matano 1992, 2001; Matano *et al.* 1985a,b; Matano & Hirasaki 1997). Their study was hampered by the limited sample of ape brains, which included only two gorillas, one chimpanzee, and one gibbon.

This limitation has been addressed in some recent studies with a wider sample size. Rilling and Insel (1998) regressed cerebellar volume against the volume of the rest of the brain in 44 anthropoid specimens and found a differential increase in cerebellar volume in hominoids over monkeys, possibly because of selection for enhanced suspensory locomotion in early hominoids. Semendeferi and Damasio (2000) interpreted the smaller percentage of total brain size occupied by the cerebellum in humans compared with the rest of the hominoids to mean that the cerebellum did not expand to the same extent as the cerebrum during hominid evolution. Weaver (2001) estimated cerebellar volumes from hominid endocasts, and concluded that the hominid cerebellum underwent a mosaic expansion in the last two million years, only reaching its present size in absolute and relative terms in recent humans. Measures of hominoid fossil endocasts might portray a less uniform picture of brain allometry than that provided by data from extant species (see Begun & Kordos, Chapter 14, this volume).

My own study combined the Yerkes and Duesseldorf samples, with some additional ape specimens from the Stephan collection, for a sample size of 97 specimens, including 42 ape brains, although only the histological sections from Duesseldorf could be used for the measure of the nuclei (MacLeod 2000; MacLeod *et al.* 2001a). The study measured the whole brain and cerebellum, distinguished the lateral cerebellum (hemispheres) from the medial cerebellum (vermis), and measured the dentate nucleus and principal inferior olive (PIO). The cerebellum has a clear zonal organization, with the oldest part of the cerebellum most medial, and the newest, or neocerebellum, the most lateral. The

HEMISPHERE TO VERMIS

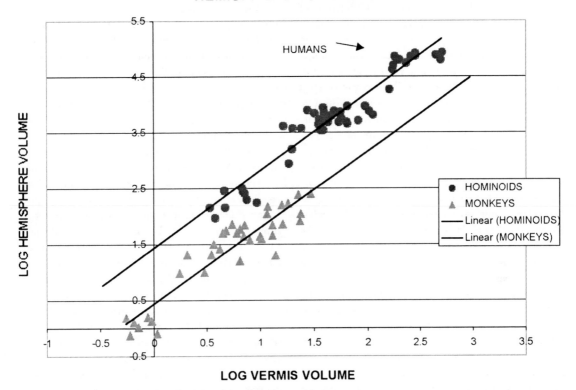

Figure 7.4. Logged cerebellar hemisphere to vermis volumes for the combined Yerkes and Hirnforschung samples. Regression was done as described in Figure 7.2. The SE is 0.268, with an r^2 value of 0.968. Regression formula for monkeys is $y' = 0.367 + 1.4588x$, and for hominoids is $y' = 1.465 + 1.365x$. The hominoid regression line, which includes humans in the regression, is significantly different from the monkey regression line.

dentate nucleus is the output nucleus for the neocerebellum; its outgoing fibers project to higher centers, mainly the cerebral cortex via the thalamus, but it also sends collaterals to the PIO. The PIO, in turn, projects specifically to the dentate nucleus and to the neocerebellum. Thus, the targeted volumes present a rough picture of an integrated cerebellar circuitry.

When the volumes were treated with multiple regression analysis that tested for best fit with either a single anthropoid regression line or a double line that distinguished hominoids from monkeys, cerebellar structures showed differential expansion beyond expected allometry. The lateral cerebellum was much larger in hominoids over monkeys. When the cerebellar hemispheres were regressed against the vermis, this expansion was 2.7 times that expected in monkeys. There was no significant difference in the slopes of monkey and hominoid regression lines, but there were

significant differences in their intercepts, i.e., a grade shift (Figure 7.4). Within cerebellar circuitry the PIO increased with the cerebellar hemispheres, but the dentate nucleus expanded only with the rest of the brain and did not participate in the grade shift of the neocerebellar structures. Cerebellar circuitry did not then expand as an integrated unit, as would be expected in functionally integrated structures (Barton & Harvey 2000), but instead evolved in a selectively mosaic fashion in hominoids (MacLeod, Schlercher & Zilles 2001b), implying neural reorganization.

Earlier insights that the cerebellum is important to many cognitive activities (Leiner, Leiner & Dow 1986; Schmahmann 1991) have been reinforced in the last decade through experimental scrutiny. The cerebellum is not simply an organ of coordination and balance but participates in an array of cognitive activities that include the planning of complex motor patterns (Thach 1996), rhythmic and sequential patterning (Keele & Ivry 1990), sensory discrimination (Gao *et al.* 1996), switching of attention (Allen *et al.* 1997), visuo-spatial problem solving (Kim, Ugurbil & Strick 1994), procedural learning (Doyon 1997; Fiez & Raichle 1997), and various linguistic operations such as generating verbs from nouns and word choice (Fiez & Raichle 1997). In cognitive tasks where regional blood flow has been measured through PET and fMRI scans, activity is most marked in the lateral cerebellum. This explains why the neocerebellum communicates so extensively with the neocortex; it receives a massive neocortical input through the pontine nuclei, and sends back information not only to the sensory-motor strip but to widespread areas of the cerebral cortex. Thus, my finding that the cerebellar hemispheres and the PIO increased disproportionately in the hominoids indicates that there was a selection for those parts of cerebellar circuitry active in cognition among the hominoids.

The hemisphere to vermis regression shows no disparity between great ape volumes and those of either humans or gibbons. When the cerebellar hemispheres are regressed against the rest of the brain the parallelism between the regression lines is lost when humans are grouped with the rest of the hominoids. This is because of a differential increase of the neocortex in humans, not because humans have smaller cerebellar hemispheres *per se* (MacLeod *et al.* 2003). Removal of humans from the regression line restores the parallel lines.

The uniformity of the ape and monkey slopes in the regressions performed on the data suggest that the differential expansion of the cerebellar hemispheres was not a species-specific event, but one which took place in the common ancestor to the hominoids in the early Miocene, before the separation of lesser and great apes. The pattern of suspensory locomotion was not yet clearly established, although some early hominoids showed evidence of suspensory feeding (Gebo, Chapter 17, this volume; Larson 1998). More importantly, it appears that the early hominoids were frugivorous (Fleagle 1999; Potts, Chapter 13, Singleton, Chapter 16, this volume). A frugivorous diet requires visuo-spatial memory and mapping skills because of patchily distributed resources scattered over large ranges (Clutton-Brock and Harvey 1979; Milton 1981). The cerebellum receives substantial input from the parietal and occipital areas of the brain (Stein, Miall & Weir 1987), important

areas in mapping skills (Kolb & Whishaw 1990). As well, the dentate nucleus projects information to the frontal and prefrontal lobes, areas important in strategy and choice. A circuitry involving the parieto-occipital areas, cerebellum, and frontal lobes might have been advantageous to a more efficient feeding strategy for the early hominoids. The ability of the lateral cerebellum to plan complex movements (Thach 1996) would minimize extraneous effort, especially in the three-dimensional world of suspensory feeding.

Cerebellar participation in procedural learning, sensory discrimination, and visuo-spatial activities that have a cognitive component could underlie some of the derived feeding adaptations seen in extant great apes. Mountain gorillas must navigate a tactile maze when eating some of their well-protected foods, and must proceed in a logical sequence of hierarchically embedded steps (Byrne 1995 and, Chapter 3, this volume) that requires procedural learning, or learning how to do a task. Procedural learning is necessary to successful extractive foraging, tool use, and other complex behavior routines (Gibson 1999; Parker & Milbraith 1993), especially in suspensory conditions (Russon 1998). Neocerebellar skills could account for great apes' superior performance in cognitive activities that require sustained attention, especially attention that focuses on relevant cues, in keeping with the directed attention skills known to be vital to human children learning language (Allen & Courchesne 2001).

If a substantial increase in the lateral cerebellum could account for the superior cognitive performance of great apes over monkeys, then why do we not see those same skills expressed in gibbons? Great ape brains are substantially larger than gibbon brains (Table 7.1). They have not only greater computing power, but also more tissue to carry out cerebellar functions. The larger brain size of great apes and humans has magnified the cognitive advantages accrued from a selective increase of the neocerebellum. Nonetheless, gibbons still share the augmented lateral cerebellum that enables the intricate choreography of brachiation, and perhaps other cognitive skills yet to be uncovered.

DISCUSSION AND CONCLUSION

Many of the distinctive aspects of great ape brain anatomy outlined in the text are explicable by absolute brain volume. Larger brains have more convolutions and a more extensive cortical mantle that facilitate interconnections. The localized functional areas that are a distinguishing feature of human brains may be largely a phenomenon of size, in which cerebral cortical connectivity becomes more and more demanding as the brain expands, forcing the brain to organize itself into more locally specialized units (Deacon 1990; Hopkins & Rilling 2000). Larger brains also have implications for the complexity of cortical processing. Gibson (1990) argues that large absolute size enables parallel processing and distributed networks; a problem can be resolved by simultaneous processing in different areas of the cortex (and, it follows, in those structures connected to the cortex such as the cerebellum). Rumbaugh (1995) and Gibson *et al.* (2001) argue that great ape superiority in transfer tasks, in which a subject must learn a correct response to a problem then unlearn that response in favor of the opposite choice in order to receive a reward, may be attributable to their large brains. The EQ alone does not explain their results.

The discrepancy between ape and monkey in volumes of cerebellar structures means that a greater percentage of the brain is devoted to lateral cerebellar function, and hominoid brains are organized differently in consequence. The continuities between human and great ape frontal lobe proportions also argue for shared neurological substrates to common ape and human cognitive abilities, and there appears to be no grade shift between great and lesser apes in frontal lobe expansion. The temporal lobes show homogeneity within the apes, with a possible differential expansion in humans, although there is disagreement on interpretation of the data. The thalamic nuclei also show continuity, but humans appear to have a more highly developed anterior thalamus. The findings of Nimchinsky *et al.* (1999) point towards an increased integration and corticalization of emotions and communication in the great apes. Zilles and colleagues also discerned this tendency towards increased corticalization of emotions for neocortical cytoarchitecture and thalamic volumes, but for the anthropoids as a whole. Semendeferi detects subtle differences within the great ape clade in areas 10 and 13, but the biological variation of brain structures acts as a caveat to interpreting brain anatomy at the level of the genus. The larger picture is one of continuity of structures within the hominoids that are sometimes distinct

from other anthropoids, with the lesser ape brains in a somewhat intermediate position.

Shape differences within the great ape clade and between apes and monkeys are difficult to interpret. The patterns of sulcal and gyral morphology reveal phylogenetic continuities, but these are not so easily translated functionally. Asymmetries are present in great ape brains, and some would interpret these asymmetries as common prelinguistic substrates, especially in chimpanzees. These shared asymmetries will be understood more fully when we have more data on finer levels of neuroanatomical organization as revealed by cytoarchitectural patterns, but structural affinities imply that human and great ape brains are working with the same *Bauplan*. Although we find continuity between human and all ape brains, particularly with regard to the organization of the cerebellum, we also find trends in shape and cytoarchitecture that distinguish great ape from lesser ape neuroanatomy. We have precious few ape brains. Answers to the question of a distinctive great ape anatomy will come with careful and dedicated coaxing, but with great reluctance.

ACKNOWLEDGMENTS

I thank the editors, Anne Russon and David Begun, for their invitation to participate in this volume, and for their invaluable editorial skills.

REFERENCES

Allen, G. & Courchesne, E. (2001). Attention function and dysfunction in autism. *Frontiers in Bioscience*, **6**, 105–19.

Allen, G., Buxton, R. B., Wong, E. C. & Courchesne, E. (1997). Attentional activation of the cerebellum independent of motor involvement. *Science*, **275**, 1940–3.

Armstrong, E. (1986). Enlarged limbic structures in the human brain: the anterior thalamus and medial mamillary body. *Brain Research*, **362**, 394–7.

(1991). The limbic system and culture: An allometric analysis of the neocortex and limbic nuclei. *Human Nature*, **2**(2), 117–36.

Armstrong, E. & Frost, G. T. (1988). The diencephalon: a comparative review. In *Orangutan Biology*, ed. J. H. Schwartz, pp. 177–88. New York: Oxford University Press.

Armstrong, E., Clarke, M. R. & Hill, E. M. (1987). Relative size of the anterior thalamic nuclei differentiates anthropoids by social system. *Brain, Behavior and Evolution*, **30**, 263–71.

Armstrong, E., Zilles, K., Schlaug, G. & Schleicher, A. (1986). Comparative aspects of the primate posterior cingulate cortex. *The Journal of Comparative Neurology*, **253**, 539–48.

Armstrong, E., Zilles, K., Curtis, M. & Schleicher, A. (1991). Cortical folding, the lunate sulcus and the evolution of the human brain. *Journal of Human Evolution*, **20**, 341–8.

Barton, R. A. & Harvey, P. H. (2000). Mosaic evolution of brain structure in mammals. *Nature*, **405**, 1055–8.

Blinkov, S. M. & Glezer, I. I. (1968). *The Human Brain in Figures and Tables: A Quantitative Handbook*. New York: Plenum Press.

Bolk, L. (1902). Beiträge zur Affenanatomie. 2. Über das Gehirn von Orang-Utan. *Peterus Camper Nederlandsche Bijdragen tot se Anatomie*, **1**, 25–84.

Brodmann, K. (1912). Neue Ergenbnisse über die vergleichende histologische Lokalisation der Grosshirnrinde mit besonderer Berücksicktigung des Stirnhirns. *Anatomischer Anzeiger*, **41**, 157–216.

Byrne, R. (1995). *The Thinking Ape: Evolutionary Origins of Intelligence*. Oxford: Oxford University Press.

Campbell, A. W. (1905). *Histological Studies on the Localisation of Cerebral Function*. Cambridge, UK: Cambridge University Press.

Cantalupo, C. & Hopkins, W. D. (2001). Asymmetric Broca's area in great apes. *Nature*, **414**, 505.

Clutton-Brock, T. H. & Harvey, P. H. (1979). Home range size, population density and phylogeny in primates. In *Primate Ecology and Human Origins*, ed. I. S. Bernstein & E. O. Smith, pp. 201–14. New York: Garland Press.

Connolly, C. J. (1950). *External Morphology of the Primate Brain*. Springfield, IL: Charles C. Thomas.

Deacon, T. W. (1990). Rethinking mammalian brain evolution. *American Zoologist*, **30**, 629–705.

  (1997). *The Symbolic Species: The Co-evolution of Language and the Brain*. New York: W. W. Norton and Company.

de Winter, W. & Oxnard, C. E. (2001). Evolutionary radiations and convergences in the structural organization of mammalian brains. *Nature*, **409**, 710–13.

Doyon, J. (1997). Skill learning. In *The Cerebellum and Cognition* (*International Review of Neurobiology*, vol. 41), ed. J. D. Schmahmann, pp. 273–94. San Diego, CA: Academic Press.

Dunbar, R. I. M. (1992). Neocortex size as a constraint on group size in primates. *Journal of Human Evolution*, **20**, 469–93.

Falk, D. (1978). Cerebral asymmetry in Old World monkeys. *Acta Anatomica (Basel)*, **101**, 334–9.

Fiez, J. A. & Raichle, M. E. (1997). Linguistic processing. In *The Cerebellum and Cognition* (*International Review of Neurobiology*, vol. 41), ed. J. D. Schmahmann, pp. 233–54. San Diego, CA: Academic Press.

Finlay, B. L. & Darlington, R. B. (1995). Linked regularities in the development and evolution of mammalian brains. *Science*, **268**, 1578–4.

Fleagle, J. C. (1999). *Primate Adaptation and Evolution*, 2nd edn. Toronto: Academic Press.

Frahm, H. D., Stephan, H. & Stephan, M. (1982). Comparison of brain structure volumes in Insectivora and Primates. I. Neocortex. *Journal für Hirnforschung*, **23**, 375–89.

Gannon, P. (1995). *Asymmetry in the Cerebral Cortex of Macaca fascicularis: A Basal Substrate for the Evolution of Brain Mechanisms Underlying Language*. New York: City University of New York.

Gannon, P. J., Holloway, R. L., Broadfield, D. C. & Braun, A. R. (1998). Asymmetry of chimpanzee planum temporale: humanlike pattern of Wernicke's brain language area homolog. *Science*, **279**, 220–2.

Gannon, P. J., Kheck, N. M. & Hof, P. R. (2001). Language areas of the hominoid brain: a dynamic communicative shift on the upper east side planum. In *Evolutionary Anatomy of the Primate Cerebral Cortex*, ed. D. Falk & K. R. Gibson, pp. 216–40. Cambridge, UK: Cambridge University Press.

Gao, J.-H., Parsons, L. M., Bower, J. M., Xiong, J., Li, J. & Fox, P. T. (1996). Cerebellum implicated in sensory acquisition and discrimination rather than motor control. *Science*, **272**, 545–7.

Geschwind, N. & Levitsky, W. (1968). Human brain: left–right asymmetries in temporal speech region. *Science*, **161**, 186–7.

Gibson, K. R. (1990). New perspectives on instincts and intelligence: brain size and the emergence of hierarchical mental constructional skills. In *"Language" and Intelligence in Monkeys and Apes: Comparative Developmental Perspectives*, ed. S. T. Parker & K. R. Gibson, pp. 97–128. New York: Cambridge University Press.

(1999). Social transmission of facts and skills in the human species: neural mechanisms. In *Mammalian Social Learning: Comparative and Ecological Perspectives* (*Symposia of the Zoological Society of London* 72), ed. H. O. Box & K. R. Gibson, pp. 351–66. Cambridge, UK: Cambridge University Press.

Gibson, K. R., Rumbaugh, D. & Beran, M. (2001). Bigger is better: primate brain size in relationship to cognition. In *Evolutionary Anatomy of the Primate Cerebral Cortex*, ed. D. Falk & K. R. Gibson, pp. 79–97. Cambridge, UK: Cambridge University Press.

Gilissen, E. (2001). Structural symmetries and asymmetries in human and chimpanzee brains. In *Evolutionary Anatomy of the Primate Cerebral Cortex*, ed. D. Falk & K. R. Gibson, pp. 187–215. Cambridge, UK: Cambridge University Press.

Gilissen, E., Amunts, K., Schlaug, G. & Zilles, K. (1998). Left–right asymmetries in the temporoparietal intrasylvian cortex of common chimpanzees. *American Journal of Physical Anthropology*, Supplement **26**, 86.

Heilbroner, P. L. & Holloway, R. L. (1988). Anatomical brain asymmetries in New World and Old World monkeys: stages of temporal lobe development in primate evolution. *American Journal of Physical Anthropology*, **76**, 39–48.

Hochberg, F. H. & LeMay, M. (1975). Arteriographic correlates of handedness. *Neurology*, **25**, 218–22.

Holloway, R. L. & de la Coste-Lareymondie, M. C. (1982). Brain endocast asymmetry in pongids and hominids: some preliminary findings on the paleontology of cerebral dominance. *American Journal of Physical Anthropology*, **58**, 101–10.

Hopkins, W. D. & Rilling, J. K. (2000). A comparative MRI study of the relationship between neuroanatomical asymmetry and interhemispheric connectivity in primates: implication for the evolution of functional asymmetries. *Behavioral Neuroscience*, **114**(4), 739–48.

Hopkins, W. D., Marino, L., Rilling, J. K. & MacGregor, L. A. (1998). Planum temporale asymmetries in great apes as revealed by magnetic resonance imaging (MRI). *NeuroReport*, **9**, 2913–8.

Jerison, H. J. (1973). *Evolution of the Brain and Intelligence*. New York: Academic Press.

Keele, S. W. & Ivry, R. (1990). Does the cerebellum provide a common computation for diverse tasks? A timing hypothesis. *Annals of the New York Academy of Sciences*, **608**, 179–211.

Kim, S.-G., Ugurbil, K. & Strick, P. L. (1994). Activation of a cerebellar output nucleus during cognitive processing. *Science*, **265**, 949–51.

Kolb, B. & Whishaw, I. Q. (1990). *Fundamentals of Human Neuropsychology*, 3rd edn. New York: W. H. Freeman and Company.

Larsell, O. & Jansen, J. (1970). *The Comparative Anatomy and Histology of the Cerebellum from Monotremes through Apes*. Minneapolis, MN: University of Minnesota Press.

Larson, S. G. (1998). Parallel evolution in the hominoid trunk and forelimb. *Evolutionary Anthropology*, **6**(3), 87–99.

Leiner, H. C., Leiner, A. L. & Dow, R. S. (1986). Does the cerebellum contribute to mental skills? *Behavioral Neuroscience*, **100**(4), 443–54.

LeMay, M. (1976). Morphological cerebral asymmetries of modern man, fossil man, and nonhuman primate. *Annals of the New York Academy of Sciences*, **280**, 349–66.

(1985). Asymmetries of the brains and skulls of nonhuman primates. In *Cerebral Lateralization in Nonhuman Species*, ed. S. D. Glick, pp. 233–45. Toronto: Academic Press.

LeMay, M. & Culebras, A. (1972). Human brain-morphologic differences in the hemispheres demonstrable by carotid arteriography. *The New England Journal of Medicine*, **267**(4), 168–70.

Le May, M. & Geschwind, N. (1975). Hemispheric differences in the brains of great apes. *Brain, Behavior and Evolution*, **11**, 48–52.

MacLeod, C. E. (2000). The Cerebellum and its Part in the Evolution of the Hominoid Brain. Ph.D. dissertation, Simon Fraser University, Burnaby, BC.

MacLeod, C. E., Schleicher, A. & Zilles, K. (2001b). Natural selection and fine neurological tuning. *American Journal of Physical Anthropology*, Supplement **32**, 100.

MacLeod, C. E., Zilles, K., Schleicher, A. & Gibson, K. R. (2001a). The cerebellum: an asset to hominoid cognition. In *All Apes Great and Small*. Vol. 1. *African Apes*, ed. B. M. F. Galdikas, N. E. Briggs, L. K. Sheeran, G. L. Shapiro & J. Goodall, pp. 35–53. New York: Kluwer Academic/Plenum Publishers.

MacLeod, C. E., Zilles, K., Schleicher, A., Rilling, J. K. & Gibson, K. R. (2003). Expansion of the neocerebellum in Hominoidea. *Journal of Human Evolution*, **44**, 401–29.

Martin, R. D. (1980). Adaptation and body size in primates. *Zeitschrift für Morphologie und Anthropologie*, **71**, 115–24.

Matano, S. (1992). A comparative neuroprimatological study on the inferior olivary nuclei (from the Stephan's collection). *Journal of Anthropology and Sociology Nippon*, **100**(1), 69–82.

(2001). Brief communication. Proportions of the ventral half of the cerebellar dentate nucleus in humans and great apes. *American Journal of Physical Anthropology*, **114**(2), 163–5.

Matano, S. & Hirasaki, E. (1997). Volumetric comparisons in the cerebellar complex of anthropoids, with special reference to locomotor types. *American Journal of Physical Anthropology*, **103**, 173–83.

Matano, S., Baron, G., Stephan, H. & Frahm, H. D. (1985b). Volume comparisons in the cerebellar complex of primates. II. Cerebellar nuclei. *Folia Primatologica*, **44**, 182–203.

Matano, S., Stephan, H. & Baron, G. (1985a). Volume comparisons in the cerebellar complex of primates. I. Ventral pons. *Folia Primatologica*, **44**, 171–81.

McGrew, W. C. & Marchant, L. F. (1996). On which side of the apes? Ethological study of laterality of hand use. In *Great Ape Societies*, ed. W. C. McGrew, L. F. Marchant, & T. Nishida, pp. 255–72. Cambridge, UK: Cambridge University Press.

Milton, K. (1981). Distribution patterns of tropical plant foods as an evolutionary stimulus to primate mental development. *American Anthropologist*, **83**, 534–48.

Nimchinsky, E. A., Gilissen, E., Allman, J. M., Perl, D. P., Erwin, J. M. & Hof, P. R. (1999). A neuronal morphologic type unique to humans and great apes. *Proceedings of the National Academy of Science*, **96**, 5268–73.

Parker, S. T. & Milbrath, C. (1993). Higher intelligence, propositional language, and culture as adaptations for planning. In *Tools, Language and Cognition in Human Evolution*, ed. K. R. Gibson & T. Ingold, pp. 314–33. Cambridge, UK: Cambridge University Press.

Preuss, T. M. (2000). What's human about the human brain? In *The New Cognitive Neurosciences*, 2nd edn., ed. M. S. Gazzaniga, pp. 1219–34. Cambridge, MA: MIT Press.

(2001). The discovery of cerebral diversity: an unwelcome scientific revolution. In *Evolutionary Anatomy of the Primate Cerebral Cortex*, ed. D. Falk & K. R. Gibson, pp. 138–64. Cambridge, UK: Cambridge University Press.

Retzius, G. (1906). *Das Affenhirn*. Jena: Gustav Fischer.

Richman, D. P., Stewart, R. M., Hutchinson, J. W., & Caviness, Jr., V. S. (1975). Mechanical model of brain convolutional development. *Science*, **189**, 19–21.

Rilling, J. K. (1998). Comparative Neuroanatomy of Anthropoid Primates using Magnetic Resonance Imaging: Insights into Human and Non-human Primate Brain Evolution. Ph.D. dissertation, Emory University, Atlanta, GA.

Rilling, J. K. & Insel, T. R. (1998). Evolution of the cerebellum in primates: differences in relative volume among monkeys, apes and humans. *Brain, Behavior and Evolution*, **52**, 308–14.

(1999). The primate neocortex in comparative perspective using magnetic resonance imaging. *Journal of Human Evolution*, **37**, 191–223.

Rilling, J. K. & Seligman, R. A. (2002). A quantitative morphometric comparative analysis of the primate temporal lobe. *Journal of Human Evolution*, **42**, 505–33.

Rogers, L. J. (1982). Light experience and asymmetry of brain function in chickens. *Nature*, **297**, 223–5.

(2000). Evolution of hemispheric specialization: advantages and disadvantages. *Brain and Language*, **73**, 236–253.

Rumbaugh, D. (1995). Primate language and cognition: common ground. *Social Research*, **62**, 711–30.

Russon, A. E. (1998). The nature and evolution of orangutan intelligence. *Primates*, **39**, 485–503.

Schleicher, A., Zilles, K. & Kretschmann, H.-J. (1978) Automatische Registrierung und Auswertung eines Grauwertindex in histologischen Schnitten. *Anatomischer Anzeiger*, **114**, 413–15.

Schmahmann, J. D. (1991). An emerging concept: the cerebellar contribution to higher function. *Archives of Neurology*, **48**, 1178–87.

Semendeferi, K. (1994). Evolution of the Hominoid Prefrontal Cortex: A Quantitative and Image Analysis of Areas 13 and 10. Ph.D. Thesis, University of Iowa, Iowa City, Iowa.

(1999). The frontal lobes of the great apes with a focus on the gorilla and the orangutan. In *The Mentalities of Gorillas and Orangutans: Comparative Perspectives*, ed. S. T. Parker, R. W. Mitchell & H. L. Miles, pp. 70–95. Cambridge, UK: Cambridge University Press.

Semendeferi, K. & Damasio, H. (2000). The brain and its main anatomical subdivisions in living hominoids using magnetic resonance imaging. *Journal of Human Evolution*, **38**, 317–32.

Semendeferi, K., Armstrong, E., Schleicher, A., Zilles, K. & Van Hoesen, G. W. (1998). Limbic frontal cortex in hominoids: a comparative study of area 13. *American Journal of Physical Anthropology*, **106**, 129–55.

(2001). Prefrontal cortex in humans and apes: a comparative study of area 10. *American Journal of Physical Anthropology*, **114**, 224–41.

Semendeferi, K., Damasio, H., Frank, R. & Van Hoesen, G. W. (1997). The evolution of the frontal lobes: a volumetric analysis based on three-dimensional reconstructions of magnetic resonance scans of human and ape brains. *Journal of Human Evolution*, **32**, 375–88.

Semendeferi, K., Lu, A., Schenker, N. & Damasio, H. (2002). Humans and great apes share a large frontal cortex. *Nature Neuroscience*, **5**(3), 272–6.

Stein, J. F., Miall, R. C. & Weir, D. J. (1987). The role of the cerebellum in the visual guidance of movement. In *Cerebellum and Neuronal Plasticity*, ed. M. Glickstein, C. Yeo & J. Stein, pp. 175–92. New York: Plenum Press.

Stephan, H., Bauchot, R. & Andy, O. J. (1970). Data on size of the brain and of various brain parts in insectivores and primates. In *The Primate Brain: Advances in Primatology*, Vol. I, ed. C. R. Noback & W. Montagna, pp. 289–98. New York: Appleton-Century Crofts.

Stephan, H., Frahm, H. & Baron, G. (1981). New and revised data on volumes of brain structures in insectivores and primates. *Folia Primatologica*, **35**, 1–29.

Thach, W. T. (1996). On the specific role of the cerebellum in motor learning and cognition: clues from PET activation and lesion studies in man. *Behavioral and Brain Sciences*, **19**, 411–31.

Tilney, F. & Riley, H. A. (1928). *The Brain from Ape to Man: A Contribution to the Study of the Evolution and Development of the Human Brain, Volumes I & II*. New York: Paul B. Hoeber.

Tomasello, M. & Call, J. (1994). Social cognition of monkeys and apes. *Yearbook of Physical Anthropology*, **37**, 273–305.

Uylings, H. B. M. & Van Eden, C. G. (1990). Qualitative and quantitative comparison of the prefrontal cortex in rat and in primates, including humans. *Progress in Brain Research*, **85**, 31–62.

Vallortigara, G., Rogers, L. J. & Bisazza, A. (1999). Possible evolutionary origins of cognitive brain lateralization. *Brain Research Reviews*, **30**, 164–75.

Weaver, A. H. (2001). Cerebellar volumes in Pliocene and Pleistocene *Homo*. Ph.D. Thesis, University of New Mexico, Albuquerque, NM.

Witelson, S. F. & Pallie, W. (1973). Left hemisphere specialization for language in the newborn: neuroanatomical evidence of asymmetry. *Brain*, **96**, 641–6.

Yeni-Komshian, G. H. & Benson, D. A. (1976). Anatomical study of cerebral asymmetry in the temporal lobe of humans, chimpanzees, and rhesus monkeys. *Science*, **192**, 387–9.

Zilles, K. & Rehkämper, G. (1988). The brain, with special reference to the telencephalon. In *Orangutan Biology*, ed. J. H. Schwartz, pp. 157–76. New York: Oxford University Press.

Zilles, K., Armstrong, E., Moser, K. H., Schleicher, A. & Stephan, H. (1989). Gyrification in the cerebral cortex of primates. *Brain, Behavior and Evolution*, **34**, 143–50.

Zilles, K., Armstrong, E., Schleicher, A. & Kretschmann, H.-J. (1988). The human pattern of gyrification in the cerebral cortex. *Anatomical Embryology*, **179**, 173–9.

Zilles, K., Schleicher, A., & Pehlemann, E. W. (1982). How many sections must be measured in order to reconstruct the volume of a structure using serial sections? *Microscopica Acta*, **86**, 339–46.

8 · Life histories and the evolution of large brain size in great apes

CAROLINE ROSS

School of Life & Sport Sciences, University of Surrey, London

INTRODUCTION

Our preoccupation with large brain size and why it evolved originally arose from interest in the evolution of traits that we consider to be special in our own species. Humans are generally thought to be more intelligent, behaviorally flexible, and culturally complex than other primate species. In humans, these complex behaviors are usually linked to the relatively large brain size that is also a key feature of *Homo sapiens*. Understanding the evolution of large brains may therefore help us to understand the evolution of complex behaviors and abilities. This requires identifying the costs and benefits of large brain size, in order to explain why large brains evolved in a few primate taxa but not in most, or in other mammal groups.

Compared to other mammals, haplorhine primates also have large brains (relative to body size), the relative brain size of humans is still larger, and those of the great apes are not exceptional within the group (Figure 8.1). However, a range of studies has found that the intellectual capacities of the great apes are significantly different from those of other haplorhine primates (e.g., Byrne 1995, 1999a; Russon, Bard & Parker 1996; and Blake, Chapter 5, Byrne, Chapter 3, Parker, Chapter 4, Russon, Chapter 6, this volume). This suggests that a large brain size relative to body size is not the only factor that may be important in determining levels of intelligence in primates. Some research suggests that absolute brain size may also be important (Beran, Gibson & Rumbaugh 1999; Byrne 1999b; Gibson 1990; MacLeod, Chapter 7, this volume). For example, some learning abilities (e.g., transfer of learning across contexts) are more closely correlated with absolute than with relative brain size (Beran *et al.* 1999). Other work has suggested that overall changes in brain size may not be as important as reorganization of the brain's information processing functions (Byrne 1995, 1999b) or the evolution of specific parts of the brain (Barton & Dunbar 1997; Barton, Purvis & Harvey 1995; Holloway 1996).

This chapter explores several issues related to the evolution of large brain size and life-history parameters. After outlining some of the life-history characteristics of the great apes, the life-history correlates of brain size in haplorhine primates are investigated and these results are compared with those from other studies. I also investigate possible explanations for these relationships, including confounding variables. I finish by looking at the great apes as a separate group, to compare patterns found within this group with those found in the haplorhine primates overall and to investigate the possibility that the evolution of complex intelligence and large brain size has correlates with life-history parameters.

BRAIN SIZE, LIFE HISTORY, AND ENVIRONMENTAL PRESSURES

There is a history of research into links between brain size and other life-history variables in primates and other taxa, beginning with Portmann (1962) and Sacher (1959). Sacher (1959, 1975) noted that in primates, brain size was a better predictor of recorded longevity than body size; this is also found after controlling for the effects of body weight (Allman, McLaughlin & Hakeem 1993) and phylogeny (Barton 1999; Judge & Carey 2000). Age at maturation and brain size in primates have also been found to be strongly related (Harvey, Martin & Clutton-Brock 1987) (Figures 8.2, 8.3). The relatively slow growth rates of primates relative to other mammals might be linked to their relatively large brain size, with energetically expensive large brains limiting the rates of growth that could be maintained by juvenile primates (Charnov & Berrigan 1993). In fact, Martin (1983, 1996)

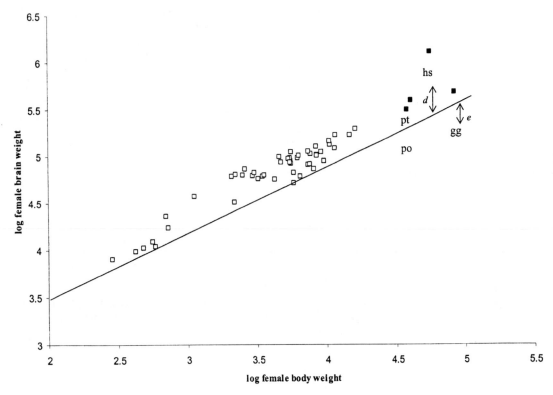

Figure 8.1. The relationship between female body weight and female brain weight in anthropoid primates. Open squares are monkeys and lesser apes; closed squares are great apes and humans (po = *Pongo*; pt = *Pan troglodytes*; hs = *Homo*; gg = *Gorilla*). A line of slope 0.64, calculated using CAIC analysis (see pages 128–9) for anthropoid primate species, is shown (Barton 1999). Relative brain size has been calculated by taking the distance of the observed brain size from this line, parallel to the *y* axis (i.e., for species hs, this is the positive value *d* whereas for species gg, it is the negative value *e*).

linked fetal brain growth with maternal energetic input in placental mammals.

Together these studies suggest that, after body size effects are accounted for, there are links between slow maturation rate, slow growth rate, long lifespan, and large adult brain. Many researchers have discussed reasons for links between a "slow life" and large brain size. Some of their hypotheses are discussed and tested below.

The question "Why evolve a large brain?" has frequently been answered by hypothesizing that some aspect of a species' physical or social environment has selected for individuals with abilities to process large amounts of information. For example, some foods may be more difficult to find or process than others, whether they be animal prey (Jerison 1973), or foods that must be extracted from a substrate (Parker & Gibson 1977, 1979) or otherwise prepared before eating (Byrne & Byrne 1993). A large home range size may require a correspondingly large memory (Milton 1981), certain

substrates may be difficult to negotiate (Povinelli & Cant 1995; Russon 1998), or complex social systems may be difficult to understand and manipulate (Dunbar 1992; Humphrey 1976; Jolly 1966). In all cases, the assumption is that individuals with greater information processing power (i.e., larger brains) will be able cope better with the environment than will those with less. Clearly, the benefits of this increased brain size must result in increased survival and reproduction in comparison with less well-endowed individuals, or they will not outweigh the costs.

There is considerable evidence that a variety of environmental variables correlate with brain size. Clutton Brock and Harvey (1980) showed that folivorous primates have smaller brain sizes than frugivorous primates of similar size. They, and others, have suggested that this may be due to the difficulty of finding fruit versus leaves or because frugivorous primates have larger home ranges (Clutton Brock & Harvey 1980; Milton 1988). Dunbar (1992, 1995) also found a positive

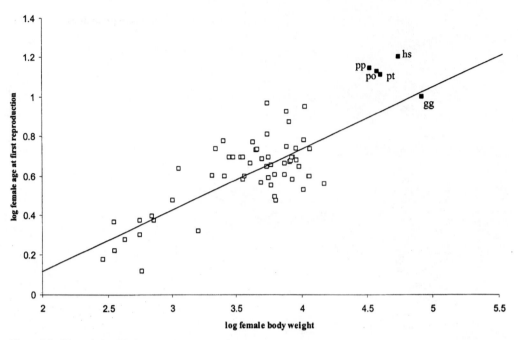

Figure 8.2. The relationship between female body weight and female age at first reproduction in anthropoid primates. Open squares are monkeys and lesser apes; closed squares are great apes and humans, (po = *Pongo*; pp = *Pan paniscus*; pt = *Pan troglodytes*; hs = *Homo*; gg = *Gorilla*). A line of slope 0.34, calculated using CAIC analysis (see pages 128–9) for anthropoid primate species, is shown (Ross & Jones 1999).

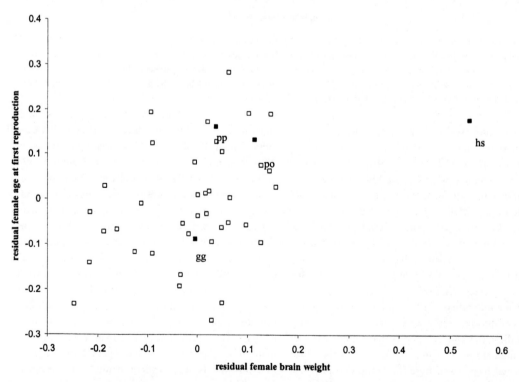

Figure 8.3. The relationship between relative female age at first reproduction and relative brain size in anthropoid primate species. Open squares are monkeys and lesser apes; closed squares are great apes and humans, (po = *Pongo*; pt = *Pan troglodytes*; hs = *Homo*; gg = *Gorilla*). Data shown are contrasts calculated using the CAIC program (see pages 128–9).

Table 8.1. *Life-history parameters used in analyses*

Parameter	Symbol	Definition
Body weight	W	Mean adult female body weight (grams)
Brain weight	Brn	Mean adult female brain weight (grams)
Age at first reproduction	AR	Mean female age at first reproduction (years)
Juvenile period	Juv	Mean female age at first reproduction (years) minus weaning age (years)
Female birth rate	b	Mean female birth rate (offspring/year). Calculated assuming 0.5 primary sex ratio
Maximum longevity	L	Maximum recorded longevity for the species (years)
r_{max}	r_m	Intrinsic rate of natural increase (a measure of a population's ability to grow)
Growth rate	GR	Growth rate from birth to AR (grams/day)
Infant growth rate	IGR	Growth rate from birth to weaning (grams/day)
Juvenile growth rate	JGR	Growth rate from weaning to AR (grams/day)

correlation between home range size and the ratio of neocortex size to brain size in haplorhine primates, but showed that this may owe primarily to both variables being independently related to body size. A number of studies have shown that relative brain size or some alternative measure of brain enlargement is positively correlated with the complexity of the social environment in haplorhine primates, as measured by social group size (e.g., Barton & Dunbar 1997; Dunbar 1992, 1995) and the prevalence of tactical deception in primates (Byrne 1995).

However, some studies show that environmental variables may also correlate with life-history parameters, particular those relating to reproductive rates (see reviews by Ross 1998; Ross & Jones 1999). Hence, the relationships that have been found between life-history variables and brain size may owe to confounding influences of the environment on both. For example, frugivorous species might have a relatively late age at first reproduction and a relatively large brain size. If so, a correlation between late age at first reproduction and relatively large brain size might be expected even if there were no direct link between the two. Below, I apply several methods to investigate the possibility that such confounding environmental variables influence connections between brain size and life-history parameters.

METHODS

Life history and brain size data

The variables used are listed and defined in Table 8.1. Most of the data used here have been published

previously (Barton 1999; Ross & Jones 1999) and the criteria used to include data are described in detail in these papers. The data set for the great apes has been expanded and updated for this work, as shown in Table 8.2. The measure of brain size used here is mean female brain weight; this is preferred to mean adult brain weight as most other parameters refer to female characteristics, for example, female body weight, female age at first reproduction. The length of the juvenile period is taken as the time between weaning and female age at first reproduction. I have not separated this period into juvenility and adolescence because the break between juvenility and adolescence is often hard to define, particularly in species that show no growth spurt at adolescence.

Controlling for body size

As brain size and life-history variables correlate highly with body size (Harvey *et al.* 1987; Peters 1983; Ross 1988; Western 1979), all comparative analyses carried out here controlled for the confounding influence of body size. This is done in two ways: (i) using multiple regression to investigate the importance of body size relative to other independent variable(s) on dependent variables, and (ii) using *relative* brain size.

The measures of relative brain size used here are the residuals of the regression of brain size on body size, which remove the effect of body size on brain size (Figure 8.1). Several different regressions with different slopes, ranging from 0.64 to 0.75, have been proposed in the literature (Barton 1999; Jerison 1973; Martin 1990). Here, a slope of 0.64 is used (Barton 1999), though the

Table 8.2. *Great ape data*

(a) Life-history data

Taxon	Body weight (kg)[1] (n)	Cranial capacity (cc)[1] (n)	AR (yr)[2] (n)	Wean (yr)[2]	IBI (yr)[2,3] (n females/n intervals)	References
Gorilla gorilla (unknown subspecies)			c: 9.85 (2)[4]			Rudder (1979)
G. g. gorilla (W. lowland gorilla)	m: 170.4 (10) f: 71.5 (3)	m: 561.2 (27) f: 477.3 (13)				Angst (1976)
G. g. beringei (mountain gorilla)	m: 162.5 (5) f: 97.5 (1)	m: 574.6 (11) f: 487.4 (14)	w: 10.42 (4)		w: 3.83 (5/8)	Angst (1976), Harcourt *et al.* (1981)
G. g. graueri (E. lowland gorilla)	m: 175.2 (4) f: 71.0 (2)					Angst (1976)
Pan paniscus	m: 45.0 (7) f: 33.2 (6)		c: 9.5 (2) w: 13–14	w: 4.0	c: 4.75 (2/2) w: 4.8 (17/28)	Furuichi *et al.* (1998), Jantschke (1975), Kano (1992), Neugebauer (1981), Nishida & Hiraiwi-Hasegawa (1987)
					c: 0.62 (49)	Nissen & Yerkes (1943)
Pan troglodytes (unknown subspecies)						
P. t. schweinfurthii (E. chimpanzee)	m: 42.7 (21) f: 33.7 (26)	m: 420.2 (9) f: 414.8 (9)	w: 14.6		w: 5.5 (13/21) w: 6.17	Angst (1976), Goodall (1986), Nishida & Hiraiwi-Hasegawa (1987), Nishida *et al.* (1990)
P. t. troglodytes	m: 59.7 (5) f: 45.8 (4)	m: 400.6 (10) f: 376.9 (9)				Angst (1976)
P. t. verus (W. chimpanzee)	m: 46.3 (1) f: 41.6 (3)	m: 404.2 (11) f: 371.7 (13)	w: 14.3 (8)	w: 5	w: 5.76 (19/33)	Angst (1976), Boesch & Boesch-Achermann (2000)
Pongo pygmaeus (unknown subspecies)			c: 9.7[4] (12–15)	w: 5–7		Jones (1982)
P. p. abelii (Sumatran orangutan)	m: 77.9 (3) f: 35.6 (4)	m: 408.5 (20) f: 380.0 (5)				Röhrer-Ertl (1988)
P. p. pygmaeus (Bornean orangutan)	m: 78.5 (7) f: 35.8 (13)	m: 431.2 (26) f: 384.4 (8)	w: 12 + (est.15)		w: 7–8	Galdikas (1981), Galdikas & Wood (1990), Röhrer-Ertl (1988)

Notes:

[1] All body weight data are from Smith and Jungers (1997). Brain sizes are cranial capacities (cc), as in the original papers. For analyses, brain weights were estimated using a regression equation linking cranial capacity to brain weight (Martin 1990).

[2] IBI = infant birth interval, c = data from captive animals, w = data from wild animals.

[3] Using only birth intervals following the survival of the first infant, when this is stated. Numbers of females and numbers of birth intervals are shown in brackets.

[4] Age at first conception given as 9 years, age at first reproduction taken as 9 years + 8 months gestation.

(b) Dietary data

	Dietary intake (% of total)								
Species/population	fruit	seeds	leaves	flowers	bark, wood, cambium	animal	other	Method	Reference
Pongo pygmaeus pygmaeus Kutai, E. Kalimantan	53.8		29.0	2.2	14.2	0.8		1	Rodman (1977)
Pongo pygmaeus pygmaeus Tanjung Puting, C. Kalimantan	61.0		15.0	4.0	11.0	4.0		1	Galdikas (1988)
Gorilla gorilla beringei Virunga	1.7		85.8	2.3	6.9		3.3	1	Fossey & Harcourt (1977)
Gorilla gorilla graueri Kahuzi-Biega, Zaire (lowland area)	55.0							2	Yamagiwa et al. (1996)
Gorilla gorilla graueri Kahuzi-Biega, Zaire (highland area)	16.0							2	Yamagiwa et al. (1996)
Gorilla gorilla gorilla Lope, Gabon	76.0							1	Tutin & Fernandez (1993)
Pan troglodytes schweinfurthii Gombe	61.0		26.0	4.0		4.0		1	Goodall (1986)
Pan troglodytes schweinfurthii Kibale	78.0	1.0	14.7	4.2	2.2			1	Ghiglieri (1984)
Pan troglodytes troglodytes Gabon	68.0		28.0					3	Hladik (1977)
Pan troglodytes troglodytes Kahuzi-Biega, Zaire (lowland area)	76.0					4.0		2	Yamagiwa et al. (1996)
Pan troglodytes troglodytes	38.0							2	Yamagiwa et al. (1996)
Pan troglodytes troglodytes Lope, Gabon	76.0							1	Tutin & Fernandez (1993)
Pan troglodytes verus Taï forest, Ivory coast			3.0 (THV)			8.0		1	Boesch & Boesch-Achermann (2000)
Pan paniscus Wamba, Zaire	83.4		15.2			1.5		1	Kano (1992)
Pan paniscus Lomako, Zaire	54.0	7.1	27.4	4.4	6.2 (pith)			1	Badrian & Malenky (1984)

Notes:

THV = terrestrial herbaceous vegetation.

Methods of data collection are: 1 = time spent feeding; 2 = fecal analysis; 3 = annual intake. Some totals do not add to 100% because of rounding or because all dietary items were not quantified.

pattern of results of this analysis is unchanged if a regression slope of 0.75 is used.

Residuals are preferable to ratios (neocortex: Dunbar 1992; brain to body size: Joffe 1997), which do not completely control for body size because brain size does not scale isometrically (1:1) with body size (Bauchot & Stephan 1969; Jerison 1973; Martin 1990). Neocortex size is not used in this study as the neocortex is not the only part of the brain involved in learning and intelligence and hence it is hard to justify its use in studies concerned with general intelligence (see also MacLeod, Chapter 7, this volume).

Measures of environmental complexity

Three measures of environmental complexity were chosen for these analyses.

Diet

Several previous studies have suggested that diet is linked to brain size in primates and other mammals. Diet is measured here by the mean percentage of leaves in the species' diet. As previous studies have suggested that a highly folivorous diet is generally easier to obtain than a more frugivorous diet or a diet high in animal food, I assume throughout that increasing the amount of leaves in the diet decreases the complexity of the environment that a species experiences. This is clearly a very simplistic assumption and does not account for other influences on dietary complexity, in particular the problems of extracting embedded foods or searching for foods that are seasonally variable in their distribution. Unfortunately, finding a simple measure of dietary complexity that could deal with these variables proved beyond the scope of this study. The composition of each great ape species' diet was taken from data in Smuts et al. (1987) and the sources shown in Table 8.2. All analyses included both the percentage of leaves and the percentage of fruit in the diet. Although other materials such as cambium, wood, and bark were also measured in some studies, these were not included in the analyses as not all studies recorded the eating of these materials separately but often lumped them under "other."

Group size

As primates are social animals that often live in stable social groups, an individual's group can be considered as a part of its environment. Group size was defined as in Dunbar (1991, 1995) as the mean size of a stable social group. For species with a fission–fusion group structure (*Pan* and *Ateles* species) the size of the community was used (Dunbar 1995). I also followed Dunbar (1995) by using data from Smuts et al. (1987) to define group size; group size for each population was taken as the mean size of breeding groups from Smuts et al. (1987) plus sources shown in Table 8.2. Where a range of group sizes was given, the mid-value was used.

Habitat

Habitat type was used as a parameter because it may be correlated with life-history variation (Ross 1988; Ross & Jones 1999). Habitat was classified as "forest" or "open." Forest species included those that typically spend all, or most, of their time living in forest habitats. "Non-forest" species included those living primarily in woodland, savannah, or grassland. Clearly, a number of confounding environmental variables may make one habitat more complex than another (including all those considered here) and it is not clear which habitat type should be considered as more complex.

Arboreality has been considered in the context of environmental complexity (Povinelli & Cant 1995). However, the degree of challenge also depends on positional behavior (e.g., suspensory or above-branch quadrupedalism) and features of the substrate (e.g., secondary or primary forest; mountain or savannah). As the comparative analyses used here do not deal well with more than two categorical variables and as positional and substrate complexity are not easily quantified as continuous variables, arboreality was not used in this analysis.

Comparative methods

Analyses shown here used two data sets: (1) data on haplorhine species, with each species treated as an independent data point (for sources, see Key & Ross 1999; Ross & Jones 1999), and (2) contrast data produced by the Comparative Analysis by Independent Contrasts program (CAIC) (Purvis & Rambaut 1995). Use of CAIC removes phylogenetic bias from analyses by transforming species data into differences or "contrasts" between clades (Harvey & Pagel 1991; Purvis & Rambaut 1995). This analysis uses a comparative method based on Felsenstein's (1985) method of independent contrasts and was carried out as detailed in Purvis and Rambaut

(1995). Potential advantages and disadvantages of CAIC are discussed by Martin (1996), MacLarnon (1999), and Purvis and Webster (1999). All analyses using CAIC were carried out using the composite primate phylogeny, including branch lengths, produced by Purvis (1995). Repeating analyses with a slightly different phylogeny (Purvis & Webster 1999) made no difference to the results and conclusions reached here.

The methods used here are almost entirely those of correlation and regression. Least squares regression examined whether adult female body weight and/or life-history parameters are good predictors of brain weight. Further multiple regressions investigated the relationship between brain weight, life history parameters, diet, and group size. Multiple regressions offer a powerful method of revealing correlation patterns between life history and intelligence measures but do not, of course, reveal cause and effect. Regression analyses could not be used for dichotomous variables so differences in relative brain weight between species with differing habitats were tested using t-tests. To control for phylogenetic bias, analyses comparing dichotomous variables were repeated using CAIC, using its "Brunch" option. This method takes the dichotomous variable as the independent variable (X) and tests the null hypothesis that change in X (e.g., forest to non-forest or *vice versa*) produces no change in the dependent variable (Y) (e.g., brain weight). If the null hypothesis is correct, results will show that a change in X from state 1 to state 2 is as likely to produce an increase in Y as it is to produce a decrease. If change in X is correlated with change in Y, then this change in X will produce either an increase or a decrease in Y.

"Contrasts" for relative female brain weight were generated by CAIC (per Purvis & Rambaut 1995). A t-test was carried out on the mean of the contrasts to test whether the mean was significantly greater than zero (indicating that higher Y evolves with higher X) or less than zero (indicating that smaller Y evolves with higher X). Also t-tests were used to test for differences between relative brain weight for "forest" versus "non-forest" species.

Although several environmental variables could operate together to increase the complexity a species experiences, the use of dichotomous data meant it was not possible to combine all measures in a multiple regression. Links between brain size, group size, and diet were explored using multivariate analyses, but the

Table 8.3. *Reproductive rate parameters versus body size and brain size. Multiple regression through the origin carried out using CAIC data (n = 23 older contrast values of log (parameter) vs. log W and log Brn). For details on calculating these equations, see Ross and Jones (1999).*

Parameter	r	p	x	p
			\multicolumn{2}{c}{Multiple regression statistics}
Juv	0.770	<0.0001	W	0.1853
			Brn	0.0113
AR	0.826	<0.0001	W	0.4191
			Brn	0.0167
L	0.717	0.0005	W	0.7626
			Brn	0.1308
b	0.687	0.0012	W	0.6175
			Brn	0.5020
r_m	0.750	0.0002	W	0.9429
			Brn	0.1535

dichotomous habitat measure could not be included. To control for confounding effects of habitat type, analyses investigating links between other measures of environmental complexity and brain size were repeated separately for forest and non-forest species.

RESULTS

Brain weight and life history in primates

Table 8.3 shows the relationship between brain weight and a range of life history traits in primate species. It provides evidence that, once body size is controlled, primates with a large brain weight (Brn) also have a relatively late age at first reproduction (AR) and a relatively long juvenile period (Juv). This finding is extremely robust for primates and supports previous studies that used different comparative methods and included different species in the data set (e.g., Harvey *et al.* 1987). Other life-history parameters are not clearly linked to brain weight in this way. Although simple regressions of other life-history variables against brain weight show the expected significant correlations, these disappear when a multiple regression including body weight is performed (Table 8.3). Longevity, birth rate, and rate of increase are all significantly linked to body weight but not to brain weight. Although longevity has sometimes been reported to have a close relationship with brain weight, these analyses suggest that this relationship owes primarily to body weight correlating with both.

Confounding variables: the Economos problem

Economos (1980) suggested that brain weight may be a more accurate measure of body size than is body weight, if body weight measures are more subject to error than brain weight measures. Body weight may fluctuate with an animal's health, nutritional status, and reproductive condition. If dead animals are measured, their weights may vary according to their state of preservation. Economos argued that body size measures that serve as proxies for body weight, including brain weight, may be more consistent, less error-prone, and hence better predictors of life-history variables.

This may explain why brain weight correlates more highly with age at first reproduction than does body weight. However, two lines of evidence suggest that the relationship between brain weight and age at first reproduction is not a statistical artifact. First, correlations with brain weight have been investigated for other reproductive parameters. When body weight effects are removed, these reproductive parameters are not significantly related to brain weight, as they should be if Economos's theory is correct (Table 8.4). Second, another way of avoiding the Economos problem is to use body weight measures from different populations to calculate residuals for the other two variables, brain weight and age at first reproduction (Barton 1999; Harvey & Krebs 1990). Using Smith and Jungers' (1997) body weight data, I took body weight data for the two largest populations in 25 primate species, assigning the first listed body weight to the regression including brain weight and the second to the regression with age at first reproduction. This approach is likely to introduce random error into the analyses, and hence reduce the probability of significant relationships between the variables being found.

Results of this study are shown in Table 8.4. Clearly, even when body weight data are taken from separate populations and body weight influences are held constant, the link between age at first reproduction and brain weight is still significant. This again indicates that the link between brain weight and age at first reproduction represents a "real" relationship rather than simply a statistical artifact. The relationship between longevity

and brain weight is significant with this data set, suggesting the results reported in Table 8.3 may not hold with all data sets.

Confounding variables: environmental complexity and brain size

The relationship between brain weight and age at maturation, after controlling for body weight, may owe to both being separately linked to environmental variables. My results showed no link between relative brain weight and habitat type ($p > 0.05$ for both contrast and species data). Accordingly, the relationship between brain weight and age at maturation is not a product of both being correlated with habitat type.

Results of multivariate analyses including the other measures of environmental complexity (diet, group size), brain weight, and age at maturity are shown in Table 8.5a. These results suggest that age at maturity is linked to brain weight, even after effects of body weight, group size, and diet are removed via multiple regression. Table 8.5b shows that the length of the juvenile period is also linked to brain weight, even after effects of body weight, group size, and diet are removed via multiple regression.

Why is brain size linked to age at maturation?

Ross and Jones (1999) tested three commonly used models that attempt to explain why species with relatively large brains should also have relatively late ages of first reproduction. These models all suggest that large brain size is linked to age of first reproduction via one or more of three intervening variables, complex environment, slow infant and/or juvenile growth rate, and long juvenile period, as illustrated in Figure 8.4. As each model suggests a different causal link between brain size and age at maturity, each generates different predictions regarding links among the various parameters.

Brain growth constraint model

Large brains are energetically costly, so relatively large-brained species must have high nutritional intake and/or save energy by cutting down on other body organs or activities. Primates may make energetic savings by decreasing energy expenditure on other organs, particularly guts (Aiello & Wheeler 1995); they might also divert energy from overall body growth into brain growth and maintenance. Thus slow postnatal growth rates could

Table 8.4. *Testing the Economos problem (see text for full details). Results show the relationship between brain size and life-history variables when: (a) relative values are calculated using the same measure of body weight; (b) relative values are calculated using the two different measures of body weight. All values are CAIC-generated contrasts, using older contrasts only*

(a) Relative brain weight and relative life history parameter values taken from the same body weight data set (fbwt1 then fbwt2)

Parameter	fbwt1			fbwt2		
	N	r	p	N	r	p
Juvenile period (Juv)	10	0.63	0.0384	10	0.310	0.3538
Age rep (AR)	12	0.69	0.0088	12	0.52	0.0710
Longevity (L)	12	0.74	0.0039	18	0.78	0.0001

(b) Relative brain weight and relative life-history parameter values taken from different body weight data sets (Brain weight residuals taken from fbwt2, life-history residuals taken from fbwt1)

Parameter	N	r	p
Juvenile period (Juv)	10	0.70	0.0165
Age rep (AR)	12	0.66	0.0148
Longevity (L)	12	0.73	0.0050

Notes: fbwt1 is the first listed body weight, fbwt2 is the second listed body weight.

Table 8.5. *Brain weight and maturation age, controlling for group size and diet as measures of environmental complexity. Analyses were done using contracts generated from the CAIC program (see pages 128–9) using the oldest 50% of contrasts only. Multiple regression through the origin (i.e., with no slope intercept) is used throughout*

(a) Brain size (female) (Brn) versus female body weight (W), age at first reproduction (AR), group size, and diet (% folivory and % frugivory)

df	r	p	x	p (direction of relationship)
8	0.99	0.0015	W	0.019 (+)
			AR	0.070 (+)
			Group size	0.031 (+)
			% folivory	0.999
			% frugivory	0.171

(b) Brain size (female) (Brn) versus female body weight (W), length of juvenile period (Juv), group size and diet (% folivory and % frugivory)

df	r	p	x	p (direction of relationship)
8	0.99	0.0002	W	0.002 (+)
			Juv	0.058 (+)
			Group size	0.049 (+)
			% folivory	0.506
			% frugivory	0.350

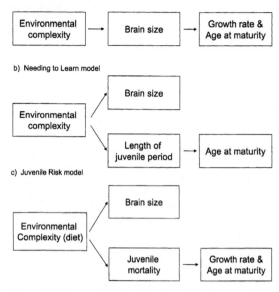

a) Brain Growth Constraint model

Environmental complexity → Brain size → Growth rate & Age at maturity

b) Needing to Learn model

Environmental complexity → Brain size

Environmental complexity → Length of juvenile period → Age at maturity

c) Juvenile Risk model

Environmental Complexity (diet) → Brain size

Environmental Complexity (diet) → Juvenile mortality → Growth rate & Age at maturity

Figure 8.4. Suggested relationships between brain size, maturation age, and environmental complexity in primates (see text for details).

result from environmental selection pressures that favor large brains (Figure 8.4). Because primate brains grow mainly before and soon after birth (Martin 1983), the relative brain weight of immature primates is particularly large. The energetic costs of maintaining large immature brains may impose slow growth during both infant and juvenile periods, which may explain the association between late maturation and large brain size. Hence this model predicts that:

(1) relative brain weight should correlate positively with environmental complexity (here measured by diet and group size), even if postnatal growth rate is held constant;
(2) relative brain weight should correlate negatively with postnatal growth rate, even if environmental complexity is held constant;
(3) there should be no direct link between environmental complexity and postnatal growth rate if brain weight is held constant.

Needing to learn model

Large brains are often thought to occur in species that need sophisticated information processing capabilities to cope with complex social and/or physical environments. As information processing capabilities depend on

learning, such species may have to delay maturity until they have learned enough to be behaviorally mature. Hence, large-brained species should have a long immaturity period to allow learning takes place (Figure 8.4). This could result in an extension of infancy and/or the juvenile period. As it is the juvenile period that is particularly extended in primates, this model predicts a link between brain weight and the length of the juvenile period, but one that owes primarily to both factors being selected for by a complex environment. This model predicts that:

(1) relative brain weight should correlate positively with environmental complexity (i.e., diet and group size), even when postnatal growth rates are held constant;
(2) relative brain weight should correlate positively with age at first reproduction but this relationship should disappear when environmental complexity is held constant;
(3) delayed maturation could occur by prolonging the juvenile period after reaching adult size (no relationship between brain weight and growth rate) or by slowing growth rates (negative relationship between brain weight and growth rate);
(4) there should be a positive correlation between environmental complexity and age at first reproduction even if brain weight is held constant. Delayed maturation could occur by prolonging the juvenile period after reaching adult size (no relationship between environmental complexity and growth rate) or by slowing growth rates (negative relationship between environmental complexity and growth rate).

Juvenile risk model

Juvenile primates are smaller as well as less skilled than adults (Janson & van Schaik 1993). Juveniles often have lower foraging success than adults in the same population, leading to increased time foraging, greater susceptibility to malnutrition or starvation during food shortages, and perhaps greater susceptibility to predators (although survival rates may be increased by living in larger groups).

Janson and van Schaik argue that juvenile primates' problems owe primarily to their relatively small size so their difficulties could be overcome by growing *faster* to adult size. However, juvenile primates appear to grow *slower* than their physiological capacity allows, so slow growth may have adaptive advantages. Janson and van Schaik suggest that juvenile primates grow slowly to reduce their metabolic needs. Living in larger groups may also increase food competition with conspecifics, so it may also slow growth rates, but in this case the disadvantages of slow growth rates may be outweighed by the advantage of lowered predation risk and correspondingly reduced mortality rates.

This model differs from the brain growth constraint model in considering slow growth rates as directly adaptive and interpreting the link between relative brain weight and relative age at first reproduction as caused by a third variable, diet, that influences both. Folivorous diets should select for both small brains and high postnatal growth rates (leaves are easy to find, so increased foraging for leaves has a smaller influence on predation risk than increased foraging for fruit). This model predicts that:

(1) relative brain weight should correlate negatively with diet (percentage folivory), even when postnatal growth rate is held constant;
(2) relative brain weight should correlate positively with postnatal growth rate but this correlation should disappear when dietary measures are held constant;
(3) percentage folivory should correlate positively with growth rate even if brain weight is held constant;
(4) there should a group size influence on postnatal growth rates, although its direction is unclear. Large groups may reduce mortality risks by providing protection from predators, but increased mortality may result from increasing competition levels.

Which model fits with the data?

Ross and Jones' (1999) analyses used data on age at first reproduction to explore the links illustrated in Figure 8.4. They assumed that slow growth rates would lead to late reproduction and, looking at data from 34 strepsirhine and haplorhine primate species, concluded that the brain growth constraint model best explains the observed patterns. However, Ross and Jones' analyses do not specify where in the life cycle slow growth occurs, i.e., during infancy (between birth and weaning) or during juvenility (between weaning and reproductive age). Here I examine infant and juvenile growth rates in anthropoid primates to explore further why reproduction age and brain weight appear to be linked and to test the predictions of the three models (analyses are restricted to anthropoid primates as patterns of growth and relative brain size are very different in strepsirhines and tarsiers).

Table 8.6a shows results from a multiple regression of brain weight against body weight and mean infant growth rate and mean juvenile growth rate. This suggests that juvenile growth rate is negatively correlated with brain weight. This is as predicted by the brain growth constraint model and does not contradict the needing to learn model, but does not support the juvenile risk model. In order to test whether the predictions of any the three models were met, I carried out further multiple regression analyses of brain weight against body weight and juvenile growth rate with the environmental measures of group size and diet included.

The results of these multiple regressions (Table 8.6b,c) indicate that:

(1) There is no direct link between any measure of environmental complexity and brain weight when juvenile growth rate is held constant. This is not consistent with the brain growth constraint or needing to learn models, both of which predict a direct positive link between environmental complexity and brain weight even if growth rate is held constant (Figure 8.4). The negative link between percentage folivory and brain weight predicted by the juvenile risk model is also not found here.
(2) Body weight and juvenile growth rate predict brain weight (although juvenile growth weight is only significant at $p = 0.08$), even if environmental factors are held constant. Together, these results indicate that, when other variables are taken into account, brain weight is negatively correlated with juvenile growth rate, i.e., large adult brain weight is related to slow juvenile growth rates. This supports the predictions of the brain growth constraint model but not those of the other models.
(3) The multiple regression of juvenile growth rate against brain weight and environmental variables indicates that, after controlling for brain weight, there is no link between the measures of environmental complexity and growth rates. This is not consistent with the juvenile risk or needing to learn models, both of which predict a positive link between environmental complexity and growth rate even if brain weight is held constant (Figure 8.4).

These new results give tentative support to the brain growth constraint model but the link between juvenile growth rates and brain weight are unexpected. The majority of brain growth in primates occurs prenatally and early in life: in the juvenile phase of growth

Table 8.6. *The relationship between brain size, growth rates and environmental complexity. Analyses were done using contracts generated from the CAIC program (see pages 128–9) using the oldest 50% of contrasts only. Multiple regression through the origin (i.e., with no slope intercept) is used throughout*

(a) Brain size (female) (Brn) versus female body weight (W), infant growth from birth to weaning age (IGR) and juvenile growth rate from weaning to maturity (JGR)

df	r	p	x	p (direction of relationship)
27	0.91	0.0001	W	0.0001 (+)
			IGR	0.757
			JGR	0.009 (−)

(b) Brain weight (female) (Brn) versus female body weight (W), juvenile growth rate (JGR), group size, and diet (% folivory and % frugivory)

df	r	p	x	p (direction of relationship)
10	0.99	0.0003	W	0.001 (+)
			JGR	0.075 (−)
			Group size	0.504
			% folivory	0.393
			% frugivory	0.350

(c) Juvenile growth rate (JGR) versus female body weight (W), Brain weight (female) (Brn), group size, and diet (% folivory and % frugivory)

df	r	p	x	p (direction of relationship)
10	0.99	0.0023	W	0.009 (+)
			Brn	0.075 (−)
			Group size	0.404
			% folivory	0.809
			% frugivory	0.28

brain growth slows and the brain is relatively smaller (Martin, 1983). The brain growth constraint model would predict that, as large adult brain weight is linked to relatively large neonatal brain weight and rapid early brain growth, it should be infant growth rates that are primarily affected by relatively large brain size.

The link between juvenile growth rates and adult brain weight may still result from constraints imposed by relatively large brains, if these are particularly hard to overcome during juvenile life. The juvenile risk model suggests that independent juvenile primates may face particularly difficult challenges that are not faced by dependent infants. If input from mothers and other caregivers mitigates the costs of large brain weight in infants, these costs may start to constrain somatic growth only at independence. Although the analyses shown above do not show direct support for the juvenile risk model, this could owe to the crude measure of dietary challenge used, percentage folivory. Testing this model further requires more sophisticated measures of diet and experimental studies on brain growth in juvenile primates under a range of dietary regimes.

THE EVOLUTION OF BRAIN SIZE AND LIFE HISTORY IN THE GREAT APES

Although brain weight and postnatal growth rates appear to be closely linked in the anthropoid primates generally, the above analyses do not investigate patterns within smaller groups. Previous studies have suggested that relative brain weight within the great apes cannot be considered directly adaptive (Pilbeam & Gould 1974; Shea 1983) but instead may result from evolution acting on body size. Shea (1983) concluded that body size variation in African great apes is related to changes in late postnatal growth after brain growth is completed. *Gorilla* has an unusual life history for a mammal of its body size, with a small neonate (Leutenegger 1973) and relatively small brains (Jerison 1973; Shea 1983). *Gorilla*'s large body size thus results from an extended period of juvenile postnatal growth, where body size but not brain size increases, leading to a relatively small adult brain weight. If so, we might expect gorillas not to have a relatively early age at first reproduction (or rapid early postnatal growth rates) despite their relatively small brain weights.

Investigating relationships between brain weight and juvenile growth rates within the great apes is difficult because data are available for only three species. Data on age at first reproduction are available for more species, so its relationships with other parameters can be examined. Brain weight and age at first reproduction are positively correlated in primates as a whole and within the two primate suborders, but these links are less clear within nonhuman apes (greater and lesser). Brain weight and age at first reproduction in apes are both significantly positively correlated with body weight (Figures 8.1 and 8.2). However, a multiple regression of brain weight against both body weight and age at first reproduction in apes does not show a significant relationship between brain weight and age at first reproduction, after body weight effects are removed ($p > 0.05$ for both species data and contrasts, Figure 8.3). Given the very small sample size ($n =$ six species, five contrasts), it is difficult to interpret this finding and to be confident that these characteristics are unrelated in the apes.

The small sample size makes it impossible to carry out statistical analyses treating the great apes as four species, but differences among great apes can be seen. *Gorilla* has approximately the age of first reproduction predicted for a haplorhine primate of its size, whereas orangutans and chimpanzees begin to reproduce relatively late (Figure 8.2). This suggests that the smaller relative brain weight of *Gorilla* is not due to constraints on growth produced by rapid maturation and that the extreme body size of gorillas is the main cause of their

Table 8.7. *Relative values for life history parameters of the great apes and* Hylobates *(data for some parameters were available for more than one species of* Hylobates *but all had the relationship to body weight shown here)*

Species	Brn	AR	Juv	L	Wn	g
Hylobates spp.	E	+	+	E	+	+
Gorilla gorilla	E	E	E	E	E	+
Pan paniscus	nd	+	+	nd	+	+
Pan troglodytes	E	+	+	+	+	+
Pongo pygmaeus	E	+	+	+	+	+
Homo sapiens	+	+	+	+	+	+

Relative values are calculated form the regression line generated by the CAIC program (slope values used are: Brain mass (Brn) – 0.64, age at female first reproduction (AR) – 0.25, length of juvenile period (Juv) – 0.24, longevity (L) – 0.22, weaning age (Wn) – 0.36, gestation length (g) – 0.09). E = expected value for body mass, + = relatively high for body mass, nd = no data available).

relatively small brains. Extending the postnatal growth period to reach a large body size leads to the expected age of first reproduction. Conversely, the late age at first reproduction of orangutans and chimpanzees may well be adaptive because it does not appear to be linked to the constraints of large brain size: both *Pan* and *Pongo* have about the expected brain weights for an anthropoid of their body weight.

Data shown in Table 8.7 allow further interpretation of ape life-history evolution. The relative values for a range of life-history parameters show that although absolute values of many gorilla life history variables are very similar to those of chimpanzees and orangutans, relative values often differ. This suggests that the common great ape–human ancestor would have had very similar absolute values to those found in modern great apes and that the relative values of gorilla life histories have been produced by the evolution of large body size without concomitant changes in life-history parameters. This table also indicates that apes and humans share a relatively long gestation period, suggesting that this characteristic is ancestral to the clade. Relatively large brain weight is seen only in humans but none of the other life-history characteristics investigated are unique to humans: an extended juvenile period occurs in humans,

Hylobates, *Pan*, and *Pongo*, and relatively long lifespan occurs in humans, *Pan*, and *Pongo*. This suggests an extended juvenile period arose early in ape evolution and was retained in *Hylobates* after this group evolved a smaller body size. These data also suggest that the relatively large brain size of *Homo* evolved after the evolution of the extended juvenile period and long lifespan. It is possible that the evolution of an extended juvenile period acted as a preadaptation to both the very large body size of *Gorilla* and the large brain size of *Homo*. Once the extended juvenile period had evolved, "spare time" was available for further growth to occur.

These comparisons indicate that, within the apes, relative brain weight may not be as important as other aspects of learning and intelligence when investigating the relationship between intelligence and life history parameters. Great apes' relative brain weight is not greater than that of other haplorhine primates, despite their capacity for more complex intelligent task solving. Similarly, the differences between species that can solve complex social and physical problems (humans and great apes) and those that cannot (monkeys and gibbons) cannot be explained by relative brain size.

CONCLUSIONS

It is clear from these analyses that relatively large brain weight in anthropoid primates is linked to slow juvenile growth rate and late age at first reproduction. It also appears that having a large brain may constrain some primate species to having a slow growth rate during their juvenile period. These results support those of Ross and Jones (1999), which also suggest that brain size is constrained by the energetics of primate growth. If this is the case, the rate of growth of a primate's brain may be limited by the energetic costs of growing to adult size. The results presented here suggest that this limitation is primarily imposed during juvenile growth, as neither fetal nor infant growth rate are significant predictors of adult brain size. At first sight this result is somewhat surprising as it is generally thought that the critical periods of primate brain growth occur before or soon after birth in primates (Martin 1983, 1996). Although some subsequent work has questioned the validity of the link between maternal metabolic rate and brain size (Barton 1999), this does not refute the idea that postnatal brain growth may be influenced by energetic constraints. The results presented here do not indicate that extended juvenile periods always have a strong adaptive purpose, but that they may also arise as a consequence of the energetic constraints of having large brains.

Within the great apes, however, an extended juvenile period has evolved without a correlated increase in relative brain size, so simple allometric scaling may account for brain expansion. Similarly, great apes' extended ontogenies are usually but not always longer than expected for primates of their body size. When the great apes are examined as a separate group it is difficult to see any clear link between life-history evolution and the evolution of relative brain weight. This suggests that it may have been the evolution of extended juvenile periods that was adaptive in the great ape lineage, and that this may have allowed the evolution of both the large body size of *Gorilla* and the relatively large brain size of *Homo*.

It appears that a study of comparative life history may offer insights into the evolution of brain size in the primates overall, but does not offer a good understanding of why apes evolved more human-like intellect than monkeys. This, together with other comparative studies, indicates that for the great apes, relative brain size may not be everything (see also MacLeod, Chapter 7, this volume). The absolute size of the brain, the organization of brain structures and/or the expansion of some parts of the brain (e.g., the neocortex) may be more important than a simple increase in relative brain size.

ACKNOWLEDGMENTS

I must thank the following people for help with this paper. Anne Russon and David Begun both offered very helpful comments on a first draft of this chapter and have also been incredibly patient with me whilst I took far too long in revising it. The chapter was also improved by comments from Kathleen Gibson and two anonymous reviewers. Thanks also to Colin Groves, who pointed me in the right direction when I was looking for data on great ape brain sizes. This work has built on previous work I have done with Kate Jones and Ann MacLarnon – I thank both of them for all their help.

REFERENCES

Aiello, L. C. & Wheeler, P. (1995). The expensive tissue hypothesis: the brain and digestive system in humans and primate evolution. *Current Anthropology*, **36**, 199–221.

Allman, J. M., McLaughlin T. & Hakeem A. (1993). Brain structures and life-span in primate species. *Proceedings of The National Academy of Sciences of the United States of America*, **90**, 3559–63.

Angst, R. (1976). Das Endocranialvolumen der Pongiden (Mammalia: Primates). *Beitraege Zur Naturkundlichen Forschung Suedwestdeutschland*, **35**, 181–8.

Badrian, N. & Malenky, R. K. (1984). Feeding ecology of *Pan paniscus* in the Lomako Forest, Zaire. In *The Pygmy Chimpanzee. Evolutionary Biology and Behavior*. ed. R. L. Susman, pp. 275–99. New York: Plenum Press.

Barton, R. A. (1999). The evolutionary ecology of the primate brain. In *Comparative Primate Socioecology*, ed. P. C. Lee, pp. 167–203. Cambridge, UK: Cambridge University Press.

Barton, R. A. & Dunbar, R. I. M. (1997). Evolution of the social brain. In *Machiavellian Intelligence II: Extensions and Evaluations*, ed. A. Whiten & R. W. Byrne, pp. 240–63. Cambridge, UK: Cambridge University Press.

Barton, R. A., Purvis, A. & Harvey, P. H. (1995). Evolutionary radiation of visual and olfactory brain systems in primates, bats and insectivores. *Philosophical Transactions of the Royal Society (Biological Sciences)*, **348**, 381–92.

Bauchot, R. & Stephan H. (1969). Encephalisation et niveau évolutif chez les simiens. *Mammalia*, **3**, 225–75.

Beran, M. J., Gibson, K. & Rumbaugh, D. (1999). Predicting hominid intelligence from brain size. In *The Descent of Mind: Psychological perspectives on Hominid Evolution*, ed. M. C. Corballis & S. E. G. Lea, pp. 88–97. Oxford: Oxford University Press.

Boesch, C. & Boesch-Achermann, H. (2000). *The Chimpanzees Of The Taï Forest: Behavioural Ecology and Evolution*. New York: Oxford University Press.

Byrne, R. W. (1995). *The Thinking Ape*. Oxford: Oxford University Press.

(1999a). Primate cognition: evidence for the ethical treatment of primates. In *Attitudes to Animals: Views in Animal Welfare*, ed. F. L. Dolins, pp. 114–25. Cambridge, UK: Cambridge University Press.

(1999b). Human cognitive evolution. In *The Descent of Mind: Psychological Perspectives on Hominid Evolution*, ed. M. C. Corballis & S. E. G. Lea, pp. 71–88. Oxford: Oxford University Press.

Byrne, R. W. & Byrne, J. M. E. (1993). Complex leaf-gathering skills of mountain gorillas (*Gorilla g. beringei*) variability and standardization. *American Journal of Primatology*, **31**, 241–61.

Charnov, E. L. & Berrigan, D. (1993). Why do female primates have such long lifespans *and* so few babies? Or life in the slow lane. *Evolutionary Anthropology*, **1**, 191–4.

Clutton Brock, T. H. & Harvey, P. H. (1980). Primates, brains and ecology. *Journal of Zoology, London*, **207**, 151–69.

Dunbar, R. I. M. (1991). Functional significance of social grooming in primates. *Folia Primatologica*, **57**, 121–31.

(1992). Neocortex size as a constraint on group size in primates. *Journal of Human Evolution*, **20**, 469–93.

(1995). Neocortex size and group size in primates: a test of the hypothesis. *Journal of Human Evolution*, **28**, 287–96.

Economos, A. C. (1980). Brain-life conjecture: a re-evaluation of the evidence. *Gerontology*, **26**, 82–9.

Felsenstein, J. (1985). Phylogenies and the comparative method. *American Naturalist*, **125**, 1–15

Fossey, D. & Harcourt, A. H. (1977). Feeding ecology of free-ranging mountain gorilla (*Gorilla gorilla beringei*). In *Primate Ecology: Studies of Feeding and Ranging Behaviour in Lemurs, Monkeys and Apes*, ed. T. H. Clutton-Brock, pp. 415–47. New York: Academic Press.

Furuichi, T., Idani, G., Ihobe, H., Kuroda, S., Kitamura, K., Mori, A., Enomoto, T., Okayasu, N., Hashimoto, C. & Kano, T. (1998). Population dynamics of wild bonobos (*Pan paniscus*) at Wamba. *International Journal of Primatology*, **19**, 1029–43.

Galdikas B. M. F. (1981). Orangutan reproduction in the wild. In *Reproductive Biology of The Great Apes: Comparative and Biomedical Perspectives*, ed. C. E. Graham, pp. 281–300. New York: Academic Press.

(1988). Orangutan diet, range, and activity at Tanjung Puting, Central Borneo. *International Journal of Primatology*, **9**, 1–35.

Galdikas, B. M. F. & Wood, J. W. (1990). Birth spacing patterns in humans and apes. *American Journal of Physical Anthropology*, **83**, 185–91.

Ghiglieri, M. P. (1984). *The Chimpanzees of Kibale Forest: A Field Study of Ecology and Social Structure*. New York: Columbia University Press.

Gibson, K. R. (1990). New perspective on instincts and intelligence: brain size and the emergence of hierarchical constructional skills. In *"Language" and Intelligence in Monkeys and Apes*, ed. S. T. Parker & K. R. Gibson, pp. 97–128. New York: Cambridge University Press.

Goodall, J. (1986). *The Chimpanzees of Gombe: Patterns of Behavior*. Cambridge, MA: Harvard University Press.

Harcourt, A. H., Stewart, K. J. & Fossey, D. (1981). Gorilla reproduction in the wild. In *Reproductive Biology of The Great Apes: Comparative and Biomedical Perspectives*, ed. C. E. Graham, pp. 265–79. New York: Academic Press.

Harvey, P. H. & Krebs, J. R. (1990). Comparing brains. *Science*, **249**, 140–6.

Harvey, P. H. & Pagel, M. D. (1991). *The Comparative Method in Evolutionary Biology*. Oxford: Oxford University Press.

Harvey, P. H., Martin, R. D. & Clutton-Brock, T. H. (1987). Life histories in comparative perspective. In *Primate Societies*, ed. B. B. Smuts, D. L. Cheney, R. M. Seyfarth, R. W. Wrangham & T. T. Struhsaker, pp. 181–96. Chicago, IL: University of Chicago Press.

Hladik, C. M. (1977). Chimpanzees of Gabon and chimpanzees of Gombe: some comparative data on the diet. In *Primate Ecology: Studies of Feeding and Ranging Behaviour in Lemurs, Monkeys and Apes*, ed. T. H. Clutton-Brock, pp. 481–501. New York: Academic Press.

Holloway, R. (1996). Evolution of the human brain. In *Handbook of Symbolic Evolution*, ed. A. Lock & C. R. Peters, pp. 74–115. Oxford: Oxford University Press.

Humphrey, N. K. (1976). The social function of the intellect. In *Growing Points in Ethology*, ed. P. P. G. Bateson & R. A. Hinde, pp. 303–17. Cambridge, UK: Cambridge University Press.

Janson, C. H. & van Schaik, C. P. (1993). Ecological risk aversion in juvenile primates: slow and steady wins the race. In *Juvenile Primates: Life History, Development and Behaviour*, ed. M. D. Pereira & L. A. Fairbanks, pp. 57–74. Oxford: Oxford University Press.

Jantschke, F. (1975). The maintenance and breeding of pygmy chimpanzees. In *Breeding Endangered Species in Captivity*, ed. R. D. Martin, pp. 245–51. London: Academic Press.

Jerison, H. J. (1973). *Evolution of Brain and Intelligence*. New York: Academic Press.

Joffe, T. H. (1997). Social pressures have selected for an extended juvenile period in primates. *Journal of Human Evolution*, **32**, 593–605.

Jolly, A. (1966). Lemur social behavior and primate intelligence. *Science*, **153**, 501–6.

Jones M. L. (1982). The orang utan in captivity. In *The Orang Utan: Its Biology and Conservation*. ed. L. E. M. de Boer, pp. 17–37. The Hague: Dr. W. Junk Publishers.

Judge, D. S. & Carey, J. R. (2000). Postreproductive life predicted by primate patterns. *Journals of Gerontology Series A*, **55**, 201–9.

Kano T. (1992). *The Last Ape: Pygmy Chimpanzee Behavior and Ecology*. Stanford: Stanford University Press.

Key, C. & Ross, C. (1999). Sex differences in energy expenditure in non-human primates. *Proceedings of The Royal Society of London B*, **266**, 2479–85.

Leutenegger, W. (1973). Maternal–fetal weight relationships in primates. *Folia Primatologica*, **20**, 280–93.

MacLarnon, A. M. (1999). The comparative method: principles and illustrations from primate socioecology. In *Comparative Primate Socioecology*, ed. P. C. Lee, pp. 5–24. Cambridge, UK: Cambridge University Press.

Martin, R. D. (1983). *Human Brain Size in an Ecological Context*. (52nd James Arther Lecture on the evolution of the brain). New York: American Museum of Natural History.

(1990). *Primate Origins and Evolution: A Phylogenetic Reconstruction*. Cambridge, UK: Cambridge University Press.

(1996). Scaling of the mammalian brain: the maternal energy hypothesis. *News in Physiological Science*, **11**, 149–56.

Milton, K. (1981). Distribution patterns of tropical plant foods as a stimulus to primate mental development. *American Anthropologist*, **83**, 534–48.

(1988). Foraging behaviour and the evolution of primate intelligence. In *Machiavellian Intelligence*, ed. A. Whiten & R. W. Byrne, pp. 285–306. Oxford: Clarendon Press.

Neugebauer, W. (1981). The rearing of the pygmy chimpanzee (*Pan paniscus*). *International Zoo Yearbook*, **21**, 64–8.

Nishida, T. & Hiraiwa-Hasegawa, M. (1987). Chimpanzees and bonobos: cooperative relationships among males. In *Primate Societies*, ed. B. B. Smuts, D. L. Cheney, R. M. Seyfarth, R. W. Wrangham & T. T. Struhsaker, pp. 165–77. Chicago, IL: University of Chicago Press.

Nishida, T., Takasaki, H. & Takahata, Y. (1990). Demography and reproductive profiles. In *The Chimpanzees of The Mahale Mountains. Sexual and Life History Strategies*. ed. T. Nishida, pp. 63–97. Tokyo: University of Tokyo.

Nissen H. W. & Yerkes R. M. (1943). Reproduction in the chimpanzee: Report on forty-nine births. *Anatomical Record*, **86**, 567–78.

Parker, S. T. & Gibson, K. R. (1977). Object manipulation, tool use and sensorimotor intelligence as feeding adaptations in cebus monkeys and great apes. *Journal of Human Evolution*, **6**, 623–41.

(1979). A developmental model for the evolution of language and intelligence in early hominids. *Behavioral and Brain Sciences*, **2**, 367–408.

Peters, R. H. (1983). *The Ecological Implications of Body Size.* Cambridge, UK: Cambridge University Press.

Pilbeam, D. & Gould, S. J. (1974). Size and scaling in human evolution. *Science*, **186**, 892–901.

Povinelli, D. J. & Cant, J. G. H. (1995). Orangutan climbing and the evolutionary origins of self-conception. *Quarterly Review of Biology*, **40**, 393–421.

Purvis, A. (1995). A composite estimate of primate phylogeny. *Philosophical Transactions of the Royal Society of London B*, **348**, 405–21.

Purvis, A. & Rambaut, A. (1995). Comparative analysis by independent contrasts (CAIC): an Apple Macintosh application for analysing comparative data. *Computer Apple Biosciences*, **11**, 247–51.

Purvis, A. & Webster, A. J. (1999). Phylogenetically independent comparisons and primate phylogeny. In *Comparative Primate Socioecology*, ed. P. C. Lee, pp. 44–70. Cambridge, UK: Cambridge University Press.

Portmann, A. (1962). Cerebralisation und Ontogenese. *Medizin. Grundlagenforsch*, **4**, 1–62.

Rodman, P. S. (1977). Feeding behaviour of orangutans of the Kutai Nature Reserve, East Kalimantan. In *Primate Ecology: Studies of Feeding and Ranging Behaviour in Lemurs, Monkeys and Apes*, ed. T. H. Clutton-Brock, pp. 383–413. New York: Academic Press.

Röhrer-Ertl, O. (1988). Cranial growth. In *Orang-Utan Biology*, ed. J. H. Schwartz, pp. 201–23. New York: Oxford University Press.

Ross, C. (1988). The intrinsic rate of natural increase and reproductive effort in primates. *Journal of the Zoological Society of London*, **214**, 199–219.

(1998). Primate life histories. *Evolutionary Anthropology*, **6**, 54–63.

Ross, C. & Jones, K. E. (1999). The evolution of primate reproductive rates. In *Comparative Primate Socioecology*, ed. P. C. Lee, pp. 73–110. Cambridge, UK: Cambridge University Press.

Rudder, B. C. C. (1979). The Allometry of Primate Reproductive Patterns. Ph.D. thesis, University of London.

Russon, A. E. (1998). The nature and evolution of intelligence in orangutans (*Pongo pygmaeus*). *Primates*, **39**, 485–503.

Russon, A. E., Bard, K. A. & Parker, S. T. (ed.) (1996). *Reaching into Thought: The Minds of the Great Apes.* Cambridge, UK: Cambridge University Press.

Sacher, G. A. (1959). Relationship of lifespan to brain weight and body weight in mammals. In *CIBA Foundation Symposium on the Lifespan of Animals*, ed. G. E. W. Wolstenholme & M. O'Connor, pp. 115–33. Boston, MA: Little Brown.

(1975). Maturation and longevity in relation to cranial capacity in hominid evolution. In *Primate Functional Morphology and Evolution*, ed. R. H. Tuttle, pp. 417–41. The Hague: Mouton and Co.

Shea, B. T. (1983). Phyletic size change and brain/body allometry: A consideration based on the African pongids and other primates. *International Journal of Primatology*, **4**, 33–61.

Smith, R. J. & Jungers, W. L. (1997). Body mass in comparative primatology. *Journal of Human Evolution*, **32**, 523–59.

Smuts, B. B., Cheney, D. L., Seyfarth, R. M., Wrangham, R. W. & Struhsaker, T. T. (ed.) (1987). *Primate Societies.* Chicago, IL: University of Chicago Press.

Tutin, C. E. G. & Fernandez, M. (1993). Composition of the diet of chimpanzees and comparisons with that of sympatric lowland gorillas in the Lope Reserve, Gabon. *American Journal of Primatology*, **30**, 195–211.

Western, D. (1979). Size, life-history and ecology in mammals. *African Journal of Ecology*, **17**, 185–204.

Yamagiwa, J., Maruhashi, T., Yumoto, T. & Mwanza, N. (1996). Dietary and ranging overlap in sympatric gorillas and chimpanzees in Kahuzi-Biega National Park, Zaire. In *Great Ape Societies*, ed. W. C. McGrew, L. F. Marchant, T. Nishida, pp. 82–98. Cambridge, UK: Cambridge University Press.

9 · Evolution of complex feeding techniques in primates: is this the origin of great ape intelligence?

GEN YAMAKOSHI

Graduate School of Asian and African Studies, Kyoto University, Kyoto

DOES GREAT APE INTELLIGENCE DIFFER FROM THAT OF MONKEYS?

There is growing consensus that great apes' intellectual abilities are qualitatively distinct from those of other primate taxa, as seen in their mirror self-recognition (e.g., Gallup, 1970) causal understanding of tool-using tasks without trial and error (Visalberghi, Fragaszy & Savage-Rumbaugh 1995), and imitative ability (e.g., Custance, Whiten & Bard 1995), among other traits and abilities (see Russon, Bard & Parker 1996; other chapters this volume). This raises the important question: In what ecological and social environments did this distinct intellectual capacity evolve?

Potential answers have been much discussed in recent years. Using brain parameters (e.g., absolute or relative brain size, neocortex ratio) as proxies for the rather amorphous concept of "intelligence," comparative studies (Dunbar 1992, 1995) have found that the size of the social network (represented by group size) better explains variation in the neocortex ratio among primate taxa than any of the ecological parameters considered thus far, such as degree of frugivory, range size, or presence/absence of "extractive foraging." This suggests that the social complexity resulting from primate-style group living is more likely to be behind variations in primate intelligence, as the so-called "social intellect hypothesis" sets out (Chance & Mead 1953; Humphrey 1976), than the ecological complexity arising from foraging problems, as some others have suggested (Menzel 1997; Milton 1981; Parker & Gibson 1979). However, these studies do not provide satisfactory explanations for the difference between great apes and other nonhuman primates, because they were aimed at discovering general tendencies across the primate order and did not focus on this specific difference.

Byrne (1997) pointed out that there are no systematic differences in group size or neocortex ratio between great apes and haplorhine monkeys, so differences in the intellectually governed behavior of great apes and monkeys cannot be straightforwardly explained by the social intellect hypothesis. He proposed an alternative "technical intelligence hypothesis" to explain the specific, supposedly qualitative change that occurred in the common ancestor of all great apes and humans. This common ancestor must have faced some sorts of ecological pressures that required more complex and efficient technical skills, foraging pressures among them, which probably then became organized hierarchically. The need for such complex behavioral structures must have been an important factor leading to the appearance of abstract cognitive abilities such as planning and mental representation.

Since "hierarchically organized" feeding techniques have been poorly described among wild primates in general (but see Byrne, Chapter 3, this volume; Byrne & Byrne 1991; Matsuzawa 1996; Russon 1998; Stokes & Byrne 2001; Yamakoshi & Sugiyama 1995), it remains difficult to test the "technical intelligence hypothesis." What does seem worth testing is the idea that foraging complexity, broadly interpreted, was behind the hypothesized cognitive leap between monkeys and great apes. My aim in this chapter is to review complex feeding techniques presented in the literature for possible selective pressures that may have differentiated great ape intelligence from that of other nonhuman primates. I focus on identifying patterns of food processing techniques, unique to great apes, that might have been critical to the evolution of great-ape-type intelligence. For this reason, I reviewed the order Primata: to isolate traits that are unique to a clade, we must ensure that its outgroup does not possess these same traits (e.g., Begun 1999).

COMPLEX FEEDING TECHNIQUES IN PRIMATES

Although there have been experimental studies of object manipulation across many primate species in laboratory settings (Glickman & Sroges 1966; Jolly 1964; Parker 1974a,b; Torigoe 1985) and some comparative theoretical models with an evolutionary perspective (Alcock 1972; Beck 1980; Parker & Gibson 1977, 1979), very little attention has been paid to comparing the complex feeding techniques of wild primates.

The exception is tool use in wild primates, which has been extensively reported and intensively reviewed (Beck 1980; Candland 1987; Goodall 1970; Hall 1963; Kortlandt & Kooij 1963; Tomasello & Call 1997; van Schaik, Deaner & Merrill 1999; and see Byrne, Chapter 3, this volume). These reports illustrate that (1) although there have been some anecdotal reports of tool use in a wide range of primate taxa, the majority of cases consist of unhabituated primates throwing objects toward human observers; (2) the more reliable examples, such as tool use in feeding contexts, were rarely documented in taxa other than capuchins, baboons, macaques, or great apes; and (3) if we focus on the habitual or customary use of tools (*sensu* McGrew & Marchant 1997, i.e., use by many individuals, regularly or predictably), only chimpanzees (McGrew 1992; Yamakoshi 2001) and Sumatran orangutans (van Schaik & Knott 2001) meet the standard. This suggests that there are large differences in tool-use frequency between at least two great ape species and other monkeys and lemurs.

However, simply showing that the outgroup taxa do not use tools as habitually or customarily as do the great apes provides little material for reconstructing cognitive evolution among primates. Parker and Gibson (1977, 1979) suggested narrowing the focus to "intelligent" tool use (i.e., flexible use of detached objects to alter the state of target objects, with understanding of the causal dynamics involved in the tool's relation with the target object) as a way of isolating features relevant to evolutionary reconstructions. An alternative approach with broader comparative scope, which shows more about the feeding techniques of non-great-ape primate taxa, is first to expand our focus and consider other complex foraging skills that could constitute a prototype of, or an alternative to, tool use (see also van Schaik *et al.* 1999).

In this regard, at least four other types of complex feeding techniques merit consideration. All contrast with tool use, in which a detached "agent-of-change" is manipulated to alter the state of an "object-of-change" (Parker & Gibson 1977). The first type includes behavior that has been described as "proto tool use" (Parker & Gibson 1977), "object use" (Panger 1998), or "substrate use" (Boinski, Quatrone & Swartz 2000). In such manipulations, an individual manipulates a detached "object-of-change" directly against a fixed substrate. The best known example may be capuchin monkeys banging palm nuts against tree trunks (Izawa & Mizuno 1977; Thorington 1967). This type of technique, henceforth "substrate use," is relatively well documented in the literature but has not been extensively investigated from a comparative perspective.

The second type is manual food processing that involves "bimanual asymmetric coordination" (*sensu* van Schaik *et al.* 1999; Byrne, Chapter 3, this volume, Byrne & Byrne 1993). It involves complementary performance of two distinct motor actions, such as holding a twig with fruit in one hand while picking the fruit with the other. This is much discussed in the context of hand preference (e.g., McNeilage, Studert-Kennedy & Lindblom 1987). It is apparently more complex than simple reaching for or picking up food, but its relation to cognition is unclear. It could be a product of relational cognition, i.e., cognition whose structures govern understanding and manipulating relations between several entities (e.g., object–object relations, coordinating multiple actions) rather than individual items or actions (Byrne & Russon 1998; Case 1985; Russon 1998). Unfortunately, few studies have focused on this particular behavior in the wild, making identification of a general trend across primates difficult. Only 15 cases among all primates were identified in van Schaik *et al.*'s (1999) extensive review.

The third type is hierarchically organized food processing, an idea that has been recently proposed and demonstrated from observations of mountain gorillas eating herbaceous pith (Byrne, Chapter 3, this volume; Byrne & Byrne 1991; Byrne & Russon 1998). To consume the noxious, prickly pith, the gorillas must implement an organized sequence of inter-coordinated steps. The process does not involve tool or additional object use at all, but the entire sequence is hierarchically organized and complex enough to suggest high-level, hierarchical cognitive processes, such as insightful comprehension of the "program-level" structure of the technique. This hierarchical cognition may be the same as that which governs bimanual asymmetric coordination: relational cognition is a rudimentary form of hierarchical cognition (Byrne & Russon 1998; Russon 1998, Chapter 6, this volume). This type of behavior has not yet been demonstrated, in detail, in primate species other than gorillas (Byrne & Byrne 1993), chimpanzees (Yamakoshi & Sugiyama 1995), and orangutans (Russon 1998). A recent study of vervet monkeys (Harrison & Byrne 2000), however, suggests the absence of such behavior so this may shed light on the great ape–monkey difference.

Given the paucity of available data on bimanual asymmetric coordination and on hierarchically or even sequentially organized food processing, I do not deal with these types in this chapter. I concentrate instead on surveying evidence of tool use proper and substrate use, which is relatively readily available in the existing literature, to illuminate their distribution across primates and investigate the basic operation patterns involved. In addition, I investigate the ecological parameters that form the contexts of both tool and substrate use. Information on the category and morphology of target foods as well as on the types and materials of tools or substrates actually used would help in reconstructing the evolutionary ecology of the common ancestors of extant great apes, ancestors that are little known and presently controversial (see Yamakoshi 2001, concerning the chimpanzee–human ancestor; see also Begun, Chapter 2, Begun & Kordos, Chapter 14, Gebo, Chapter 17, Kelley, Chapter 15, Singleton, Chapter 16, this volume).

APPROACH

I undertook an extensive review of published primate studies for observations on *tool use* and *substrate use* for food processing using the following principles:

(1) I collected observations of both types of manipulation, in feeding contexts only. This was partly because I was particularly interested in testing the technical intelligence hypothesis (Byrne 1997), which focuses more on feeding techniques than on object manipulation itself, and partly because this limitation ruled out many ambiguous cases of object manipulation that occurred in other contexts, such as play (e.g., Sabater Pi *et al.* 1993; Starin 1990).

(2) I limited data to "native" primate groups, excluding captive or reintroduced primates to avoid possible contamination by human-induced behaviors (e.g., Fitch-Snyder & Carter 1993; Hannah & McGrew 1987). I included studies on free-ranging, provisioned groups but excluded behaviors used to obtain provisioned foods (e.g., Suzuki 1965; Wheatley 1999: 57–8) because some researchers claim that biased provisioning behaviors by human caretakers could have directly influenced the behavior of the target primates (e.g., Green 1975). Consequently, the famous "sweet potato washing" behavior of Japanese macaques (Kawamura 1954, 1959) was excluded. Observations on feeding behaviors involving human crops or other human-oriented foods were included if these behaviors occurred spontaneously during normal foraging activity in free-ranging situations (e.g., Wheatley 1988).

(3) I counted only direct observations that provided details on the behavior itself. I omitted cases based on indirect evidence, such as inference from tool collection (e.g., Hashimoto, Furuichi & Tashiro 2000; Stanford *et al.* 2000), confirmation by sounds (e.g., Langguth & Alonso 1997; Whitesides 1985), and incomplete observations that did not confirm the end result (eating a target food) (e.g., Chevalier-Skolnikoff 1990), amongst others.

(4) I did not include secondhand information. For instance, Hill's description (1960: 427) of capuchins cracking open an oyster with a stone has occasionally been cited, but Hill only relayed Buffon's quotation of Dampier (1697). Moreover, Dampier's *A New Voyage round the World* does not contain such a description; he actually wrote, ". . . the monkeys come down by the sea-side and catch them; digging them out of their shells with their claws" (Dampier 1927: 123).

(5) I excluded behavioral observations in which either the species or study site was impossible to confirm (e.g., Beatty 1951).

When counting cases, I defined a case as one behavioral pattern toward one food category, observed within one particular population, not a species or a genus. I chose population as the basis for counting cases rather than species (1) to avoid confusion caused by species classification, which changes over time, and (2) to take into account that chimpanzee tool-use repertoires vary between populations (McGrew 1992; Whiten *et al.* 1999). When comparing numbers of cases between taxa, however, I also lumped population-based data into species- and genus-based numbers to avoid possible biases for well-studied species with larger numbers of study sites and for behavioral patterns maintained by learning rather than by a "hard-wired" process.

"Food categories" were broadly defined according to their physical structures, as leaves, fruits, nuts, insects, eggs, etc. This is because observations of species other than chimpanzees tend not to be intensive and detailed enough to compare with the very detailed chimpanzee observations. For instance, Taï Forest chimpanzees (Côte d'Ivoire) crack open five different species of nuts with either stone or stick hammers (Boesch & Boesch 1983); here, the five species of nuts were treated as one food category (i.e., nuts).

"Behavioral patterns" were also crudely defined. Chimpanzee researchers normally distinguish "ant-dipping" from "ant-fishing" because of subtle differences in behavioral patterns and tool materials (McGrew 1974; Nishida 1973). Here, I combined these two into one behavior to balance the intensity of observations between chimpanzees and other species. In the case of "tool composites" (i.e., two or more types of tool used sequentially or in association to achieve a single goal; *sensu* Sugiyama 1997), I counted each component of a composite as a single case. This means that if chimpanzees used digging sticks to perforate a termite mound and then fishing probes to fish termites in succession (see Suzuki, Kuroda & Nishihara 1995), I treated them as two independent tool-using cases.

In surveying the literature, I did not use a systematic method (e.g., selecting a limited number of articles from four international primatology journals; see Reader & Laland 2001). Since primate journals now rarely accept simple anecdotal descriptions of feeding techniques, particularly for non-tool behaviors, such a systematic method could result in substantial bias for tool use. Descriptions of non-tool behaviors are also likely to be found in other media, such as newsletters or books. Furthermore, detailed descriptions of food processing are likely to be found in earlier monograph-style studies, which often preceded the establishment of primatology journals.

I employed non-systematic (*ad libitum*) sampling, therefore, with help from (1) already published review articles and books (Beck 1980; Candland 1987; Goodall 1970; Hall 1963; Kortlandt & Kooij 1963; McGrew 1992; Tomasello & Call 1997; van Schaik *et al.* 1999; Williams 1984, 1992; Yamakoshi 2001), and (2) available databases (Biological Abstracts, University of Washington Primate Literature Database, Primate Literature Database of Primate Research Institute Kyoto University), which I searched with key words such as "tool use," "object use," "substrate use," "proto tool use," "object manipulation," "feeding technique/skill," "food processing," "extractive foraging," "manual dexterity," "fine manipulation," and some variations (e.g., plurals). I further expanded the database to include additional works referenced in these articles. Given the non-systematic method of developing this dataset, I limited my analyses primarily to qualitative aspects.

PHYLOGENETIC DISTRIBUTION

In total, 76 cases of tool use, 74 cases of substrate use, and 4 ambiguous cases were collected (Tables 9.1, 9.2). These were reported from only a small number of primate genera. All instances of tool use came from 5/66 extant nonhuman primate genera (42 nonhuman anthropoid genera), and all instances of substrate use from 7 genera (numbers of genera were calculated from Fleagle 1999: 6–7).

Moreover, tool-using and substrate-using genera largely overlapped. All five tool-using genera also showed substrate use, while squirrel monkeys (*Saimiri*) and mangabeys (*Lophocebus*) showed only substrate use (Table 9.2). In other words, tool use, substrate use, or ambiguous cases were not observed in 58/66 nonhuman primate genera, or in 34/42 nonhuman anthropoid genera in the wild. In addition, the phylogenetic distribution was fairly limited. All the reported cases were from the Cebinae, Cercopithecinae, and Pongidae. No case was reported from prosimians, Atelidae, Colobinae, or Hylobatidae. The overall picture of phylogenetic distribution is almost identical to that generated by van Schaik *et al.*'s (1999) review.

Between tool and substrate use, there was a sharp contrast in the observed numbers of cases in each taxa. For substrate use, most cases were reported in capuchins (*Cebus*) (55%, 52%, and 46% of all cases on a population, species, and genus basis, respectively) and for tool use, in chimpanzees (*Pan*) (64%, 52%, and 52% respectively) (Table 9.2, Figure 9.1). These findings represent an amalgam of contributing factors, including the range of tool/substrate use techniques, the number of populations examined, and the intensity of research. They none

Table 9.1. *Reported evidences of tool use and substrate use in primates*

(a) Tool use

Case no.	Species	Behavioral descriptions	Observation sites	Target foods[1]	Protection modes[3]	Tools/ substrates used	Operation patterns	References
1	*Cebus albifrons*	Use leaves to collect and drink water	Bush Bush W.S., Trinidad	Water	Fluid	Leaf	Soak	Phillips (1998)
2	*Cebus apella*	Crack open hard fruits with branch segment	Raleighvallen, Suriname	Fruit	Hard	Branch	Bang	Boinski *et al.* (2000)
3	*Cebus apella*	Pound open oyster with oyster colony	Canelatiua, Brazil	Mollusk	Hard	Oyster colony	Bang	Fernandes (1991)
4	*Cebus capucinus*	Use leaves to obtain water from tree-holes	Santa Rosa, Costa Rica	Water	Fluid	Leaf	Soak	Rose (2001)
5	*Cebus capucinus*	Protect hands while processing hairy caterpillars	Santa Rosa, Costa Rica	Insect	Noxious	Leaf	Rub	Rose (2001)
6	*Cebus capucinus*	Protect hands while processing hairy fruits	Santa Rosa, Costa Rica	Fruit	Noxious	Leaf	Rub	Rose (2001)
7	*Cebus capucinus*	Protect hands while processing hairy caterpillars	Lomas Barbudal, Costa Rica	Insect	Noxious	Leaf	Rub	Panger *et al.* (2002)
8	*Macaca fascicularis*	Break open oysters with a stone	Mergui Archipelago, Myanmar	Mollusk	Hard	Stone/rock	Bang	Carpenter (1887)
9	*Macaca fascicularis*	Rub small fruits (with ants?) on the ground with leaves	Botanical Garden, Singapore	Fruit	Dirty	Leaf	Rub	Chiang (1967)
10	*Macaca fascicularis*	Rub rubber bands on the ground with leaves	Botanical Garden, Singapore	Rubber band	Dirty	Leaf	Rub	Chiang (1967)
11	*Macaca fascicularis*	Wash seeds with leaves or piece of paper	Ubud, Indonesia	Fruit	Dirty	Leaf, paper	Rub	Wheatley (1988)
12	*Macaca fascicularis*	Wash sweet potato rhizome with leaves or piece of paper	Ubud, Indonesia	USO[2]	Dirty	Leaf, paper	Rub	Wheatley (1988)
13	*Macaca fascicularis*	Wash caterpillar or worms with leaves	Ubud, Indonesia	Insect	Noxious	Leaf	Rub	Wheatley (1988)

No.	Species	Description	Location	Object	Property	Tool	Action	Reference
14	*Macaca silenus*	Roll and remove pilose from chrysalis with a leaf	Anaimalai, India	Insect	Noxious	Leaf	Rub	Hohmann (1988)
15	*Papio cynocephalus*	Use stick to scatter or sort small stones or pry pebbles from clay	Nairobi N. P., Kenya	Stone	Hidden	Branch/twig/stalk	Pry	Oyen (1978)
16	*Papio ursinus*	Pound baobab fruits with stones	Magalakwên Valley, South Africa	Fruit	Hard	Stone/rock	Bang	Marais (1969: 56)
17	*Pongo pygmaeus*	Wrap ant nest on twig with leaves	Ketambe, Indonesia	Insect	Mobile	Leaf	Wipe	Rijksen (1978: 89)
18	*Pongo pygmaeus*	Hold a spiny durian fruit into a crevice with a piece of dead wood	Ketambe, Indonesia	Fruit	Unstable	Branch/twig/stalk	Press	Rijksen (1978: 84)
19	*Pongo pygmaeus*	"Hammer" or poke insect nest (hand held)	Suaq Balimbing, Indonesia	Insect	Hidden	Branch/twig/stalk	Dig	van Schaik et al. (1996), Fox et al. (1999)
20	*Pongo pygmaeus*	"Hammer" or poke insect nest (mouth held)	Suaq Balimbing, Indonesia	Insect	Hidden	Branch/twig/stalk	Dig	van Schaik et al. (1996), Fox et al. (1999)
21	*Pongo pygmaeus*	Probe honey ants and termites (hand held)	Suaq Balimbing, Indonesia	Insect	Hidden	Branch/twig/stalk	Probe	van Schaik et al. (1996), Fox et al. (1999)
22	*Pongo pygmaeus*	Probe honey ants and termites (mouth held)	Suaq Balimbing, Indonesia	Insect	Hidden	Branch/twig/stalk	Probe	van Schaik et al. (1996), Fox et al. (1999)
23	*Pongo pygmaeus*	Scrape out *Neesia* seeds (mouth held)	Suaq Balimbing, Indonesia	Fruit	Noxious	Branch/twig/stalk	Scrape	van Schaik et al. (1996)
24	*Pongo pygmaeus*	Scoop out *Neesia* seeds from husks (hand held)	Suaq Balimbing, Indonesia	Fruit	Hard	Branch/twig/stalk	Probe	van Schaik et al. (1996)
25	*Pongo pygmaeus*	Scrape out irritant hairs of *Neesia* seeds (mouth held)	Ie Mdamai, Indonesia	Fruit	Noxious	Branch/twig/stalk	Scrape	van Schaik & Knott (2001)
26	*Pongo pygmaeus*	Scoop out *Neesia* seeds from husks (hand held)	Ie Mdamai, Indonesia	Fruit	Hard	Branch/twig/stalk	Probe	van Schaik & Knott (2001)
27	*Pongo pygmaeus*	Use leaves as scoops for drinking	Gunung Palung, Indonesia	Water	Fluid	Leaf	Soak	van Schaik & Knott (2001)

(cont.)

Table 9.1. (cont.)

Case no.	Species	Behavioral descriptions	Observation sites	Target foods[1]	Protection modes[3]	Tools/substrates used	Operation patterns	References
28	*Pan troglodytes*	Fish/dip ants and termites with a stick	Mahale Mts., Tanzania	Insect	Hidden	Branch/twig/stalk	Probe	Nishida (1973), Nishida & Uehara (1980), Uehara (1982)
29	*Pan troglodytes*	Wipe ants from a tree trunk with a clump of leafy twigs	Mahale Mts., Tanzania	Insect	Mobile	Leafy twig	Wipe	Nishida (1973)
30	*Pan troglodytes*	Dip a stick into a hole to extract honey	Mahale Mts., Tanzania	Insect	Hidden	Branch/twig/stalk	Probe	Nishida & Hiraiwa (1982)
31	*Pan troglodytes*	Insert a stick forcefully into a tree hole to expel squirrel	Mahale Mts., Tanzania	Mammal	Hidden	Branch/twig/stalk	Dig	Huffman & Kalunde (1993)
32	*Pan troglodytes*	Extract water from tree hole or stream with leaves	Mahale Mts., Tanzania	Water	Fluid	Leaf/leafy twig	Soak	Matsusaka & Kutsukake (2002)
33	*Pan troglodytes*	Fish termites with a stick	Kasakati Basin, Tanzania	Insect	Hidden	Branch/twig/stalk	Probe	Suzuki (1966)
34	*Pan troglodytes*	Dip honey from arboreal bee nest with a stick	Kasakati Basin, Tanzania	Insect	Hidden	Branch/twig/stalk	Probe	Izawa & Itani (1966)
35	*Pan troglodytes*	Fish/dip ants and termites with a stick	Gombe, Tanzania	Insect	Hidden	Branch/twig/stalk	Probe	Goodall (1963, 1986: 252–3), McGrew (1974)
36	*Pan troglodytes*	Extract water from a tree hole or a stream with leaf sponges	Gombe, Tanzania	Water	Fluid	Leaf	Soak	Goodall (1964)
37	*Pan troglodytes*	Dig or pry open subterranean bee nests with sticks	Gombe, Tanzania	Insect	Hidden	Branch/twig/stalk	Dig/pry	Goodall (1970)
38	*Pan troglodytes*	Sponge out the inside of skull cavity with a wad of leaves	Gombe, Tanzania	Mammal	Hard	Leaf	Soak	Teleki (1973a)
39	*Pan troglodytes*	Sponge out the inside of *Strychnos* fruits with a wad of leaves	Gombe, Tanzania	Fruit	Hard	Leaf	Soak	Wrangham (1977)

No.	Species	Description	Location	Mammal	Mobile	Stone/rock	Throw	Reference
40	*Pan troglodytes*	Throw a rock toward a bushpig or baboons during hunting	Gombe, Tanzania	Mammal	Mobile	Stone/rock	Throw	Plooij (1978)
41	*Pan troglodytes*	Pry open a tree hole to extract a fledgling	Gombe, Tanzania	Bird	Hidden	Branch/twig/stalk	Pry	Goodall (1986: 540)
42	*Pan troglodytes*	Insert a stick forcefully into a tree cavity to expel ants/termites	Gombe, Tanzania	Insect	Hidden	Branch/twig/stalk	Dig	Goodall (1986: 541–542)
43	*Pan troglodytes*	Wipe bees or ants with a handful of leaves	Gombe, Tanzania	Insect	Mobile	Leaf	Wipe	Goodall (1986: 542)
44	*Pan troglodytes*	Extract water from the ground hole with leaf sponges	Semliki, Uganda	Water	Fluid	Leaf	Soak	K. D. Hunt (2000)
45	*Pan troglodytes*	Extract water from a tree hole with leaf sponges	Budongo, Uganda	Water	Fluid	Leaf	Soak	Quiatt & Kiwede (1994)
46	*Pan troglodytes*	Extract water from a tree hole with a moss sponge	Tongo, D. R. Congo	Water	Fluid	Moss	Soak	Lanjouw (2002)
47	*Pan troglodytes*	Pound an arboreal bee hive with a dead piece of branch	Ndakan, Central Africa	Insect	Hidden	Branch/twig/stalk	Dig	Fay & Carroll (1994)
48	*Pan troglodytes*	Pound an arboreal bee hive with a dead piece of branch	Bai Hokou, Central Africa	Insect	Hidden	Branch/twig/stalk	Dig	Fay & Carroll (1994)
49	*Pan troglodytes*	Fish termites with a stick	Ndoki, Congo	Insect	Hidden	Branch/twig/stalk	Probe	Suzuki *et al.* (1995)
50	*Pan troglodytes*	Dig a hole with a stick for later termite fishing	Lossi, Congo	Insect	Hidden	Branch/twig/stalk	Dig	Bermejo & Illera (1999)
51	*Pan troglodytes*	Fish for termites with a stick	Lossi, Congo	Insect	Hidden	Branch/twig/stalk	Probe	Bermejo & Illera (1999)
52	*Pan troglodytes*	Pound an arboreal bee hive with a dead piece of branch	Lossi, Congo	Insect	Hidden	Branch/twig/stalk	Dig	Bermejo & Illera (1999)
53	*Pan troglodytes*	Puncture an arboreal bee hive with a dead piece of branch	Lossi, Congo	Insect	Hidden	Branch/twig/stalk	Dig	Bermejo & Illera (1999)

(cont.)

Table 9.1. (cont.)

Case no.	Species	Behavioral descriptions	Observation sites	Target foods[1]	Protection modes[3]	Tools/substrates used	Operation patterns	References
54	Pan troglodytes	Dip honey from an arboreal bee hive with a stick	Lossi, Congo	Insect	Hidden	Branch/twig/stalk	Probe	Bermejo & Illera (1999)
55	Pan troglodytes	Dip honey from subterranean beehive with a long twig	Yaounde Forests, Cameroon	Insect	Hidden	Branch/twig/stalk	Probe	Merfield & Miller (1956: 43–44)
56	Pan troglodytes	Extract water from a tree hole with a leaf as a sponge	Lopé, Gabon	Water	Fluid	Leaf	Soak	Tutin et al. (1995)
57	Pan troglodytes	Dip honey from arboreal bee nest with a stick	Lopé, Gabon	Insect	Hidden	Branch/twig/stalk	Probe	Tutin et al. (1995)
58	Pan troglodytes	Crack hard nuts with stones and/or branches	Taï, Côte d'Ivoire	Fruit	Hard	Stone/rock branch	Bang	Boesch & Boesch (1981, 1983)
59	Pan troglodytes	Dip bone marrow with a stick	Taï, Côte d'Ivoire	Mammal	Hard	Branch/twig/stalk	Probe	Boesch & Boesch (1989)
60	Pan troglodytes	Dip/extract ants/larvae with a stick from the nest	Taï, Côte d'Ivoire	Insect	Hidden	Branch/twig/stalk	Probe	Boesch & Boesch (1990), Boesch (1995)
61	Pan troglodytes	Pull out wood-boring bee with a stick	Taï, Côte d'Ivoire	Insect	Hard	Branch/twig/stalk	Probe	Boesch & Boesch (1990)
62	Pan troglodytes	Dip honey from bee nest with a stick	Taï, Côte d'Ivoire	Insect	Hidden	Branch/twig/stalk	Probe	Boesch & Boesch (1990)
63	Pan troglodytes	Dip remains from colobus skull with a stick	Taï, Côte d'Ivoire	Mammal	Hard	Branch/twig/stalk	Probe	Boesch & Boesch (1990)
64	Pan troglodytes	Remove remains from cracked hard nuts with a stick	Taï, Côte d'Ivoire	Fruit	Hard	Branch/twig/stalk	Pry	Boesch & Boesch (1990)
65	Pan troglodytes	Extract mushroom from termite nest with a stick	Taï, Côte d'Ivoire	Mushroom	Hidden	Branch/twig/stalk	Probe	Boesch (1995)
66	Pan troglodytes	Pound nuts with stones	Cape Palmas, Côte d'Ivoire	Fruit	Hard	Stone/rock	Bang	Savage & Wyman (1843–44)

67	*Pan troglodytes*	Crack hard nuts with stones	Bossou, Guinea	Fruit	Hard	Stone/rock	Bang	Sugiyama & Koman (1979)
68	*Pan troglodytes*	Extract water from a tree hole with a leaf as a sponge	Bossou, Guinea	Water	Fluid	Leaf	Soak	Sugiyama & Koman (1979)
69	*Pan troglodytes*	Pound termites in a tree hole with a stick	Bossou, Guinea	Insect	Hidden	Branch/twig/stalk	Dig	Sugiyama & Koman (1979)
70	*Pan troglodytes*	Pound resin in a tree hole with a stick	Bossou, Guinea	Resin	Hidden	Branch/twig/stalk	Dig	Sugiyama & Koman (1979)
71	*Pan troglodytes*	Hook a fruiting branch with a twig	Bossou, Guinea	Fruit	Far	Branch/twig/stalk	Hook	Sugiyama & Koman (1979)
72	*Pan troglodytes*	Fish/dip ants and termites with a stick	Bossou, Guinea	Insect	Hidden	Branch/twig/stalk	Probe	Humle (1999), Sugiyama et al. (1988)
73	*Pan troglodytes*	Stabilize an anvil with another stone during nut-cracking	Bossou, Guinea	Fruit	Unstable	Stone/rock	Wedge	Matsuzawa (1991a)
74	*Pan troglodytes*	Push a "leaf sponge" deeper into a tree hole with a twig	Bossou, Guinea	Water	Hidden	Branch/twig/stalk	Probe	Matsuzawa (1991b: 256–57)
75	*Pan troglodytes*	Pound apical meristem of palm with a palm frond pestle	Bossou, Guinea	Pith	Hidden	Branch/twig/stalk	Dig	Sugiyama (1994), Yamakoshi & Sugiyama (1995)
76	*Pan troglodytes*	Scoop algae from water with a wand	Bossou, Guinea	Algae	Far	Branch/twig/stalk	Scoop	Matsuzawa et al. (1996)

(cont.)

Table 9.1. (cont.)
(b) Substrate use

Case no.	Species	Behavioral descriptions	Observation sites	Target foods[1]	Protection modes[3]	Tools/substrates used	Operation patterns	References
1	*Saimiri oerstedi*	Rub caterpillars roughly on a stem	Corcovado, Costa Rica	Insect	Noxious	Branch/trunk	Rub	Boinski & Fragaszy (1989)
2	*Saimiri sciureus*	Rub caterpillars against a branch	Manu, Peru	Insect	Noxious	Branch/trunk	Rub	Janson & Boinski (1992)
3	*Cebus albifrons*	Pound nuts against substrate	El Tuparro, Colombia	Fruit	Hard	Branch/trunk	Bang	Defler (1979)
4	*Cebus albifrons*	Tap palm nuts against a branch or another nut to assess	Manu, Peru	Fruit	Hard	Branch/trunk	Tap	Terborgh (1983)
5	*Cebus albifrons*	Bash palm nuts against a branch	Manu, Peru	Fruit	Hard	Branch/trunk	Bang	Terborgh (1983)
6	*Cebus albifrons*	Smash hard fruits against substrate	Dardanelos, Brazil	Fruit	Hard	Unident.	Bang	Rylands (1987)
7	*Cebus apella*	Pound open hard fruits against a trunk	La Macarena, Colombia	Fruit	Hard	Branch/trunk	Bang	Izawa (1979), Izawa & Mizuno (1977)
8	*Cebus apella*	Strike snail against tree branch	La Macarena, Colombia	Mollusk	Hard	Branch/trunk	Bang	Izawa & Mizuno (1977)
9	*Cebus apella*	Smash dead branch against a trunk to extract larvae	La Macarena, Colombia	Insect	Hard	Branch/trunk	Bang	Izawa (1979), Izawa & Mizuno (1977)
10	*Cebus apella*	Rub frog against branch bark	La Macarena, Colombia	Amphibian	Noxious	Branch/trunk	Rub	Izawa (1978)
11	*Cebus apella*	Pound open palm nuts against bamboo trunk	La Macarena, Colombia	Fruit	Hard	Branch/trunk	Bang	Struhsaker & Leland (1977)
12	*Cebus apella*	Strike hard fruit against a branch	River Peneya, Colombia	Fruit	Hard	Branch/trunk	Bang	Izawa & Mizuno (1977)
13	*Cebus apella*	Pound fruits or nuts on branches	Monte Seco, Colombia	Fruit	Hard	Branch/trunk	Bang	Thorington (1967)
14	*Cebus apella*	Rub fruits or nuts on branches	Monte Seco, Colombia	Fruit	Hard	Branch/trunk	Rub	Thorington (1967)

#	Species	Description	Location	Food type	Hardness	Substrate	Action	Reference
15	*Cebus apella*	Hit pods against a trunk	Santa Genebra, Brazil	Fruit	Hard	Branch/trunk	Bang	Galetti & Pedroni (1994)
16	*Cebus apella*	Bash open husked fruits against branch	Raleighvallen, Suriname	Fruit	Hard	Branch/trunk	Bang	Boinski et al. (2000)
17	*Cebus capucinus*	Pound dried fruit on the branch	Santa Rosa, Costa Rica	Fruit	Hard	Branch/trunk	Bang	Freese (1977)
18	*Cebus capucinus*	Roll or rub coati pups and squirrels against branch	Santa Rosa, Costa Rica	Mammal	Hard	Branch/trunk	Rub	Rose (2001)
19	*Cebus capucinus*	Rub fruits to damage hard outer coatings	Santa Rosa, Costa Rica	Fruit	Hard	Branch/trunk	Rub	Panger et al. (2002)
20	*Cebus capucinus*	Rub *Acacia* thorns to damage hard outer coatings	Santa Rosa, Costa Rica	thorn	Hard	Branch/trunk	Rub	Panger et al. (2002)
21	*Cebus capucinus*	Rub caterpillars to remove noxious substance	Santa Rosa, Costa Rica	Insect	Noxious	Branch/trunk	Rub	Panger et al. (2002)
22	*Cebus capucinus*	Pound vertebrate prey against branch	Santa Rosa, Costa Rica	Mammal	Hard	Branch/trunk	Bang	Panger et al. (2002)
23	*Cebus capucinus*	Fulcrum fruits to damage hard outer coatings	Santa Rosa, Costa Rica	Fruit	Hard	Branch/trunk	Fulcrum[4]	Panger et al. (2002)
24	*Cebus capucinus*	Rub fruits to damage hard outer coatings	Lomas Barbudal, Costa Rica	Fruit	Hard	Branch/trunk	Rub	Panger et al. (2002)
25	*Cebus capucinus*	Rub caterpillars to remove noxious substance	Lomas Barbudal, Costa Rica	Insect	Noxious	Branch/trunk	Rub	Panger et al. (2002)
26	*Cebus capucinus*	Rub vertebrate prey against branch	Lomas Barbudal, Costa Rica	Mammal	Hard	Branch/trunk	Rub	Panger et al. (2002)
27	*Cebus capucinus*	Pound fruits to damage hard outer coatings	Lomas Barbudal, Costa Rica	Fruit	Hard	Branch/trunk	Bang	Panger et al. (2002)
28	*Cebus capucinus*	Pound vertebrate prey against branch	Lomas Barbudal, Costa Rica	Mammal	Hard	Branch/trunk	Bang	Panger et al. (2002)
29	*Cebus capucinus*	Rub snail to damage hard outer coatings	Palo Verde, Costa Rica	Mollusk	Hard	Unident.	Rub	Panger (1998)
30	*Cebus capucinus*	Rub duck egg to damage hard outer coatings	Palo Verde, Costa Rica	Egg	Hard	Unident.	Rub	Panger (1998)
31	*Cebus capucinus*	Rub fruits to damage hard outer coatings to soften	Palo Verde, Costa Rica	Fruit	Hard	Unident.	Rub	Panger (1998)

(cont.)

Table 9.1. (cont.)

Case no.	Species	Behavioral descriptions	Observation sites	Target foods[1]	Protection modes[3]	Tools/substrates used	Operation patterns	References
32	*Cebus capucinus*	Rub fruits to remove wind-dispersed seeds	Palo Verde, Costa Rica	Fruit	Noxious	Unident.	Rub	Panger (1998)
33	*Cebus capucinus*	Rub caterpillars to remove noxious or stinging substance	Palo Verde, Costa Rica	Insect	Noxious	Unident.	Rub	Panger (1998)
34	*Cebus capucinus*	Pound snail to damage hard outer coatings	Palo Verde, Costa Rica	Mollusk	Hard	Unident.	Bang	Panger (1998)
35	*Cebus capucinus*	Pound duck egg to damage hard outer coatings	Palo Verde, Costa Rica	Egg	Hard	Unident.	Bang	Panger (1998)
36	*Cebus capucinus*	Pound clay wasp hives to damage hard outer coatings	Palo Verde, Costa Rica	Wasp hive	Hard	Unident.	Bang	Panger (1998)
37	*Cebus capucinus*	Pound sticks to damage hard outer coatings to extract insects	Palo Verde, Costa Rica	Insect	Hard	Unident.	Bang	Panger (1998)
38	*Cebus capucinus*	Pound fruits to damage hard outer coatings to soften	Palo Verde, Costa Rica	Fruit	Hard	Unident.	Bang	Panger (1998)
39	*Cebus capucinus*	Pound fruits to remove seeds' noxious or stinging substance	Palo Verde, Costa Rica	Fruit	Noxious	Unident.	Bang	Panger (1998)
40	*Cebus capucinus*	Fulcrum fruits to damage hard outer coatings	Palo Verde, Costa Rica	Fruit	Hard	Unident.	Fulcrum	Panger (1998)
41	*Cebus olivaceus*	Pound open hard fruits	Fundo Pecuario Masaguaral, Venezuela	Fruit	Hard	Unident.	Bang	Robinson (1986)
42	*Cebus olivaceus*	Bang snail on a tree trunk or branch	Fundo Pecuario Masaguaral, Venezuela	Mollusk	Hard	Branch/trunk	Bang	Fragaszy (1986), Robinson (1986)
43	*Cebus olivaceus*	Bang twigs or branches on another surface to extract insects	Fundo Pecuario Masaguaral, Venezuela	Insect	Hard	Branch/trunk	Bang	Fragaszy (1986)
44	*Macaca fascicularis*	Rub small seeds and fruits (with ants?) on the ground	Botanical Garden, Singapore	Fruit	Dirty	Ground	Rub	Chiang (1967)

#	Species	Behavior	Location	Item	Property	Substrate	Action	Reference
45	*Macaca fascicularis*	Wash sweet potatoes or cassava root in water	Ubud, Indonesia	USO	Dirty	Water	Wash	Wheatley (1988)
46	*Macaca fascicularis*	Wash papaya leaves in water	Ubud, Indonesia	Leaf	Dirty	Water	Wash	Wheatley (1988)
47	*Macaca fascicularis*	Dip a fruit (maybe with sand on it) into a river	Kutai, Indonesia	Fruit	Dirty	Water	Wash	Wheatley (1980)
48	*Macaca fuscata*	Wash dirty plant root into stream water	Hakusan, Japan	USO	Dirty	Water	Wash	Izawa (1982: 188)
49	*Macaca fuscata*	Wash grass roots in the water (on a flat stone)	Katsuyama, Japan	USO	Dirty	Water	Wash	Nakamichi et al. (1998)
50	*Macaca fuscata*	Rub fruits against tree trunk to remove bitter pulp	Koshima, Japan	Fruit	Noxious	Branch/trunk	Rub	Yamakoshi (pers. obs. 1992)
51	*Macaca fuscata*	Roll a frog on fallen tree trunk before eating	Yakushima, Japan	Amphibian	Noxious	Branch/trunk	Rub	Suzuki et al. (1990)
52	*Macaca radiata*	Rub to clean fruit on rough tree surface	Elephanta, India	Fruit	Dirty	Branch/trunk	Rub	Kuruvilla (1980)
53	*Macaca radiata*	Rub to open fruit on rough tree surface	Elephanta, India	Fruit	Hard	Branch/trunk	Rub	Kuruvilla (1980)
54	*Lophocebus albigena*	Rub fruits on branches	Kibale, Uganda	Fruit	Hard	Branch/trunk	Rub	Waser (1977)
55	*Lophocebus albigena*	Rub large gryllacrid orthopteran on branches	Kibale, Uganda	Insect	?	Branch/trunk	Rub	Waser (1977)
56	*Lophocebus albigena*	Rub large fruits on a branch	Bujuko, Uganda	Fruit	Hard	Branch/trunk	Rub	Chalmers (1968)
57	*Papio anubis*	Roll and remove prickles from *Opuntia* sp. fruit on the earth	Debra Libanos, Ethiopia	Fruit	Noxious	Ground	Rub	Crook & Aidrich-Blake (1968)
58	*Papio ursinus*	Rub fish in the sand to remove mucus layer of scales	Kuiseb River Canyon, Namibia	Fish	Noxious	Ground	Rub	Hamilton et al. (1975)
59	*Papio ursinus*	Hammer baobab fruits on the rock by hand	Magalakwén Valley, South Africa	Fruit	Hard	Ground	Bang	Marais (1969: 56)
60	*Pongo pygmaeus*	Use tree trunk to stabilize a fruit held between teeth	Tanjung Puting, Indonesia	Fruit	Unstable	Branch/trunk	Press	Chevalier-Skolnikoff et al. (1982)
61	*Pongo pygmaeus*	Rub burr-covered fruits against a branch	Tanjung Puting, Indonesia	Fruit	Noxious	Branch/trunk	Rub	Galdikas & Vasey (1992)

(cont.)

Table 9.1. (cont.)

Case no.	Species	Behavioral descriptions	Observation sites	Target foods[1]	Protection modes[3]	Tools/substrates used	Operation patterns	References
62	*Pan paniscus*	Wash aquatic herb in the water	Lilungu, D. R. Congo	Root	Dirty	Water	Wash	Bermejo *et al.* (1994)
63	*Pan paniscus*	Wash earthworms in the water	Lilungu, D. R. Congo	Insect	Dirty	Water	Wash	Bermejo *et al.* (1994)
64	*Pan troglodytes*	Smash a captured colobus against riverbed rocks	Mahale Mts., Tanzania	Mammal	Mobile	Ground	Bang	Takahata *et al.* (1984)
65	*Pan troglodytes*	Bang *Strychnos* and other fruits against tree trunk or a rock	Gombe, Tanzania	Fruit	Hard	Branch/trunk ground	Bang	Goodall (1963), Nishida *et al.* (1983)
66	*Pan troglodytes*	Flail animal prey's head against ground or tree trunk	Gombe, Tanzania	Mammal	Mobile	Branch/trunk ground	Bang	Teleki (1973b: 135)
67	*Pan troglodytes*	Smash open *Monodora* fruit against a hard branch	Kibale, Uganda	Fruit	Hard	Branch/trunk	Bang	Lambert (1999)
68	*Pan troglodytes*	Rub and roll fruits against a branch to remove hairs	Lopé, Gabon	Fruit	Noxious	Branch/trunk	Rub	Tutin *et al.* (1996)
69	*Pan troglodytes*	Pound hard fruits against a root or a tree trunk	Taï, Côte d'Ivoire	Fruit	Hard	Branch/trunk	Bang	Boesch & Boesch (1989)
70	*Pan troglodytes*	Pound a colobus skull against a root or a tree trunk	Taï, Côte d'Ivoire	Mammal	Mobile	Branch/trunk	Bang	Boesch & Boesch (1989)
71	*Pan troglodytes*	Rub hairy fruits against hard surface	Taï, Côte d'Ivoire	Fruit	Noxious	Unident.	Rub	Boesch (1996)
72	*Pan troglodytes*	Pound a pangolin against branch on the ground during hunt	Bossou, Guinea	Mammal	Mobile	Ground	Bang	Sugiyama (pers. com.)
73	*Pan troglodytes*	Pound fruit against branch on the ground	Bossou, Guinea	Fruit	Hard	Ground	Bang	Sugiyama (pers. com.)
74	*Pan troglodytes*	Pound baobab fruits on hard objects	Mt. Assirik, Senegal	Fruit	Hard	Ground	Bang	Hunt & McGrew (2002)

(c) Ambiguous cases

Case no.	Species	Behavioral descriptions	Observation sites	Target foods[1]	Protection modes[3]	Tools/ substrates used	Operation patterns	References
1	*Cebus albifrons*	Knock two *Astrocaryum* nuts forcibly together with both hands	Manu, Peru	Fruit	Hard	Hard nuts	Bang	Terborgh (1983: 83)
2	*Cebus apella*	Pound two palm nuts against each other (scissors-like manner)	La Macarena, Colombia	Fruit	Hard	Hard nuts	Bang	Struhsaker & Leland (1977)
3	*Chlorocebus aethiops*	Dip *Acacia tortilis* pods into tree exudates	Amboseli, Kenya	Fruit/water	Hard/fluid	Pod water	Soak	Hauser (1988)
4	*Pan troglodytes*	Dip chewed *Sacoglottis* fruits into water	Taï, Côte d'Ivoire	Fruit/water	?/fluid	Fruit water	Soak	Boesch (1991)

Notes:

[1] Classification based on food items for plants on broad taxonomic groups for animals.

[2] Underground storage organ (see Peters & O'Brien 1981).

[3] See text for classification.

[4] Defined as "an individual applies force on an object working against a substrate (which was used as a fulcrum)" (Panger *et al.* 2002).

Table 9.2. *Numbers of reported tool-/substrate-use cases among primate genera*

	Tool use			Substrate use			Ambiguous		
	Population	Species	Genus	Population	Species	Genus	Population	Species	Genus
Saimiri	0	0	0	2	2	1	0	0	0
Cebus	7	6	6	41	25	17	2	2	1
Chlorocebus	0	0	0	0	0	0	1	1	1
Macaca	7	6	6	10	9	7	0	0	0
Lophocebus	0	0	0	3	2	2	0	0	0
Papio	2	2	2	3	3	3	0	0	0
Pongo	11	9	9	2	2	2	0	0	0
Pan	49	25	25	13	5	5	1	1	1
Total	76	48	48	74	48	37	4	4	3

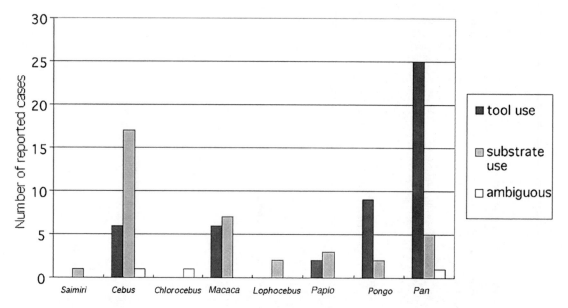

Figure 9.1. Numbers of reported cases of tool/substrate use among primate genera. The numbers were counted on a genus basis (see text and Table 9.2 for details).

the less support the standard view that tool and substrate use are unequally distributed across nonhuman primate taxa.

TARGET FOODS

The kinds of food exploited by tool and substrate techniques represent a very important source of information for reconstructing the evolutionary origins and functional importance of these techniques (e.g., Parker & Gibson 1979). I classified the observed cases in relation to target food categories (Table 9.3).

Tool use and substrate use were applied mostly to fruits (predominantly nuts) and insects. For tool use, insects were the targets in 43%/33%/33% of observed cases on a population/species/genus basis, respectively, and fruits in 22%/25%/25% of cases. For substrate use, fruits dominated (54%/46%/46%),

Table 9.3. *Categories of target foods in tool/substrate use*

	Tool use						Substrate use					
	Fruit	Other plant matter	Insect	Mammal	Other animal matter	Misc.	Fruit	Other plant matter	Insect	Mammal	Other animal matter	Misc.
Saimiri	0	0	0	0	0	0	0	0	2/2/1	0	0	0
Cebus	2/2/2	0	2/1/1	0	1/1/1	2/2/2	22/10/6	1/1/1	6/4/2	4/2/2	7/7/5	1/1/1
Macaca	2/1/1	1/1/1	2/2/2	0	1/1/1	1/1/1	5/5/4	4/3/2	0	0	1/1/1	0
Lophocebus	0	0	0	0	0	0	2/1/1	0	1/1/1	0	0	0
Papio	1/1/1	0	0	0	0	1/1/1	2/2/2	0	0	0	1/1/1	0
Pongo	5/3/3	0	5/5/5	0	0	1/1/1	2/2/2	0	0	0	0	0
Pan	7/5/5	3/3/3	24/8/8	5/5/5	1/1/1	9/3/3	7/3/3	1/1/1	1/1/1	4/1/1	0	0
Total	17/12/12	4/4/4	33/16/16	5/5/5	3/3/3	14/8/8	40/23/18	6/5/4	10/8/5	8/3/3	9/9/7	1/1/1

Numbers are in order of: number of cases in a population basis/species basis/genus basis.

Table 9.4. *Protection modes of target foods in tool/substrate use*

	Tool use						Substrate use					
	Hard	Hidden	Noxious	Fluid	Dirty	Misc.	Hard	Hidden	Noxious	Fluid	Dirty	Misc.
Saimiri	0	0	0	0	0	0	0	0	2/2/1	0	0	0
Cebus	2/2/2	0	3/2/2	2/2/2	0	0	35/21/13	0	6/4/4	0	0	0
Macaca	1/1/1	0	2/2/2	0	4/3/3	0	1/1/1	0	2/2/2	0	7/6/4	0
Lophocebus	0	0	0	0	0	0	3/2/2	0	0	0	0	0
Papio	1/1/1	0	0	0	0	1/1/1	1/1/1	0	2/2/2	0	0	0
Pongo	2/1/1	4/4/4	2/1/1	1/1/1	0	2/2/2	0	0	1/1/1	0	0	1/1/1
Pan	9/7/7	27/12/12	0	7/1/1	0	6/5/5	5/1/1	0	2/1/1	0	2/2/2	4/1/1
Total	15/12/12	31/16/16	7/5/5	10/4/4	4/3/3	9/8/8	45/26/18	0	15/12/11	0	9/8/6	5/2/2

Numbers are in order of: number of cases in a population basis/species basis/genus basis.

whereas insect cases were substantial, but less significant (14%/17%/14%). These patterns were heavily influenced by the fact that there were many observations of tool use for insects (mostly nesting ones) in chimpanzees, and of substrate use for fruits (mostly hard-shelled ones) in capuchins. Other ripe fruits and leaves, the staple food categories for many primate species, were rarely the targets of these feeding techniques, probably because of their physically unprotected forms.

PROTECTION MODES: WHAT ARE THE OBSTACLES?

Logically, target foods for both tool and substrate use should be relatively difficult to prepare and consume; otherwise, these techniques would be unnecessary. I roughly classified observed target foods according to their protection modes (Table 9.4). In many cases, the technique used contributed to identifying the protection mode (e.g., foods were classified as dirty when foragers cleaned them), so classification reflects how nonhuman primate foragers perceived these target foods.

"Hard" foods (detachable items encased by hard materials, e.g., nutshell) were relatively common targets of technically complex feeding, both tool use (20%/25%/25% of the cases) and substrate use (61%/54%/49%). Foods "hidden" within or under immovable, protective matrices that functioned as substrates (e.g., termite mound), were obtained exclusively by tool use (41%/33%/33%), probably because of their topographical requirement. Notably, exploiters of hidden foods were exclusively great apes. A similar conspicuous difference was found for "fluid" foods, also obtained solely by tool use. "Noxious" foods were foods protected with thorns, hairy covers, disagreeable or painful chemicals, scales, etc. This type of food involved tool use infrequently (9%/10%/10%) but substrate use commonly, in various taxa (20%/25%/30%). "Dirty" foods were foods that were cleaned in any way with the use of tools or substrates before ingestion. Dirty foods were infrequent targets, but had a highly uneven phylogenetic distribution. Macaques were responsible for most of the cases of dirty-food processing by tool use and all the cases of substrate use.

Theoretically, target food protection modes should not be limited to physical and behavioral (e.g., agility or aggressiveness of hunted species) defenses, but should also include chemical ones (i.e., toxic compounds),

which can be surmounted with help from specialized techniques (e.g., cooking; see Wrangham et al. 1999 for early humans). In reality, this study identified few cases involving other protection modes (Table 9.1). Some cases such as "removing poisonous hair from foods" (see tool-use case 14, Table 9.1) seem to involve chemical protection, but appear to work as a part of physical protection; others involve avoiding contact with the chemical (e.g., tool-use cases 5–7, 23, 25, Table 9.1). The main reason that no tool or substrate use was reported from Colobinae must be that they do not need specialized feeding techniques to process their staple foods (chemically but not physically protected leaves) before ingestion, because they have specialized digestive tracts for cellulose fermentation.

TOOLS AND SUBSTRATES USED

To use tools, a primate has to find appropriate material from its surrounding environment and occasionally has to modify the material into a proper tool. In substrate use cases, it needs to utilize a particular function of a substrate (e.g., hard, coarse, liquid, etc.) to change the physical condition of the target food. Differences among taxonomic groups in the choice of tools or substrate materials may help us understand each group's use of its environments or ecological niches (Table 9.5).

For tools, stick-shaped items of various sizes made of tree branches, twigs, or the stalks of herbs were most commonly used (58%/56%/56%). This pattern was observed almost exclusively in great apes and rarely in monkeys. Second in frequency were leaves, widely used as tools across taxonomic groups in various contexts (29%/29%/29%). Third, stones/rocks were used as hammers/anvils/missiles in 90%/100%/10% of cases. These three categories of tool resources accounted for almost all observed cases (96%/96%/96%), implying a limitation to the materials that nonhuman primates see as available in the natural environment. Overall, tools were usually made from vegetation (88%/85%/85%); animal matter was used in only one exceptional case (tool-use case No. 3, Table 9.1).

Similarly, almost all documented cases of substrate use involved only three substrate types: tree branch or trunk (58%/56%/49%), water (9%/13%/14%), and the ground (11%/8%/11%). Understandably, water and the ground were used only by more terrestrial species (Macaca, Papio, Pan).

Table 9.5. *Categories of tools and substrates used*

	Tool use				Substrate use			
	Branch/twig/stalk	Leaf	Stone/rock	Misc.	Branch/trunk	Water	Ground	Misc.
Saimiri	0	0	0	0	2/2/1	0	0	0
Cebus	1/1/1	5/4/4	0	1/1/1	27/16/9	0	0	14/9/8
Macaca	0	6/5/5	1/1/1	0	4/4/3	5/4/3	1/1/1	0
Lophocebus	0	0	0	0	3/2/2	0	0	0
Papio	1/1/1	0	1/1/1	0	0	0	2/2/2	1/1/1
Pongo	9/7/7	2/2/2	0	0	2/2/2	0	0	0
Pan	33/18/18	10/4/4	5/3/3	1/0/0	5/1/1	2/2/2	5/1/1	1/1/1
Total	44/27/27	23/15/15	7/5/5	2/1/1	43/27/18	7/6/5	8/4/4	16/11/10

Numbers are in order of: number of cases in a population basis/species basis/genus basis.

Table 9.6. *Essential operation patterns involved*

	Tool use						Substrate use			
	Probe	Dig	Soak	Rub	Bang	Misc.	Bang	Rub	Wash	Misc.
Saimiri	0	0	0	0	0	0	0	2/2/1	0	0
Cebus	0	0	2/2/2	3/2/2	2/2/2	0	24/14/7	14/9/8	0	3/2/2
Macaca	0	0	0	6/5/5	1/1/1	0	0	5/5/4	5/4/3	0
Lophocebus	0	0	0	0	0	0	0	3/2/2	0	0
Papio	0	0	0	0	1/1/1	1/1/1	1/1/1	2/2/2	0	0
Pongo	4/3/3	2/2/2	1/1/1	0	0	4/3/3	0	1/1/1	0	1/1/1
Pan	18/7/7	11/7/7	9/3/3	0	3/1/1	8/7/7	9/2/2	2/1/1	2/2/2	0
Total	22/10/10	13/9/9	12/6/6	9/7/7	7/5/5	13/11/11	34/17/10	29/22/19	7/6/5	4/3/3

Numbers are in order of: number of cases in a population basis/species basis/genus basis.

ESSENTIAL OPERATION PATTERNS

All cases were sorted according to the operation patterns involved (Table 9.6). Most cases of substrate use could be classified into three operation patterns: banging, rubbing, and washing (95%/94%/92%). Operations were more varied in tool-use cases and differences between taxonomic groups seemed to emerge much more clearly in tool use.

In tool use, great apes (*Pongo* and *Pan*) clearly had the most diverse operation patterns. The two most dominant operations (probing and digging) were also unique to great apes. Importantly, these patterns typically involved "hidden foods," also consumed almost exclusively by great apes. In contrast, rubbing tool use was unique to capuchins and macaques. In substrate use, rubbing was common across taxonomic groups. Capuchins tended to bang, macaques tended to wash, and great apes did both. Irrespective of tool use or substrate use, typical species manipulation patterns emerged clearly. Capuchins were inclined to bang or rub, macaques appeared to specialize in rubbing or washing, and great apes had much more diverse repertories.

TRAITS OF PHYLOGENETIC GROUPS

Although the literature survey was based largely on simple descriptions, several interesting differences were detected among phylogenetic groups. Here I focus on these group-specific characteristics, with special emphasis on identifying great ape uniqueness. Since the cases obtained were from very limited groups and data from some groups were insufficient to generalize, I will summarize and compare results for three major groups: *Cebus*, *Macaca*, and great apes (*Pongo* and *Pan*).

Cebus

It has been suggested that capuchins' fine manipulative ability represents an independent evolutionary event, having no direct homological relation with that of humans and great apes (Parker & Gibson 1979; van Schaik *et al.* 1999). In this study, capuchins, coupled with *Saimiri*, formed one independent phylogenetic cluster that performed complex feeding techniques.

Capuchin feeding techniques can be characterized by, (1) relative dominance of substrate over tool use in terms of number of cases (Figure 9.1); (2) a clear tendency for using substrates and on rare occasions tools to obtain hard foods, such as nuts (Tables 9.3, 9.4); (3) use of an arboreal environment, reflected in preferences for tree branches as substrates (Table 9.5); and (4) an overwhelming majority of "banging" operation patterns (Table 9.6). These four characteristics, obviously interrelated, indicate the importance of extracting material from hard fruits by banging them against a substrate, usually a tree branch, in their feeding ecology. At Coca Cash, Peru, although capuchins are ripe-fruit eaters in quantitative terms, they specialize in hard palm nuts that other monkeys cannot utilize during seasonal fruit scarcities (Terborgh 1983). In addition, from the viewpoint of functional morphology, capuchins are considered adapted to hard nut feeding, although their hard nut feeding occurs only during limited periods (Rosenberger 1992).

The relative paucity of tool use compared with the rich repertoire of substrate use in these highly dexterous species may be partly explained by their feeding arboreally. Capuchins commonly select tree trunks or branches as substrates for their banging or rubbing, in sharp contrast to more terrestrial species, such as macaques, baboons, and chimpanzees, which often use the ground as a substrate (Table 9.5). As was suggested in explaining the rarity of tool use by orangutans (e.g., Russon 1998), it may be difficult or dangerous to manipulate detached objects in highly arboreal contexts.

Capuchins' tendency to bang objects has also been well documented in experimental studies, suggesting that it is "hard-wired." Torigoe (1985) gave a nylon rope or a wooden cube to subjects from 74 primate species to observe species differences in object manipulation patterns. Capuchins (20 individuals from four species: *C. nigrivittatus*, *C. capucinus*, *C. apella*, *C. albifrons*) tended to "roll," "rub," and "slide" the manipulandum, as did Old World monkeys (macaques, guenons, mangabeys, baboons). Capuchins were also quite specialized in "striking" the objects. With the exception of great apes, this operation pattern was unique to capuchins.

Macaca

Conspicuous tendencies in the complex feeding techniques of macaques were: (1) utilization of "dirty" foods (Table 9.4); (2) use of water and the ground as substrates (Table 9.5); and (3) operation patterns such as "rubbing" and "washing" (Table 9.6). As with capuchins, these tendencies can be reasonably explained by their foraging ecology.

Macaques are terrestrial to some degree and highly omnivorous. Terrestrial activities, including searching for foods on the ground, constitute a large proportion of their foraging activity (e.g., Nakagawa 1990). Foods on the ground can be "dirty," and may contain sandy, hard materials that cause frictional wear to their teeth, as was suggested in *Paranthropus* (Jolly 1970). Therefore, it could be adaptive for macaques to rub or wash dirty foods before ingestion.

The predisposition of macaques to rub and wash objects is also considered "hard-wired." Their "rubbing" and "washing" nature has been demonstrated in experimental studies (Glickman & Sroges 1966; Torigoe 1985), in observations on their reaction to provisioned foods (Kawamura 1954; Suzuki 1965), and in their behavior in non-feeding, playful contexts (Huffman 1984).

Pongo and Pan

The most notable characteristic of great ape feeding techniques must be their variability and flexibility. Great apes consume a wide range of foods that are protected in a variety of ways (Tables 9.2, 9.3) by using various tools and substrates (Table 9.4). Compared with monkeys, they employ an especially diverse repertoire of manipulative operations (Table 9.6). This apparent correlation between complex feeding ecology and complex object manipulation, which may lead to complex information processing, has been pointed out repeatedly (e.g., Byrne 1997; Byrne & Russon 1998, Parker & Gibson 1979; Russon 2003; van Schaik *et al.* 1999). Variability and flexibility are very important in their own right but it is difficult to determine whether these general complexities occur only in great apes on an evolutionary timescale. To infer specific evolutionary events that have shaped great ape cognition, we need not only these general characteristics but also any specific characteristics that are unique to great apes.

Besides variability and flexibility, additional traits that characterize great ape feeding techniques should be noted. These include: (1) dominance of tool over substrate use in a number of cases (Figure 9.1); (2) exploitation of social insects and their products (Table 9.3); (3) utilization of "hidden" foods embedded within substrate-like protective matrices (Table 9.4); (4) unique operation patterns, such as "probing" and "digging" (Table 9.6); and (5) tool and substrate use in hunting mammals (Tables 9.1, 9.3).

Traits 1 to 4 are, for the most part, interrelated. Social insects (ants, bees, termites) and their products are often hidden deep in nests, which are often difficult and ineffective to access without "probing" and/or "digging" tools. Interestingly, this pattern of traits in is both dominant and almost unique to great apes. Monkeys consume insects relatively infrequently and rarely utilize hidden foods, use stick-shaped tools, and "probe" or "dig" (Tables 9.3–9.6).

Importantly, recently reported wild orangutan tool use (Fox, Sitompul & van Schaik 1999; van Schaik, Fox & Sitompul 1996) includes the same types of tools and tool use that appear to be unique to great apes. That is, their tool-using behavior is directed at social insects hidden within substrate-like matrices and they use stick-type tools in probing operations to obtain them. These patterns were previously known only in chimpanzees and have never been reported in monkeys. Orangutans and chimpanzees appear to share great ape specific feeding techniques and consequently form a distinct phylogenetic cluster associated with the evolution of primate feeding techniques.

Distinctive use of complex techniques in animal hunting by great apes (or more accurately, by chimpanzees and occasionally capuchins, Tables 9.1, 9.3) is also interesting because meat eating, with the aid of tools, has been thought to play an important role in some human evolution scenarios (Lee & DeVore 1968; Stanford & Bunn 2001). However, the number of instances of great ape hunting is small and the pattern is not well defined compared with the insect-probing patterns mentioned above (Table 9.1). It has also been suggested that insectivory in chimpanzees is ecologically far more important than meat eating, and, as such, provides references that are more (or at least equally) promising than meat eating for understanding hominin evolution (McGrew 2001).

"INSERTION FEEDING": THE ORIGIN OF GREAT APE INTELLIGENCE?

This literature review suggested that one specific feeding technique, using stick-type tools to probe and extract foods hidden in substrate-like matrices, such as social insects, may be unique to great apes and is virtually absent in monkeys in the wild. This is referred to here as "insertion feeding." What is the significance of insertion feeding in the evolution of great ape cognition? If this specific technique requires qualitatively or quantitatively different cognitive mechanisms, it may have been associated with the cognitive leap that is thought to have occurred in the common ancestor of extant great apes (Byrne 1997) and it may help explain the origin of the unique cognitive abilities observed in living great apes. Four possible scenarios for cognitive dimensions that might have been enhanced with the emergence of insertion feeding are discussed here.

Precision grip scenario

Insertion feeding techniques appear to require precision grips (broadly defined; Marzke & Wullstein 1996) for tool manipulation, more so than other major techniques (e.g., banging). The precision grip is considered a critical morphological feature that enables humans to perform tool behaviors distinguishable from those of other primates (e.g., Napier 1962). It is not clear, however, whether there is any direct causal relationship between fine manipulation and general intellectual ability, or how the establishment of techniques involving precise manipulation could have been linked with the emergence of a uniquely great ape intelligence (but see Byrne, Chapter 3, this volume).

The supposed increase of precision in manipulation may not directly explain great ape technical intelligence, but it may provide a basis for it. The ability to use a tool as an extension of one's fingers (or lips, in the case of orangutans; see Fox *et al.* 1999) rather than one's hand or arm may have expanded the repertoire of possible action units for processing foods. The series of fine finger-level actions that mountain gorillas use to process herbaceous vegetation protected by barbs or stinging hairs, among the most complex feeding techniques known in nonhuman primates, not only consists of many action units but is also organized hierarchically even though no tool is involved (Byrne, Chapter 3, this volume; Byrne & Byrne 1993). Motor dexterity no doubt provides a morphological basis for this behavioral complexity.

Difficulty-in-learning scenario

Feeding techniques are prime candidates for social learning in primates (e.g., Itani & Nishimura 1973; Lefebvre 1995). When a technique is shared within a social group and maintained via social learning, the complexity of the feeding technique should be evaluated by the task difficulty not only for performers, but also for learners (e.g., Russon 1997). If a particular feeding technique demands high cognition for discovery or learning, but not necessarily for performing, the technique may provoke the evolution of a particular type of cognition.

Insertion feeding techniques might be more difficult to learn than other types of tool or substrate use. In typical insertion feeding (e.g., termite fishing), tools are inserted through the protective matrix, which functions as a substrate (ground, tree branch, etc.), and delicately manipulated there. This means that an essential part of the whole sequence of the feeding behavior is invisible to conspecific learners. The hidden part concerns how the tool is acting on the target food and this may increase the difficulty of independently understanding the critical causal relationships in the task.

Since the ability to understand causality in tool-use tasks is claimed to differ significantly between great apes and capuchin monkeys in experimental settings (Visalberghi *et al.* 1995), the presence or absence of insertion feeding in nature may contribute to explaining the origin of great ape intelligence. There might have been selection pressures for the common great ape ancestor to achieve good understanding of the causal relationships associated with feeding tasks in order to learn insertion techniques for effectively exploiting hidden food resources. This idea needs further testing to determine whether there are differences in social learning difficulty between the insertion type of tool-using behavior and types in which the entire procedure is visible to learners (e.g., nut cracking).

Tool-use scenario

One distinctive characteristic of insertion feeding is that the task can be solved only by tool use, and not by substrate use. Theoretically, other types of tasks, such as cracking nuts, can employ both tool and substrate use (i.e., nuts can be cracked open by hitting them with a hammer tool or against a hard substrate). The difference lies in the fact that "hidden foods" – the targets of insertion feeding – are in substrate-like matrices that cannot be detached so they cannot be directly manipulated themselves, while foods like nuts are detachable objects that are easily manipulated. This means that these hidden foods cannot be targets of substrate use. In other words, to consume foods that are hidden within substrate-like protective matrices, only tool use seems to be effective. Therefore, the need to exploit some "hidden" food resources could have pushed the common ancestor of extant great apes to evolve cognitive capacities related to tool use.

Although it seems to make sense to reason that insertion feeding promoted wider tool-use practice and that this consequently promoted higher intelligence, the latter part of such reasoning is not self-evident. Currently, further theoretical and empirical developments are needed to allow us to determine whether these tool-using techniques are more cognitively demanding than other feeding techniques (e.g., substrate use) in terms of discovering, performing, or learning them.

"Level 2" tree-structure scenario

In addition to the topological requirement for tool use discussed above, insertion feeding appears to have a unique task composition that merits consideration. A typical tool-use task (e.g., a chimpanzee dipping a leaf sponge into stream water to drink) involves a subject (a chimpanzee), a tool (a sponge), and a target food (water). An insertion-type tool-use task (e.g., a chimpanzee inserting a twig into the entrance of a termite mound to fish termites) has one additional active component, a substrate-like protective matrix (a termite mound), that could increase task complexity. The number of active task components is increased because, in insertion feeding, the target is not simply an independent detached object (e.g., a hard nut), but an independent protecting object (e.g., a termite mound) plus a food (e.g., a termite). In other words, in insertion feeding, the subject must manipulate a tool relative to a target food and its protecting substrate.

In classifying various tool-using behaviors of wild chimpanzees, Matsuzawa (1996) proposed a "tree-structure analysis," which arranges tool behaviors from "Level 0" to "Level 3" according to the number of tools (detached objects) actively involved in the task structure. He ranked termite fishing as "Level 1" because only one tool is involved in the task – "using *a twig* to fish for *termites*," as he described it. An example of "Level 2" is the nut-cracking behavior, which he described as "using *a hammer stone* to hit *a nut* on *an anvil stone*," in which two tools, the hammer and the anvil, are actively involved in solving the task. This analysis detects no likely meaningful difference between a simple tool-using task and an insertion feeding task because it considers only the number of tools, not the number of active components of the task.

In a modified version of Matsuzawa's tree-structure analysis (Table 9.7), which counts both the number of tools and other protecting objects as active components in determining task complexity, such as substrate-like matrices, insertional feeding tasks qualify as "Level 2" because they involve two active components, the tool and the matrix-substrate. Similarly, substrate use may qualify as "Level 1" because it involves manipulating one active component, the substrate, which places it on the same level as simple tool use. Using this modified classification, the difference between great apes and other monkeys in tool use and substrate use, as detected in my review, may lie between "Level 1" and "Level 2."

It is still necessary to examine further why the distinction between "Level 1" and "Level 2" is important in cognitive and evolutionary terms. The answer may lie in the number of object–object relations that must be handled in coordinated fashion to solve the task. In this view, manipulating relations between food items and (non-detachable) protecting objects plays a cognitive role equivalent to that of manipulating relations between food and (detachable) tools. However, the increase in cognitive complexity induced by adding one protecting object/tool goes beyond adding one more object–object relation: it increases the number of

Table 9.7. *Modified version of Matsuzawa's tree-structure analysis*

Level 0
Pick up *a nut*
Pick up *a termite*

Level 1
Dip *a sponge* to drink *water* (tool use)
Bang *a nut* against *a tree branch* (substrate use)

Level 2
Insert *a twig* into *a hole of termite mound* to fish for *termites*
Use *a hammer stone* to crack *a nut* on *an anvil stone*

essential relations multiplicatively. Feeding techniques that involve one tool or non-detachable matrix ("Level 1") must manage one object–object relation (tool–food or substrate–food). Techniques that involve two tools and/or substrate-like protecting objects ("Level 2") must manage three such relations (food–tool, food–protecting substrate, tool–protecting substrate) in coordinated fashion. The ability to coordinate multiple object–object relations (i.e., relations between relations) may be the critical cognitive advance. Some cognitive studies suggest that like great apes, monkeys master tasks that require manipulating a single object–object relation; differently than great apes, monkeys do not master tasks that require manipulating relations-between-relations (Parker, Chapter 4, Russon, Chapter 6, this volume).

FUTURE PERSPECTIVES

Insertion feeding, which this review determined is unique to great apes, could have been a trigger in promoting the package of cognitive abilities now found only in great apes, via the evolution of (1) motor precision in object handling; (2) deep understanding of object–object relationships in feeding tasks; (3) tool use; and (4) complex techniques that involve two or more active components of the task. These four scenarios are obviously interrelated and not mutually exclusive. All or some of them could have acted together to favor the intelligence that is unique to great apes, under the ecological conditions that favored insertion feeding. Future experimental and field studies that are designed to compare insertion with other feeding techniques will hopefully elucidate the overall validity of the hypothesis and the relative significance of the four (or more) scenarios.

It is also necessary to date and place the emergence of insertion feeding in evolutionary perspective. The common ancestor of living great apes is thought to have lived about 14 million years ago, but the likely fossil species and its location (Africa vs. Europe; e.g., Stewart & Disotell 1998) are very controversial (Begun 2001; Begun, Ward & Rose 1997). By inference from current evidence of insertion feeding, the likely ancestors must have lived in a termite-rich environment and relied to some extent on termite consumption in their feeding ecology. Future studies reconstructing the feeding ecology of fossil Miocene apes, particularly those focusing on termite feeding, are sorely needed (see evocative evidence for probable termite feeding using bone tools by *Australopithecus robustus*; Backwell & d'Errico 2001).

This study stressed the uniqueness of "insertion feeding" in great apes among nonhuman primates in the wild, and its cognitive and evolutionary implications. Outside the Primate order, however, some nonprimate species seem to master similar insertion techniques. Two avian species have been reported to use stick tools to extract insects from their hiding places: woodpecker finches in Galápagos Islands (Gifford 1919; Hundrey 1963) and New Caledonian crows (Orenstein 1972). Recent studies claimed similarities between their tool use and that of great apes in terms of tool standardization (Hunt 1996), local differences (G. R. Hunt 2000), and ecology (Tebbich *et al.* 2002). It would be valuable to articulate the differences between these tool-using bird species and closely related non-tool-using species, particularly in their learning mechanisms and feeding ecology, as an analogue of the difference between great apes and other nonhuman primate species.

ACKNOWLEDGMENTS

I am especially obliged to the two editors of this book, A. E. Russon and D. R. Begun. Without their extraordinary patience and encouragement, my primitive ideas would not be brought to fruition. I also thank D. M. Fragaszy, H. Ihobe, T. Matsuzawa, M. Myowa-Yamakoshi, T. Nishida, Y. Sugiyama, S. Suzuki, J. Yamagiwa, and the members of Laboratory of Human Evolution Studies, Kyoto University for their helpful comments and encouragements. This study was financed by the grant under Research Fellowships of the Japan Society for the Promotion of Science for Young Scientists to the author (No. 2670).

REFERENCES

Alcock, J. (1972). The evolution of the use of tools by feeding animals. *Evolution*, 26, 464–73.

Backwell, L. R. & d'Errico, F. (2001). Evidence of termite foraging by Swartkrans early hominids. *Proceedings of the National Academy of Science of the United States of America*, 98, 1358–63.

Beatty, H. (1951). A note on the behavior of the chimpanzee. *Journal of Mammalogy*, 32, 118.

Beck, B. B. (1980). *Animal Tool Behavior: The Use and Manufacture of Tools by Animals.* New York: Garland STPM Press.

Begun, D. R. (1999). Hominid family values: morphological and molecular data on relations among the great apes and humans. In *The Mentalities of Gorillas and Orangutans: Comparative Perspectives*, ed. S. T. Parker, R. W. Mitchell & H. L. Miles, pp. 3–42. Cambridge, UK: Cambridge University Press.

(2001). African and Eurasian Miocene hominoids and the origins of the Hominidae. In *Hominoid Evolution and Environmental Change in the Neogene of Europe*, ed. P. Andrews, G. Koufos & L. de Bonis, pp. 231–53. Cambridge, UK: Cambridge University Press.

Begun, D. R., Ward, C. V. & Rose, M. D. (ed.) (1997). *Function, Phylogeny, and Fossils: Miocene Hominoid Evolution and Adaptations.* New York: Plenum Press.

Bermejo, M. & Illera, G. (1999). Tool-set for termite-fishing and honey extraction by wild chimpanzees in the Lossi Forest, Congo. *Primates*, 40, 619–27.

Bermejo, M., Illera, G. & Sabater Pí, J. (1994). Animals and mushrooms consumed by bonobos (*Pan paniscus*): new records from Lilungu (Ikela), Zaire. *International Journal of Primatology*, 15, 879–98.

Boesch, C. (1991). Handedness in wild chimpanzees. *International Journal of Primatology*, 12, 541–58.

(1995). Innovation in wild chimpanzees (*Pan troglodytes*). *International Journal of Primatology*, 16, 1–16.

(1996). Three approaches for assessing chimpanzee culture. In *Reaching into Thought: The Minds of the Great Apes*, ed. A. E. Russon, K. A. Bard & S. T. Parker, pp. 404–29. Cambridge, UK: Cambridge University Press.

Boesch, C. & Boesch, H. (1981). Sex differences in the use of natural hammers by wild chimpanzees: a preliminary report. *Journal of Human Evolution*, **10**, 585–93.

(1983). Optimization of nut-cracking with natural hammers by wild chimpanzees. *Behaviour*, **83**, 265–86.

(1989). Hunting behavior of wild chimpanzees in the Taï National Park. *American Journal of Physical Anthropology*, **78**, 547–73.

(1990). Tool use and tool making in wild chimpanzees. *Folia Primatologica*, **54**, 86–99.

Boinski, S. & Fragaszy, D. M. (1989). The ontogeny of foraging in squirrel monkeys, *Saimiri oestedi*. *Animal Behaviour*, **37**, 415–28.

Boinski, S., Quatrone, R. P. & Swartz, H. (2000). Substrate and tool use by brown capuchins in Suriname: ecological contexts and cognitive bases. *American Anthropologist*, **102**, 741–61.

Byrne, R. W. (1997). The Technical Intelligence hypothesis: an additional evolutionary stimulus to intelligence? In *Machiavellian Intelligence II: Extensions and Evaluations*, ed. A. Whiten & R. W. Byrne, pp. 289–311. Cambridge, UK: Cambridge University Press.

Byrne, R. W. & Byrne, J. M. E. (1991). Hand preferences in the skilled gathering tasks of mountain gorillas (*Gorilla g. beringei*). *Cortex*, **27**, 521–46.

(1993). Complex leaf-gathering skills of mountain gorillas (*Gorilla g. beringei*): variability and standardization. *American Journal of Primatology*, **31**, 241–61.

Byrne, R. W. & Russon, A. E. (1998). Learning by imitation: a hierarchical approach. *Behavioral and Brain Sciences*, **21**, 667–721.

Candland, D. K. (1987). Tool use. In *Comparative Primate Biology*. Vol. 2B. *Behavior, Cognition, and Motivation*, ed. G. Mitchell & J. Erwin, pp. 85–103. New York: Alan R. Liss.

Carpenter, A. (1887). Monkeys opening oysters. *Nature*, **36**, 53.

Case, R. (1985). *Intellectual Development: Birth to Adulthood*. New York: Academic Press.

Chalmers, N. R. (1968). Group composition, ecology and daily activities of free living mangabeys in Uganda. *Folia Primatologica*, **8**, 247–62.

Chance, M. R. A. & Mead, A. P. (1953). Social behavior and primate evolution. *Symposia of the Society for Experimental Biology*, **7**, 395–439.

Chevalier-Skolnikoff, S. (1990). Tool use by wild *Cebus* monkeys at Santa Rosa National Park, Costa Rica. *Primates*, **31**, 375–83.

Chevalier-Skolnikoff, S., Galdikas, B. M. F. & Skolnikoff, A. Z. (1982). The adaptive significance of higher intelligence in wild orang-utans: a preliminary report. *Journal of Human Evolution*, **11**, 639–52.

Chiang, M. (1967). Use of tools by wild macaque monkeys in Singapore. *Nature*, **214**, 1258–59.

Crook, J. H. & Aidrich-Blake, P. (1968). Ecological and behavioural contrasts between sympatric ground dwelling primates in Ethiopia. *Folia Primatologica*, **8**, 192–227.

Custance D. M., Whiten, A. & Bard, K. A. (1995). Can young chimpanzees imitate arbitrary actions? Hayes and Hayes (1952) revisited. *Behaviour*, **132**, 839–58.

Dampier, W. (1927). *A New Voyage round the World*. London: Argonaut Press.

Defler, T. R. (1979). On the ecology and behavior of *Cebus albifrons* in eastern Colombia: I. Ecology. *Primates*, **20**, 475–90.

Dunbar, R. I. M. (1992). Neocortex size as a constraint on group size in primates. *Journal of Human Evolution*, **22**, 469–93.

(1995). Neocortex size and group size in primates: a test of the hypothesis. *Journal of Human Evolution*, **28**, 287–96.

Fay, J. M. & Carroll, R. W. (1994). Chimpanzee tool use for honey and termite extraction in Central Africa. *American Journal of Primatology*, **34**, 309–17.

Fernandes, M. E. B. (1991). Tool use and predation of oysters (*Crassostrea rhizophorae*) by the tufted capuchin, *Cebus apella apella*, in brackish water mangrove swamp. *Primates*, **32**, 529–31.

Fitch-Snyder, H. & Carter, J. (1993). Tool use to acquire drinking water by free-ranging lion-tailed macaques (*Macaca silenus*). *Laboratory Primate Newsletter*, **32**, 1–2.

Fox, E. A., Sitompul, A. F. & van Schaik, C. P. (1999). Intelligent tool use in wild Sumatran orangutans. In *The Mentalities of Gorillas and Orangutans: Comparative Perspectives*. ed. S. T. Parker, R. W. Mitchell, & H. L. Miles, pp. 99–116. Cambridge, UK: Cambridge University Press.

Fragaszy, D. M. (1986). Time budgets and foraging behavior in wedge-capped capuchins (*Cebus olivaceus*): age and sex differences. In *Current Perspectives in Primate Social Dynamics*, ed. D. M. Taub & F. A. King, pp. 159–74. New York: Van Nostrand Reinhold.

Freese, C. H. (1977). Food habits of white-faced capuchins *Cebus capucinus* L. (Primates: Cebidae) in Santa Rosa National Park, Costa Rica. *Brenesia*, **10/11**, 43–56.

Galetti, M. & Pedroni, F. (1994). Seasonal diet of capuchin monkeys (*Cebus apella*) in a semideciduous forest in south-east Brasil. *Journal of Tropical Ecology*, **10**, 27–39.

Galdikas, B. M. F. & Vasey, P. (1992). Why are orangutans so smart? Ecological and social hypotheses. In *Social Processes and Mental Abilities in Non-Human Primates: Evidences from Longitudinal Field Studies*, ed. F. D. Burton, pp. 183–224. New York: Edwin Mellen Press.

Gallup, G. Jr. (1970). Chimpanzees: self-recognition. *Science*, **167**, 86–7.

Gifford, E. (1919). Field notes on the land birds of the Galapagos Islands and Cocos Islands, Costa Rica. *Proceedings of the California Academy of Sciences*, **2**, 189–258.

Glickman, S. E. & Sroges, R. W. (1966). Curiosity in zoo animals. *Behaviour*, **26**, 151–88.

Goodall, J. (1963). Feeding behaviour of wild chimpanzees: a preliminary report. *Symposia of the Zoological Society of London*, **10**, 39–48.

    (1964). Tool-using and aimed throwing in a community of free-living chimpanzees. *Nature*, **201**, 1264–6.

Goodall, J. van Lawick (1970). Tool-using in primates and other vertebrates. In *Advances in the Study of Behavior*, Vol. 3, ed. D. S. Lehrman, R. A. Hinde & E. Shaw, pp. 195–249. London: Academic Press.

Goodall, J. (1986). *The Chimpanzees of Gombe: Patterns of Behavior*. Cambridge, MA: Harvard University Press.

Green, S. (1975). Dialects in Japanese monkeys: vocal learning and cultural transmission of local-specific vocal behavior? *Zeitschrift für Tierpsychologie*, **38**, 304–14.

Hall, K. R. L. (1963). Tool-using performances as indicators of behavioral adaptability. *Current Anthropology*, **4**, 479–94.

Hamilton, W. J. Jr., Buskirk, R. E. & Buskirk, W. H. (1975). Environmental determinants of object manipulation by chacma baboons (*Papio ursinus*) in two southern African environments. *Journal of Human Evolution*, **7**, 205–16.

Hannah, A. C. & McGrew, W. C. (1987). Chimpanzees using stones to crack open oil palm nuts in Liberia. *Primates*, **28**, 31–46.

Harrison, K. E. & Byrne, R. W. (2000). Hand preferences in unimanual and bimanual feeding by wild vervet monkeys (*Cercopithecus aethiops*). *Journal of Comparative Psychology*, **114**, 13–21.

Hashimoto, C., Furuichi, T. & Tashiro, Y. (2000). Ant dipping and meat eating by wild chimpanzees in the Kalinzu Forest, Uganda. *Primates*, **41**, 103–8.

Hauser, M. D. (1988). Invention and social transmission: new data from wild vervet monkeys. In *Machiavellian Intelligence: Social Expertise and the Evolution of Intellect in Monkeys, Apes, and Humans*, ed. R. Byrne & A. Whiten, pp. 327–43. Oxford: Clarendon Press.

Hill, W. C. O. (1960). *Primates: Comparative Anatomy and Taxonomy, IV, Cebidae, Part A*. Edinburgh: Edinburgh University Press.

Hohmann, G. (1988). A case of simple tool use in wild liontailed macaques (*Macaca silenus*). *Primates*, **29**, 565–7.

Huffman, M. A. (1984). Stone-play of *Macaca fuscata* in Arashiyama B troop: transmission of a non-adaptive behavior. *Journal of Human Evolution*, **13**, 725–35.

Huffman, M. A. & Kalunde, M. S. (1993). Tool-assisted predation on a squirrel by a female chimpanzee in the Mahale Mountains, Tanzania. *Primates*, **34**, 93–8.

Humle, T. (1999). New record of fishing for termites (*Macrotermes*) by the chimpanzees of Bossou (*Pan troglodytes verus*), Guinea. *Pan Africa News*, **6**, 3–4.

Humphrey, N. K. (1976). The social function of intellect. In *Growing Points in Ethology*, ed. P. Bateson & R. A. Hinde, pp. 303–21. Cambridge, UK: Cambridge University Press.

Hundrey, M. (1963). Notes on methods of feeding and the use of tools in the Geospizinae. *Auk*, **80**, 372–3.

Hunt, G. R. (1996). Manufacture and use of hook-tools by New Caledonian crows. *Nature*, **379**, 249–51.

    (2000). Human-like, population-level specialization in the manufacture of pandanus tools by New Caledonian crows *Corvus moneduloides*. *Proceedings of the Royal Society of London, Series B, Biological Sciences*, **267**, 403–13.

Hunt, K. D. (2000). Initiation of a new chimpanzee study site at Semliki-Toro Wildlife Reserve, Uganda. *Pan Africa News*, **7**, 14–6.

Hunt, K. D. & McGrew, W. C. (2002). Chimpanzees in the dry habitats of Assirik, Senegal and Semliki Wildlife Reserve, Uganda. In *Behavioural Diversity in Chimpanzees and Bonobos*, ed. C. Boesch, G. Hohmann & L. F. Marchant, pp. 35–51. Cambridge, UK: Cambridge University Press.

Itani, J. & Nishimura, A. (1973). The study of infra-human culture in Japan. In *Precultural Primate Behavior*, ed. E. Menzel, pp. 26–50. Basel: Karger.

Izawa, K. (1978). Frog-eating behavior of wild black-capped capuchin (*Cebus apella*). *Primates*, **19**, 633–42.

(1979). Foods and feeding behavior of wild black-capped capuchin (*Cebus apella*). *Primates*, **20**, 57–76.

(1982). *Nihonzaru no Seitai: Gousetsu no Hakusan ni Yasei wo Tou.* [*Ecology of Japanese Macaques: Considering the Wilderness in the Heavy-Snowing Hakusan Mountains*]. (In Japanese.) Tokyo: Doubutsu-Sha.

Izawa, K. & Itani, J. (1966). Chimpanzees in Kasakati Basin, Tanganyika. (1): Ecological study in the rainy season 1963–1964. *Kyoto University African Studies*, **1**, 73–156.

Izawa, K. & Mizuno, A. (1977). Palm-fruit cracking behavior of wild black-capped capuchin (*Cebus apella*). *Primates*, **18**, 773–92.

Janson, C. H. & Boinski, S. (1992). Morphological and behavioral adaptations for foraging in generalist primates: the case of the cebines. *American Journal of Physical Anthropology*, **88**, 483–98.

Jolly, A. (1964). Prosimians' manipulation of simple object problems. *Animal Behaviour*, **12**, 560–77.

Jolly, C. F. (1970). The seed-eaters: a new model of hominid differentiation based on a baboon analogy. *Man*, **5**, 5–28.

Kawamura, S. (1954). Yasei nihon-zaru no shokuji-koudou ni arawareta atarashii koudou-kei: Doubutsu ni okeru culture no bunseki. [Newly-emerged behavioural patterns in feeding behaviour of wild Japanese macaques: an analysis of a culture in animals.] (In Japanese.) *Seibutsu Shinka*, **2**(1), 10–13.

(1959). The process of sub-culture propagation among Japanese macaques. *Primates*, **2**, 43–60.

Kortlandt, A. & Kooij, M. (1963). Protohominid behaviour in primates (preliminary communication). *Symposia of the Zoological Society of London*, **10**, 61–88.

Kuruvilla, G. P. (1980). Ecology of the bonnet macaque (*Macaca radiata* Geoffroy) with special reference to feeding habits. *Journal of Bombay Natural History Society*, **75**, 976–88.

Lambert, J. E. (1999). Seed handling in chimpanzees (*Pan troglodytes*) and redtail monkeys (*Cercopithecus ascanius*): implications for understanding hominoid and cercopithecine fruit-processing strategies and seed dispersal. *American Journal of Physical Anthropology*, **109**, 365–86.

Langguth, A & Alonso, C. (1997). Capuchin monkeys in the Caatinga: tool use and food habits during drought. *Neotropical Primates*, **5**, 77–8.

Lanjouw, A. (2002). Behavioural adaptations to water scarcity in Tongo chimpanzees. In *Behavioural Diversity in Chimpanzees and Bonobos*, ed. C. Boesch, G.

Hohmann & L. F. Marchant, pp. 52–60. Cambridge, UK: Cambridge University Press.

Lee, R. B. & DeVore, I. (eds.) (1968). *Man the Hunter*. New York: Aldine de Gruyter.

Lefebvre, L. (1995). Culturally-transmitted feeding behaviour in primates: evidence for accelerating learning rates. *Primates*, **36**, 227–39.

McGrew, W. C. (1974). Tool use by wild chimpanzees in feeding upon driver ants. *Journal of Human Evolution*, **3**, 501–8.

(1992). *Chimpanzee Material Culture: Implications for Human Evolution*. Cambridge, UK: Cambridge University Press.

(2001). The other faunivory: primate early human diet. In *Meat-eating and Human Evolution*, ed. C. B. Stanford & H. T. Bunn, pp. 160–78. Oxford: Oxford University Press.

McGrew, W. C. & Marchant, L. F. (1997). Using the tools at hand: manual laterality and elementary technology in *Cebus* spp. and *Pan* spp. *International Journal of Primatology*, **18**, 787–810.

McNeilage, P. F., Studert-Kennedy, M. & Lindblom, B. (1987). Primate handedness reconsidered. *Behavioral and Brain Sciences*, **10**, 247–303.

Marais, E. (1969). *The Soul of the Ape*. Harmondsworth, UK: Penguin Books.

Marzke, M. W. & Wullstein, K. L. (1996). Chimpanzee and human grips: a new classification with a focus on evolutionary morphology. *International Journal of Primatology*, **17**, 117–39.

Matsusaka, T. & Kutsukake, N. (2002). Use of leaf-sponge and leaf spoon by juvenile chimpanzees at Mahale. *Pan Africa News*, **9**, 6–9.

Matsuzawa, T. (1991a). Nesting cups and metatools in chimpanzees. *Behavioral and Brain Sciences*, **14**, 570–1.

(1991b). *Chimpanzee Mind*. (In Japanese.) Tokyo: Iwanami-Shoten.

(1996). Chimpanzee intelligence in nature and in captivity: isomorphism of symbol use and tool use. In *Great Ape Societies*, ed. W. C. McGrew, L. F. Marchant & T. Nishida, pp. 196–209. Cambridge, UK: Cambridge University Press.

Matsuzawa, T., Yamakoshi, G. & Humle, T. (1996). A newly found tool-use by wild chimpanzees: algae scooping. (In Japanese.) *Primate Research*, **12**, 283.

Menzel, C. R. (1997). Primates' knowledge of their natural habitat: as indicated in foraging. In *Machiavellian Intelligence II: Extensions and Evaluations*, ed. A.

Whiten & R. W. Byrne, pp. 207–39. Cambridge, UK: Cambridge University Press.

Merfield, F. G. & Miller, H. (1956). *Gorilla Hunter: The African Adventures of a Hunter Extraordinary*. New York: Farrar, Straus & Cudahy.

Milton, K. (1981). Distribution patterns of tropical plant foods as a stimulus to primate mental development. *American Anthropologist*, **83**, 534–48.

Nakagawa, N. (1990). Choice of food patches by Japanese monkeys (*Macaca fuscata*). *American Journal of Primatology*, **21**, 17–29.

Nakamichi, M., Kato, E., Kojima, Y. & Itoigawa, N. (1998). Carrying and washing of grass roots by free-ranging Japanese macaques at Katsuyama. *Folia Primatologica*, **69**, 35–40.

Napier, J. R. (1962). The evolution of the hand. *Scientific American*, **207**, 56–62.

Nishida, T. (1973). The ant-gathering behaviour by the use of tools among wild chimpanzees of the Mahale Mountains. *Journal of Human Evolution*, **2**, 357–70.

Nishida, T. & Hiraiwa, M. (1982). Natural history of a tool-using behavior by wild chimpanzees in feeding upon wood-boring ants. *Journal of Human Evolution*, **11**, 73–99.

Nishida, T. & Uehara, S. (1980). Chimpanzees, tools, and termites: another example from Tanzania. *Current Anthropology*, **21**, 671–2.

Nishida, T., Wrangham, R. W., Goodall, J. & Uehara, S. (1983). Local differences in plant-feeding habits of chimpanzees between the Mahale Mountains and Gombe National Park, Tanzania. *Journal of Human Evolution*, **12**, 467–80.

Orenstein, R. (1972). Tool use by the New Caledonian crow (*Corvus moneduloides*). Auk, **89**, 674–6.

Oyen, O. J. (1978). Stone-eating and tool-use among olive baboons. *Texas Journal of Science*, **30**, 295.

Panger, M. A. (1998). Object-use in free-ranging white-faced capuchins (*Cebus capucinus*) in Costa Rica. *American Journal of Physical Anthropology*, **106**, 311–21.

Panger, M. A., Perry, S., Rose, L., Gros-Louis, J., Vogel, E., Mackinnon, K. C. & Baker, M. (2002). Cross-site differences in foraging behavior of white-faced capuchins (*Cebus capucinus*). *American Journal of Physical Anthropology*, **119**, 52–66.

Parker, C. E. (1974a). Behavioral diversity in ten species of nonhuman primates. *Journal of Comparative and Physiological Psychology*, **87**, 930–7.

(1974b). The antecedents of man the manipulator. *Journal of Human Evolution*, **3**, 493–500.

Parker, S. T. & Gibson, K. R. (1977). Object manipulation, tool use, and sensorimotor intelligence as adaptations in *Cebus* monkeys and great apes. *Journal of Human Evolution*, **6**, 623–41.

(1979). A developmental model for the evolution of language and intelligence in early hominids. *Behavioral and Brain Sciences*, **2**, 367–408.

Peters, C. R. & O'Brien, E. M. (1981). The early hominid plant-food niche: insights from an analysis of plant exploitation by *Homo*, *Pan*, and *Papio* in Eastern and Southern Africa. *Current Anthropology*, **22**, 127–40.

Phillips, K. A. (1998). Tool use in wild capuchin monkeys (*Cebus albifrons trinitatis*). *American Journal of Primatology*, **46**, 259–61.

Plooij, F. X. (1978). Tool-use during chimpanzees' bushpig hunt. *Carnivore*, **1**, 103–6.

Quiatt, D. & Kiwede, Z. T. (1994). Leaf sponge drinking by a Budongo Forest chimpanzee. *American Journal of Primatology*, **33**, 236.

Reader, S. M. & Laland, K. N. (2001). Primate innovation: sex, age and social rank differences. *International Journal of Primatology*, **22**, 787–805.

Rijksen, H. D. (1978). *A Field Study on Sumatran Orang Utans* (Pongo pygmaeus abelii *Lesson 1827*): *Ecology, Behaviour and Conservation*. Wageningen: H. Veenman and Zonen B. V.

Robinson, J. G. (1986). Seasonal variation in use of time and space by the wedge-capped capuchin monkey, *Cebus olivaceus*: implications for foraging theory. *Smithsonian Contributions to Zoology*, **431**, 1–60.

Rose, L. M. (2001). Meat and the early human diet: insight from Neotropical primate studies. In *Meat-eating and Human Evolution*, ed. C. B. Stanford & H. T. Bunn, pp. 141–59. Oxford: Oxford University Press.

Rosenberger, A. L. (1992). Evolution of feeding niches in New World monkeys. *American Journal of Physical Anthropology*, **88**, 525–62.

Russon, A. E. (1997). Exploiting the expertise of others. In *Machiavellian Intelligence II: Extensions and Evaluations*, ed. A. Whiten & R. W. Byrne, pp. 174–206. Cambridge, UK: Cambridge University Press.

(1998). The nature and evolution of intelligence in orangutans (*Pongo pygmaeus*). *Primates*, **39**, 485–503.

(2003). Developmental perspectives on great ape traditions. In *Towards a Biology of Traditions: Models and Evidence*, ed. D. Fragaszy & S. Perry, pp. 329–64. Cambridge, UK: Cambridge University Press.

Russon, A. E., Bard, K. A. & Parker, S. T. (ed.) (1996). *Reaching into Thought: The Minds of the Great Apes*. Cambridge, UK: Cambridge University Press.

Rylands, A. B. (1987). Primate communities in Amazonian forests: their habitats and food resources. *Experientia*, 43, 265–79.

Sabater Pi, J., Bermejo, M., Illera, G & Vea, J. J. (1993). Behavior of bonobos (*Pan paniscus*) following their capture of monkeys in Zaire. *International Journal of Primatology*, 14, 797–804.

Savage, T. S. & Wyman, J. (1843–44). Observation on the external characters and habits of the *Troglodytes niger* Geoff. *Boston Journal of Natural History*, 4, 362–86.

Stanford, C. B. & Bunn, H. T. (ed.) (2001). *Meat-Eating and Human Evolution*. New York: Oxford University Press.

Stanford, C. B., Gambaneza, C., Nkurunungi, J. B. & Goldsmith, M. L. (2000). Chimpanzees in Bwindi-Impenetrable National Park, Uganda, use different tools to obtain different types of honey. *Primates*, 41, 337–41.

Starin, E. D. (1990). Object manipulation by wild red colobus monkeys living in the Abuko Nature Reserve, the Gambia. *Primates*, 31, 385–91.

Stewart, C.-B. & Disotell, T. R. (1998). Primate evolution: in and out of Africa. *Current Biology*, 8, R582–88.

Stokes, E. J. & Byrne, R. W. (2001). Cognitive capacities for behavioural flexibility in wild chimpanzees (*Pan troglodytes*): the effect of snare injury on complex manual food processing. *Animal Cognition*, 4, 11–28.

Struhsaker, T. T. & Leland, L. (1977). Palm-nut smashing by *Cebus apella* in Colombia. *Biotropica*, 9, 124–6.

Sugiyama, Y. (1994). Tool use by wild chimpanzees. *Nature*, 367, 327.

(1997). Social tradition and the use of tool-composites by wild chimpanzees. *Evolutionary Anthropology*, 6, 23–7.

Sugiyama, Y. & Koman, J. (1979). Tool-using and making behavior in wild chimpanzees at Bossou, Guinea. *Primates*, 20, 513–24.

Sugiyama, Y., Koman, J. & Sow, M. B. (1988). Ant-catching wands of wild chimpanzees at Bossou, Guinea. *Folia Primatologica*, 51, 56–60.

Suzuki, A. (1965). An ecological study of wild Japanese monkeys in snowy areas: focused on their food habits. *Primates*, 6, 31–72.

(1966). On the insect-eating habits among wild chimpanzees living in the savanna woodland of Western Tanzania. *Primates*, 7, 481–7.

Suzuki, S., Hill, D. A., Maruhashi, T. & Tsukahara, T. (1990). Frog- and lizard-eating behaviour of wild Japanese macaques in Yakushima, Japan. *Primates*, 31, 421–26.

Suzuki, S., Kuroda, S. & Nishihara, T. (1995). Tool-set for termite-fishing by chimpanzees in the Ndoki Forest, Congo. *Behaviour*, 132, 219–35.

Takahata, Y., Hasegawa, T. & Nishida, T. (1984). Chimpanzee predation in the Mahale Mountains from August 1979 to May 1982. *International Journal of Primatology*, 5, 213–33.

Tebbich, S., Taborsky, M., Fessl, B. & Dvorak, M. (2002). The ecology of tool-use in the woodpecker finch *Cactospiza pallida*. *Ecology Letters*, 5, 656–664.

Teleki, G. (1973a). The omnivorous chimpanzee. *Scientific American*, 228, 32–42.

(1973b). *The Predatory Behavior of Wild Chimpanzees*. Lewisburg, PA: Bucknell University Press.

Terborgh, J. (1983). *Five New World Primates: A Study in Comparative Ecology*. Princeton, MA: Princeton University Press.

Thorington, R. W. Jr. (1967). Feeding and activity of *Cebus* and *Saimiri* in a Colombian forest. In *Progress in Primatology*, ed. D. Starck, R. Schneider & H. J. Kuhn, pp. 180–4. Stuttgart: Gustav Fischer Verlag.

Tomasello, M. & Call, J. (1997). *Primate Cognition*. New York: Oxford University Press.

Torigoe, T. (1985). Comparison of object manipulation among 74 species of non-human primates. *Primates*, 26, 182–94.

Tutin, C. E. G., Ham, R. & Wrogemann, D. (1995). Tool-use by chimpanzees (*Pan t. troglodytes*) in the Lope Reserve, Gabon. *Primates*, 42, 1–24.

Tutin, C. E. G., Parnell, R. J. & White, F. (1996). Protecting seeds from primates: Examples from *Diospyros* spp. in the Lopé Reserve, Gabon. *Journal of Tropical Ecology*, 12, 371–84.

Uehara, S. (1982). Seasonal changes in the techniques employed by wild chimpanzees in the Mahale Mountains, Tanzania, to feed on termites (*Pseudacanthotermes spiniger*). *Folia Primatologica*, 37, 44–76.

van Schaik, C. P. & Knott, C. D. (2001). Geographic variation in tool use on *Neesia* fruits in orangutans. *American Journal of Physical Anthropology*, 114, 331–42.

van Schaik, C. P., Deaner, R. O. & Merrill, M. Y. (1999). The conditions for tool use in primates: implications for the evolution of material culture. *Journal of Human Evolution*, **36**, 719–41.

van Schaik, C. P., Fox, E. A. & Sitompul, A. F. (1996). Manufacture and use of tools in wild Sumatran orangutans. *Naturwissenschaften*, **83**, 186–8.

Visalberghi, E., Fragaszy, D. M. & Savage-Rumbaugh, E. S. (1995). Performance in a tool-using task by common chimpanzees (*Pan troglodytes*), bonobos (*Pan paniscus*), an orangutan (*Pongo pygmaeus*), and capuchin monkeys (*Cebus apella*). *Journal of Comparative Psychology*, **109**, 52–60.

Waser, P. (1977). Feeding, ranging and group size in the mangabey *Cercocebus albigena*. In *Primate Ecology: Studies of Feeding and Ranging Behaviour in Lemurs, Monkeys and Apes*, ed. T. H. Clutton-Brock, pp. 504–38. London: Academic Press.

Wheatley, B. P. (1980). Feeding and ranging of East Bornean *Macaca fascicularis*. In *The Macaques: Studies in Ecology, Behavior, and Evolution*, ed. D. G. Lindburg, pp. 215–46. New York: Van Nostrand Reinhold.

(1988). Cultural behavior and extractive foraging in *Macaca fascicularis*. *Current Anthropology*, **29**, 516–19.

(1999). *The Sacred Monkeys of Bali*. Prospect Heights, IL: Waveland Press.

Whitesides, G. H. (1985). Nut cracking by wild chimpanzees in Sierra Leone, West Africa. *Primates*, **26**, 91–4.

Whiten, A., Goodall, J., McGrew, W. C., Nishida, T., Reynolds, V., Sugiyama, Y., Tutin, C. E. G., Wrangham, R. W. & Boesch, C. (1999). Cultures in chimpanzees. *Nature*, **399**, 682–5.

Williams, J. B. (1984). *Tool Use by Nonhuman Primates: A Bibliography, 1975–1983*. Seattle, WA: Primate Information Center, University of Washington.

(1992). *Tool Use by Nonhuman Primates: A Bibliography, 1983–1992*. Seattle, WA: Primate Information Center, University of Washington.

Wrangham, R. W. (1977). Feeding behaviour of chimpanzees in Gombe National Park, Tanzania. In *Primate Ecology: Studies of Feeding and Ranging Behaviour in Lemurs, Monkeys and Apes*, ed. T. H. Clutton-Brock, pp 504–38. London: Academic Press.

Wrangham, R. W., Holland Jones, J., Laden, G., Pilbeam, D. & Conklin-Brittain, N. L. (1999). The raw and stolen: cooking and the ecology of human origins. *Current Anthropology*, **40**, 567–94.

Yamakoshi, G. (2001). Ecology of tool use in wild chimpanzees: toward reconstruction of early hominid evolution. In *Primate Origins of Human Cognition and Behavior*, ed. T. Matsuzawa, pp. 537–56. Tokyo: Springer-Verlag.

Yamakoshi, G. & Sugiyama, Y. (1995). Pestle-pounding behavior of wild chimpanzees at Bossou, Guinea: a newly observed tool-using behavior. *Primates*, **36**, 489–500.

10 • The special demands of great ape locomotion and posture

KEVIN D. HUNT

Department of Anthropology, Indiana University, Bloomington

INTRODUCTION

Amidst the welter of competencies that could be labeled "intelligence," the great apes repeatedly demonstrate numerous high-level abilities that distinguish them from other mammals and ally them with humans (Griffin 1982; Parker & Gibson 1990; Russon, Bard & Parker 1996; Suddendorf & Whiten 2001). Self-concept is argued to be among this set of distinctive abilities. It is often viewed as an integral aspect of advanced intelligence, one that some have argued allows great apes to have a theory of mind (Heyes 1998 and references therein). Among the abilities that co-occur with it in humans are symbolic play, simple altruism, reciprocal relationships, a concept of planning, and pleasure in completion of complex tasks (Povinelli & Cant 1995).

Until recently, the demands of locomotion and posture, together referred to as positional behavior (Prost 1965), were not explicitly considered to correlate with any aspect of primate intelligence or its evolution, self-concept included. Primate intelligence is most often hypothesized to have evolved either for negotiating complex social problems, or for mapping and resolving complicated foraging challenges (for an overview, see Russon, Chapter 1, this volume). Chevalier-Skolnikoff, Galdikas and Skolnikoff (1982: 650) suggested instead that, at least for orangutans, locomotor demands were "the single major function for which the advanced cognitive abilities . . . evolved." Povinelli and Cant (1995) subsequently refined and expanded this hypothesis, asserting that self-concept in orangutans evolved to enable these large-bodied apes to negotiate thin, compliant (i.e., flexible) branches during suspensory locomotor bouts, particularly when crossing gaps in the canopy. They hypothesized that the unpredictable response of compliant, weight-bearing structures when weight is transferred onto them, the need for several such structures to

support the weight of a single individual, and the erratic orientation of supports together require that large primates such as great apes have an "ability to engage in a type of mental experimentation or simulation in which one is able to plan actions and predict their likely consequence before acting" (Povinelli & Cant 1995: 409). In order to move safely in the forest canopy, orangutans and perhaps other great apes must be able to step outside themselves and imagine how their body and its movements will affect fragile, easily deformable branches and twigs. I will refer to these argument as the "Povinelli and Cant hypothesis," cognizant of Chevalier-Skolnikoff *et al.*'s contribution.

This hypothesis is consistent with evidence that only massive primates, the great apes, have a concept of self. Evidence rests heavily on one measure, mirror self-recognition (MSR), which is often taken as particularly informative about self-concept. Gallup (1970, 1982, 1991) forcefully argued that MSR is found only in species that possess a self-concept, and Parker (1996) contended it is displayed only in species that also display high-level imitation. Chimpanzees and orangutans consistently recognize themselves in mirrors, as do a few gorillas, whereas other nonhuman primates do not (Gallup 1970; Lethmate & Ducker 1973; Miles 1994; Nicholson & Gould 1995; Patterson 1984; Patterson & Cohn 1994; Suarez & Gallup 1981; Swartz *et al.* 1999; see reviews by Gallup 1991; Inoue-Nakamura 1997).[1] Although other capacities that co-occur with self-concept, such as symbolic play, simple altruism, reciprocal relationships, a concept of planning, and pleasure in completion of complex tasks, are not clearly identifiable in any great ape, narratives of their daily lives in captivity and in the wild convince me they have these capacities.

From the positional side, this hypothesis has not been systematically evaluated. This chapter attempts to craft informed estimates of locomotor and postural frequencies for each of the apes in order to place positional behavior in the context of Povinelli and Cant hypothesis, as well as other prominent hypotheses on the evolution of great ape intelligence, namely foraging-related ecological pressures and social pressures.

BACKGROUND

The connection between primate positional behavior and self-concept or other higher cognitive capabilities receives *prima facie* support from research on great apes – they are unusually suspensory. However, quantitative studies of apes' positional behavior are relatively recent and the meaning of these data is still in contention. Perhaps one source of the contention is that positional behavior theory has a long history, and thus a deep timescale to add heft to opposing hypotheses. Currently, two distinct positional modes (or categories – modes will be used here) are most often argued to be responsible for ape anatomy: vertical climbing and arm-hanging. The two modes have quite different demands relative to the Povinelli and Cant hypothesis.

Early research on ape functional anatomy was grounded in anatomical research, a field already well developed by the nineteenth century (Owen 1835; Savage & Wyman 1847; Tyson 1699), rather than in ape positional behavior study, which began in earnest only in the 1960s. Keith's (1891) contention that brachiation was the behavior for which ape specializations were evolved permeated early research on ape positional behavior. Keith and other anatomists argued that adaptation to hand-over-hand under-branch suspensory locomotion ("brachiation") selected for shared ape traits such as long forelimbs, long, curved digits, mobile shoulders, elongated scapulae, broad (i.e., human-like) torsos, short, stiff backs, taillessness, and a predominance of muscles that flex the elbow, extend the humerus, and raise the arm. Comparison of ape and monkey muscle weights largely supported Keith's hypothesis (Ashton & Oxnard 1964).

Data on wild ape behavior failed to corroborate the brachiation hypothesis. Mountain gorillas (Tuttle & Watts 1985 and references therein), chimpanzees (Goodall 1968; Reynolds 1965) and even orangutans (Harrison 1962) brachiated less than theory demanded. Although brachiation made up more than 50% of locomotion among hylobatids (Fleagle 1980), 20% among bonobos (Susman 1984), and more than 10% in orangutans (Cant 1987a), another mode, "quadrumanous climbing" (i.e., "four-handed" movement in which feet and hands grip a support), was even more common among great apes: 31% in orangutans, and 31% in bonobos. Quadrumanous climbing quickly replaced brachiation as the positional mode for which ape "brachiating" characters were considered to have evolved (Cartmill & Milton 1977; Fleagle 1976; Kortlandt 1974; Mendel 1976; Tuttle 1975; Tuttle, Basmajran & Ishida. 1979). The mode lacked a widely agreed upon, rigorous definition, but it has encompassed, among other behaviors, brachiation, quadrupedal walking on slightly inclined boughs, irregular-gait walking on thin supports, vertical climbing, gap crossing suspensory behaviors, clambering (a hindlimb assisted brachiation), and forelimb-assisted bipedalism. The more suspensory of these behaviors are those that Povinelli and Cant hypothesize to be related to self-concept in orangutans, but other behaviors are more similar to quadrupedal walking or bipedalism. Because quadrumanous climbing conflates kinematically different behaviors that require different anatomical adaptations, it seems to have outlived its usefulness. Hunt *et al.* (1996) strongly recommended discarding the term entirely and instead reporting its constituent modes separately.

Of the component positional modes in quadrumanous climbing, vertical climbing was often singled out as the most important shared ape locomotor mode. Long arms were hypothesized to facilitate ascending large diameter trunks (Cartmill 1974; Kortlandt 1974), and vertical climbing on smaller diameter supports was argued to require shoulder mobility to allow alternate reaching for new handholds. Large muscles that retract the humerus and flex the elbow were seen as vertical climbing propulsors (Fleagle *et al.* 1981; Jungers, Fleagle & Simons 1982).

Notably, vertical climbing does not pose the sorts of intellectual demands that Povinelli and Cant link to suspension. Vertical supports are not compliant, either because they are large (hence the need for a robust, divergent great toe in apes) and do not deform under weight, or because smaller supports are stabilized by the weight of the suspended climber, in particular by weight depending on the trailing hindfoot, which makes deformation minor and predictable.

Quantitative positional behavior data on chimpanzees (Hunt 1989, 1991a,b) provided only partial support for a vertical climbing hypothesis. Hunt's data showed that vertical climbing was not dramatically more common in apes than monkeys (0.9% of behavior versus 0.5%), and large diameter vertical climbing was rare. Unimanual forelimb suspension (arm-hanging) was more common than anticipated, and much more common among chimpanzee than monkeys (4.4% versus 0.0%). Hunt suggested that ape shoulder mobility allows much greater joint excursion than is necessary for vertical climbing. He suggested that shoulder mobility, scapula shape, torso shape, wrist mobility, and some muscular specializations are adaptations to arm-hanging, but most ape muscular specializations and their gripping great toe fit a vertical climbing hypothesis. Finger curvature and length were suggested to be adaptations to arm-hanging and vertical climbing. Hunt's (1991a) review of ape positional behavior studies then available concluded that arm-hanging and vertical climbing were the behaviors most clearly identifiable as shared among all apes.

Doran (1989, 1996) disagreed. She argued for a return to a vertical-climbing-only hypothesis, since her data showed that "climbing" was more common than suspensory behaviors among Taï, Ivory Coast, chimpanzees. Her evidence in support of the vertical climbing hypothesis is weak, most importantly because vertical climbing was not one of her locomotor categories. As currently conceived (most eloquently by Fleagle *et al.* 1981), the climbing hypothesis is a *vertical* climbing hypothesis. The mode Doran sometimes refers to as "climbing" (e.g., Doran 1996) is not vertical climbing, but short-hand for the catch-all mode "quadrumanous climbing and scrambling" (Doran 1989: 328). Whereas most anatomists read "vertical climbing" when Doran writes "climbing," her climbing mode pooled suspensory modes (such as clambering, bridging, tree swaying), quadrupedalism (scrambling), and an unknowable proportion of true vertical climbing. In contrast to this liberality, her suspensory mode was narrowly defined to include only "alternating hand to hand progression beneath substrate" (Doran 1989: 328).

In this chapter I attempt to adjust for this and other biases to craft informed estimates of locomotor and postural frequencies for each ape species, after which I place positional behavior in the context of the Povinelli and Cant and other hypotheses on great ape intelligence and its evolution. I standardized and recalculated available data to allow comparability. Rather than providing ranges of possible frequencies or qualitative estimates, I

provide exact values, but offer reliability judgments to offset this false accuracy. I formulate predictions drawn from Povinelli and Cant's hypothesis, and then test them against positional behavior estimates. My aims are to work towards resolving debates over how great ape positional behavior should be characterized, and to apply these findings to the question of whether some distinctively great ape forms of arboreal positional behavior demand high-level intelligence that may take the form of a self-concept.

Like others, I assume that cognitive capacities, which rely on expensive brain tissue, are unlikely to have evolved or to be maintained unless they serve important functions (see Russon, Chapter 1, this volume), and therefore that living species that have a self-concept use it.

POVINELLI AND CANT PREDICTIONS

It is the non-stereotyped, figure-it-out-as-you-go nature of some locomotor or postural modes that is central to Povinelli and Cant's argument. They argue that primates that locomote on stable supports, which are stable either because the animal is light or the support is large, locomote using stereotyped, preprogrammed movements (cognitively simple action schemata). These movements are less cognitively challenging than those on unstable supports. Movement on compliant or fragile supports must be planned, and plans must be adjusted moment to moment as supports are found to be more or less compliant than estimated. Highly intelligent primates may be those that must locomote in a more moment-to-moment, calculating, context-contingent manner. I will call these cognitively challenging positional repertoires *self-concept eliciting positional regimes* (SCEPRs), and I will refer to individual modes as SCEP modes.

Chevalier-Skolnikoff *et al.* (1982) and Povinelli and Cant (1995) conceived of the SCEPR as a locomotor repertoire. I argue that postures can require a work-it-out-as-you-go approach as well. An orangutan may walk on a large support to the periphery of a tree, but reaching out, grasping a small support among the terminal branches, and assuming an arm-hanging posture requires the consideration of the compliance and fragility of supports and an accommodation to unexpected compliance. Arm-hanging chimpanzees may make a number of small adjustments to posture (e.g., gripping a different support with one foot, but leaving the other grips unchanged) that can leave them, over a period of minutes, meters from their starting point and suspended from completely different supports, without ever locomoting. These postural behaviors require individuals to be aware of and respond to various degrees of compliance.

The following testable predictions grow out of the Povinelli and Cant hypothesis:

(1) Great apes that have demonstrated the ability to form self-concepts will have SCEPRs, and vice versa.
(2) If the 11 kg siamang has a SCEPR compared to the anatomically near-identical 6 kg gibbon, the siamang should have a more cognitively sophisticated self-concept than gibbons.
(3) Species with great body weight dimorphism and similar SCEPRs, or with great differences in SCEPR between the sexes, should exhibit sex differences in self-concept.
(4) In comparisons among species, the more common SCEP modes are in a species' positional repertoire, the more compliant supports are, and/or the more critical SCEP modes are to survival, the more robust and sophisticated should be self-conception.

POSITIONAL MODE DEFINITIONS

I followed Hunt et al.'s (1996) positional mode definitions, and greater detail is presented there. Here, categories such as "sit" and "lie" need no elaboration. Other modes that have been defined differently in different studies require some explanation.

"Stand" is quadrupedal or tripedal posture (P4 in Hunt et al.). In the "biped" mode weight is borne by hindlimbs, usually without significant assistance from the forelimbs (Hunt et al. mode P5). In the "squat" (P2) mode the heels only contact the support. "Cling" is a torso orthograde (i.e., erect) posture where hands and feet grip a relatively vertical support; the elbows and knees are quite flexed (P3). "Arm-hang" (= forelimb-suspend, P8) is a one- or very rarely two-handed forelimb suspension, typically engaged in on small-diameter and therefore compliant supports, sometimes assisted by a hindlimb (P8a). "Arm–foot hang" (P9a,b) is suspension from a foot and a hand; the torso is parallel to the ground, usually engaged in on relatively small supports. Both postures are argued to exert the same sorts of selective pressures as suspensory locomotion. Both apply to

the forest's horizontal structure, where Povinelli and Cant argue the greatest locomotory difficulties occur.

Among locomotor modes "walk" (L1), "leap" (L12), and "run" (L5) are straightforward. "Climbing" throughout means "vertical climbing" (L8). It refers to a behavior wherein the individual ascends or descends a vertical or near-vertical support much as a person would ascend or descend on a ladder. "Bipedal" includes both walking and running, using hindlimbs alone and forelimbs only for incidental support. Chimpanzees use it on relatively large supports (Hunt 1989). "Scramble" (L1c(1)) is quadrupedal walking on small, often flexible, approximately horizontal supports. Orientation of supports is irregular, and the gait itself looks irregular in consequence. Scrambling requires some appreciation of compliance. "Brachiate" refers to hand-over-hand suspensory movement underneath branches, and includes the rapid, stereotyped ricochetal brachiation of gibbons. "Clamber" is a torso-upright suspensory locomotion different from brachiation in that the hindlimbs also provide support, with their grip above the center of gravity of the individual, in orangutans, often near the ear (Cant 1987a). "Suspensory" is a miscellaneous category that encompasses below-branch behaviors that cannot be considered brachiation or clamber, such as tree sway. "Transfer" (L9f) often begins with bimanual forelimb suspension, and may contain a brachiation-like gap-closing motion (a "lunge"), wherein a hand grasps a small support in an adjacent tree, after which a branch is pulled toward the animal with a hand-over-hand or hand-over-foot motion. Weight is gradually transferred to the adjacent tree. The torso remains more or less orthograde throughout; more weight is born by the forelimbs than the hindlimbs.

These last five modes, scramble, brachiate, clamber, suspensory movement, and transfer are all used on small, flexible supports and require awareness of support compliance and fragility. These modes, along with the two postural modes (arm-hanging and arm–foot hanging), form the core of a SCEPR.

Biases

Studies reviewed here utilized four sampling modes, instantaneous (focal), instantaneous (scan), continuous (bout) (Altmann 1974), and continuous (meters/kilometer) (Tuttle & Watts 1985). Recent work suggests these sampling methods are rather comparable (Doran 1992). Instantaneous scan sampling theoretically yields positional mode frequencies that are quite similar to those produced by instantaneous focal sampling (Altmann 1974). Continuous bout sampling under-represents long-duration bouts and over-represents short-duration bouts. In theory, comparability between instantaneous sampling and bout sampling is not expected. In practice, the two sampling regimes yield quite similar positional mode frequencies because bout lengths vary little (Doran 1992). Meters/kilometers and bout sampling regimes would yield identical figures if velocity were constant, and it is rather constant in chimpanzees (Hunt 1989) and probably other species. I will assume figures based on meters/kilometer and bout sampling are roughly equivalent, based in part on the comparability of instantaneous and bout sampling.

As positional data have accumulated, it has become apparent that positional mode frequency estimations for regimes with only five or ten modes are relatively robust with respect to sampling differences. Table 10.1 includes two studies of different hylobatids that yielded quite similar mode frequencies, despite having been conducted by different researchers on different species, at different times, and at different sites. Three studies of bonobo locomotion had sample sizes that varied by an order of magnitude, yet they yielded quite similar mode frequencies (Table 10.4). It seems that when Ns reach 100 or so, mode frequencies are rather reliable even in the face of large sample size differences.

A second bias is introduced by differences in the level of habituation to human observation. Poorly habituated individuals tend to run, leap and brachiate at unnaturally high frequencies. Unhabituated individuals are less likely to flee when arboreal, leading to oversampling of arboreal behaviors, while terrestrial behaviors are often undersampled because targets are obscured by foliage. Habituated individuals have higher frequencies of walking versus running, transferring versus leaping, posture versus locomotion, and terrestriality versus arboreality.

A common compromise when reporting data on poorly habituated subjects is reporting arboreal and terrestrial observations separately, under the assumption that even though terrestrial behaviors may be undersampled, the relative proportions of terrestrial modes to one another will be accurate. With a similar rationale, locomotion and posture are often reported separately, assuming that even if unhabituated animals locomote more often, the relative proportions of individual

locomotor modes is representative. Unfortunately, these divisions are sometimes perpetuated in later studies after subjects are habituated, in order to allow comparability.

There is little question that the best comparisons between species will be made on habituated subjects using methods that record relative frequencies of every positional mode in the study population's entire positional repertoire, whether locomotor or postural, and in both arboreal and terrestrial contexts. It is no surprise that studies with large sample sizes were conducted on populations habituated for a decade or more. Four pioneers, Goodall, Nishida, Boesch, and Fossey, habituated populations on which more than two-thirds of the observations below are based. Of course, short studies on unhabituated populations are vastly better than nothing. Here, I consider these potential biases before including data in tables. Sometimes I report data from short-term studies for the sake of completeness, but exclude them from calculations and discussion. To allow comparability, I calculated locomotor and postural mode frequencies separately.

The most serious bias in positional study is using non-comparable positional mode definitions. I attempt to compensate for this bias with adjustments explained below.

CALCULATIONS OF POSTURAL MODE FREQUENCIES

Hylobatids

Four studies have reported hylobatid postural mode frequencies (Table 10.1). I divided hylobatids into two groups, the siamang (*Hylobates syndactylus*) and other gibbon species. While anatomically similar, siamangs weigh approximately 11 kg, whereas gibbons average only 6 kg (Plavcan & van Schaik 1997; Smith & Jungers 1997). Larger primates leap less and climb quadrumanously more (Fleagle 1976).

Gibbons

Two gibbon studies observed subjects in all behavioral contexts, rather than, for example, only during feeding or travel, and sample sizes, while small, are well above 100 (322 and 655). However, these data included only two postural modes, sit and arm-hang; I assume postural modes other than sit and arm-hang were rare. The average of the two studies is reported in Table 10.1.

Table 10.1. *Hylobatid postural modes (percentages)*

	Sit	Lie	Stand	Squat	Cling	Biped	Arm-hang	Hand–foot hang
Hylobates agilis[1]	65.5	0.0	0.0	0.0	0.0	0.0	34.5	0.0
Hylobates pileatus[2]	61.7	0.0	0.0	0.0	0.0	0.0	38.3	0.0
Gibbon average	**63.6**	**0.0**	**0.0**	**0.0**	**0.0**	**0.0**	**36.4**	**0.0**
Hylobates syndactylus[3]	47.0	0.0	0.0	0.0	0.0	0.0	53.0	0.0
Hylobates syndactylus[4]	38.3	0.0	0.0	0.0	0.0	0.0	61.7	0.0
Siamang best est.	**47.0**	**0.0**	**0.0**	**0.0**	**0.0**	**0.0**	**53.0**	**0.0**

Notes:

[1] Gittins (1983). Percentage of 322 bouts sampled by 10-minute scan surveys.

[2] Srikosamatara (1984). Percentage of 655 5-minute scan surveys.

[3] Chivers (1972). Percentage of 234 5-second instantaneous focal surveys.

[4] Fleagle (1976). Percentage of 1376 postural bouts during feeding.

Bold indicates those values are the best estimate for the taxon indicated. SCEP modes are shaded.

Siamang

One siamang study observed individuals only when feeding; a second recorded all behavioral contexts. Feeding observations undersample sitting and oversample arm-hanging (i.e., suspension), since frugivores arm-hang most often when gathering fruits. Only two postural modes (sit, arm-hang) were recorded, and sample sizes were small. I assume the broader study offers the better estimate, despite its small sample size.

Great apes

Orangutan

Three positional studies on orangutans yielded over 6000 observations. However, observations were limited to arboreal feeding in two studies, and to arboreal travel and resting in a third. The arboreal limitation likely introduces little bias because Bornean orangutans are highly arboreal (females nearly 100%, males 80%; Rodman 1979) and Sumatran orangutans are completely arboreal (Povinelli & Cant 1995). Context, however, may introduce bias. Standing and arm-hanging were much more common during travel and resting, whereas arm–foot hang was much more common during feeding. To adjust for this bias, frequencies were weighted by context (Table 10.2). Five studies have reported activity budgets (Galdikas 1978; MacKinnon 1977; Rijksen 1978; Rodman 1979; Wheatley 1982), from which I calculated an average activity budget of 42.7% feed, 39.6% rest, and 17.4% travel. I multiplied postural mode

frequencies during feeding by 0.427, and resting + travel by 0.396 + 0.174. Given the similarity of values between studies before weighting, the weighted average in Table 10.2 is a good estimate.

Bonobo

Bonobos are poorly habituated and therefore their posture is poorly characterized. The only study to date (Table 10.2) yielded 132 observations made on subjects feeding arboreally on fruit. Bonobos have terrestrial knuckle-walking adaptations virtually identical to those of chimpanzees, and their diets include significant amounts of terrestrial herbaceous vegetation (Malenky *et al.* 1994), suggesting they spend a significant amount of time on the ground. Since arboreal and terrestrial postures differ dramatically in apes, the absence of terrestrial observations likely introduces significant bias. These biases and the low sample size make this estimate poor.

Chimpanzee

Three studies of chimpanzee posture have yielded over 20 000 observations (Table 10.2). Although one study was limited to three postural modes, the unsampled modes represent only 5% of posture in the other studies. Frequencies for all three studies, even with this bias, are quite similar. Studies by Doran (1989) and Hunt (1989) yielded much larger sample sizes; these were used to generate a best estimate. The biggest difference between the two studies is less frequent suspensory behavior in West than East African chimpanzees.

Table 10.2. *Great ape postural modes (percentages)*

	Sit	Lie	Stand	Squat	Cling	Biped	Arm–hang	Hand–foot hang
Pongo[1]	46.0	0.0	24.4	0.0	0.0	0.0	29.7	0.0
Pongo[2]	42.1	0.0	6.7	0.0	0.0	3.8	17.8	30.0
Pongo[3]	49.0	0.0	1.0	0.0	0.0	2.0	12.0	36.0
Pongo **weighted avg.**	**45.6**	**0.0**	**15.5**	**0.0**	**0.0**	**1.1**	**23.3**	**14.1**
Bonobo[4]	**90.0**	**3.0**	**2.0**	**0.0**	**0.0**	**0.0**	**5.0**	**0.0**
P. t. verus[5]	80.0	5.0	15.0	0.0	0.0	0.0	0.0	0.0
P. t. verus[6]	75.8	16.8	5.8	0.0	0.0	0.0	1.6	0.0
P. t. schweinfurthii[7]	75.2	15.1	3.0	0.8	0.4	0.4	5.3	0.0
Chimpanzee best est.	**75.5**	**16.0**	**4.4**	**0.4**	**0.2**	**0.2**	**3.5**	**0.0**
Mountain gorilla[8]	60.0	1.3	2.7	35.4	0.0	0.2	0.0	0.0
Mountain gorilla[9]	73.4	20.1	6.5	0.0	0.0	0.0	0.1	0.1
Mtn. gorilla average	**66.9**	**10.7**	**4.6**	**17.7**	**0.0**	**0.1**	**0.1**	**0.0**
Lowland gorilla[10]	48.3	8.3	4.6	31.5	0.0	5.1	1.9	0.0
Lowland gorilla est.[11]	**59.3**	**9.7**	**4.6**	**23.3**	**0.0**	**2.2**	**0.8**	**0.0**

Notes:

[1] Sugardjito & van Hooff (1986). Percentage of 5836 bouts during arboreal travel and resting, Sumatran orangutans.

[2] Cant (1987a). Percentage of 350 bouts while feeding on figs, Bornean females.

[3] Cant (1987b). Percentage of time spent in each bout during 1682 minutes of focal arboreal feeding observations, Sumatran females.

[4] Kano & Mulavwa (1984). Percentage of 132 instantaneous time-point surveys during arboreal feeding on fruit.

[5] Sabater Pi (1979). Percentage of bouts during 186 hours of continuous sampling.

[6] Doran (1989). Percentage of 8660 1-minute time-point samples.

[7] Hunt (1989, 1992a). Percentage of 11 848 2-minute time-point samples.

[8] Tuttle & Watts (1985). Percentages each bout makes up of total bouts observed in 2300 h of continuous bout sampling.

[9] Doran (1996). Percentage of 10 674 1-minute instantaneous focal samples on Karisoke gorillas.

[10] Calculated from Remis (1995, table 9).

[11] Calculated assuming terrestrial postures of lowland and mountain gorillas are similar; weighted following Remis' (1995) estimate that lowland gorillas are 41% arboreal and 59% terrestrial (see text).

Bold indicates those values are the best estimate for the taxon indicated. SCEP modes are shaded.

Gorilla

Because mountain gorillas live in montane habitats nearly devoid of climbable trees, whereas lowland gorillas live in rainforest, postural profiles might be expected to differ considerably. Data support that expectation. A study of the Karisoke mountain gorillas yielded a prodigious 2300 hours of observation; another study generated 10 674 observations. I averaged values from both studies to produce the estimates in Table 10.2.

Lowland gorillas remain poorly habituated. The terrestrial positional behavior of this presumably quite terrestrial subspecies is largely unknown. Remis (1995) reported that for 382 first sightings (the most objective measure of terrestriality for poorly habituated subjects), 59% were terrestrial and 41% were arboreal. Data were limited to wet-season observations. Remis tabulated arboreal postural data for females, group males, and lone males. I pooled male data, then averaged male and female frequencies to get mid-sex averages (Table 10.2). I estimated lowland gorilla terrestrial behavior assuming that wet and dry season behavior differ little. This assumption seems reasonably sound because the proportion of time spent on the ground is similar in wet and dry seasons (Remis 1999). I estimated lowland

Table 10.3. *Gibbon locomotor modes (percentages)*

	Walk	Climb	Leap	Run	Biped	Scramble	Brachiate	Clamber	Suspensory	Transfer
H. agilis[1]	3.5	6.3	23.9	0.0	0.0	0.0	66.3	0.0	0.0	0.0
H. lar[2]	0.0	34.2	9.3	0.0	5.2	0.0	51.2	0.0	0.0	0.0
H. pileatus[3]	0.0	6.0	8.7	0.0	0.9	0.0	84.4	0.0	0.0	0.0
Gibbon avg.	**1.2**	**15.5**	**14.0**	**0.0**	**2.0**	**0.0**	**67.3**	**0.0**	**0.0**	**0.0**
H. syndactylus[4]	0.0	10.0	0.0	0.0	11.0	0.0	80.0	0.0	0.0	0.0
H. syndactylus[5]	0.0	54.3	3.2	0.0	4.1	0.0	37.9	0.0	0.0	0.0
Siamang avg.	**0.0**	**32.2**	**1.6**	**0.0**	**7.6**	**0.0**	**59.0**	**0.0**	**0.0**	**0.0**

[1] Gittins (1983). Percentage of 255 10-minute scan surveys.
[2] Fleagle (1980). Percentage of 211 pooled feeding and travel bouts; continuous focal sampling.
[3] Srikosamatara (1984). Percentage of 218 5-minute scan surveys.
[4] Gittins (1983). Percentage of 208 10-minute scan surveys.
[5] Fleagle (1980). Percentage of 1206 pooled feeding and travel bouts; continuous focal sampling.
Bold indicates those values are the best estimate for the taxon indicated. SCEP modes are shaded.

gorilla terrestrial plus arboreal postural frequencies using mountain gorilla terrestrial behavior to estimate the missing lowland gorilla terrestrial data, then weighting terrestrial (i.e. mountain gorilla) frequencies by 0.59 (the proportion of time spent in terrestrial behavior in the lowland gorilla) and arboreal frequencies by 0.41 (proportion of arboreality).

CALCULATIONS OF LOCOMOTOR MODE FREQUENCIES

Hylobatids

Gibbon

Locomotor mode frequencies are available for three gibbon species ($N = 684$; Table 10.3). *Hylobates lar* were observed during feeding and travel modes, contexts that presumably sample most gibbon locomotor activity. I pooled travel and feeding observations to make this study comparable to others. The three species differed. *H. agilis* displayed more leaping than other species, *H. lar* much more climbing activity, and *H. pileatus* more brachiation. I averaged the three studies to produce the gibbon positional profile in Table 10.3.

Siamang

Two studies totaling 1414 observations document siamang locomotor behavior (Table 10.3). In one study, siamangs were observed during feeding and travel contexts.

I pooled these observations to afford comparability. Gittins (1983) reported more brachiation, Fleagle (1980) found more climbing. These differences could reflect mode definition biases, in which case averaging ameliorates the bias.

Great apes

Orangutan

In two studies, male and female orangutans were observed during travel only (Sugardjito 1982; Sugardjito & van Hooff 1986). A third study observed females during feeding and travel (Cant 1987a), but only in arboreal contexts. Travel-only data overestimate walking, and female-only data underestimate quadrupedalism. In other words, these two studies' biases offset one another. Assuming no locomotion occurs during resting, travel plus resting contexts account for over 97% of orangutan locomotion. The remainder is building sleeping nests (0.8%) and social display (1.5%). Nest building is mostly postural (all my chimpanzee nest building observations were). No data exist for social display. I averaged the two travel studies then averaged these values with travel + feeding values to yield a best estimate (Table 10.4).

Bonobo

Three bonobo studies provided similar numbers of observations, but only Doran (1989) observed partly

Table 10.4. *Great ape locomotor modes (percentages)*

	Walk	Climb	Leap	Run	Biped	Scramble	Brachiate	Clamber	Suspensory	Transfer
Orangutan[1]	13.0	10.0	0.0	0.0	0.0	0.0	21.0	41.0	0.0	15.0
Orangutan[2]	10.8	9.8	0.0	0.0	0.0	0.0	19.8	43.0	0.0	16.8
Orangutan[3]	12.0	31.3	0.0	0.0	0.0	0.0	10.6	39.4	1.2	5.6
Orangutan est.	**12.0**	**20.6**	**0.0**	**0.0**	**>0.0**	**0.0**	**15.5**	**40.7**	**0.6**	**10.8**
Bonobo[4]	34.0	20.0	18.0	0.0	8.0	0.0	20.0	0.0	0.0	0.0
Bonobo[5]	31.0	31.0	10.0	0.0	6.0	0.0	21.0	0.0	0.0	0.0
Bonobo[6]	35.3	50.4	3.1	0.0	1.5	0.0	8.9	0.0	0.0	0.0
Bonobo est.	**35.3**	**50.4**	**3.1**	**0.0**	**1.5**	**0.0**	**8.9**	**0.0**	**0.0**	**0.0**
P. t. verus[7]	86.1	11.0	0.3	0.0	1.2	0.0	1.3	0.0	0.0	0.0
P. t. verus (est.)[8]	86.1	9.6	0.3	0.0	1.2	0.5	1.3	0.0	0.1	0.8
P. t. schweinfurthii[9]	91.8	5.1	0.2	0.8	0.4	0.1	0.8	0.0	0.1	0.6
P. t. schweinfurthii[10]	91.8	4.8	0.0	1.4	0.4	0.4	0.2	0.0	0.0	0.2
P. t. s. average[11]	91.8	5.0	0.1	1.1	0.4	0.3	0.5	0.0	0.1	0.4
Chimpanzee est.[12]	**89.9**	**6.5**	**0.2**	**0.7**	**0.7**	**0.3**	**0.8**	**0.0**	**0.1**	**0.5**
Mountain gorilla[13]	95.6	0.2	0.0	1.9	0.0	0.0	0.0	0.0	0.0	0.0
Mountain gorilla[14]	96.5	<1.7	0.0	0.0	1.6	>0.0	0.1	0.0	0.0	0.0
Mountain gorilla est.	**96.0**	**<1.0**	**0.0**	**1.0**	**0.8**	**>0.0**	**0.1**	**0.0**	**0.0**	**0.0**
Lowland gorilla[15]	18.8	46.6	0.0	0.0	13.7	0.0	8.7	0.0	3.2	8.0
Lowland gorilla est.	**64.4**	**19.7**	**0.0**	**0.6**	**6.1**	**0.0**	**3.6**	**0.0**	**1.3**	**3.3**

Notes:

[1] Sugardjito (1982). Percentage each mode makes up of all bouts observed during 219 h of continuous bout sampling; Sumatran orangutans; during travel only.

[2] Sugardjito & van Hooff (1986). Percentage each mode makes up of 10 601 bouts observed; Sumatran orangutans; continuous bout sampling for travel only.

[3] Cant (1987a). Percentage each mode makes up of all bouts observed during 4360 minutes of continuous bout sampling. Bornean females only were observed during feeding and travel.

[4] Susman, Badrian & Badrian (1980). Percentage each mode makes up of 131 arboreal feeding bouts.

[5] Susman (1984). Percentage each mode makes up of 1722 arboreal bouts, mostly during feeding.

[6] Doran (1996). Percentage each mode makes up of 1461 1-minute time-point samples. Arboreal locomotion only; mid-sex average.

[7] Doran (1996, table 16.3). Mid-sex averages of percentages of 1417 1-minute instantaneous time-point samples.

[8] Doran values recalculated, assuming the proportion that scramble, tree sway and transfer making up "climbing" is the same as at Mahale and Gombe. Percentages of each mode constituting climbing taken from Table 10.5.

[9] Percentages of 1751 2-minute instantaneous time-point samples at Mahale Mountains; mid-sex averages. Reanalyzed data originally presented in Hunt (1992a).

[10] Percentages of 484 2-minute instantaneous time-point samples at Mahale Mountains; mid-sex averages. Reanalyzed data from Hunt (1992a).

[11] Average of Gombe and Mahale data. Note that values are virtually identical to Hunt (1991a).

[12] Average of *P.t. verus* estimate, Gombe frequencies, and Mahale frequencies.

[13] Tuttle & Watts (1985). Percentage of each kilometer constituted by each mode in 2300 h of continuous bout sampling; mid-sex average for four adults.

[14] Doran (1996). Percentage each mode makes up of 1848 1-minute time-point samples; mid-sex average.

[15] Remis (1995). Percentage of 122 1-minute instantaneous time-point samples; arboreal, wet season observations only; midsex average. Calculated from Remis (1995, table 11).

Bold indicates those values are the best estimate for the taxon indicated. SCEP modes are shaded.

Table 10.5. *Percentage of each constituent locomotor mode in Doran's "climbing" category, for chimpanzees*

Mode	Mahale[1]	Gombe[1]	Mean
Vertical climbing	86.4	88.9	87.7
Scramble	1.7	7.4	4.6
Suspensory (tree sway)	1.7	0.0	0.8
Transfer (= bridge)	10.2	3.7	6.9

Note: [1] Data from Hunt (1992a).

habituated individuals; her values are reported in Table 10.4. Unhabituated bonobos leaped and brachiated as they fled observers. Doran found bonobos too poorly habituated to make terrestrial observations. No estimate of the relative frequency of arboreal versus terrestrial behavior is available, so it is unclear how representative of the bonobos' entire locomotor repertoire these data are. They seem unlikely to offer more than a crude estimate.

Chimpanzee

Two studies offer chimpanzee arboreal locomotor data (Table 10.4). Comparability between the two studies is problematic. Hunt (1992a) defined vertical climbing as hand-over-hand ascents on supports angled greater than 45°, whereas Doran (1996) pooled vertical climbing with other modes in a quadrumanous climbing category. This is critical to the current discussion because her data do not distinguish SCEP modes, i.e., those typically used on compliant supports such as transfer, tree sway or clamber, from modes used on stable supports. To estimate compliant-support modes in *P. t. verus*, I estimated the proportion of each of the constituent modes in Doran's climbing category (Table 10.4) by assuming that her quadrumanous climbing and scrambling mode contained proportions of transferring, vertical climbing and other modes in the same proportions found in *P. t. schweinfurthii*. Vertical climbing was indeed the largest component of "climbing" (nearly 90%), but other modes were significant at both East African sites. I multiplied these proportions by 11% (Doran's value for "climbing," see Table 10.4) to yield the *P. t. verus* estimate in Table 10.5. I calculated the chimpanzee locomotor profile by averaging values for Gombe, Mahale and the *P. t. verus* estimate (Table 10.5).

Mountain gorilla

Tuttle and Watts (1985) provided frequencies from a 2300-hour study. Doran (1996) recorded 1848 instantaneous samples. Although Doran again pooled scramble with vertical climbing, these modes are uncommon in the mountain gorilla and therefore probably bias these observations little. I averaged these two locomotor profiles to provide an estimate (Table 10.5).

Lowland gorilla

I recalculated Remis' (1995) data to produce a mid-sex average. One difficulty is that Remis' "scramble" involved "suspension by forelimbs with substantial support from hindlimbs (in compression)" wherein "weight was distributed relatively evenly across four limbs" (1995: 417). The "scramble" mode is more commonly defined as torso-pronograde quadrupedal walking, distinguished by its unpatterned gait (Hunt *et al.* 1996). Scramble *sensu* Remis is a mode that ranges between forelimb-assisted bipedalism and hindlimb-assisted brachiation. I divided her "scrambling" value, placing half in brachiation and half in bipedalism, to yield the approximation in Table 10.5. As above, I then used terrestrial mountain gorilla data to produce a weighted lowland gorilla estimate, assuming 59% terrestrial and 41% arboreal behavior.

DISCUSSION

Postural profiles (Table 10.6) for the seven ape taxa reviewed here provide one profile that is probably biased (the arboreal bonobo study), two profiles that are merely estimates but have no identified biases, and four profiles derived from long-term studies for which known biases have been corrected or that suffer no known biases. Locomotor profiles (Table 10.7) are derived from limited, biased studies in two cases, estimated in three species, and derived from long-term studies on well-habituated populations in two cases. We expect primates with a self-concept, great apes, to have SCEPRs compared to primates without self-concept, for example monkeys. Baboon positional frequencies provide this contrast. Data were collected using identical methods to those for Mahale and Gombe chimpanzees (Hunt 1991b).

Table 10.6. *Summary postural mode frequencies percentages*

	Sit	Lie	Stand	Squat	Cling	Biped stand	Arm-hang	Hand–foot hang	Quality of profile[1]
Gibbon	63.6	0.0	0.0	0.0	0.0	0.0	36.4	0.0	Reliable
Siamang	47.0	0.0	0.0	0.0	0.0	0.0	53.0	0.0	Estimate
Orangutan	44.8	0.0	14.6	0.0	0.0	1.3	22.3	15.0	Reliable
Bonobo[2]	90.0	3.0	2.0	0.0	0.0	0.0	5.0	0.0	Arboreal
Chimpanzee	75.5	16.0	4.4	0.4	0.2	0.2	3.5	0.0	Reliable
Mtn. Gorilla	66.9	10.7	4.6	17.7	0.0	0.1	0.1	0.0	Reliable
L. Gorilla	59.3	9.7	4.6	23.3	0.0	2.2	0.8	0.0	Estimate
Papio anubis[3]	75.3	4.0	19.7	0.2	0.3	0.1	0.2	0.0	Reliable

Notes:
[1] Values categorized as "estimate" are considered approximate frequencies.
[2] Bonobo estimates are shown for completeness; they are not discussed because they reflect arboreal feeding only.
[3] Percentage of 1555 2-minute instantaneous focal observations; mid-sex average. From Hunt (1991a).
SCEP modes are shaded.

Table 10.7. *Summary locomotor mode percentages*

	Walk	Climb	Leap/ hop	Run	Biped walk	Scramble	Brachiate	Clamber	Other susp.	Transfer	Quality of profile
Gibbon	1.2	15.5	14.0	0.0	2.0	0.0	67.3	0.0	0.0	0.0	Small NS
Siamang	0.0	32.2	1.6	0.0	7.6	0.0	59.0	0.0	0.0	0.0	Estimate
Orangutan	12.0	20.6	0.0	0.0	0.0	0.0	15.5	40.7	0.6	10.8	Estimate
Bonobo	35.3	50.4	3.1	0.0	1.5	0.0	8.9	0.0	0.0	0.0	Arboreal
Chimpanzee	89.9	6.5	0.2	0.7	0.7	0.3	0.8	0.0	0.1	0.5	Reliable
Mtn. Gorilla	96.0	<1.0	0.0	1.0	0.8	>0.0	0.1	0.0	0.0	0.0	Reliable
L. Gorilla	64.4	19.7	0.0	0.6	6.1	0.0	3.6	0.0	1.3	3.3	Estimate
Papio anubis[1]	97.0	0.7	0.5	1.6	0.0	0.0	0.0	0.0	0.0	0.0	Reliable

Notes:
[1] Percentage of 497 2-minute instantaneous focal observations; mid-sex average. From Hunt (1991b).
SCEP modes are shaded.

Posture

Compared to baboons, SCEP postures (arm-hang, arm–foot hang) occurred more often in all apes except the mountain gorilla. Gibbons and siamangs frequently use SCEP modes during posture. Cannon and Leighton (1994) found that gibbon supports during locomotion are quite stable even compared to those of macaques, just as Povinelli and Cant note, but suspensory postures are engaged in on small, compliant supports (Grand 1972; Gittins 1982 illustrates this spectacularly). The Povinelli and Cant hypothesis predicts that gibbons and siamangs will have self-conception, though perhaps less so than arboreal great apes. The larger siamang engaged in arm-hanging more often than gibbons, suggesting siamangs must accommodate more to compliant supports, and therefore have a more SCEPR than gibbons.

Among great apes, orangutans demonstrated the highest frequency of the SCEP modes arm-hang and arm–foot hang. They also stood the most. Suspensory postures among chimpanzees were only a tenth as common, despite similar body weights. Chimpanzees emerged overall as generalists. Mountain gorillas were

Table 10.8. *Percentages of arboreal locomotor modes in bonobos and other great apes*

	Bonobo[1]	Mahale chimpanzee[2]	Gombe chimpanzee[3]	Orangutan[4]	Lowland gorilla[5]
Quadrupedal walk	35.3	31.1	38.0	12.0	18.8
"Quadrumanous climb"	50.4	51.7	55.8	31.4	46.6
Suspension	8.9	14.4	3.1	56.8	19.9
Bipedalism	1.5	1.7	3.1	0.0	13.7
Leap	3.1	1.1	0.0	0.0	0.0
N	1461	178	45	4360 min.	122

[1] After Doran (1996, table 16.5). 1-minute instantaneous focal observations; mid-sex average.

[2] 2-minute instantaneous focal observations; mid-sex average.

[3] 2-minute instantaneous focal observations; mid-sex average.

[4] Values for "quadrumanous climbing" were calculated by pooling values for climb, scramble and transfer. Values for suspension were obtained by adding brachiation, clamber and miscellaneous suspensory modes.

[5] Calculated from Remis (1995, table 11). 1-minute instantaneous focal observations; mid-sex average. See discussion above for discussion of regularization of Remis' locomotor modes.

distinctive only for their high frequency of squatting and lying. Lowland gorillas had a distinctively high frequency of bipedalism. Bonobo profiles are not compared because they reflect arboreal feeding only.

Of all posture among gibbons, siamangs and orangutans, SCEPR postures constituted ≥35%. Among chimpanzees, mountain gorillas, lowland gorillas, baboons and perhaps bonobos, SCEP modes made up less than 5% of all postures. Posture typically makes up the vast majority of positional behavior (e.g., 85% in chimpanzees, Hunt 1989). Some experts suggest that relatively immobile postures produce too little stress on the musculoskeletal system to demand morphological adaptations. My view is that while locomotion is more stressful and dangerous because falls are more likely, posture is five times more common. If posture exerts significant selective pressures, all Asian apes have profoundly greater SCEPRs than African apes or baboons.

Locomotion

Brachiation, clamber, transfer, and miscellaneous suspensory modes constituted 59% or more of all Asian ape locomotor behavior. As Povinelli and Cant maintained, orangutans have high frequencies of locomotor SCEP modes, such as clamber and transfer. African apes, compared with Asian apes, are quadrupedal walkers. Walking, a distinctly un-SCEP mode, made up >60% of all locomotion in African apes, but constituted <15% in all Asian apes. Even scrambling, a walking-like compliant support mode, was uncommon among African apes. While African apes do not have a SCEPR compared with orangutans, they may still be SCEPR-selected compared with monkeys. Walking constituted 97% of baboon locomotor behavior. In the same forested habitat, walking constituted 91.8% of chimpanzee behavior. Walking made up only 64.4% of lowland gorilla behavior. Mountain gorillas are distinctive for their high frequencies of squatting and running, neither part of a SCEPR. *In toto*, SCEP modes made up less than 4% of all locomotor modes among the African apes. These locomotor data suggest that among the great apes, orangutans alone exhibit a distinct SCEPR.

Although the bonobo data are not directly comparable to the complete ape data set, arboreal-only behavior can be compared (Table 10.8). Bonobos and chimpanzees, in this limited comparison, are nearly indistinguishable; suspension represents <15% in both. Walking, likewise, is seen in similar frequencies in the two species. It is considerably less common among orangutans and gorillas. Although the catch-all category "quadrumanous climbing" makes comparisons tentative, gorillas appear much more Asian in this comparison than either *Pan* species. Suspensory mode frequencies in the lowland gorilla are exceeded among the great apes only by the orangutan, a quite unexpected result. They also exhibited distinctively high frequencies of bipedal posture, bipedal locomotion, and squatting. The lowland gorilla data are reliable in this comparison, since the missing terrestrial data are not a factor. These data leave the status of lowland gorillas as likely exhibitors of a SCEPR, but the case is equivocal.

In summary, Tables 10.6, 10.7, and 10.8 suggest that suspensory positional modes such as arm–foot hang, arm-hang, orthograde clamber, transfer and brachiate are more common in orangutans than other great apes, and more common in all apes than in monkeys. Sitting and quadrupedal walking, distinctively un-SCEP modes, were considerably more common among African apes than orangutans.

Among chimpanzees, unimanual forelimb-suspension (arm-hanging) and vertical climbing were distinctively common, compared with baboons, but their positional regime was unremarkable compared with other great apes. Bonobos, at least from evidence in Table 10.8, are indistinguishable from chimpanzees. Their high proportion of leaping in Table 10.7 is likely a reflection of poor habituation, and the seemingly distinctive level of climbing is an artifact of arboreal-only observations.

Gibbons have the highest frequency of leaping among the apes. Gibbons and siamangs, not surprisingly, are brachiation and arm-hanging specialists, but only postural modes show evidence of a need to accommodate compliant supports, and even this evidence is circumstantial.

Predictions

None of the predictions growing out of Povinelli and Cant's hypothesis were corroborated unequivocally, though some evidence is supportive.

(1) Apes demonstrating self-concepts were predicted to have SCEPRs. Only orangutans clearly exhibit a SCEPR, but other apes have varying expressions of a SCEPR compared with monkeys. Estimates presented here suggest that great apes' SCEPRs rank: orangutan $\gg$ lowland gorilla > chimpanzee (= bonobo) > hylobatids $\gg$ mountain gorilla. Povinelli and Cant might predict lowland gorillas to have a self-concept, but mountain gorillas, for which we have little laboratory cognitive evidence, should not. Chimpanzees have a less demanding SCEPR than lowland gorillas, yet they appear to express self-concept equal to that of orangutans, and

have been among the most successful on MSR tests (Gallup 1970; Povinelli *et al.* 1997). Equivocal evidence suggests that bonobos have a chimpanzee-like low-level SCEPR, yet they, too, pass the MSR mark test (Walraven, Van Eslsacker & Vesheyen 1995) and exhibit symbolic behavior perhaps beyond that of common chimpanzees (Savage-Rumbaugh *et al.* 1993). Hylobatids have a postural but not a locomotor SCEPR, but offer little evidence of self-concept (Hyatt 1998; Inoue-Nakamura 1997). Some gibbons exhibit evidence of passing the mark test (Ujhelyi *et al.* 2000), and others examine body parts in mirrors (Hyatt 1998). Other indications of symbolic behavior or self-concept are lacking. While positional behavior suggests that self-concept should roughly follow the pattern of orangutan $\gg$ lowland gorilla > chimpanzee = bonobo > hylobatids $\gg$ mountain gorilla, MSR results and other self-concept indicators suggest orangutan = chimpanzee = bonobo $\geq$ mountain gorilla $\gg$ hylobatids, with lowland gorillas unknown. This evidence does not support the Povinelli and Cant hypothesis.

(2) Siamangs have a SCEPR in their high frequency of arm-hanging, and are therefore predicted to have more sophisticated self-conception than closely related gibbons. No siamang has yet passed the MSR mark test (Hyatt 1998), but the contrast in SCEPR among the hylobatids suggests that as a program to test the compliant support hypothesis, further research is warranted.

(3) If SCEPRs are comparable, the heavier gorilla and orangutan males should display more sophisticated self-concepts than females. Gorillas did not meet the prerequisite comparability of male and female SCEPRs. Although Remis (1995) found very little difference in male and female positional mode frequencies, her observations were arboreal only, and females are much more arboreal than males (58% vs. 24%). Orangutan results are negative. Female orangutans engage in more clambering (47.8% vs. 38%) but males engage in more tree swaying (24% vs. 9.7%) (Table 10.9). Both behaviors should require a self-concept, so overall male and female SCEPRs appear comparable. No sex differences in self-concept have yet been noted in orangutans (Inoue-Nakamura 1997 and references therein). This result is consistent with the compliant support hypothesis, but is not support for it.

Table 10.9. *Sex differences in orangutan locomotor behavior (percentages)*[1]

	Walk	Climb	Brachiate	Clamber	Tree sway[2]
Male	8.0	9.0	21.0	38.0	24.0
Female	13.3	10.3	18.5	47.8	9.7

Notes:
[1] From Sugardjito & van Hooff (1986), Table II. Percentage each mode makes up of 10 601 bouts observed; continuous bout sampling for travel only.
[2] Pooled with "transfer" in other tables.

In chimpanzees, females have a more pronounced SCEPR than males (Hunt 1992b). Females arm-hang more often and from smaller supports, and females brachiate more than males (Hunt, 1992b). Males have high frequencies of un-SCEP postures such as sit (Hunt 1992b). The Povinelli and Cant hypothesis predicts that female chimpanzees should exhibit a more sophisticated self-concept; no such difference has been observed. This observation is at odds with the compliant support hypothesis.

(4) The more profound the SCEPR, the more robust and sophisticated self-concepts should be. No indices of self-concept sophistication exist, but robustness can be indexed by the proportion of individuals within a species that exhibit it and how early in development it appears. The consistency of success on self-concept measures is orangutan = chimpanzee = bonobo ≥ lowland gorilla ≫ hylobatids, with mountain gorillas unknown and hylobatid data contested. Their SCEPRs rank orangutan ≫ lowland gorilla > chimpanzee (= bonobo) > hylobatids ≫ mountain gorilla. No age differences in self-concept acquisition are yet apparent (Inoue-Nakamura 1997). The compliant support hypothesis is not supported by these data.

CONCLUSIONS

A comparison of ape positional behavior repertoires confirms Povinelli and Cant's contention that orangutans position themselves among compliant and unpredictable supports, but the positional behavior of other apes does not clearly support their hypothesis. Positional mode frequencies presented here support only one of four predictions developed from the compliant support hypothesis. Apes with a self concept were predicted to have self-concept eliciting positional regimes, but only orangutans clearly demonstrated a SCEPR. The compliant support hypothesis predicts that siamangs will evince greater evidence of self-concept than gibbons or mountain gorillas. No such difference has been observed, but further investigation seems warranted. Orangutans possess far more elements of a SCEPR than other great apes, which predicts more advanced self-conception in orangutans, but this has not been observed. Mountain gorillas do not have a SCEPR, yet there seems to be no sentiment among ape researchers that their cognitive sophistication or concept of self is different from that of lowland gorillas. Female chimpanzees should show greater expression of self-concept than males, but there is no objective evidence for such a sex difference, and my subjective opinion is that there is not one.

Orangutans offer a challenge to the social brain hypothesis in that their society is simple, yet they are cognitively complex. African apes offer a challenge to the compliant support hypothesis, as perhaps do hylobatids. Gorillas, with their simple foraging regime compared with other apes, offer a challenge to the foraging complexity hypothesis. Casting the net more widely, spider monkeys (*Ateles* spp.) offer a challenge to both the social complexity and foraging demands hypotheses. Spider monkeys have social relationships, group sizes and composition, and diet similar to those of chimpanzees. Social complexity and foraging hypotheses would predict their concept of self and other cognitive abilities should rival those of chimpanzees, yet *Ateles* have shown no evidence of a self-concept or any other form of high-level intelligence comparable to great apes, or even to *Cebus* (Chevalier-Skolnikoff 1991).

It might be argued that self-concept evolved in one of the common ancestors of apes due to SCEPRs, as the compliant support hypothesis suggests, and has been retained for use in other contexts. This seems unlikely, since self-concept is presumably dependent on large, metabolically expensive brains, and it would disappear without selective pressure to maintain it. If it were to be retained, a non-SCEPR selective pressure for self-concept must have appeared just as African apes were losing their ancestral SCEPR. This coincidence seems unlikely.

Resolution of the evolutionary origins of great ape self-concept and other evidence of higher intelligence, therefore, awaits further study of positional behavior as well as of the complexity of social relationships, diet, food resource distribution, food chemistry, and their intelligence itself. The best conclusion concerning the compliant support hypothesis is at present a tentative one: if foraging demands explain intelligence little compared with the demands of sociality, and if our understanding of orangutans as rather anti-social apes holds, and if phylogenetic inertia is insufficient to explain the retention of orangutan intelligence, then a locomotor origin for self-conception in orangutans is possible, but its origin in other apes is unexplained.

A broader conclusion concerning the evolution of self-concept and other higher cognitive abilities among other apes is similarly tentative. Among the apes, species with massive bodies have a concept of self, and smaller primates do not, even when they have SCEPRs, complex foraging regimes, and/or demanding social lives. Great apes may have larger brains not because the have unique selective pressures impinging on them, but because they can. Perhaps we must fall back on the hypothesis that organisms with larger bodies have lower costs for maintaining relatively large brains (Jerison, 1973), and therefore "intelligence" (including cognition involved in self conception) is found among the great apes simply because it is less expensive for massive primates than it is for other primates. From this perspective, increased locomotion among compliant supports derives from the same cause as presence of self-concept – great body weight – but the two are not causally connected.

ACKNOWLEDGMENTS

I am grateful to the editors, D. Begun and A. E. Russon, for inviting me to contribute a chapter to this volume. Melissa Remis helped me revise an early version. I am particularly indebted to Russon for her heroic efforts to educate me on cognition research, and for her incredible patience in awaiting the manuscript and its revisions. I thank J. Goodall and T. Nishida for inviting me to work at their field sites. I am indebted to R. W. Wrangham for support during the period of time during which much of the original research presented here was conducted. J. G. H. Cant graciously offered numerous suggestions and improvements, without which this chapter would not have been completed. He has come to my aid other times as well. I thank the staff of the Mahale Mountains Wildlife Research Center, and the Gombe Stream Research Center for their help and support. Research was aided by grants from the Leakey Foundation and a National Science Foundation grant BNS-86-09869 to R. W. Wrangham.

ENDNOTE

1 Povinelli and Cant suggest that most gorillas have lost their capacity for self-recognition secondarily, as part of an adaptation to terrestriality, maintaining that the ability of the lowland gorilla Koko to recognize herself in a mirror (Patterson 1984) is an unrepresentative exception. Recent work, however, suggests that gorillas do exhibit MSR (Swartz *et al.* 1999). This seems in keeping with other evidence of self-concept implicit in Koko's signing ability.

REFERENCES

Altmann, J. (1974). Observational study of behavior: sampling methods. *Behaviour*, **49**, 227–67.

Ashton, E. H. & Oxnard, C. E. (1964). Functional adaptations in the primate shoulder girdle. *Proceedings of the Zoological Society of London*, **142**, 49–66.

Cannon, C. H. & Leighton, M. (1994). Comparative locomotor ecology of gibbons and macaques: selection of canopy elements for crossing gaps. *American Journal of Physical Anthropology*, **93**, 505–24.

Cant, J. G. H. (1987a). Positional behavior of female Bornean orangutans (*Pongo pygmaeus*). *American Journal of Primatology*, **12**, 71–90.

(1987b). Effects of sexual dimorphism in body size on feeding postural behavior of Sumatran orangutans (*Pongo pygmaeus*). *American Journal of Physical Anthropology*, **74**, 143–8.

Cartmill, M. (1974). Pads and claws in arboreal locomotion. In *Primate Locomotion*, ed. F. A. Jenkins, Jr., pp. 45–83. New York: Academic Press.

Cartmill, M. & Milton, K. (1977). The lorisiform wrist joint and the evolution of "brachiating" adaptations in the Hominoidea. *American Journal of Physical Anthropology*, **47**, 249–72.

Chevalier-Skolnikoff, S. (1991). Spontaneous tool use and sensorimotor intelligence in *Cebus* compared with other monkeys and apes. *Behavioral and Brain Sciences*, **14**, 368.

Chevalier-Skolnikoff, S., Galdikas, B. M. F. & Skolnikoff, A. (1982). The adaptive significance of higher intelligence in wild orangutans, a preliminary report. *Journal of Human Evolution*, **11**, 639–52.

Chivers, D. J. (1972). The siamang and the gibbon in the Malay peninsula. In *The Gibbon and the Siamang,* Vol. 1, ed. D. M. Rumbaugh, pp. 103–35. Basel: Karger.

Doran, D. M. (1989). Chimpanzee and Pygmy Chimpanzee Positional Behavior: The Influence of Environment, Body Size, Morphology, and Ontogeny on Locomotion and Posture. Unpublished Ph.D. dissertation. SUNY, Stony Brook.

(1992). A comparison of instantaneous and locomotor bout sampling methods: a case study of adult male chimpanzee locomotor behavior and substrate use. *American Journal of Physical Anthropology*, **89**, 85–99.

(1996). Comparative positional behavior of the African apes. In *Great Ape Societies*, ed. W. C. McGrew, L. F. Marchant & T. Nishida, pp. 213–24. Cambridge, UK: Cambridge University Press.

Fleagle, J. G. (1976). Locomotion and posture of the Malayan siamang and implications for hominid evolution. *Folia Primatologica*, **26**, 245–69.

(1980). Locomotion and posture. In *Malayan Forest Primates: Ten Year's Study in Tropical Rain Forest*, ed. D. J. Chivers, pp. 191–207. New York: Plenum Press.

Fleagle, J. G., Stern, J. T., Jungers, W. L., Susman, R. L., Vangor, A. K. & Wells, J. P. (1981). Climbing: a biomechanical link with brachiation and with bipedalism. In *Vertebrate Locomotion*, ed. M. H. Day, pp. 359–75. New York: Academic Press.

Galdikas, B. M. F. (1978). Orangutan Adaptation at Tanjung Puting Reserve, Central Borneo. Unpublished Ph.D. thesis. University of California, Los Angeles.

Gallup, G. G. (1970). Chimpanzees: self-recognition. *Science*, **167**, 86–7.

(1982). Self-awareness and the emergence of mind in primates. *American Journal of Primatology*, **2**, 237–48.

(1991). Toward a comparative psychology of self-awareness: species limitations and cognitive consequences. In *The Self: An Interdisciplinary Approach*, ed. G. R. Goethals & J. Strauss, pp. 121–35. New York: Springer-Verlag.

Gibson, K. R. (1990). New perspectives on instincts and intelligence: brain size and the emergence of hierarchical mental constructional skills. In *"Language" and Intelligence in Monkeys and Apes*, ed. S. T. Parker & K. R. Gibson, pp. 97–128. New York: Cambridge University Press.

Gittins, S. P. (1982). Feeding and ranging in the agile gibbon. *Folia Primatologica*, **38**, 39–71.

(1983). Use of the forest canopy by the agile gibbon. *Folia Primatologica*, **40**, 134–44.

Goodall, J. van Lawick (1968). The behavior of free-living chimpanzees in the Gombe Stream Reserve. *Animal Behaviour Monographs*, **1**(3), 165–311.

Grand, T. I. (1972). A mechanical interpretation of terminal branch feeding. *Journal of Mammalogy*, **53**, 198–201.

Griffin, D. R. (1982). *Animal Mind – Human Mind*. New York: Springer-Verlag.

Harrison, B. (1962). *Orangutan*. London: Collins.

Heyes, C. M. (1998). Theory of mind in nonhuman primates. *Behavioral and Brain Sciences*, **21**, 101–48.

Hunt, K. D. (1989). Positional behavior in *Pan troglodytes* at the Mahale Mountains and Gombe Stream National Parks, Tanzania. Ph.D. dissertation, University of Michigan. Ann Arbor: University Microfilms.

(1991a). Positional behavior in the Hominoidea. *International Journal of Primatology*, **12**, 95–118.

(1991b). Mechanical implications of chimpanzee positional behavior. *American Journal of Physical Anthropology*, **86**, 521–36.

(1992a). Positional behavior of *Pan troglodytes* in the Mahale Mountains and Gombe Stream National Parks, Tanzania. *American Journal of Physical Anthropology*, **87**, 83–107.

(1992b). Sex differences in chimpanzee positional behavior, activity budget and diet. *Bulletin of the Chicago Academy of Sciences*, **15**, 4.

Hunt, K. D., Cant, J. G. H., Gebo, D. L., Rose, M. D., Walker, S. E. & Youlatos, D. (1996). Standardized descriptions of primate locomotor and postural modes. *Primates*, **37**, 363–87.

Hyatt, C. W. (1998). Responses of gibbons (*Hylobates lar*) to their mirror images. *American Journal of Primatology*, **45**, 307–11.

Inoue-Nakamura, N. (1997). Mirror self-recognition in nonhuman primates: a phylogenetic approach. *Japanese Psychological Research*, **39**, 266–75.

Jerison, H. J. (1973). *The Evolution of the Brain and Intelligence*. New York: Academic Press.

Jungers, W. L., Fleagle, J. G. & Simons, E. L. (1982). Limb proportions and skeletal allometry is fossil catarrhine primates. *American Journal of Physical Anthropology*, **57**, 200–19.

Kano, T. & Mulavwa, M. (1984). Feeding ecology of the pygmy chimpanzees *Pan paniscus* of Wamba. In *The Pygmy Chimpanzee*, ed. R. L. Susman, pp. 233–74. New York: Plenum Press.

Keith, A. (1891). Anatomical notes on Malay apes. *Journal of the Straits Branch of the Royal Asiatic Society*, **23**, 77–94.

Kortlandt, A. (1974). Ecology and paleoecology of ape locomotion. *Symposia of the 5th Congress of the International Primate Society*, pp. 361–4. Tokyo: Japan Science Press.

Lethmate, J. & Ducker, G. (1973). Untersuchungen am sebsterkennen im spiegel bei orangutans einigen anderen affenarten [Self-recognition by orangutans and some other primates]. *Zeitschrift für Tierpsychologie*, **33**, 248–69.

MacKinnon, J. R. (1977). A comparative ecology of Asian apes. *Primates*, **18**, 747–72.

Malenky, R. K., Kuroda, S., Vineberg, E. O. & Wrangham, R. W. (1994). The significance of terrestrial herbaceous foods for bonobos, chimpanzees and gorillas. In *Chimpanzee Cultures*, ed. R. W. Wrangham, W. C. McGrew, F. B. de Waal & P. G. Heltne, pp. 59–75. Cambridge, MA: Harvard University Press.

Mendel, F. (1976). Postural and locomotor behavior of *Alouatta palliata* on various substrata. *Folia Primatologica*, **26**, 36–53.

Miles, H. L. (1994). Me Chantek: the development of self-awareness in a signing gorilla. In *Self-awareness in Animals and Humans*, ed. S. T. Parker, R. W. Mitchell & M. L. Boccia, pp. 254–72. New York: Cambridge University Press.

Nicholson, I. S. & Gould, J. E. (1995). Mirror mediated object discrimination and self-directed behavior in a female gorilla. *Primates*, **36**, 515–21.

Owen, R. (1835). On the osteology of the chimpanzee and orang-utan. *Transactions of the Zoological Society of London*, **1**, 343–79.

Parker, S. T. (1996). Apprenticeship in tool-mediated extractive foraging: the origins of imitation, teaching, and self-awareness in great apes. In *Reaching into Thought: The Minds of the Great Apes*, ed. A. E. Russon, K. A. Bard & S. T. Parker, pp. 348–70. Cambridge, UK: Cambridge University Press.

Parker, S. T. & Gibson, K. R. (ed.) (1990). *"Language" and Intelligence in Monkeys and Apes: Comparative Developmental Perspectives*. New York: Cambridge University Press.

Patterson, F. G. (1984). Self-recognition of *Gorilla gorilla gorilla*. *Gorilla*, **7**, 2–3.

Patterson, F. G. & Cohn, R. (1994). Self-recognition and self-awareness in lowland gorillas. *In Self-awareness in Animals and Humans*, ed. S. T. Parker, R. W. Mitchell & M. L. Boccia, pp. 273–90. New York: Cambridge University Press.

Plavcan, J. M. & van Schaik, C. P. (1997). Intrasexual competition and body weight dimorphism in anthropoid primates. *American Journal of Physical Anthropology*, **103**, 37–68.

Povinelli, D. J. & Cant, J. G. H. (1995). Arboreal clambering and the evolution of self-conception. *Quarterly Review of Biology*, **70**, 393–421.

Povinelli, D. J., Gallup, G. G., Eddy, T. J. *et al.* (1997). Chimpanzees recognize themselves in mirrors. *Animal Behaviour*, **53**, 1083–8.

Prost, J. (1965). A definitional system for the classification of primate locomotion. *American Anthropologist*, **67**, 1198–214.

Remis, M. J. (1995). Effects of body size and social context on the arboreal activities of lowland gorillas in the Central African Republic. *American Journal of Physical Anthropology*, **97**, 413–33.

(1999). Tree structure and sex differences in arboreality among western lowland gorillas (*Gorilla gorilla gorilla*) at Bai Hokou, Central African Republic. *Primates*, **40**, 383–96.

Reynolds, V. F. (1965). *Budongo: A Forest and Its Chimpanzees*. London: Methuen.

Rijksen, H. D. (1978). *A Field Study on Sumatran Orangutans Pongo pygmaeus abelii (Lesson): Ecology, Behavior and Conservation*. Wageningen, the Netherlands: H. Veenen and Zonen, B. V.

Rodman, P. S. (1979). Individual activity patterns and the solitary nature of orangutans. In *The Great Apes*, ed. D. A. Hamburg & E. R. McCown, pp. 235–256. Menlo Park: Benjamin/Cummings.

Russon, A. E., Bard, K. A. & Parker, S. T. (ed.) (1996). *Reaching into Thought: The Minds of the Great Apes*. Cambridge, UK: Cambridge University Press.

Sabater Pi, J. (1979). Feeding behavior and diet of the chimpanzees in the Okorobiko Mountains, Rio Muni, Republic of Equatorial Guinea (West Africa). *Primates*, **18**, 183–204.

Savage, T. S. & Wyman, J. (1847). Notice of the external characters and habits of *Troglodytes gorilla*, a new species of orang from the Gabon River. *Boston Journal of Natural History*, **5**, 28–43.

Savage-Rumbaugh, E. S., Murphy, J., Savcik, R. A., Brakke, K. E., Williams, S. L. & Rumbaugh, D. M. (1993). *Language Comprehension in Ape and Child*. Chicago, IL: University of Chicago Press.

Smith, R. J. & Jungers, W. L. (1997). Body mass in comparative primatology. *Journal of Human Evolution*, **32**, 523–59.

Srikosamatara, S. (1984). Notes on the ecology and behavior of the hoolock gibbon. In *The Lesser Apes*, ed. H. Preuschoft, D. J. Chivers, W. Y. Brockelman & N. Creel, pp. 242–57. Edinburgh: Edinburgh University Press.

Suarez, S. & Gallup, G. G. (1981). Self-recognition in chimpanzees and orangutans, but not gorillas. *Journal of Human Evolution*, **10**, 175–88.

Suddendorf, T. & Whiten, A. (2001). Mental evolution and development: evidence for secondary representation in children, great apes, and other animals. *Psychological Bulletin*, **127**, 629–50.

Sugardjito, J. (1982). Locomotor behavior of the Sumatran orangutan *Pongo pygmaeus abelii* at Ketambe, Gunung Leuser National Park. *Malay Naturalist Journal*, **35**, 57–64.

Sugardjito, J. & van Hooff, J. A. R. A. M. (1986). Age–sex class differences in the positional behavior of the Sumatran orang-utan *Pongo pygmaeus abelii* in the Gunung Leuser National Park, Indonesia. *Folia Primatologica*, **47**, 14–25.

Susman, R. L. (1984). The locomotor behavior of *Pan paniscus* in the Lomako forest. In *The Pygmy Chimpanzee*, ed. R. L. Susman, pp. 369–94. New York: Plenum Press.

Susman, R. L., Badrian, N. L. & Badrian, A. J. (1980). Locomotor behavior of *Pan paniscus* in Zaire. *American Journal of Physical Anthropology*, **53**, 69–80.

Swartz, K., Evans, S., Mollerus, T. & Sarauw, D. (1999). Species differences in mirror behavior among gorillas, orangutans and chimpanzees. In *The Mentalities of Gorillas and Orangutans: Comparative Perspectives*, ed. S. T. Parker, R. W. Mitchell & H. L. Miles, pp. 283–94. Cambridge, UK: Cambridge University Press.

Tuttle, R. H. (1975). Parallelism, brachiation and hominoid phylogeny. In *The Phylogeny of the Primates: A Multidisciplinary Approach*. ed. W. P. Luckett & F. Szalay, pp. 447–80. New York: Plenum.

Tuttle, R. H. & Watts, D. P. (1985). The positional behavior and adaptive complexes of *Pan gorilla*. In *Primate Morphophysiology, Locomotor Analysis and Human Bipedalism*, ed. S. Kondo, pp. 261–88. Tokyo: Tokyo University Press.

Tuttle, R. H., Basmajian, J. V. & Ishida, H. (1979). Activities of pongid thigh muscles during bipedal behavior. *American Journal of Physical Anthropology*, **50**, 123–36.

Tyson, E. (1699). *Orang-outan, sive Homo sylvestris: or the Anatomy of a Pygmie Compared with that of a Monkey, an Ape and a Man*. London: Osborne.

Ujhelyi, M., Merker, B., Buk, P. & Geissmann, T. (2000). Observations on the behavior of gibbons (*Hylobates leucogenys*, *H. gabriellae*, and *H. lar*) in the presence of mirrors. *Journal of Comparative Psychology*, **114**, 253–62.

Walraven, V., Van Eslsacker, L. & Verheyen, R. (1995). Reactions of a group of pygmy chimpanzees (*Pan paniscus*) to their mirror images: evidence of self-recognition. *Primates*, **36**, 145–50.

Wheatley, B. (1982). Energetics of foraging in *Macaca fascicularis* and *Pongo pygmaeus* and a selective advantage of large body size in the orang-utan. *Primates*, **23**(3), 348–63.

11 · Great ape social systems

CAREL P. VAN SCHAIK,[1] SIGNE PREUSCHOFT,[2] AND
DAVID P. WATTS[3]

[1] *Department of Biological Anthropology and Anatomy, Duke University, Durham*
[2] *Living Links, Yerkes Primate Center, Emory University, Atlanta*
[3] *Department of Anthropology, Yale University, New Haven*

INTRODUCTION

Cognitive capacities may be more highly developed in most primates than among mammals in general (Tomasello & Call 1997), although other mammalian radiations such as cetaceans (e.g., Connor, Smolker & Richards 1992) and birds (e.g., Hunt 1996; Marler 1996) may have evolved similar capacities independently. Numerous studies have also suggested to some that great apes stand out among nonhuman primates in achieving more advanced cognitive abilities (e.g., Byrne 1995; Parker & Gibson 1990; Rumbaugh Savage-Rumbaugh & Washburn 1996; Russon, Bard & Parker 1996). Phenomena such as mirror self-recognition, imitation, pretend play, teaching, and manufacture and flexible use of tools have been cited as evidence that great apes, but not other nonhuman primates, have some form of self-concept, some ability to attribute mental states to others, and greater understanding of physical causality (Byrne 1995, 1997a; Byrne & Whiten 1997; Parker, Chapter 4, this volume; Russon 1997, Chapter 6, this volume; Russon & Bard 1996). Even skeptics note that great apes learn more rapidly than monkeys (Tomasello & Call 1997).

Our own recent meta-analysis of published studies on nonhuman primate cognition confirmed this assessment, that great apes are more intelligent than other nonhuman primates (Deaner *et al.* unpublished). It found that primate cognition is distinguished by some generalized capacity rather than a collection of narrow, problem- or domain-specific abilities, supporting the view that great apes constitute a homogeneous group that outranks other primates in cognitive performance.

The inevitable question, and one that inspired this book, is "Which selective pressures have been responsible for the evolution of these unusual cognitive abilities?" In this chapter, we first briefly review existing ideas, then characterize the social systems of great apes in order to evaluate the most prominent among them, the social intelligence hypothesis. Neither this nor any other current hypothesis unambiguously accounts for the unusual cognitive abilities of great apes, largely because it ignores the costs of cognitive adaptations, which are closely linked to a taxon's life history (Deaner, Barton & van Schaik 2003; van Schaik & Deaner 2003). We therefore attempt to distill the commonalities of great ape social characteristics and identify factors that have produced opportunities for the evolution of more advanced cognitive abilities in the socioecological realm.

Hypotheses for cognitive evolution in primates

Over the past three decades, primate researchers have proposed various hypotheses to account for the apparently exceptional cognitive capacities of primates as a whole and for cognitive variation within primates (see Russon, Chapter 1, this volume). These focus on three main classes of selective demands, all thought to have selected for advanced cognitive capabilities of varying degrees of generality (Table 11.1):

- challenges of interacting with conspecifics in permanent social groups; these interactions vary from highly competitive and manipulative to highly cooperative (Byrne & Whiten 1988, 1997; Cheney, Seyfarth & Smuts 1986; Humphrey 1976; Whiten 2000);
- foraging challenges, either tracking spatio-temporal variation in food distribution or processing food (Byrne 1997a; Clutton-Brock & Harvey 1980; Milton 1988; Parker & Gibson 1977);
- locomotor demands on large-bodied quadrupeds moving in the three-dimensional forest canopy (Povinelli & Cant 1995).

Table 11.1. *Overview of hypotheses for the evolution of cognitive capacities, especially among primates*

Hypothesis	Selective demand
Social strategizing (a.k.a. Machiavellian Intelligence)	1. Competitive advantages from exploiting and manipulating others
	2. Improved competitive ability due to coalitions and alliances
	3. Benefits from awareness of third-party relationships
	4. Benefits arising from social exchange, including reciprocity and exchange of different behavioral modalities
	5. Benefits from conflict mitigation and resolution mechanisms
Ecological demands	1. Spatio-temporal mapping: monitoring food availability in space and time
	2. Extractive foraging and food processing: ability to understand physical causality; manual dexterity and bi-manual coordination to process foods
Arboreal clambering	1. Non-stereotyped, quadrumanous movement by large-bodied animals in three-dimensional habitats

Testing these hypotheses is not easy (van Schaik & Deaner 2003). Direct interspecific comparisons of cognitive abilities are difficult, so comparative analyses often use neuroanatomical measures as proxies for cognitive ability. Unfortunately, none of the existing neuroanatomical scaling techniques is inherently superior to the others, and all have potential drawbacks (Deaner, Nunn & van Schaik 2000). Socioecological demands are also usually estimated through proxy variables, which are equally beset with difficulties. Thus, group or clique size is often used as a proxy for social complexity (Dunbar 1992, 1998; Kudo & Dunbar 2001), but group size especially expresses the potential for social complexity at best rather than any of the five selective demands listed in Table 11.1. Degree of frugivory, home range size, and day journey length are often used as proxies for ecological demands, but they are highly correlated with group size and body size, which compromises the resolving power of comparative tests (Deaner *et al.* 2000; Dunbar 1992).

None the less, several tests have led to the emergence of the social (or Machiavellian) intelligence hypothesis as the leading contender for explaining cognitive evolution within the primates (Barton & Dunbar 1997; Byrne & Whiten 1988, 1997; Cummins 1998; Dunbar 1998; Whiten 2000). This is not unexpected. Most primates, unlike the majority of mammals, live in stable mixed-sex groups (e.g., van Schaik 1996), which allows for establishing long-term and dynamic social relationships within and between sexes. Hence, primate social systems have a high potential for social complexity.

Cognition manifests itself in the social domain in many abilities: individual recognition; conflict management through such means as awareness of rank relations and social relationships between self and others and among others; tactical alliance formation; attempts to manipulate the behavior or the relationships of other group members, sometimes deceptively; and exchanges of services such as grooming and agonistic support within long-term social relationships. More controversially, some nonhuman primates may be able to take the perspective of conspecifics and to attribute mental states to them (Hare *et al.* 2000; Whiten 1998).

Recent work has not been entirely favorable to this hypothesis, especially with respect to explaining the cognitive capabilities that distinguish the great apes as a group (e.g., Byrne 1997a,b). We review the social lives of living great apes, to provide a more careful evaluation of the social intelligence hypothesis as an explanation for the evolution of great ape intelligence.

SOCIAL SYSTEMS OF GREAT APES

Although only four species survive today, the great apes show dramatic variation in their social systems despite their cognitive homogeneity. Here, we briefly explain the key concepts used to organize the species vignettes presented below.

Social systems can be described in terms of social organization, i.e., the characteristic grouping patterns and mating system, and social structure, i.e., the nature of the social relationships among individuals. Social relationships can be understood to reflect the strategic goals of individuals, and the nature of dyadic relationships depends on the nature of competition and the extent of cooperation between dyad members (Cords 1997; Kummer 1978). Thus, social relationships can have agonistic and affiliative components, reflecting competition and cooperation, respectively, which are expressed in special signals.

First, consider competition. When contest competition for monopolizable resources (food or mates) can significantly influence fitness, consistent asymmetries between individuals in competitive power yield predictable agonistic outcomes and translate into dominance–subordination relationships that are stable across contexts (Preuschoft & van Schaik 2000; van Schaik 1989, 1996). Dyadic dominance relationships often translate into linear dominance hierarchies. In contrast, clear dominance relationships are absent when contest competition is weak, i.e., when competition is mainly by scramble or largely absent.

Within dominance relationships, individuals may use formal *status signals* to communicate their assessment of their relative status and thereby prevent escalation of contests (cf. de Waal 1986). The establishment and communication of dominance relationships thus benefits both dyad members, even if they have no affiliative bond. Status signals can indicate either dominance or subordination, and are shown only by one member of the dyad, regardless of the context (Preuschoft & van Schaik 2000). Subordinate status indicators function like ritualized signals of submission: they signify absence of aggressive tendencies in the sender and correlate with yielding to the interaction partner (cf. Schenkel 1967). Dominance status indicators function like ritualized signals of assertion: they correlate with assertive or aggressive behavior in the sender and induce yielding in the receiver (Preuschoft 1999). Either member of the dyad may perform other assertive and submissive signals, however, depending on the context (e.g., Preuschoft 1992).

Affiliative bonds reflect active cooperation. They exist when receivers meet senders' affiliative initiatives with tolerance, reciprocate them, or exchange other kinds of socio–positive acts for them. In primates, active cooperation includes behaviors such as allogrooming, agonistic coalitions and alliances (consistent formation of coalitions within the same dyads), protection against harassment or infanticide, and food sharing (e.g., Dugatkin 1997). The bonds reflect long-term investments that individuals need to maintain and defend against possible disruptions caused by conflicts of interest (Cords 1997; Kummer 1978). Post-conflict behaviors, such as reconciliation, and various conflict-prevention behaviors (Aureli, Cords & van Schaik 2002; Aureli & de Waal 2000), confirm the presence and the value of these affiliative bonds.

The need for cooperation, especially when it involves unrelated partners, can profoundly alter the power dynamics within dyads. If individual A is dominant to individual B, but B can withhold a commodity or service that A needs, B has leverage power over A (Hand 1986). This dilutes the effects of dominance. Power depends not only on an individual's capacity for physical coercion, but also on the extent to which it controls commodities on which others rely (Lewis 2002). When contest competition is important, but subordinates can have considerable leverage, dyads tend to form affiliative bonds and to have relaxed dominance relationships. This can reach the point where status signals disappear and social relations become egalitarian (Preuschoft & van Schaik 2000), as when contest potential is weak. Thus, the absence of formal status signals indicates either the absence of strong contest competition or the presence of contest potential in a situation with high subordinate leverage.

The cognitive demands that these social interactions and relationships impose are not fully understood but may increase as social decisions become flexible (cf. Byrne & Whiten 1997; Cheney & Seyfarth 1990; Harcourt 1992). Thus, members of fission–fusion communities must re-assess their relationships every time they reunite after separation. Likewise, in some species, affiliative bonds occur not only between relatives, but also between non-kin. In the latter, the mutualism or social exchange requires frequent re-assessment and hence more flexible decision making (e.g., mutualistic agonistic support that helps female macaques to acquire and maintain their dominance ranks: Chapais, 1992).

Using this framework, we describe the social organization and social structure of the four extant great apes. Their social systems and social structures are summarized in Figure 11.1 and Table 11.2, respectively.

Table 11.2. *The coordinates of power: dominance and leverage derived from affiliative bonds and cooperation*

	Dominance			Bonding			Cooperation		
	F–M	F–F	M–M	F–M	F–F	M–M	F–M	F–F	M–M
Pongo pygmaeus	Y, SS	Y, SS	Y, AS or DI	(Y, mating)	mo–da	N	N	N	N
Gorilla gorilla	Y, ?	N, AS	Y, AS or DI	Y	N	N (fa–so?)	Little	N	fa–so
Pan troglodytes	Y, SI	rare, SI	Y, SI	Y	N/Y*	Y	N/Y, esp. mo–so*	N/Y*	Y
Pan paniscus	Y (F>M), ?	Y?, SS?	Y?	Y, mainly mo–so	Y	loose	mo–so	Y	N

Notes:

Dominance: approach–retreat and winning of conflicts. SI: Subordination indicator, SS: submissive signal, DI: dominance indicator, AS: assertive signal. Agonistic signals (AS/SS) are reported only when no status indicator (SI/DI) is documented.
Bonding: As evident from patterns of reconciliation and other affiliative interactions (grooming, communication sex).
Cooperation: agonistic support against group members, intergroup aggression, hunting, food sharing.
Y: yes, N: no, mo: mother, da: daughter, fa: father, so: son, *: site dependent (yes in West Africa and in captivity).

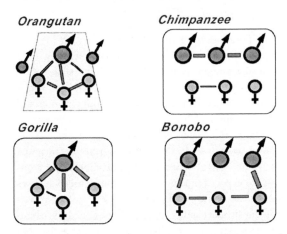

Figure 11.1. Great ape social systems. Pictorial representation of the social systems of the four extant great ape species (modified after de Waal 1995). Each pictogram represents one social unit. Solid lines surround socially cohesive groupings that share a common range; dotted lines surround less stable groupings. Bars of varying thickness between the symbols of males and females indicate social bonds of variable strength.

Orangutans (*Pongo pygmaeus*)

Social organization and mating

Orangutans are largely solitary, although association between individuals other than females with dependent offspring can vary seasonally and is more pronounced overall in Sumatran than Bornean study sites (Delgado & van Schaik 2000; van Schaik 1999). Females have overlapping individual home ranges, with a tendency for daughters to remain close to their natal range (Galdikas 1984; Singleton & van Schaik 2001). Clearly delimited social units probably do not exist, but female clusters, like prosimian "noyaux," are known at one site. These females have highly overlapping ranges and preferentially associate, co-feed and even share food, and are probably relatives (Singleton & van Schaik 2002).

Females lack sexual swellings. Males roam more widely than females, and their ranges overlap even more. Sexually mature males come in two morphs. "Flanged" males, far bigger than females, have fully developed secondary sexual characteristics – cheek pads or flanges, long hair, throat sack – and emit characteristic loud vocalizations. Non-flanged males, smaller, lack these features and are often younger, although the age at which males acquire secondary sexual characteristics varies greatly (Utami Atmoko 2000). Non-flanged males are far more active and more sociable than most flanged males, and seek and follow potentially fertile females (those without infants or with large infants). Sexually active females are extremely rare: interbirth intervals last 8 years on average (Galdikas & Wood 1990), but parous females may be sexually active on and off for about a year before conception (Fox 1998; van Schaik unpublished).

Parties may contain all age–sex classes, but flanged males who are not dominant in their regular home range are the least sociable (van Schaik 1999). Where parties are rare, they usually consist of male–female mating associations. Where they are common, as in swamps or rich alluvial forests rich in strangling figs, many other kinds of associations form, including those containing multiple females with infants and older offspring. Sexually active females that share the same area seek out the same locally dominant flanged male for near-exclusive consortships, avoid other flanged males, and vehemently resist mating with tenaciously associating non-flanged males. At the Sumatran sites, voluntary consortships with a dominant flanged male can last weeks (Delgado & van Schaik 2000); in Borneo they more typically last for days at most. Flanged males are highly intolerant of each other, but unflanged ones sometimes associate, usually when they follow the same females.

Within a local population, various females and numerous males, both flanged and unflanged, have widely overlapping ranges. Each individual has established dominance relationships with many other familiar ones (Rijksen 1978), although the mobility of males other than local dominants is so high that some may not be as familiar with the locals.

Dominance

Flanged adult males compete vigorously with other flanged males and form dominance relationships, not necessarily transitive (Utami & Mitra Setia 1995), with those they regularly encounter (up to about a dozen: Singleton & van Schaik 2001). Low vulnerability to predation alleviates the need for social cohesiveness so despite extreme intolerance, formal subordination signals, which mediate peaceful coexistence, are superfluous. Unrelated adult females have dominance relationships tending toward uni-directionality (Utami *et al.* 1997), expressed most commonly in one-sided avoidance behavior and submissive vocalizations (e.g., Rijksen 1978). At sites where females commonly associate, females may rarely express their dominance relationships, and even those dominant within their own regular ranges are uncertain and avoid confrontations when outside these areas (Singleton & van Schaik 2002). Winners of decided agonistic interactions, within and between age–sex classes, are predictably bigger and/or in better condition than losers (van Schaik unpublished). No coalitions have been observed.

Bonding and cooperation

Adult females in the same cluster associate preferentially and share food (Singleton & van Schaik 2002). These associations provide contexts for play among immatures and for the transfer of learned skills, such as tool use (e.g., van Schaik 2003; van Schaik, Fox & Sitompul 1996). Otherwise, clear social bonds are not found among adults: consistent spatial proximity is absent and grooming virtually absent, except between mothers and infants. Closer bonds occur among adolescents of both sexes, but these dissolve at adulthood (e.g., Rijksen 1978). The strong female mating preferences for the dominant flanged males do not translate into consistent active affiliation, because these males rarely associate with females outside mating consortships. However, some affiliative behavior and food sharing occurs during consortships (van Schaik unpublished). Neither reconciliation nor other tension-reduction behaviors have been studied in orangutans.

Gorillas (*Gorilla gorilla*)

Social organization and mating

Gorillas live in groups of several adult females, their offspring, and one or more mature adult ("silverback") males, with a modal size of about 12 members. Males are much larger than females. Most of our data on social relationships and life histories come from mountain gorillas in the Virungas population. However, despite known ecological variation across subspecies (reviewed in Doran & McNeilage 1998), no compelling reasons exist to expect fundamental variation in social systems (Watts 2003). Lowland gorillas are highly frugivorous in many habitats and groups sometimes divide into temporary subgroups to exploit scattered fruit patches, but subgroups apparently always contain at least one adult male (Doran & McNeilage 1998).

Most female mountain gorillas transfer from their natal groups to other groups or to solitary males at sexual maturity, and secondary female transfer is common (Harcourt 1978; Watts 1996). Most male mountain gorillas also disperse and become solitary, perhaps after some time in all-male groups; solitary males try to attract females. Other males become followers in established bisexual groups (usually their natal groups) where they have good chances to displace or replace aging leaders as dominant, breeding males (Harcourt 1978; Watts 1996, 2000a). Female transfer also occurs in both lowland subspecies (Tutin 1996; Yamagiwa & Kahekwa 2001) and solitary males are present in populations of all subspecies.

Several lines of evidence support the argument that gorilla groups form because multiple females converge around males to gain male protection, especially against other infanticidal males (Watts 1990a, 1996, 2003; Wrangham 1979). Infanticide is a major threat to female reproductive success in mountain gorillas and probably in other subspecies (Doran & McNeilage 1998; Watts 1989; but see Yamagiwa & Kahekwa 2001). As in langurs (Steenbeek 2000), groups without mature follower males dissolve on the death or disappearance of leader males (but see Yamagiwa & Kahekwa 2001). In such circumstances, females with small infants are almost certain to lose them to infanticide by outside males (Watts 1989). Infanticide risk is lower in multi-male than in single-male groups (Robbins 1995; Watts 2000a). This helps explain why females transfer disproportionately often from single-male to multi-male groups, despite the fact that multi-male groups are usually larger; we would expect the reverse if transfer served primarily to reduce feeding competition.

Gorillas have the shortest interbirth intervals among great apes (Watts 1996). Estrus lasts only about two days and females lack conspicuous swellings (Harcourt et al. 1980). Dominant males in groups with multiple sexually mature males do most mating and often try to prevent mating by subordinates, but females usually mate with most or all available males, which can induce male protection against outside males and forestall within-group infanticide (Watts 1989, 1991).

Dominance

Researchers have made contrasting statements about whether gorillas have a vocalization homologous to chimpanzee pant-grunts, which are formal subordination signals (below). Marler (1976) tentatively suggested that the mountain gorilla "pant series" (Fossey 1972), a response to mild threats, is morphologically equivalent to pant-grunts. However, Harcourt, Stewart and Hauser (1993) treated pant series and "mild cough-grunts," which are mild threats, as identical. They also noted that cough-grunting often goes both ways in dyads, albeit at different frequencies, and so does not signify dominance. Watts (1995) noted that females typically "grumble" to males when males are aggressive to them and sometimes when males make non-aggressive approaches, but rarely grumble to other females, even after aggression. Thus, he suggested that grumbles are formal signals of female subordination to males. However, Harcourt et al. (1993) argued that individuals simultaneously feeding in close proximity often engage in choruses of "non-syllabled" vocalizations that include grumbles, none of which formally indicate status.

Absence of formal status signals between males and females would be interesting given that males are twice as heavy as females and always dominant to them, and that female mountain gorillas almost never cough-grunt or otherwise behave aggressively to silverbacks (Watts 1995, 1997, pers. obs.). Females presumably have some leverage over males because they can transfer at low cost, when they do not have dependent infants. This may give them room for some negotiation via context-dependent agonistic signals in relationships with males.

Absence of unconditional signals of status in female–female relationships is easier to understand. Harcourt et al.'s (1993) data for female–female dyads show only a weak imbalance in the distribution of cough-grunts. This fits with the observation that aggression is commonly bi-directional in female dyads and that neither female shows submission to the other in many dyads (Watts 1994). Size and age differences can lead to differences in fighting ability between females (e.g., very old females threaten others infrequently), but three important contravening factors contribute to symmetry in female–female agonistic relationships. First, mountain gorillas are largely folivorous, and females have little to gain from contest competition over food. They engage in contests, including fights, in other contexts – notably over proximity to males – but most of these are undecided as well (Watts 1994). Second, females often have no female relatives available to serve as allies, so most contests between females are dyadic. Perhaps most importantly, males intervene in many contests between females and usually stop these without either opponent winning (Harcourt & Stewart 1989; Watts 1994, 1997); these interventions negate asymmetries in fighting ability among females and facilitate assertiveness and retaliation by weaker opponents.

Male gorillas in multi-male groups establish dominance relationships (Harcourt & Stewart 1981; Robbins 1996; Watts 1995). Aggression, including cough-grunts, is almost entirely unidirectional in male–male dyads, except when a younger male is trying to reverse rank with an older (Robbins 2001; Watts 1995).

Bonding and cooperation

Unless closely related, female mountain gorillas do not form strong, lasting affiliative bonds with each other even though they may live together for decades (Harcourt 1979a; Stewart & Harcourt 1987; Watts 1994, 2001). Grooming among females is rare and mostly restricted to kin. Unrelated females occasionally form coalitions, but alliances are also mostly restricted to kin. Alliance formation offers females few benefits in terms of access to the densely distributed plants that are the gorillas' main foods. This helps to explain why female kin often do not stay together. Also, males limit the advantages of alliance formation because their interventions in polyadic conflicts among females usually prevent allies from winning against their opponents (Watts 1997). Affiliative interactions are virtually absent among males, but those in multi-male groups cooperate to defend their groups against outside males (Robbins 2001; Sicotte 1993; Watts 2000a, 2003). Most males who cooperate in this manner are relatives, but this is not a prerequisite for cooperation (Robbins 2001; Watts 2000a). Adult females have strong bonds with their groups' dominant males, while some differentiation of male–female relationships occurs in multi-male groups (Harcourt 1979b; Sicotte 1994; Stewart & Harcourt 1987; Watts 1992, 2003). Reconciliation is restricted to male–female dyads (Watts 1995), emphasizing the importance of male–female bonds.

Chimpanzees (*Pan troglodytes*)

Social organization and mating

Chimpanzees form fission–fusion social units called communities, with up to 150 members. Members travel and forage in constantly changing subgroups within the community range. Males cover the community's entire range, while females generally stay within smaller parts of the range except when in estrus (but see Hasegawa 1990). Most females leave their natal communities as sexually active adolescents, whereas males stay in their natal communities for life. Males are generally more gregarious than females, but female sociability and social relationships vary across populations, perhaps in association with variation in habitat productivity (Doran *et al.* unpublished). Females commonly associate with other adults in West Africa (Boesch & Boesch-Achermann 2000), but are more solitary in East African populations (Goodall 1986; Wrangham 2000). Relations between

neighboring communities are hostile, and males engage in coalitionary aggression, which is sometimes lethal, against neighbors (Boesch & Boesch-Achermann 2000; Goodall 1986; Manson & Wrangham 1991; Wrangham 1999; Watts & Mitani 2001).

Females begin to develop conspicuous sexual swellings during adolescence and are sexually very active for several years prior to first parturition; most emigrate from their natal communities during this time. For parous females, sexual activity is largely limited to the three to four cycles between conceptions, which occur about every 6 years (this is variable; see Boesch & Boesch-Achermann 2000). Estrus lasts about 12–15 days, and females are most sexually active during the period of maximum tumescence, about a third of the menstrual cycle. Males are most interested in females on the days immediately preceding detumescence, when ovulation, and thus fertilization, is most likely (Wrangham 1993). Females typically mate opportunistically with multiple males, and sperm competition among males is presumably high. High-ranking males sometimes try to monopolize fertile females when other males are present, with varying success, and males sometimes persuade females to mate exclusively with them during consortships, on which they avoid other males (Hasegawa & Hiraiwa-Hasegawa 1983; Tutin 1979). Female compliance is important for exclusive mating, although males sometimes enforce exclusivity coercively (Goodall 1986; Hemelrijk, Van Laere & van Hooff 1992; Watts 1998).

Dominance

Dominance ranks tend to be difficult to discern among females, and many female dyads lack dominance relationships (but see Pusey, Williams & Goodall 1997). In contrast, males usually have dyadic dominance relationships and often form clear dominance hierarchies, although not necessarily in large communities (Bygott 1979; Goodall 1986; Nishida & Hosaka 1996; Watts 2000b). Access to fertile females depends partly on male rank (de Waal 1982; Tutin 1979; Watts 1998). For some males, often including alphas, rank depends on alliances (de Waal 1982; Goodall 1986; Nishida & Hosaka 1996). Males are heavier than females and all fully adult males dominate all adult females.

Chimpanzees use a variety of signals in agonistic contexts (van Hooff 1973). Of these, only pant-grunting is uni-directional in any given dyad (Bygott

1979; Hayaki, Huffman & Nishida 1989; Noë, de Waal & van Hooff 1980; Takahata 1990). It functions as a signal of subordination, but it is frequently volunteered in the absence of any aggression and is sometimes followed by neutral proximity or even affiliative physical contact.

While females commonly pant-grunt to males, and males often pant-grunt to other males, pant-grunting is rare between females (Preuschoft & de Waal 2001; Pusey et al. 1997). Dominance relationships between males and females and between males are sufficiently asymmetric to be formalized. Encounters between females may be too rare or agonistic outcomes too unpredictable for this to happen.

Bonding and cooperation

Females at Taï maintain long-lasting affiliative relationships with specific other females and males (Boesch & Boesch-Achermann 2000). Taï females sometimes form coalitions with males against other males; coalitions between females are also known (Boesch & Boesch-Achermann 2000), although frequency data are not available. Long-term male–female association and coalition formation are not typical in known East African populations, where relationships between females are marked by competition, except among mother–daughter dyads (Goodall 1986; Pusey et al. 1997). Affiliative interactions between males are a conspicuous feature of chimpanzee society. Males often associate with each other. Males groom other males more than females and more than females groom other females, and they form alliances that are important in within-community competition (Boesch & Boesch-Achermann 2000; Bygott 1979; de Waal 1982; Muroyama & Sugiyama 1994; Nishida 1983; Nishida & Hosaka 1996; Takahata 1990; Watts 2000b). The combination of frequent association and grooming with alliance formation justifies characterizing chimpanzee societies as "male bonded" (van Hooff & van Schaik, 1994). Males also cooperate during hunting and territory defense (Boesch & Boesch 1989; Boesch & Boesch-Achermann 2000; Stanford 1998; Watts & Mitani 2001; Wrangham 1999). Although males are philopatric, allies are not closely related on average (Goldberg & Wrangham 1997; Vigilant et al. 2001).

Reconciliation has been studied in captive and wild chimpanzees (Arnold & Whiten 2001; de Waal & van Roosmalen 1979; Preuschoft et al. 2002). Across demographic classes, reconciliation frequency varies positively with the frequency of cooperation: adult male–male dyads usually reconcile more than male–female and female–female dyads. Chimpanzees also seem to reconcile more readily with opponents with whom they groom often, and captive females may reconcile with each other more frequently than any other demographic class, even males (Preuschoft et al. in 2002).

Chimpanzees often respond to the imminent availability of food with excited displays and flurries of pant-hooting, pant-grunting, embracing, and patting, and sometimes with play (de Waal 1992; Reynolds & Reynolds 1965). This seems to reduce the social tension that accompanies the desire to feed in the presence of other community members, so that individuals can do so without much overt competition and can even share food (de Waal 1992). Socio-sexual behavior patterns like hold-bottom, genital contacts among females, and male–male mounting occur but are not prominent in these interactions.

Bonobos (Pan paniscus)

Social organization and mating

Bonobos live in fission–fusion communities of at least 20 members, but parties tend to be more cohesive than in chimpanzees and more often contain various members of both sexes (Wrangham & White 1988). Adolescent females transfer between communities, while males apparently remain in their natal communities permanently (for possible exceptions, see Hohmann et al. 1999). Females are rarely alone, spending most time in mixed-sex parties. Interactions between communities are usually hostile (Kano 1992), although they do not involve boundary patrols, coalitionary raids and lethal violence – a major contrast with chimpanzees (Wrangham 1999).

Females have large sexual swellings, are highly promiscuous, and have very long periods of sexual attractivity (Furuichi 1989; Takahata, Ihobe & Idane 1996; Wrangham 1993). After their first birth, bonobo females spend almost 50% of their time in a stage of maximal tumescence, as compared with about 4% for chimpanzees (Wrangham 1993). Along with this prolonged sexual activity, female bonobos spend close to 100% of their time in association with males (vs. < 40% in chimpanzees, Wrangham 1993).

Dominance

Whereas status typically reflects an individual's potential to coerce others (e.g., in macaques and chimpanzees), it apparently reflects the potential to provide commodities in bonobos. Adult female bonobos are lighter than males but tend to dominate them (Parish 1996; Vervaecke, de Vries & van Elsacker 2000a). However, dominance relationships are less salient among bonobos than chimpanzees, even if some individuals have decided dominance relationships, and tolerance among individuals is greater (Furuichi 1989; de Waal 1989a; de Waal & Lanting 1997; Kano 1992; Vervaecke *et al.*, 2000a).

Also, bonobos apparently lack formal status signals, at least among adults. Bonobo greeting grunts, which often occur in response to intimidation, social tension, or aggression, and which accompany post-conflict approaches and assertive play wrestling, are homologous to pant-grunts (de Waal 1988). Greeting grunts occur between males and between juvenile males and females, but not between females or between males and females. Because the sender is typically the member of the dyad that loses agonistic interactions and retreats when the other approaches, de Waal (1988) hypothesized that greeting grunts are formal signals of subordination. Vervaecke *et al.* (2000a) could not confirm this in a captive group, but could discern dominance relationships on the basis of conflict outcomes and approach–retreat interactions. They suggested that peering was entirely unidirectional, from subordinate to dominant, but other work (Johnson *et al.* 1999) does not support this suggestion.

Bonding and cooperation

Adult females, although largely unrelated, form close bonds with each other (Furuichi 1989; Idani 1991). They sometimes form coalitions that allow them to defeat males (Kano 1992; Vervaecke, de Vries & van Elsacker 2000b). In contrast, males rarely form coalitions with each other (Furuichi & Ihobe 1994; Kano 1992), despite the fact that grooming is more common between males than in other types of adult dyads except those of mothers with sons (Furuichi 1989; Muroyama & Sugiyama 1994). Communal male hunting is also absent or at least quite rare (White 1996). Males establish dominance relationships, and male rank and copulation rank were positively correlated at Wamba (Kano 1996). However, female sexual behavior in bonobos is only loosely tied to fertility and females are better able to resist male coercion than their chimpanzee counterparts. These factors may help to explain why males do not form alliances with each other. Mother–son bonds are remarkably strong, and mothers sometimes give maturing sons agonistic support (Hohmann *et al.* 1999; Kano 1992; Muroyama & Sugiyama 1994). This support may help sons to rise in rank (Kano 1992), although, if so, the non-occurrence of male–male alliances is puzzling. The extent to which females support unrelated males is unclear. Intergroup aggression consists largely of displays and chases and rarely involves contact aggression (de Waal & Lanting 1997; Kano 1992), perhaps because bonobos' greater gregariousness reduces the probability of encountering lone neighbors compared with chimpanzees (Wrangham 1986, 1999).

Among captive bonobos reconciliation occurs regularly, but no information about rates of reconciliation in different demographic classes is available (de Waal 1987). Many of the behavior patterns that bonobos, chimpanzees, and gorillas use in reconciliation are similar: facial–vocal signals of appeasement or reassurance, holding out a hand, and embracing. Alone among great apes, however, bonobos use sexual interactions in almost 50% of reconciliations, regardless of opponent age or sex (de Waal 1987).

Among wild bonobo females, genito-genital (G-G) rubbing increases after conflicts, and is more frequent when females are in large parties with high potential for conflict. In the presence of limited, monopolizable food, non-owners present to food-possessors, who usually mount them (Hohmann & Fruth 2000). Thus, G-G rubbing seems to promote tolerance and facilitate access to food. Furuichi (1989) hypothesized that adolescent immigrant females use sexual behavior to establish bonds with resident, dominant females. No data are available to test this hypothesis, but females seem either to develop grooming bonds or to engage in G-G rubbing, but not both (Hohmann & Fruth 2000). Thus, G-G rubbing may serve as a conflict management device when the potential for feeding competition is high and when relationships are endangered by conflicts or are not yet securely established (Hohmann & Fruth 2000; Hübsch 1970; Jordan 1977; Kano 1992). That sex serves this communicative function, rather than grooming or some other kind of affiliative behavior, can be seen as the outcome of the extended proceptivity and attractivity of bonobo females, which led to a high and lasting motivation for sexual behavior (de Waal & Lanting 1997).

SOCIAL DIAGNOSIS OF THE GREAT APES

Can we conclude from these descriptions that great ape sociality is indeed more complex than that of other primates? On the surface the answer is negative. The sizes of parties, groups and even communities fall comfortably within the range found among other primates, and their demographic composition is no more complex (Smuts *et al.* 1987). The kinds of social interactions associated with social complexity in other primates, for example alliance formation with non-relatives (Harcourt 1992), are not strikingly more common. In fact, alliances are absent in orangutans and limited mostly to female kin in gorillas. Nor do all great apes obviously share any complex social phenomena absent in all other nonhuman primates.

More detailed examination suggests that some great apes show greater cognitive complexity in dealing with social problems also faced by other nonhuman primates. Thus, male baboons and macaques engage in chimpanzee-like coalition formation tactics (Kuester & Paul 1992; Noë 1990; Silk 1994), but chimpanzee social decision making shows subtleties not apparent in these taxa, such as separating interventions to prevent rivals from establishing threatening alliances (de Waal 1982; Nishida & Hosaka 1996; Parker, Chapter 4, this volume). Also, male chimpanzees form both dyadic alliances for within-community status competition and community-wide alliances that compete against other communities, sometimes via territorial incursions and potentially lethal attacks of a form not seen in other nonhuman primates (Watts & Mitani 2001; Wrangham 1999).

Given the possibility that such social subtleties distinguish all the great apes from other nonhuman primates and the likelihood that great apes are more intelligent than other nonhuman primates are, we search for social commonalities that could help to explain their advanced cognitive abilities then consider the argument that advanced social cognition in extant great apes is a consequence of cognitive capacities that arose in their last common ancestor in response to non-social selective pressures.

The species vignettes allow us to recognize the following great ape commonalities that distinguish them from most other anthropoids:

(1) *A tendency toward fission–fusion social organization (or at least toward non-permanence of social units), with individuals out of contact with conspecifics for prolonged periods and with foraging females notably solitary.* Only gorillas form cohesive groups, but female membership in these groups is flexible. Flexible choice of association partners or group membership raises interesting questions about the extent to which individuals can enforce power differentials and the amount of uncertainty about these differentials (below). Most other primates with some form of fission–fusion social organization either have stable and cohesive, "modular" subgroups (e.g., Hamadryas baboons, geladas) or form short-term parties that usually maintain visual or auditory contact with other members of stable groups (e.g., long-tailed macaques) (Smuts *et al.* 1987). Only the atelines show obvious convergence with the great apes in this respect (Strier *et al.* 1993; Symington 1990).

(2) *Relatively high subordinate leverage.* Clearly signaled decided ("formal") dominance relationships among frequently associating same-sex individuals are rare in bonobos, chimpanzees (especially captives), and Sumatran orangutans (especially females) (Table 11.2). Subsets of male chimpanzee dyads form the major exception. Concomitantly, social tolerance is marked and accompanied by affiliative behaviors such as food sharing. Outside the great apes, the absence of formal dominance despite the potential for clear-cut contest competition is only found among a few species of macaques and perhaps capuchins (Preuschoft & van Schaik 2000).

(3) *Intrasexual bonds among non-relatives are as common, or more so, than bonds among relatives* (male chimpanzees; female bonobos; perhaps female chimpanzees). Intrasexual bonds with non-kin also occur among males and among females in some cercopithecines (e.g., Chapais 1992; Noë 1990), but kinship-based bonding is typically more important (Smuts *et al.* 1987; but see Chapais 2001).

(4) *Remarkably extensive intra-specific flexibility in social organization and affiliation,* in orangutans and chimpanzees (and perhaps in the other species).

	Ecology, Association	*Social Cognition, Sociability*
EVOLUTIONARY PRECONDITIONS	• large body size • slow life history • arboreal life style	• large brains
INFLUENCES ON BEHAVIOR 	• low vulnerability to predation • high vulnerability to scramble competition • (high-quality diet)	• opportunities for long-term bonds • potential for efficient social learning • scarcity of same-sex kin suitable as allies • more elaborate preparatory food processing
SOCIOECOLOGY	• tendency toward solitary foraging • fission-fusion sociality • vulnerability to lethal aggression from conspecifics	• benefits of gregariousness social rather than ecological
SOCIAL BEHAVIOR	• flexible association choices	• social leverage power and high tolerance • bonds with non-kin • social exchange relationships, like food sharing • behavioral traditions
SOCIAL INTELLIGENCE	• socially transmitted extractive foraging techniques	• strategic social decision making accounting for cooperation among non-kin • more efficient social learning • take partners' perspectives to a degree

Figure 11.2. The working hypothesis for the evolution of derived ecological, social, and cognitive features of the great apes developed in this chapter. It starts with the biological factors that provided the basic influences on their behavior, and favored the evolution of particular socioecological contexts and their social and psychological correlates. These in turn favored the evolution of several exponents of social intelligence. The black arrow indicates a direct causal link; grey arrows represent favoring the evolution of the trait at the receiving end of the arrow, whereas double-sided arrows indicate correlation or positive feedback.

Most of these features have an obvious socioecological basis, illustrated in Figure 11.2. Mostly they are indirect consequences of large body size, which leads to an increased potential for contest competition, especially for females and especially in those species unable to switch to high-fiber foods during times of fruit scarcity (see Yamagiwa, Chapter 12, this volume). However, large body size also makes great apes less vulnerable to predation than other primates. This gives females the option to forage alone when contest competition becomes too intense, turning it largely into scramble competition. Solitary foraging may make them more vulnerable to sexual coercion, including infanticide, by males of their own species, but females can seek refuge with protectors if they are aware of their approximate location. Even gorilla females without dependent infants can easily switch associates by transferring between groups.

Likewise, male chimpanzees and bonobos have the option to avoid rivals by foraging alone, although this may increase the risk of inter-community aggression for chimpanzee males. Thus, great ape life both requires and allows facultative switches between solitary and gregarious foraging as well as switches of associates.

The presence of these alternative options for subordinates means that potentially dominant individuals must curb their aggressive tendencies if they are to reap the social benefits of being in parties with subordinates. The suppression of contest tendencies favors social tolerance, which in turn facilitates social learning of manipulative skills (Coussi-Korbel & Fragaszy 1995; van Schaik 2003) and provides a substrate for food sharing and other forms of cooperation (de Waal 1989b, 2000; Preuschoft *et al.* in prep.). One important benefit for females could be exposure to better opportunities for socially learning foraging or social skills for their offspring (van Schaik 1999).

Subordinate leverage may explain the much reduced rigidity of great ape dominance relationships, compared with most cercopithecines. It also explains the absence of formal subordination signals in species or age–sex classes that would be expected to have strong contest potential (cf. Preuschoft & van Schaik 2000). Obviously, formal subordination signals are also absent where competition is largely by scramble (female gorillas) or competition is by contest and no bonding is necessary (male orangutans). Male chimpanzees are somewhat different. Subordinate leverage is less than expected. All males must be members of a large alliance in order to be successful but where there are many subordinates, males' options are limited to choices between possible allies to support. Thus, dominance relationships and status signaling are more pronounced, albeit only among a subset of the males.

The great apes' extraordinary capacity for cooperation with non-relatives, although possibly linked to their flexible association patterns, may have been imposed on them by their slow life histories, which produce demographic conditions in which close relatives of the preferred sex, age, and fighting ability are often not available. Their remarkable intra-specific social variability (e.g., in male–female and female–female social relationships in chimpanzees; Baker & Smuts 1994; Boesch & Boesch-Achermann 2000) may reflect developmentally flexible rather than canalized social decision rules.

Thus, all great ape genera share at least some aspects of social life that require greater cognitive abilities than other nonhuman primates. Do these features represent evolved responses to subtle social demands? First, fission–fusion sociality, relaxed dominance relations, and cooperation with non-relatives may have favored cognitive evolution because they require flexible tactics (with the clear exception of adult male orangutans). And at least in *Pan*, individuals must form complex social relationships, balancing rivalry and interdependence that go well beyond the alliances of monkeys. Chimpanzees may have some ability to attribute mental states to others and may use these abilities to minimize the costs of contest competition; presumably, they also use them to make other strategic social decisions. Second, more relaxed dominance may also have favored improvements in the capacity for copying the behaviors

of conspecifics, especially specialized foraging and tool skills. Thus, evidence for behavioral traditions in non-human primates, whose maintenance requires both horizontal and vertical social transmission, is by far the most extensive in two great apes species, chimpanzees and orangutans (van Schaik 2003; Whiten *et al.* 1999).

However, fission–fusion sociality, relaxed dominance, or cooperation with non-relatives are also found in some monkeys (capuchins, some macaques: Preuschoft & van Schaik 2000; Thierry, Wunderlich & Gueth 1989), without great ape level cognitive abilities. Therefore, the argument that these social features are sufficient to account for great ape–monkey cognitive differences is not compelling, and we conclude that social strategizing, whether in its traditional or its modified version, cannot directly account for the evolution of great ape cognition (cf. Byrne 1997a).

THE EVOLUTION OF GREAT APE COGNITION

Most hypotheses proposed so far (Table 11.1) face this problem: they should also apply to primates other than the great apes and/or to numerous non-primate species that live in permanent social groups and that eat food whose abundance varies in space and time or must be extracted or processed before ingestion.[1] We propose that the crucial factor needed to resolve this problem is consideration of life history variation (see also Figure 11.2).

Recently, Deaner *et al.* (2003) and van Schaik and Deaner (2003) argued that taxonomic differences in cognitive capacities often reflect life history differences (see also Parker & McKinney 1999; Ross, Chapter 8, this volume). Extensive correlated evolution between brain size and life history has apparently occurred among mammals (Deaner *et al.* 2003; van Schaik & Deaner 2003; cf. Allman 1999). Faster life history constrains cognitive evolution, whereas slower life history releases it. Taxa with relatively fast life histories cannot afford to respond to socioecological demands with cognitive adaptations; those with slower life history can and, under some conditions, perhaps they must (Potts, Chapter 13, this volume). Slow life history in itself will not lead to enhanced cognitive capacities, but it allows or perhaps requires responses to selective pressures that favor such capacities (see also Kelley, Chapter 15, Ross, Chapter 8, this volume).

Such differential response may explain the unusual cognitive position of the great apes, which stand out among primates and other mammals for the slowness of their life histories (Harvey, Martin & Clutton-Brock 1987; Kelley, 1997; Read & Harvey 1989). The combination of large body size and arboreality presumably minimized the risk of predation for the emerging great ape lineage, and consequent low extrinsic mortality risks presumably predisposed them toward the evolution of unusually slow life histories (Kelley, 1997; van Schaik & Deaner 2003; cf. Begun & Kordos, Chapter 14, Kelley, Chapter 15, this volume). This would have facilitated evolutionary increases in brain size, which are constrained by the high metabolic costs of neural tissue.

Increased brain size and prolonged neural development would have enabled responses to selective pressures that favored cognitive solutions. The combination of increased vulnerability to feeding competition and reduced predation risk, due to large body size, would have promoted flexibility in grouping (cf. Figure 11.2).

This perspective does not reveal the actual pressures that favored brain evolution, and thus cognitive evolution. However, it emphasizes that advances in general cognitive capacities could have allowed great apes to make fitness gains by improving their performance in a large array of technical, ecological, and social tasks. It also suggests the action of numerous, largely compatible and additive or interacting selective pressures. Thus, some ecological pressure may select for improved cognitive abilities for handling tasks in that domain (life history permitting), but these capacities may be exapted to improve performance in another domain, for example managing social relationships. In turn, these improvements may create new selection pressures in the social domain that brought about the particular social-cognition skills of great apes. Teasing apart the relative role of the various selective forces that gave rise to the actual historical trajectories may prove next to impossible.

The stem large hominoid, with its primarily frugivorous and presumably difficult foraging niche (see Potts, Chapter 13, Singleton, Chapter 16, this volume), would presumably have benefited from enhanced spatial memory, and improved ability to monitor food availability (cf. Milton 1988), and enhanced abilities for locating and obtaining the additional high-quality and fall-back foods needed to balance their frugivorous preferences (see Yamagiwa, Chapter 12, this volume). Improved technical foraging skills that depended on complex motor coordination, planning, and insight would also have been beneficial and presumably required prolonged learning periods (Byrne 1997a; Kaplan *et al.* 2000; Parker & Gibson 1977). Even if such foraging and locomotion pressures enhanced great ape cognition, we can still ask what cognitive abilities peculiar to the social domain develop in great apes, and which aspects of great ape life favored them. Great apes' special aptitude for observational learning (cf. Russon 1997; Russon *et al.* 1998) and their greater propensities toward establishing cooperative relationships with non-relatives both involve social exchange (cf. de Waal 2000). Also, slow life history in itself sets up various other pressures. In particular, the slow-down in female reproductive rates leads to more male-biased operational sex ratios, which in turn increases the potential of sexual coercion in the form of sexual harassment and infanticide. The need to avoid sexual coercion may have generated arms races with major behavioral and hence cognitive components, and a general need to avoid escalated fighting could have generated pressures to solve social conflicts in non-violent ways (van Schaik & Deaner 2003).

In conclusion, we argue that the original social intelligence hypothesis – that intense social life led to improved social cognition and thus to greater general intelligence – cannot explain great ape distinctiveness. However, incorporating life history variation into our explanatory paradigm leads to a plausible revision that includes various selection pressures, including (prominently) those arising from social life. Evolution of slow life histories is coupled with the evolution of large brains and large bodies (see Kelley, Chapter 15, Ross, Chapter 8, this volume). If having large bodies has major ecological consequences, it also has dramatic social ones: it increases the costs of sociality, which leads to flexible grouping patterns through increased vulnerability to competition, and it substitutes vulnerability to predators for vulnerability to hostile conspecifics. These two consequences lead to a cascade of further social consequences including increased social leverage for subordinate individuals and cooperation among non-relatives. Possession of large brains in large bodies thus indirectly set the stage for uniquely elaborate cognitive solutions of non-unique social problems.

ACKNOWLEDGMENTS

We thank the following organizations for supporting our research: Wildlife Conservation Society (CvS), L. S. B. Leakey Foundation (CvS, DW), and the Templeton Foundation (SP). We thank Rob Deaner, and the reviewers of a previous draft for a very valuable discussion.

ENDNOTE

1 The only exception is the "arboreal clambering hypothesis" (Povinelli & Cant 1995), which applies specifically to great apes and holds that ancestral apes required some form of self-concept to cope with the challenges that arboreality, especially arboreal locomotion, poses for large-bodied animals (see Gebo, Chapter 17, Hunt, Chapter 10, this volume).

REFERENCES

Allman, J. M. (1999). *Evolving Brains*. New York: Scientific American Library.

Arnold, K. & Whiten, A. (2001). Post-conflict behavior of wild chimpanzees (*Pan troglodytes schweinfurthii*) in the Budongo forest, Uganda. *Behaviour*, 138, 649–90.

Aureli, F. A. & de Waal, F. B. M. (ed.) (2000). *Natural Conflict Resolution*. Berkeley, CA: University of California Press.

Aureli, F. A., Cords, M. & van Schaik, C. P. (2002). Conflict resolution following aggression in gregarious animals: a predictive framework. *Animal Behaviour*, 64(3), 325–43.

Baker, K. C. & Smuts, B. B. (1994). Social relationships of female chimpanzees: diversity between captive social groups. In *Chimpanzee Cultures*, ed. R. W. Wrangham, W. C. McGrew, F. B. M. de Waal & P. G. Heltne, pp. 227–42. Cambridge, MA: Harvard University Press.

Barton, R. & Dunbar, R. I. M. (1997). Evolution of the social brain. In *Machiavellian Intelligence II: Extensions and Evaluations*, ed. A. Whiten and R. W. Byrne, pp. 240–63. Cambridge, UK: Cambridge University Press.

Boesch, C. & Boesch, H. (1989). Hunting behavior of wild chimpanzees in the Tai National Park. *American Journal of Physical Anthropology*, 78, 547–73.

Boesch, C. & Boesch-Achermann, H. (2000). *The Chimpanzees of the Taï Forest: Behavioural Ecology and Evolution*. Oxford: Oxford University Press.

Bygott, J. D. (1979). Agonistic behavior, dominance, and social structure in wild chimpanzees of the Gombe National Park. In *The Great Apes*, ed. D. A. Hamburg & E. R. McCown, pp. 405–28. Menlo Park, CA: Benjamin/Cummings.

Byrne, R. W. (1995). *The Thinking Ape: Evolutionary Origins of Intelligence*. Oxford: Oxford University Press.

(1997a). The technical intelligence hypothesis: an additional evolutionary stimulus to intelligence? In *Machiavellian Intelligence II: Extensions and Evaluations*, ed. A. Whiten and R. W. Byrne, pp. 289–311. Cambridge, UK: Cambridge University Press.

(1997b). Machiavellian Intelligence. *Evolutionary Anthropology*, 5, 172–80.

Byrne, R. W. & Whiten, A. (ed.) (1988). *Machiavellian Intelligence: Social Expertise and The Evolution of Intellect in Monkeys, Apes, and Humans*. Oxford: Clarendon Press.

(1997). Machiavellian intelligence. In *Machiavellian Intelligence II: Extensions and Evaluations*, ed. A. Whiten & R. W. Byrne, pp. 1–23. Cambridge, UK: Cambridge University Press.

Chapais, B. (1992). The role of alliances in social inheritance of rank among female primates. In *Coalitions and Alliances in Humans and Other Animals*, ed. A. H. Harcourt & F. B. M. de Waal, pp. 29–59. Oxford: Oxford University Press.

(2001). Primate nepotism: what is the explanatory value of kin selection? *International Journal of Primatology*, 22, 203–29.

Cheney, D. L. & Seyfarth, R. M. (1990). *How Monkeys See the World*. Chicago: University of Chicago Press.

Cheney, D. L., Seyfarth, R. M. & Smuts, B. B. (1986). Social relationships and social cognition in non-human primates. *Science*, 234, 1361–6.

Clutton-Brock, T. H. & Harvey, P. H. (1980). Primate brains and ecology. *Journal of Zoology*, 207, 151–69.

Connor, R. C., Smolker, R. A. & Richards, A. F. (1992). Dolphin alliances and coalitions. In *Coalitions and Alliances in Humans and Other Animals*, ed. A. Harcourt & F. B. M. de Waal, pp. 415–43. New York: Oxford University Press.

Cords, M. (1997). Friendships, alliances, reciprocity and repair. In *Machiavellian Intelligence II: Extensions and Evaluations*, ed. A. Whiten & R. W. Byrne, pp. 24–49. Cambridge, UK: Cambridge University Press.

Coussi-Korbel, S. & Fragaszy, D. M. (1995). On the relation between social dynamics and social learning. *Animal Behaviour*, 50, 1441–53.

Cummins, D. D. (1998). Social norms and other minds: the evolutionary roots of higher cognition. In *The Evolution of Mind*, ed. D. D. Cummins & C. Allen, pp. 30–50. Oxford: Oxford University Press.

Deaner, R. O., Barton, R. A. & van Schaik, C. P. (2003). Primate brains and life histories: renewing the connections. In *Primate Life Histories and Socioecology*, ed. P. M. Kappeler & M. E. Pereira, pp. 233–65. Chicago, IL: University of Chicago Press.

Deaner, R. O., Nunn, C. L. & van Schaik, C. P. (2000). Comparative tests of primate cognition: different scaling methods produce different results. *Brain and Behavior Evolution*, **55**, 44–52.

Delgado, R. & van Schaik, C. P. (2000). The behavioral ecology and conservation of the orangutan (*Pongo pygmaeus*): a tale of two islands. *Evolutionary Anthropology*, **9**, 201–18.

de Waal, F. B. M. (1982). *Chimpanzee Politics*. London: Jonathan Cape Ltd.

(1986). The integration of dominance and social bonding in primates. *Quarterly Review of Biology*, **61**, 459–79.

(1987). Tension regulation and nonreproductive functions of sex in captive bonobos (*Pan paniscus*). *National Geographic Research*, **3**, 318–35.

(1988). The communicative repertoire of captive bonobos (*Pan paniscus*), compared to that of chimpanzees. *Behaviour*, **106**, 183–251.

(1989a). Dominance "style" and primate social organization. In *Comparative Socioecology*, ed. V. Standen & R. A. Foley, pp. 243–63. London: Blackwell.

(1989b). Food sharing and reciprocal obligations among chimpanzees. *Journal of Human Evolution*, **18**, 433–59.

(1992). Coalitions as part of reciprocal relations in the Arnhem chimpanzee colony. In *Coalitions and Alliances in Humans and Other Animals*, ed. A. H. Harcourt & F. B. M. de Waal, pp. 233–58. Oxford: Oxford University Press.

(2000). Primates: a natural heritage of conflict resolution. *Science*, **289**, 586–90.

de Waal, F. B. M. & Lanting, F. (1997). *Bonobo: The Forgotten Ape*. Berkeley, CA: University of California Press.

de Waal, F. B. M. & Roosmalen, A. V. (1979). Reconciliation and consolation among chimpanzees. *Behavioral Ecology and Sociobiology*, **5**, 55–66.

Doran, D. & McNeilage, A. (1998). Gorilla ecology and behavior. *Evolutionary Anthropology*, **6**, 120–31.

Dugatkin, L. A. (1997). *Cooperation Among Animals: An Evolutionary Perspective*. New York: Oxford University Press.

Dunbar, R. I. M. (1992). Neocortex size as a constraint on group size in primates. *Journal of Human Evolution*, **20**, 469–93.

(1998). The social brain hypothesis. *Evolutionary Anthropology*, **6**, 178–90.

Fossey, D. (1972). Vocalizations of the mountain gorilla (*Gorilla gorilla beringei*). *Animal Behaviour*, **20**, 36–53.

Fox, E. A. (1998). The function of female mate choice in the Sumatran orangutan (*Pongo pygmaeus abelii*). Ph.D. thesis, Duke University.

Furuichi, T. (1989). Social interactions and the life history of female *Pan paniscus* in Wamba, Zaire. *International Journal of Primatology*, **10**, 173–97.

Furuichi, T. & Ihobe, H. (1994). Variation in male relationships in bonobos and chimpanzees. *Behaviour*, **130**, 211–28.

Galdikas, B. M. F. (1984). Adult female sociality among wild orangutans at Tanjung Puting Reserve. In *Female Primates: Studies by Women Primatologists*, ed. M. F. Small, pp. 217–35. New York: Alan R. Liss.

Galdikas, B. M. F. & Wood, J. W. (1990). Birth spacing patterns in humans and apes. *American Journal of Physical Anthropology*, **83**, 185–91.

Goldberg, T. L. & Wrangham, R. W. (1997). Genetic correlates of social behavior in wild chimpanzees: evidence from mitochondrial DNA. *Animal Behaviour*, **54**, 559–70.

Goodall, J. (1986). *The Chimpanzees of Gombe*. Cambridge, MA: Harvard University Press.

Hand, J. L. (1986). Resolution of social conflicts: dominance, egalitarianism, spheres of dominance and game theory. *Quarterly Review of Biology*, **61**, 201–20.

Harcourt, A. H. (1978). Strategies of emigration and transfer by primates, with particular reference to gorillas. *Zeitschrift für Tierpsychologie*, **48**, 401–20.

(1979a). Social relationships among adult female mountain gorillas. *Animal Behaviour*, **27**, 325–42.

(1979b). Social relationships between adult male and female mountain gorillas in the wild. *Animal Behaviour*, **27**, 325–42.

(1992). Coalitions and alliances: are primates more complex than non-primates. In *Coalitions and Alliances in Humans and Other Animals*, ed. A. H. Harcourt & F. B. M. de Waal, pp. 445–71. Oxford: Oxford University Press.

Harcourt, A. H. & Stewart, K. J. (1981). Gorilla male relationships: can differences during immaturity lead to contrasting reproductive tactics in adulthood? *Animal Behaviour*, **29**, 206–10.

(1989). Functions of alliances in contests within wild gorilla groups. *Behaviour*, **109**, 176–90.

Harcourt, A. H., Fossey, D., Stewart, K. & Watts, D. P. (1980). Reproduction in wild gorillas and some comparisons with chimpanzees. *Journal of Reproduction and Fertility Supplement*, **28**, 59–70.

Harcourt, A. H., Stewart, K. J. & Hauser, M. (1993). Function of wild gorilla "close" calls. Repertoire, context, and interspecific comparison. *Behaviour*, **124**, 89–122.

Hare, B., Call, J., Agnetta, B. & Tomasello, M. (2000). Chimpanzees know what conspecifics do and do not see. *Animal Behaviour*, **59**, 771–85.

Harvey, P. H., Martin, R. D. & Clutton-Brock, T. H. (1987). Life histories in comparative perspective. In *Primate Societies*, ed. B. B. Smuts, D. L. Cheney, R. M. Seyfarth, R. W. Wrangham & T. T. Struhsaker, pp. 181–96. Chicago, IL: University of Chicago Press.

Hasegawa, T. (1990). Sex differences in ranging patterns. In *The Chimpanzees of the Mahale Mountains: Sexual and Life History Strategies*, ed. T. Nishida, pp. 99–114. Tokyo: University of Tokyo Press.

Hasegawa, T. & Hiraiwa-Hasegawa, M. (1983). Opportunistic and restrictive mating among wild chimpanzees in the Mahale mountains, Tanzania. *Journal of Ethology*, **1**, 75–85.

Hayaki, H., Huffman, M. A. & Nishida, T. (1989). Dominance among male chimpanzees in the Mahale Mountains, National Park, Tanzania: a preliminary study. *Primates*, **30**, 187–97.

Hemelrijk, C. K., Van Laere, G. J. & van Hooff, J. A. R. A. M. (1992). Sexual exchange relationships in captive chimpanzees? *Behavioral Ecology and Sociobiology*, **30**, 269–75.

Hohmann, G. & Fruth, B. (2000). Use and function of genital contacts among female bonobos. *Animal Behaviour*, **60**, 107–20.

Hohmann, G., Gerloff, U., Tautz, D. & Fruth, B. (1999). Social bonds and genetic ties: kinship, association and affiliation in a community of bonobos (*Pan paniscus*). *Behaviour*, **136**, 1219–35.

Hübsch, I. (1970). Einiges zum Verhalten der Zwergschimpansen (*Pan paniscus*) und Schimpansen (*Pan troglodytes*) im Frankfurter Zoo. *Zoologischer Garten*, **38**, 107–32.

Humphrey, N. K. (1976). The social function of intellect. In *Growing Points in Ethology*, ed. P. P. G. Bateson & R. A.

Hinde, pp. 303–17. Cambridge, UK: Cambridge University Press.

Hunt, G. R. (1996). Manufacture and use of hook-tools by New Caledonian crows. *Nature*, **379**, 249–51.

Idani, G. (1991). Social relationships between immigrant and resident bonobo females (*Pan paniscus*) female at Wamba. *Folia Primatologica*, **57**, 83–95.

Johnson, V. E., Deaner, R. O. & van Schaik, C. P. (2002). Bayesian analysis of rank data with application to primate intelligence experiments. *Journal of the American Statistical Association*, **97**(457), 8–17.

Jordan, C. (1977). Das Verhalten zoolebender Zwergschimpansen. Ph.D. dissertation. Goethe Universität Frankfurt.

Kano, T. (1992). *The Last Ape: Pygmy Chimpanzee Behavior and Ecology*. Stanford, CA: Stanford University Press.

(1996). Male rank order and copulation rate in a unit-group of bonobos at Wamba, Zaïre. In *Great Ape Societies*, ed. W. C. McGrew, L. F. Marchant & T. Nishida, pp. 135–45. Cambridge, UK: Cambridge University Press.

Kaplan, H., Hill, K., Lancaster, J. & Hurtado, A. M. (2000). A theory of human life history evolution: diet, intelligence, and longevity. *Evolutionary Anthropology*, **9**, 156–85.

Kelley, J. (1997). Paleobiological and phylogenetic significance of life history in Miocene hominoids. In *Function, Phylogeny, and Fossils: Miocene Hominoid Evolution and Adaptations*, ed. D. R. Begun, C. V. Ward & M. D. Rose, pp. 173–208. New York: Plenum Press.

Kudo, H. & Dunbar, R. I. M. (2001). Neocortex size and social network size in primates. *Animal Behaviour*, **62**, 711–22.

Kuester, J. & Paul, A. (1992). Influence of male competition and female mate choice on male mating success in barbary macaques (*Macaca sylvanus*). *Behaviour*, **120**, 192–217.

Kummer, H. (1978). On the value of social relationships to nonhuman primates: a heuristic scheme. *Social Sciences Information*, **17**, 687–705.

Lewis, R. J. (2002). Beyond dominance: the importance of leverage. *Quarterly Review of Biology*, **77**, 149–64.

Manson, J. & Wrangham, R. W. (1991). Intergroup aggression in chimpanzees and humans. *Current Anthropology*, **32**, 369–90.

Marler, P. (1976). Social organization, communication and graded signals, the chimpanzee and the gorilla. In

Growing Points In Ethology, ed. P. P. G. Bateson & R. A. Hinde, pp. 239–80. Cambridge, UK: Cambridge University Press.

(1996). Social cognition: are primates smarter than birds? In *Current Ornithology*, Vol. 13, ed. V. Nolan, Jr. & E. D. Ketterson, pp. 1–32. New York: Plenum Press.

Milton, K. (1988). Foraging behaviour and the evolution of primate intelligence. In *Machiavellian Intelligence*, ed. R. W. Byrne & A. Whiten, pp. 285–306. Oxford: Clarendon Press.

Muroyama, Y. & Sugiyama, Y. (1994). Grooming relationships in two species of chimpanzees. In *Chimpanzee Cultures*, ed. R. W. Wrangham, W. C. McGrew, F. B. M. de Waal & P. G. Heltne, pp. 169–80. Cambridge, MA: Harvard University Press.

Nishida, T. (1983). Alpha status and agonistic alliance in wild chimpanzees. *Primates*, **24**, 318–36.

Nishida, T. & Hosaka, K. (1996). Coalition strategies among adult male chimpanzees of the Mahale mountains, Tanzania. In *Great Ape Societies*, ed. W. C. McGrew, L. F. Marchant & T. Nishida, pp. 114–34. Cambridge, UK: Cambridge University Press.

Noë, R. (1990). A veto game played by baboons: a challenge to the use of the prisoner's dilemma as a paradigm for reciprocity and cooperation. *Animal Behaviour*, **39**, 78–90.

Noë, R., de Waal, F. B. M. & van Hooff, J. A. R. A. M. (1980). Types of dominance in a chimpanzee colony. *Folia Primatologica*, **34**, 90–110.

Parish, A. R. (1996). Female relationships in bonobos (*Pan paniscus*). *Human Nature*, **7**, 61–96.

Parker, S. T. & Gibson, K. R. (1977). Object manipulation, tool use and sensorimotor intelligence as feeding adaptations in cebus monkeys and great apes. *Journal of Human Evolution*, **6**, 623–41.

(ed.) (1990). *"Language" and Intelligence in Monkeys and Apes*. New York: Cambridge University Press.

Parker, S. T. & McKinney, M. L. (1999). *Origins of Intelligence: The Evolution of Cognitive Development in Monkeys, Apes, and Humans*. Baltimore, MD: The Johns Hopkins Press.

Povinelli, D. J. & Cant, J. G. H. (1995). Arboreal clambering and the evolution of self-conception. *Quarterly Review of Biology*, **70**, 393–421.

Preuschoft, S. (1992). "Laughter" and "smile" in Barbary macaques (*Macaca sylvanus*). *Ethology*, **91**, 200–36.

(1999). Are primates behaviorists? Formal dominance, cognition, and free-floating rationales. *Journal of Comparative Psychology*, **113**, 91–5.

Preuschoft, S. & de Waal, F. (2001). Food sharing and competitive exclusion by chimpanzees: experimental manipulation of risks and stakes. Presented at the 18th Congress of the International Primatological Society, Adelaide.

Preuschoft, S. & van Schaik, C. P. (2000). Dominance and communication: conflict management in various social settings. In *Natural Conflict Resolution*, ed. F. Aureli & F. B. M. de Waal, pp. 77–105. Berkeley, CA: University of California Press.

Preuschoft, S., Wang, X., Aureli, F. & de Waal, F. B. M. (2002). Reconciliation in captive chimpanzees: a reevaluation with controlled methods. *International Journal of Primatology*, **23**, 29–50.

Pusey, A., Williams, J. & Goodall, J. (1997). The influence of dominance rank on the reproductive success of female chimpanzees. *Science*, **277**, 828–31.

Read, A. F. & Harvey, P. H. (1989). Life history differences among the eutherian radiations. *Journal of the Zoological Society of London*, **219**, 329–53.

Reynolds, V. & Reynolds, F. (1965). Chimpanzees of the Budongo Forest. In *Primate Behavior: Field Studies of Monkeys and Apes*, ed. I. DeVore, pp. 368–424. New York: Holt, Rinehart, and Winston.

Rijksen, H. D. (1978). *A Field Study on Sumatran Orang Utans* (*Pongo pygmaeus abelii Lesson 1827*). Wageningen, the Netherlands: H. Veenman & Zonen, B. V.

Robbins, M. M. (1995). A demographic analysis of male life history and social structure of mountain gorillas. *Behaviour*, **132**, 21–47.

(1996). Male–male interactions in heterosexual and all-male wild mountain gorilla groups. *Ethology*, **102**, 942–65.

(2001). Variation in the social system of mountain gorillas: the male perspective. In *Mountain Gorillas: Three Decades of Research at Karisoke*, ed. M. M. Robbins, P. Sicotte, & K. J. Stewart, pp. 29–58. Cambridge, UK: Cambridge University Press.

Rumbaugh, D. M., Savage-Rumbaugh, E. S. & Washburn, D. A. (1996). Toward a new outlook on primate learning and behavior: complex learning and emergent processes in comparative perspective. *Japanese Psychological Research*, **38**, 113–25.

Russon, A. E. (1997). Exploiting the expertise of others. In *Machiavellian Intelligence II: Extensions and Evaluations*, ed. A. Whiten & R. W. Byrne, pp. 174–206. Cambridge, UK: Cambridge University Press.

Russon, A. E. & Bard, K. A. (1996). Exploring the minds of great apes: issues and controversies. In *Reaching into Thought: The Minds of the Great Apes*, ed. A. E. Russon, K. A. Bard & S. T. Parker, pp. 1–20. Cambridge, UK: Cambridge University Press.

Russon, A. E., Bard, K. A. & Parker, S. T. (ed.) 1996. *Reaching into Thought: The Minds of the Great Apes*. Cambridge, UK: Cambridge University Press.

Russon, A. E., Mitchell, R. E., Lefebvre, L. & Abravanel, E. (1998). The evolution of imitation. In *Piaget, Evolution and Development*, ed. J. Langer & M. Killen, pp. 103–43. Mahwah, NJ: Lawrence Erlbaum Associates.

Schenkel, R. (1967). Submission: its features and function in the wolf and dog. *American Zoologist*, 7, 319–23.

Sicotte, P. (1993). Inter-group encounters and female transfer in mountain gorillas: influence of group composition on male behavior. *American Journal of Primatology*, 30, 21–36.

(1994). Effect of male competition on male–female relationships in bi-male groups of mountain gorillas. *Ethology*, 97, 47–64.

Silk, J. B. (1994). Social relationships of male bonnet macaques: male bonding in a matrilineal society. *Behaviour*, 130, 271–91.

Singleton, I. S. & van Schaik, C. P. (2001). Orangutan home range size and its determinants in a Sumatran swamp forest. *International Journal of Primatology*, 22, 877–911.

(2002). The social organisation of a population of Sumatran orang-utans. *Folia Primatologica*, 73, 1–20.

Smuts, B. B., Cheney, D. L., Seyfarth, R. M., Wrangham, R. W. & Struhsaker, T. T. (ed.) (1987). *Primate Societies*. Chicago, IL: University of Chicago Press.

Stanford, C. B. (1998). Predation and male bonds in primate societies. *Behaviour*, 135, 513–33.

Steenbeek, R. (2000). Infanticide by males and female choice in Thomas's langurs. In *Infanticide by Males and its Implications*, ed. C. P. van Schaik & C. H. Janson, pp. 153–77. Cambridge, UK: Cambridge University Press.

Stewart, K. J. & Harcourt, A. H. (1987). Gorillas: variation in female social relationships. In *Primate Societies*, ed. B. B. Smuts, D. L. Cheney, R. M. Seyfarth, R. W. Wrangham & T. T. Struhsaker, pp. 155–64. Chicago, IL: University of Chicago Press.

Strier, K. B., Mendes, F. D. C., Rimoli, J. & Rimoli, A. O. (1993). Demography and social structure in one group

of muriquis (*Brachyteles arachnoides*). *International Journal of Primatology*, 14, 513–26.

Symington, M. M. (1990). Fission–fusion social organization in *Ateles* and *Pan*. *International Journal of Primatology*, 11, 47–61.

Takahata, Y. (1990). Adult males' social relationships with adult females. In *The Chimpanzees of the Mahale Mountains: Sexual and Life History Strategies*, ed. T. Nishida, pp. 133–48. Tokyo: University of Tokyo Press.

Takahata, Y., Ihobe, H., & Idani. G. (1996). Comparing copulations of chimpanzees and bonobos: do females exhibit proceptivity or receptivity? In *Great Ape Societies*, ed. W. C. McGrew, L. F. Marchant & T. Nishida, pp. 146–58. Cambridge, UK: Cambridge University Press.

Thierry, B., Wunderlich, D. & Gueth, C. (1989). Possession and transfer of objects in a group of brown capuchins (*Cebus apella*). *Behaviour*, 110, 294–305.

Tomasello, M. & Call, J. (1997). *Primate Cognition*. New York: Oxford University Press.

Tutin, C. E. G. (1979). Responses of chimpanzees to copulation, with special reference to interference by immature individuals. *Animal Behaviour*, 27, 845–54.

(1996). Ranging and social structure of lowland gorillas in the Lopé Reserve, Gabon. In *Great Ape Societies*, ed. W. C. McGrew, L. F. Marchant & T. Nishida, pp. 58–70. Cambridge, UK: Cambridge University Press.

Utami Atmoko, S. S. (2000). Bimaturism in orang-utan males: reproductive and ecological strategies. Ph.D. Dissertation thesis, Utrecht University.

Utami, S. S. & Mitra Setia, T. (1995). Behavioral changes in wild male and female Sumatran orangutans (*Pongo pygmaeus abelii*) during and following a resident male take-over. In *The Neglected Ape*, ed. R. D. Nadler, B. F. M. Galdikas, L. K. Sheeran & N. Rosen, pp. 183–90. New York: Plenum Press.

Utami, S. S., Wich, S. A., Sterck, E. H. M. & van Hooff, J. A. R. A. M. (1997). Food competition between wild orangutans in large fig trees. *International Journal of Primatology*, 18, 909–27.

van Hooff, J. A. R. A. M. (1973). A structural analysis of the social behaviour of a semi-captive group of chimpanzees. In *Expressive Movement and Non-Verbal Communication*, ed. M. von Cranach & I. Vine, pp. 75–162. London: Academic Press.

van Hooff, J. A. R. A. M. & van Schaik, C. P. (1994). Male bonds: affiliative relationships among nonhuman primate males. *Behaviour*, 130, 309–37.

van Schaik, C. P. (1989). The ecology of social relationships amongst female primates. In *Comparative Socioecology*, ed. V. Standen & R. A. Foley, pp. 195–218. Oxford: Blackwell.

(1996). Social evolution in primates: the role of ecological factors and male behaviour. *Proceedings of the British Academy*, **88**, 9–31.

(1999). The socioecology of fission–fusion sociality in orangutans. *Primates*, **40**, 73–90.

(2003). Local traditions in orangutans and chimpanzees: social learning and social tolerance. In *The Biology of Animal Traditions*, ed. D. M. Fragaszy & S. Perry, pp. 297–328. Cambridge, UK: Cambridge University Press.

van Schaik, C. P. & Deaner, R. O. (2003). Life history and cognitive evolution in primates. In *Animal Social Complexity*, ed. F. B. M. de Waal & P. L. Tyack, pp. 5–25. Cambridge, MA: Harvard University Press.

van Schaik, C. P., Fox, E. A. & Sitompul, A. F. (1996). Manufacture and use of tools in wild Sumatran orangutans. *Naturwissenschaften*, **83**, 186–8.

Vervaecke, H., de Vries, H. & van Elsacker, L. (2000a). Dominance and its behavioral measures in a captive group of bonobos (*Pan paniscus*). *International Journal of Primatology*, **21**, 47–68.

(2000b). Function and distribution of coalitions in captive bonobos (*Pan paniscus*). *Primates*, **41**, 249–65.

Vigilant, L., Hofreiter, M., Siedel, H. & Boesch, C. (2001). Paternity and relatedness in wild chimpanzee communities. *Proceedings of the National Academy of Sciences*, **98**, 12 890–5.

Watts, D. P. (1989). Infanticide in mountain gorillas: new cases and a reconsideration of the evidence. *Ethology*, **81**, 1–18.

(1990a). Ecology of gorillas and its relation to female transfer in mountain gorillas. *International Journal of Primatology*, **11**, 21–45.

(1991). Mountain gorilla reproduction and sexual behavior. *American Journal of Primatology*, **24**, 211–26.

(1992). Social relationships of immigrant and resident female mountain gorillas. I. Male–female relationships. *American Journal of Primatology*, **28**, 159–81.

(1994). Agonistic relationships between female mountain gorillas (*Gorilla gorilla beringei*). *Behavioral Ecology and Sociobiology*, **34**, 347–58.

(1995). Post-conflict social events in wild gorillas, 1. Social interactions between opponents. *Ethology*, **100**, 158–74.

(1996). Comparative socio-ecology of gorillas. In *Great Ape Societies*, ed. W. C. McGrew, L. F. Marchant & T. Nishida, pp. 16–28. Cambridge, UK: Cambridge University Press.

(1997). Agonistic interventions in wild mountain gorilla groups. *Behaviour*, **134**, 23–57.

(1998). Coalitionary mate guarding by male chimpanzees at Ngogo, Kibale National Park, Uganda. *Behavioral Ecology and Sociobiology*, **44**, 43–56.

(2000a). Causes and consequences of variation in male mountain gorillas life histories and group membership. In *Primate Males: Causes and Consequences of Variation in Group Composition*, ed. P. M. Kappeler, pp. 169–79. Cambridge, UK: Cambridge University Press.

(2000b). Grooming between male chimpanzees at Ngogo, Kibale National Park, Uganda. I. Partner number and diversity and reciprocity. *International Journal of Primatology*, **21**, 189–210.

(2001). Social relationships of female mountain gorillas. In *Mountain Gorillas: Three Decades of Reasearch at Karisoke*, ed. M. M. Robbins, P. Sicotte & K. J. Stewart, pp. 215–40. New York: Cambridge University Press.

(2003). Gorilla social relationships: a comparative overview. In *Gorilla Biology: A Multidisciplinary Perspective*, ed. A. Taylor & M. Goldsmith, pp. 302–27. Cambridge, UK: Cambridge University Press.

Watts, D. P. & Mitani, J. C. (2001). Boundary patrols and intergroup encounters in wild chimpanzees. *Behaviour*, **138**, 299–327.

White, F. J. (1996). *Pan paniscus* 1973 to 1996: twenty-three years of field research. *Evolutionary Anthropology*, **5**, 11–17.

Whiten, A. (1998). Evolutionary and developmental origins of the mindreading system. In *Piaget, Evolution, and Development*, ed. J. Langer & M. Killen, pp. 73–99. Mahwah, NJ: Lawrence Erlbaum Associates.

(2000). Social complexity and social intelligence. In *The Nature of Intelligence*, Novartis Foundation Symposium 233, pp. 185–96. Chichester, UK: Wiley.

Whiten, A., Goodall, J., McGrew, W. C., Nishida, T., Reynolds, V., Sugiyama, Y., Tutin, C. E. G., Wrangham, R. W. & Boesch, C. (1999). Cultures in chimpanzees. *Nature*, **399**, 682–5.

Wrangham, R. W. (1979). On the evolution of ape social systems. *Social Science Information*, **18**, 334–68.

(1986). Ecology and social relationships in two species of chimpanzee. In *Ecological Aspects of Social Evolution: Birds and Mammals*, ed. D. I. Rubenstein & R. W.

Wrangham, pp. 352–78. Princeton, NJ: Princeton University Press.

(1993). The evolution of sexuality in chimpanzees and bonobos. *Human Nature*, **4**, 47–79.

(1999). Evolution of coalitionary killing. *Yearbook of Physical Anthropology*, **42**, 1–30.

(2000). Why are male chimpanzees more gregarious than mothers? A scramble competition hypothesis. In *The Socioecology of Primate Males*, ed. P. M. Kappeler, pp. 248–58. Cambridge, UK: Cambridge, University Press.

Wrangham, R. W. & White, F. J. (1988). Feeding competition and patch size in the chimpanzee species *Pan paniscus* and *Pan troglodytes. Behaviour*, **105**, 148–64.

Yamagiwa, J. & Kahekwa, J. (2001). Dispersal patterns, group structure, and reproductive parameters of eastern lowland gorillas at Kahuzi in the absence of infanticide. In *Mountain Gorillas. Three Decades of Research at Karisoki*, ed. M. M. Robbins, P. Sicotte & K. J. Stewart, pp. 91–122. Cambridge, UK: Cambridge University Press.

12 • Diet and foraging of the great apes: ecological constraints on their social organizations and implications for their divergence

JUICHI YAMAGIWA

Laboratory of Human Evolution Studies, Kyoto University, Kyoto

INTRODUCTION

Unlike the majority of the larger mammals, which are terrestrial herbivores, omnivores, or insectivores, non-human primates have created unique niches as arboreal insectivores, frugivores, or folivores. Primates now play important roles as fruit consumers and seed dispersers in tropical forests (Gautier-Hion *et al.* 1985; Terborgh 1986). However, food is still the primary limiting factor of primate populations because of its sparse distribution, physical protection (hard shells, spines, etc.), and toxic secondary compounds (Feeny 1976; Freeland & Janzen 1974; Milton 1984). Primates have evolved different strategies to cope with these dietary difficulties, and their specializations have influenced both anatomy and behavior.

Primates have evolved various features of gastrointestinal anatomy and the digestive system to cope with such dietary constraints. Leaves in particular are high in structural carbohydrates and are difficult to digest. Folivorous primates need more time to digest and absorb important food components to satisfy nutritional requirements. Specialization in gut morphology has raised the capacity of some primates to consume structural carbohydrates and detoxify secondary compounds (Kay & Davies 1994; Milton 1986). For example, the Colobinae have evolved a sacculated fermenting chamber in the stomach in which microbial fermentation precedes digestion and absorption (Bauchop & Martucci 1968; Chivers & Hladik 1980). Some secondary compounds are degraded during fermentation in the alkaline stomach environment before absorption. As an alternative strategy, a number of more folivorous primates, including howler monkeys, gorillas, bamboo lemurs, and sportive lemurs, have evolved an enlarged caecum or colon in which bacterial fermentation is activated (Stevens & Hume 1995). More frugivorous and faunivorous primates lack these fore- or hindgut specializations (Chivers & Hladik 1980; Chivers & Langer 1994; Parra 1978).

The dietary constraints that promote a strong relationship between diet and digestion also affect behavior. They constitute basic ecological factors influencing activity time budgets and activity rhythms in daily primate life. Since foliage is distributed more densely and evenly than fruit, for instance, folivorous primates need less time and space for searching for foods than frugivorous primates do. The larger body weight, larger biomass, and smaller home ranges of folivorous versus frugivorous primates reflect such relationships between diet and foraging strategies (Clutton-Brock & Harvey 1977; Kay 1984). However, the strongly frugivorous diets of orangutans, chimpanzees, and bonobos are not consistent with their large body weight and with their high biomass (Kano & Mulavwa 1984; Reynolds & Reynolds 1965; Rijksen 1978; Rodman 1973; Tutin & Fernandez 1984). Because of their large body weight, great apes need more foods in wider ranges than sympatric Old World monkeys. Moreover, great apes are less able to digest unripe fruit and mature leaves than Old World monkeys. Apparently for these reasons, great apes have broadened their diets to include a highly diverse and flexible range of non-fruit foods; the flexibility and breadth may have precluded their evolving specialized digestive systems and forced them to find behavioral means of coping with dietary constraints.

A strong relationship is also suggested between diet and social organization. Fission–fusion characteristics in grouping, with multi-male and multi-female group compositions, appear in chimpanzees and spider monkeys, both persistent frugivores but phylogenetically distant (Chapman, Wrangham & Chapman 1995;

Symington 1990; Wrangham 1986). There are numerous exceptions to these tendencies, however, and it
is difficult to find a simple relationship between any
diet-related ecological variable and social organization
(Wrangham 1987).

Diet-related and other ecological analyses of primate social organization focus on females because they
are based on sexual selection theory (Trivers 1972),
which holds that female behavior is adapted more
directly to ecological pressures such as food availability
while male behavior is adapted to maximizing mating
success, which depends on the distribution and behavior of females. Two competing hypotheses concerning
the ecological factors favoring female social organization
have considered feeding competition. The first argues
that females may tend to associate in extended kin groups
to defend sparsely distributed food resources against
other groups (Wrangham 1980). Between-group feeding competition would then have a greater effect on the
evolution of female sociality than within-group competition. The second argues that within-group feeding
competition increases with group size, and predation
pressure is the primary selective factor favoring female
sociality in primates (Terborgh & Janson 1986; van
Schaik 1983). Interestingly, none of the female great apes
exhibits philopatry, that is, stays in their natal groups
after maturity and forms alliances with kin-related
females. They tend to disperse from their natal range;
female African great apes usually join other groups after
emigration and female orangutans tend to stay near their
mother's range (Galdikas 1984, 1988; Harcourt, Stewart
& Fossey 1976; Kano 1992; Nishida & Kawanaka 1972;
Rijksen 1978; Rodman 1973; Singleton & van Schaik
2001). Also, all great apes are less vulnerable to predation pressure because of their large size, especially those
that are predominantly arboreal. Accordingly, dietary
and other ecological factors may shape great ape sociality in different ways than they shape sociality in other
nonhuman primates (Dunbar 1988; Watts 1996; White
1996; Yamagiwa 1999).

Among great apes, orangutans, chimpanzees, and
bonobos rely heavily on fruits (Galdikas 1988; Goodall
1968; Kano 1992; Rodman 1977). Only gorillas have
been regarded as specialized folivores (Fossey &
Harcourt 1977; Schaller 1963; Watts 1984), although
they do not have typical folivore digestive systems. However, gorilla data come primarily from studies on mountain gorillas inhabiting montane forests at high altitudes
where fruit is rare. Recent studies on western and eastern
lowland gorillas have reported frugivorous diets when
and where fruit is abundant (Kuroda et al. 1996; Tutin
& Fernandez 1993; Yamagiwa et al. 1994). In all great
apes, digestion is oriented toward frugivory and based
on the caeco-colic fermenting system (Martin 1990).
Remarkable similarities in gut morphology and gut pas-
sage time have also been reported between gorillas and
chimpanzees (Chivers & Hladik 1984; Milton 1984).

VARIATION IN GREAT APE SOCIAL ORGANIZATION

Despite similarities in great apes' diet and digestive
systems, however, marked differences are found in
their social organizations (Table 12.1; see van Schaik,
Preuschoft & Watts, Chapter 11, this volume, for
detailed descriptions).

Orangutans usually travel alone, although they
probably live in loosely organized, highly dispersed
communities (van Schaik & van Hooff 1996). There are
two known types of fully mature males: large and small
bodied, with secondary sexual characteristics versus
without, and strongly solitary versus somewhat sociable,
respectively (Boekhorst, Schurmann & Sugardjito 1990;
Sugardjito, Boekhorst & van Hooff 1987; van Schaik &
van Hooff 1996). Adult females tend to travel alone with
dependent offspring, within small ranges nested within
larger adult male ranges; ranges overlap considerably
within and between sexes (Galdikas 1988; Horr 1975;
Knott 1998a; Rodman 1973; van Schaik & van Hooff
1996). Temporal groups consist primarily of females
with offspring and smaller males (Galdikas 1988; Mac-
Kinnon 1974; Rodman 1977) and occasional, temporary
mating consortships during and outside estrus (Galdikas
1981, 1985; MacKinnon 1979; Rodman 1979; van Schaik
1999). Orangutans also aggregate occasionally in large
fruiting trees (Knott 1998a; MacKinnon 1974; Sugardjito et al. 1987). Both males and females disperse from
their natal range, although females tend to settle nearby
(Galdikas 1984; Singleton & van Schaik 2001). Little is
known of what happens between communities.

Chimpanzees and bonobos both live in large communities (or unit-groups) comprised of both females
and males in fluid fission–fusion grouping patterns
(Goodall 1968; Kano 1982; Nishida 1968; White 1988).
Chimpanzees form temporal parties of various age/sex
compositions; bonobo subgroups are usually bisexual.

Table 12.1. *Social organizations of great apes*

		Orangutan	Gorilla	Chimpanzee	Bonobo
Group size		1.0–1.9 (mean)[1]	3–17 (mean)	19–106 (range) 4.0–8.3 (mean)[1]	30–120 (range) 4.3–16.9 (mean)[1]
Age/sex composition		Solitary, temporal association of ♂♀, ♂♂, ♀♀	♂♀♀, ♂♂♀♀	♂♂♀♀	♂♂♀♀
Emigration		♂♀	♂♀	♀	♀
Immigration		—	♀	♀	♀
Foraging group	♀	Individual	Group (♂♀)	Individual	Group (♂♀)
	♂	Individual	Group (♂♀)	Group (♂♂)	Group (♂♀)
Association	♂♂	Rare Only small ♂	Rare[2] Only kin-related ♂	Frequent Strong alliance among kin-related ♂♂	Frequent Kin-related ♂♂
	♀♀	Rare	Constant	Rare	Frequent
	♂♀	Temporal[3]	Constant	Frequent ♂+ cycling ♀	Frequent mother–son
Reassurance & appeasement		Rare	Rare	Diverse & frequent	Sexual behavior
Sexual dimorphism in body weight ♂/♀ (mean range)		2.04–2.37	1.63–2.37	1.27–1.29	1.36–1.38

Notes:

[1] Mean party size: Orangutan (van Schaik 1999); Chimpanzee & Bonobo (Boesch 1996).

[2] About half of Mountain Gorilla groups include two or more adult males in the Virungas and Bwindi (Robbins 2001).

[3] Some pairs consisting of reproductive males and females, especially adolescents and subadults, last for years in Sumatra (Schurmann 1982).

Sources: Orangutans: Galdikas 1984, 1985; Rodman 1979; Rodman & Mitani 1987; Sugardjito *et al.* 1987; van Schaik 1999; *Gorillas:* Harcourt 1978; Stewart & Harcourt 1987; Yamagiwa 1983, 1987a; Yamagiwa & Kahekwa 2001; Yamagiwa *et al.* 1996a; Watts 1991, 1996; Tutin 1996; *Chimpanzees:* Goodall 1968, 1986; Nishida & Kawanaka 1972; Wrangham 1979a; Nishida & Hasegawa 1987; *Bonobos:* Kano 1980, 1982, 1992; Thompson-Handler *et al.* 1984; Furuichi 1989; Furuichi *et al.* 1998; Idani 1991; White 1996: *Body weight:* Leigh & Shea 1995.

Female chimpanzees tend to travel alone within a small range, while male chimpanzees associate with other males to range in larger areas (Wrangham 1979a). Neighboring communities partly overlap in their ranging but inter-community relationships are usually hostile and territorial, at least in forested habitats, and sometimes lethal (Chapman & Wrangham 1993; Goodall *et al.* 1979; Nishida *et al.* 1985). Female bonobos tend to form more stable associations with unrelated females than female chimpanzees, and frequent sexual interaction helps reduce social tension at aggregations (Furuichi 1989; Kuroda 1980). A group's range overlaps extensively with the ranges of neighboring groups and intergroup encounters can last for hours with no conflict (Idani 1991). Both chimpanzee and bonobo females tend to emigrate from their natal groups.

Gorillas usually form cohesive bisexual groups, but most groups contain only one mature male. Both males and females tend to emigrate from their natal groups and only females transfer into other social units

(Harcourt 1978). Gorilla groups do not show territoriality and their home ranges overlap extensively with those of neighboring groups; however, intergroup encounters are frequently accompanied by aggressive contacts between silverback (fully adult) males (Fossey 1983; Schaller 1963; Tutin 1996; Yamagiwa *et al.* 1996a).

If these variations in social organization are related to diet, they may have derived, in part, from small variations in diet and digestion but large variations in foraging strategies. Some cognitively governed abilities used for foraging, such as excellent memory for distant and highly varied food resources, tool use, rapid adaptation to novel foods, and food-sharing among conspecifics, are uniquely sophisticated in the large Hominoidea and may reflect such species differences (Byrne & Byrne 1993; Kuroda 1984; Rodman 1977; van Schaik *et al.* 1999; Whiten *et al.* 1999). Particularly important may be strategies used in times of food scarcity, when ecological pressures and feeding competition are most severe. Seasonal food scarcity has long been proposed as a key selection pressure favoring the evolution of enhanced intelligence in great apes (Parker & Gibson 1979).

How social pressures affect and are affected by these periods has not yet received serious consideration. An analysis of great ape diets and foraging behavior relative to different ecological and social environments is critical to understanding the evolutionary processes that shaped such intellectual abilities within the Hominoidea. This chapter will describe the ecological constraints that each great ape species faces in its habitat and the foraging strategies that each employs to survive times of scarcity of their primary foods. It will also discuss how the great apes' unique foraging patterns may relate to their capacity for cognitively governed behavior that is highly complex and flexible.

ECOLOGICAL CONSTRAINTS AND VARIATION IN APE DIETS

Great apes are primarily dwellers of tropical forests. Lowland moist forest is the main habitat of all four species. Only chimpanzees are distributed in dry savanna, in Senegal and Tanzania, and only gorillas are found in subalpine zones, in the Virunga Volcanoes of Central Africa. Great apes' dietary features may reflect the characteristics and diversity of their habitats. One element of their feeding strategies, their dietary flexibility in response to a scarcity of high-quality foods,

also differs between species. In order to elucidate the ecological constraints linked to these dietary and foraging differences, I will compare diet, locomotion, group size, day range, home range, home range overlap between neighboring social units, and inter-unit relationships (Table 12.2). To discriminate between flexible and stable features within species, I will compare the ecological features between subspecies of gorillas, whose variations are the most pronounced among the great apes.

Recent studies on great apes have demonstrated their general tendencies of having broad variety in their diet but a strong preference for fruits (Table 12.2). Except for mountain gorillas living in the montane forest of the Virunga Volcanoes, all four great ape species feed annually on hundreds of kinds of food, including fruits, leaves, bark, pith, flowers, roots, fungi, and invertebrates (Badrian & Malenky 1984; Galdikas 1988; Goodall 1986; Kano & Mulavwa 1984; Knott 1998b; Nishida & Uehara 1983; Tutin & Fernandez 1993; Yamagiwa *et al.* 1994). Gorillas inhabiting lowland tropical forests feed on a wide range of foods, and fecal analysis shows that the diversity of fruits they consume sometimes exceeds that of sympatric chimpanzees (Remis 1994; Tutin & Fernandez 1993). Although the Virunga mountain gorillas feed on fewer kinds of food, they inhabit a higher montane forest, including a subalpine zone, where no other primates exist (Fossey & Harcourt 1977; Watts 1984). They eat spiny nettle and galium instead of fruits, using complex food processing techniques (Byrne & Byrne 1993). Their broad diet and intellectual ability may enable gorillas to survive in such a fruitless habitat without specialized digestive systems. Except for gorillas, the great apes occasionally hunt vertebrates and eat their meat (Boesch & Boesch 1989; Hohmann & Fruth 1993; Ihobe 1992; Uehara *et al.* 1992; Utami & van Hooff 1997; Wrangham & Bergmann-Riss 1990). Chimpanzees are the most active hunters; the Gombe community was estimated to kill more than 150 colobus monkeys in peak hunting years (Stanford *et al.* 1994).

Orangutans

Orangutans spend more than half of their feeding time eating fruits, although they also feed on large amounts of flowers, leaves, shoots, barks, small amounts of ants and termites (Galdikas 1988; Knott 1998b; Rodman 1977), and occasionally meat (Utami & van Hooff 1997). They are the most active seed-eaters of the great apes

Table 12.2.

(a) Ecological features of great apes

	Orangutan	Gorilla	Chimpanzee	Bonobo
Habitat	Lowland tropical mosaic Swamp forest	Lowland tropical (L) Swamp (L) Montane (M)	Lowland tropical Montane Woodland, Savanna	Lowland tropical Woodland
Diet type	Frugivorous	Seasonal frugivorous (L) Folivorous (M)	Frugivorous	Frugivorous
Number of foods (spp.) (Observation period)	306 (229)[1] (5 years)	230 (129)[2] (15 years)	328 (198)[3] (16 years)	147 (100)[4] (7 years)
Locomotion	Arboreal $\gg$ Terrestrial	Terrestrial > Arboreal	Arboreal > Terrestrial	Arboreal > Terrestrial
Day range (mean length)	305–800 m	378–1531 m	910–5000 m	2400 m
Home range	0.40–>15 km[2]	4–31 km[2]	5–560 km[2]	22–58 km[2]
Home range overlap with neighboring units*	Extensive	Extensive	Partly	Extensive
Inter-unit relationships	Antagonistic or Peaceful	Antagonistic or Peaceful	Antagonistic	Peaceful

(b) Response to fruit scarcity

	Orangutan	Gorilla	Chimpanzee	Bonobo
Diet	Search fruit Bark, stems, pith as fallback	Shift to bark & THV (WLG, ELG) Habitual folivore (MG)	Search fruit Bark, THV, fig fruit, pith as fallback	Search fruit THV as fallback
Day range	Increase	Decrease (WLG, ELG) Constant (MG)	Increase	Constant
Party size	Decrease	Constant	Decrease	Constant
Gregariousness	Decrease	Decrease (WLG)	Decrease	Constant

Notes:

L, lowland, M, montane; WLG, western lowland gorilla, ELG, eastern lowland gorilla, MG, mountain gorilla; THV, terrestrial herbaceous vegetation. *Due to orangutans' solitary nature, one unit is defined as an individual or group ranging independently.

Sources: Orangutans: Galdikas 1978, 1988[1]; Knott 1998b; Rodman 1973, 1977; Singleton & van Schaik 2001; van Schaik 1999; van Schaik & van Hooff 1996. *Gorillas*: Fossey & Harcourt 1977; Goldsmith 1999; Remis 1997b; Remis *et al.* 2001[2]; Sabater Pi 1977; Tutin 1996; Tutin & Fernandez 1993; Watts 1996; Yamagiwa 1999; Yamagiwa *et al.* 1996a,b. *Chimpanzees*: Goodall 1968, 1986; Nishida 1976; Nishida & Hasegawa 1987; Nishida & Kawanaka 1972; Nishida & Uehara 1983[3]; Wrangham 1979a,b; Wrangham *et al.* 1996. *Bonobos*: Furuichi 1989; Idani 1990; Kano 1980, 1982, 1992[4]; Thompson-Handler *et al.* 1984; White 1992, 1996; White & Wrangham 1988.

(Galdikas 1988; Rodman 1977; Rodman & Mitani 1986). Galdikas (1988) reported 306 foods from 229 plant species from her observations of 58 orangutans for 5 years at Tanjung Puting. Preferences are for soft, pulpy ripe fruits (Leighton 1993; Rijksen 1978; van Schaik 1986). However, they shift their dietary composition in response to fruit availability. Especially in Borneo, bark and perhaps stems constitute important fallback foods when fruit is scarce (Galdikas 1988; Knott 1998b; Suzuki 1988). They are opportunistic foragers with a diet that is broad and shows marked seasonal variations in composition (Galdikas 1988; Knott 1998b). Sex differences have also been found in their diets (Galdikas & Teleki 1981; Rodman 1977). While orangutans are masterful tool users in captivity, they are not known for extensive tool use in the wild, although one community in Suaq Balimbing, N. Sumatra, habitually uses tools and tool sets to obtain foods (Fox, Sitompul & van Schaik 1999).

The mean length of day journeys is less than 1 km everywhere (Galdikas 1988; MacKinnon 1974; Rodman 1977). Home range size varies from 0.40 to over 15 km^2 (Singleton & van Schaik 2001). Given their frugivorous diet and large body weight, their nutritional needs are not normally satisfied within these small ranges (Knott 1998b). They occasionally travel outside their core areas to exploit seasonally abundant foods (Galdikas 1988; Singleton & van Schaik 2001). Seasonal changes in fruit abundance and distribution affect their day journey length and grouping. During fruit scarcity, both females and males tend to travel longer distances daily and to avoid grouping (Galdikas 1988; van Schaik 1999). Males tend to wander long distances and some shift their ranges frequently (Rodman & Mitani 1986), perhaps depending on the reproductive states of neighboring females and fruit availability (Knott 1998a; Mitani 1985).

Orangutans' less gregarious nature, compared with the other great apes, is partly explained by ecological factors. Asian tropical forests are characterized by "mast fruiting," a high synchronization in fruiting at irregular intervals of several years, and for that reason has greater fluctuations in fruit production than African tropical forest (Janzen 1974; van Schaik 1986). The fruit trees preferred by orangutans are more widely dispersed and significantly smaller in diameter than the African fruit trees used by chimpanzees and bonobos (Fleming, Brettwisch & Whitesides et al. 1987; Knott 1999). The scarcity of fruit and large fruit patches may limit orangutans' ability to forage together in groups (Galdikas 1988; MacKinnon 1974; Sugardjito et al. 1987). Local variations in diet and gregariousness may support this interpretation. Sumatran orangutans live at densities of two to three times higher than Borneans and associate more frequently (Rijksen 1978; van Schaik 1999). Sumatran forests offer large fruit patches, such as large fruiting fig trees, which Bornean forests lack in many areas. Orangutans tend to aggregate in large fig trees when their fruits are available (MacKinnon 1974; Sugardjito et al. 1987), which suggests that high fruit density and large fruit patches may allow gregariousness and sociability (Utami et al. 1997). Tigers range in orangutan habitat in Sumatra but no large predators threaten Bornean orangutans, so the risk of predation may not be very important for orangutans given their large body size and arboreal locomotion. The benefits of grouping are therefore low compared with the high costs, especially during periods of fruit scarcity (Sugardjito et al. 1987). Social tolerance among orangutans may prevail primarily when and where fruit is abundant (Boekhorst et al. 1990; Knott 1998a), but it may none the less provide the opportunity to socialize offspring and to learn foraging skills, including the tool-using techniques observed in Sumatra.

Gorillas

The dietary features of gorillas closely reflect differences in habitats (Table 12.3). Western lowland gorillas (WLG, *Gorilla gorilla gorilla*) are distributed in lowland forest, Mountain gorillas (MG, *G. g. beringei*) in the mountains at higher altitudes (>1000 m above sea level), and Eastern lowland gorillas (ELG, *G. g. graueri*) in both lowland and highland forests. According to the diversity of fauna and flora, WLG and ELG show broader diets than MG. Watts (1984) reported that MG consumed 75 foods from 38 plant species in the Virungas, from his 1.5 years' direct observations of a well-habituated group ranging at an altitude of 3000 m; McNeilage (2001) also reported low diversity of food (72 foods from 44 plant species) at a lower altitude (2000 m) from his 1-year study on a habituated group. Fruit is a minor part of total plant species in MG's diet. For unhabituated WLG, based on fecal analysis and feeding remains, Williamson et al. (1990) reported 182 foods from 134 plant species consumed over 8 years at Lopé and Remis et al. (2001) reported 230 foods from 129 species consumed over 15 years at Bai Hokou. Fruit constitutes the major part of

Table 12.3. *Ecological features of three subspecies of gorillas*

	G. g. gorilla	*G. g. graueri*	*G. g. beringei*
Habitat type	Lowland tropical forest	Lowland tropical forest Montane forest	Montane forest
Number of plant foods (spp.)	182 (134)[1] 230 (129)[2]	194 (121)[3] 129 (79)[4]	75 (38)[5] 72 (44)[6]
% fruit in plant food species	71%[1] 69%[2]	40%[3] 25%[4]	5%[5] 5%[6]
Mean length of day journey	1100–2600 m[7]–[10]	1500 m[11] 800–1300 m[12], [13]	500–1000 m[14]–[16]
Annual home range	10–20 km^2 [7], [8], [9], [10], [17]	20–50 km^2 [12], [13], [18] Unknown	4–11 km^2 [19], [20]
Home range overlap	Extensive	Extensive Extensive	Extensive
Fission–fusion	Frequent/rare	Rare Rare	Rare
Mean group size (maximum)	6–14 (32)[7], [8], [9], [10], [17]	3–6 (31)[12], [21] 11–16 (42)[22], [23]	8–17 (34)[14], [24], [25], [26]

Sources: Williamson *et al.* 1990[1]; Remis *et al.* 2001[2]; Yamagiwa *et al.* 1994[3]; 1996a,b[4]; Watts 1984[5], McNeilage 2001[6]: Tutin 1996[7]; Goldsmith 1996[8]; Doran & McNeilage 2001[9]; Bermejo 1997[10]; Yamagiwa & Mwanza 1994[11]; Yamagiwa 1999[12]; Goodall 1977[13]; Schaller 1963[14]; Elliott 1976[15]; Yamagiwa 1986[16]; Remis 1997a[17]; Casimir 1975[18]; Fossey & Harcourt 1977[19]; Watts 1998[20]; Hall *et al.* 1998[21]; Murnyak 1981[22]; Yamagiwa *et al.* 1993[23]; Weber & Vedder 1983[24]; Aveling & Aveling 1987[25]; Watts 1996[26].

WLG plant species foods (71% and 69%, respectively). Yamagiwa *et al.* (1994) reported that unhabituated ELG consumed 194 foods from 121 plant species over 3 years in lowland (600 m) habitat and semi-habituated ELG groups consumed 129 foods from 79 plant species for a single dry season (3 months) in the highland (2000 m) habitat of Kahuzi. The ELG also consume a wide variety of fruits, although fruit represents a smaller proportion of their plant species foods (25%–40%; Yamagiwa *et al.* 1991, 1994, 1996a).

Like chimpanzees, WLG daily consume various kinds of fruits and regularly feed on insects (Nishihara 1995; Remis 1997a; Tutin & Fernandez 1992, 1993). They avoid unripe, fatty fruits and prefer succulent, sweet fruits (Rogers *et al.* 1990). During periods of fruit scarcity, they increase consumption of foliage and terrestrial herbaceous vegetation (THV) (Kuroda *et al.* 1996; Remis 1997a). The WLG frequently eat some forms of aquatic herbaceous vegetation, which are high in proteins and minerals, in swamps (Nishihara 1995). The ELG also consume a large variety of fruits and often

feed on ants in the lowland forest (Yamagiwa *et al.* 1991, 1994). The ELG inhabiting the montane forest of Kahuzi (at 1800–3300 m) show frugivorous features during the dry season when succulent fruits are abundant (Yamagiwa *et al.* 1996a). They usually eat barks of various trees and woody vines, which may contribute to their diet as fallback foods (Casimir 1975; Yamagiwa *et al.* 1996b). For MG, vegetative foods make up the major portion of the diet. No seasonal change has been found in their dietary composition, except for bamboo shoots (Fossey & Harcourt 1977; Watts 1984).

Some ecological variables seem to cause variation in gorilla diets. Due to the clumped distribution of their major food (fruit) in the lowland forest, WLG and ELG show longer day journeys and larger annual home ranges than MG inhabiting high montane forest where THV is densely and evenly distributed (Table 12.3). Seasonal shift of range by WLG and ELG may be responsible for differences in annual home ranges (Casimir & Butenandt 1973; Remis 1994; Tutin 1996; Yamagiwa *et al.* 1996b). For both WLG and ELG, day journey length during the fruiting season is far longer than that during the non-fruiting season within the same habitat, which suggests that they actively prefer fruits but do not search them out when they are scarce (Goldsmith 1999; Yamagiwa & Mwanza 1994). For WLG, the small group sizes estimated in lowland habitats and the frequent sub-groupings observed may possibly be caused by high scramble feeding competition around fruiting trees and sparse distribution of fruits (Harcourt, Fossey & Sabater Pi 1981a; Remis 1994). However, the extensive overlap of home ranges among neighboring groups illustrates their apparent lack of territoriality in all types of habitat.

Chimpanzees

Chimpanzees live in the most diverse habitats of the great apes. Their distribution covers a wide area of Equatorial Africa, including lowland moist evergreen forests, semi-deciduous forests at medium altitudes (around 1000 m), montane forests, woodland, and dry savanna. Although the total number of foods eaten by chimpanzees varied with habitat types and the length of study period, Nishida and Uehara (1983) reported 328 foods from 198 plant species from their direct observations of two habituated groups at Mahale over 16 years. Their food items consist of fruits, flowers, leaves, bark, shoots, pith, gum, honey, insects, and meat of various vertebrates. However, like orangutans, fruit constitutes the major part of their diet in any type of habitat (Baldwin, McGrew & Tutin 1982; Ghiglieri 1984; Hladik 1977; Tutin & Fernandez 1993; Wrangham 1977; Yamagiwa et al. 1996b).

Unlike lowland gorillas and orangutans, chimpanzees may not markedly change their dietary composition according to seasonal fluctuation in food availability. Instead, they change grouping patterns as well as searching time and distance traveled for fruits. During periods of fruit scarcity, chimpanzees in the Kibale medium-altitude forest tend to decrease their party size (Wrangham, Clark & Isabirye-Basuta 1992) and in the Kahuzi montane forest to enlarge their monthly ranges (Yamagiwa 1999). In addition to these changes, fallback fruits such as figs or oil-palm nuts, or pith, bark, THV, and insects may supplement the lack of succulent fruits (Nishida 1976; Tutin & Fernandez 1993; Wrangham et al. 1996). The dietary composition of their fallback foods closely resembles that of orangutans

(Galdikas 1988; Knott 1998b; Leighton 1993; Sugardjito et al. 1987). Chimpanzees use various tools for collecting honey, ants, and termites, cracking hard nuts, and pestle-pounding oil-palm pith (Boesch & Boesch 1983; McGrew 1992; Sugiyama & Koman 1979; Yamakoshi & Sugiyama 1995). Such tool use may buffer seasonal scarcity of high-quality foods (Yamakoshi 1998).

Hunting vertebrates is another important feature of chimpanzee foraging. Monkeys and ungulates constitute the major prey in the three long-term study sites of Gombe (Wrangham & Bergmann-Riss 1990), Mahale (Uehara et al. 1992), and Taï (Boesch & Boesch 1989). Meat acquired by hunting constitutes a substantial part of chimpanzee diet, as with human hunter–gatherers, and chimpanzee predation pressure has a tremendous effect on the red colobus population at Gombe (Stanford 1996, 1998). Marked sex differences are found in the frequency of insect eating, hunting, and tool using. Males tend to eat more meat than females, while females more frequently feed on insects and use tools for capturing insects or cracking nuts than males (Boesch & Boesch 1981; Goodall 1986; McGrew 1979; Uehara 1984). Meat does not appear to serve as a fallback food (Mitani, Watts & Muller 2002).

Ranging also shows sex differences. Males tend to travel longer distances daily and to range more widely than females (Chapman & Wrangham 1993; Wrangham 1979a; Wrangham & Smuts 1980). Pronounced flexibilities in grouping and ranging may enable chimpanzees to live in similar sized home ranges ($11–34$ km^2) in various forest habitats (Yamagiwa 1999). However, their home ranges are extremely large in arid areas, for example $122–124$ km^2 at Kasakati (Izawa 1970), 150 km^2 at Filabanga (Kano 1971), $250–560$ km^2 at Ugalla and Wansisi (Kano 1972), and $278–333$ km^2 at Mt. Assirik (Baldwin et al. 1982), probably because of more limited food availability. Population density is very low (less than 0.2 individuals/km^2) in these dry savannas.

Bonobos

Bonobos are distributed in the lowland tropical forest of the Congo Basin, where neither gorillas nor chimpanzees live. Kano (1992) reported 147 foods from 100 plant species from his direct observations of several habituated groups of bonobos at Wamba over 7 years. Their dietary features resemble those of chimpanzees, and fruit is their major food throughout the year. They also eat a wide variety of invertebrates, such as earthworms and millipedes (Badrian & Malenky 1984; Kano 1983; Kano & Mulavwa 1984), and prey on flying squirrels, infant duikers, and bats, although the frequency of such predation is very low (Badrian & Malenky 1984; Bermejo, Liera & Sabater Pi 1994; Hohman & Fruth 1993; Ihobe 1992; White 1994). In captivity, bonobos show a variety of tool use equal to that of chimpanzees (Jordan 1982). However, in the wild, no tool using for insect eating in bonobos has been observed, although it has been in chimpanzees and orangutans. The most striking difference between bonobo and chimpanzee or orangutan diets is bonobos' frequent and constant feeding on THV (Badrian, Badrian & Sussman 1981; Kano 1983; Kuroda 1979). Their constant use of THV decreases feeding competition and may enable them to form larger foraging parties than do chimpanzees (Wrangham 1986). Large overlap of home ranges and peaceful relationships among neighboring groups can be explained by the availability of large arboreal fruit patches, which may mitigate conflicts caused by feeding competition (Kano 1992; White & Wrangham 1988).

Bonobos show small seasonal changes in their diet, day journey length, and party size, for which the presence of large food patches throughout the year may be responsible (Kano 1992; Malenky & Wrangham 1994). The patchy distribution of preferred THV in the lowland forest of Lomako is associated with dispersion rather than cohesion of bonobo parties, and the presence of larger fruit patches throughout the year may mitigate within-group feeding competition (Malenky & Stiles 1991; White & Wrangham 1988). The influence of THV as a fallback food on diet-related ecological variables may be small. No sex differences in range size or daily travel distance have been reported because bonobos usually form mixed parties.

Comparisons of diets and other ecological features among the great apes reveal marked similarities among all four species, especially orangutans and chimpanzees, as suggested by Rodman (2000). All the great apes are opportunistic foragers, showing a wide range of foods in their repertoires. Sex differences in diet, feeding techniques, and ranging are obvious in orangutans and chimpanzees. Orangutans and chimpanzees also resemble each other in dietary composition during periods of fruit scarcity and in tool using while feeding. However, sexual dimorphism in body weight is prominent for orangutans and gorillas but not for chimpanzees or

bonobos (Table 12.1). Sociality among males is strong for chimpanzees and bonobos but not for orangutans or gorillas. Both ecological and social factors may influence great ape social organization, and the combinations of these factors may differ among species. Great apes' foraging strategies in relation to fruit scarcity possibly reflect such differences.

GREAT APE FORAGING STRATEGIES AND GROUPING: THE ROLE OF DIET AND OTHER FACTORS

Concerning grouping patterns, great apes' foraging strategies can be classified into two types: individual and group (Table 12.1) (see van Schaik et al., Chapter 11, this volume, for a related classification). Females' foraging behavior clearly reflects the differences between the two types. Female orangutans and chimpanzees, whose diets have stronger frugivorous features, tend to forage individually (Chapman et al. 1995; Galdikas 1988; Goodall 1968; Sugardjito et al. 1987; Wich, Sterck & Utami 1999; Wrangham 1979a). Female gorillas and bonobos, whose diets include substantial vegetative foods, usually forage in bisexual groups or parties. The extent to which grouping patterns owe to diet can be examined through the effects of fluctuations in food availability and the probable role of other factors on grouping. Females and males are discussed separately.

Female grouping patterns

Female orangutans may be more solitary than female chimpanzees because of their more dispersed and smaller fruit food patches and their more arboreal locomotion, which may impose stronger feeding competition and reduce vulnerability to predation (Knott 1998b; Sugardjito 1983; Wich et al. 1999). However, studies in Ketambe and Suaq Balimbing, Sumatra, show that orangutans frequently form small groups according to fruit availability (Sugardjito et al. 1987; van Schaik 1999; van Schaik & van Hooff 1996). They tend to associate when fruits are abundant, or when fruit is scarce but large patches of figs are available (Sugardjito et al. 1987). Females with infants tend to travel without other adult conspecifics in both orangutans and chimpanzees (van Schaik 1999; Wrangham 1979a), probably because of the higher cost of feeding competition for mothers with dependent offspring. Female chimpanzees with dependent offspring were also less often found in groups than females without dependent offspring in Gombe and Kibale, which accords with ecological constraints (Chapman *et al.* 1995; Goodall 1986; Wrangham 1979a). Matsumoto-Oda (1999) reported that noncycling adult females were less often observed in large bisexual parties than cycling females and males in Mahale. These findings suggest that female orangutans and chimpanzees are unlikely to form groups except for reproductive purposes or in large fruit patches.

For female gorillas, the folivorous features of their diet may allow greater gregariousness by decreasing the cost of feeding competition (Wrangham 1986). However, their grouping patterns are not solely a function of ecological factors related to food availability. They do not tend to alter their grouping patterns in response to fruit availability, although WLG groups sometimes subdivide into temporary subgroups to exploit scattered fruit resources (Doran & McNeilage 1998; Remis 1994; Tutin 1996). Rather, both WLG and ELG in lowland forests tend to change their daily travel length (Goldsmith 1999; Yamagiwa & Mwanza 1994). Although WLG exhibit strong frugivorous features seasonally, their subgroups usually consist of both sexes and may not allow individual foraging like those of chimpanzees (Remis 1997b; Tutin 1996). Female gregariousness may also be caused by their vulnerability to predators and infanticide (Stewart & Harcourt 1987; Watts 1989, 1996; Wrangham 1979b; Yamagiwa & Kahekwa 2001). In the Virungas (MG), infanticide causes 37% of infant mortality and is regarded as a reproductive tactic adopted by extra-group males to hasten resumption of reproductive cyclicity in nursing females and to stimulate female transfer to them (Fossey 1984; Watts 1989, 1991). In Mt. Kahuzi (ELG), no infanticide has been reported, but females still tend to form a group, all-female, for a prolonged period after the death of a leading male (Yamagiwa & Kahekwa 2001). Such female groups prominently increase arboreal nesting during the absence of an adult male, probably to enhance their vigilance against terrestrial predators (Yamagiwa 2001). These observations suggest that female gorillas need a protector male against both predation by large terrestrial carnivores and harassment by extra-group males. The cohesiveness in their groupings may affect their foraging patterns, rather than the reverse. The ELG tend to visit fruiting trees very briefly and to avoid reusing the same ranging area repeatedly during the fruiting season (Yamagiwa *et al.* 1996b).

Such range shifts are also observed in WLG (Doran & McNeilage 1998; Tutin 1996).

Unlike gorillas, female bonobos do not change their dietary composition or grouping patterns seasonally. Two hypotheses have been devised to explain this. First, larger fruit patches are available throughout the year and abundant potential fallback foods like THV may mitigate the cost of grouping (Wrangham 1986). Second, female bonobos tend to use sexual behavior to reduce social tension caused by feeding competition (Kano 1980, 1989, 1992; Kitamura 1989; Kuroda 1980; Parish 1994, 1996; Thompson-Handler 1990). Copulation occurs frequently at the artificial feeding sites in Wamba and food sharing sometimes follows it (Kitamura 1989; Kuroda 1984). Genito-genital (G-G) rubbing (ventro-ventral embracing and rubbing sexual skins together) occurs between females in various situations during high social tension, such as aggressive encounters or potential conflicts around limited food resources or mating partners (Furuichi 1987; Kuroda 1984).

Male grouping patterns

Male grouping patterns differ considerably from those of females and appear to owe less to diet and more to mating patterns (Table 12.4). They may, however, be influenced by or influence foraging strategies.

In orangutans and chimpanzees, in contrast to females, male grouping patterns differ. Adult male orangutans do not show mutual affiliations, while adult male chimpanzees tend to associate with each other more frequently than with females (Galdikas 1985; Nishida 1979; van Schaik & van Hooff 1996; Wrangham 1979a). Although males in both orangutans and chimpanzees have larger and more complex home ranges than females, their relations with one another differ. Large adult male orangutans' home ranges extensively overlap with those of other males, but they maintain antagonistic relationships with each other, competing over access to females (Galdikas 1985; Rodman & Mitani 1986; van Schaik & van Hooff 1996). Small adult males occasionally travel in groups and force females to mate with them (Galdikas 1981, 1985; MacKinnon 1974; Mitani 1985; van Schaik & van Hooff 1996). Male chimpanzees tend to hunt colobus in groups during fruiting periods, which suggests that food availability allows forming the male groups that are needed to hunt successfully (Mitani *et al.*

Table 12.4. *Sexual activities of great apes*

	Orangutan	Gorilla	Chimpanzee	Bonobo
Seasonality	No	No	No	No
Menstrual cycle	29–30 days	31–32 days	34 days	42 days
Period of copulation	Unlimited[1]	1–3 days	7–17 days	5–40 days
Period of maximal swelling	No	No	12 days	15 days
Lactation period	3–4 years	2–3 years	3–4 years	3–4 years
Non-estrus period after birth	6–7 years[2]	2–3 years	3–4 years	1 year
Interbirth interval	7–8 years	4 years	4–7 years	4–7 years
Mating pattern	Temporal & prolonged consort	Prolonged consort	Promiscuous > possessive > temporal consort	Promiscuous
Infanticide	No	Mostly by extra-group males	Mostly by group males	No

Notes:

[1] Copulation occurs during consort lasting for days and weeks.

[2] Supposed by inter-birth interval in the wild, because of invisibility of female's estrus and non-estrous mating.

Sources: Orangutans: Galdikas 1981; Galdikas & Wood 1990; Nadler 1977; Rodman & Mitani 1987. *Gorillas:* Fossey 1984; Harcourt *et al.* 1981c; Watts 1989, 1991, 1996. *Chimpanzees:* Goodall 1986; Hiraiwa-Hasegawa 1987; Nishida & Hasegawa 1987; Takahata 1985; Tutin & McGinnis 1981; Wallis 1997. *Bonobos:* Furuichi 1987; Furuichi & Hashimoto 2002; Kano 1992; Kano 1996.

2002). Male chimpanzees occasionally form groups to patrol the boundary area of their home ranges (Chapman & Wrangham 1993; Wrangham 1979a); these groups are known to have killed conspecifics from neighboring communities in both Gombe and Mahale (Goodall *et al.* 1979; Nishida *et al.* 1985). Infanticide by males has occasionally been observed in chimpanzees (Goodall 1986; Hiraiwa-Hasegawa 1987; Takahata 1985) but never in orangutans. Female orangutans sometimes seek male protection to prevent another male's coercive mating (van Schaik & van Hooff 1996). However, female–male associations usually last only for days and females may not seek prolonged association with males (Galdikas 1981; Mitani 1985; Rodman & Mitani 1986). Orangutan and chimpanzee males appear to have evolved different tactics in their mating strategies, which may in turn affect female association patterns: female chimpanzees seek male protection against male sexual aggression more frequently than do female orangutans, who apparently do not usually need it.

Male gorillas do not usually associate with other males after maturity and tend to establish their own polygynous group, luring females from other groups (Fossey 1983; Harcourt 1978; Stewart & Harcourt 1987; Yamagiwa 1987a). Although the home ranges of these groups overlap extensively, adult males, including solitary males, maintain antagonistic relationships among each other (Caro 1976; Fossey 1974; Yamagiwa 1986). However, in the Virungas related MG males tend to associate in groups after maturity (Harcourt 1978; Robbins 1995, 2001). This is probably caused by female preferences in their choice of groups to join, and about half of the groups have recently shifted to multi-male composition in the Virunga and Bwindi populations (Robbins 2001; Stewart & Harcourt 1987; Watts 1996). Female MG may seek more protection from males to avoid infanticide by extra-group males and prefer to join large multi-male groups if they are available (Robbins 1995; Watts 1989, 1996). Their folivorous diet may enable them to transfer into large groups while maintaining a lower level of feeding competition. Subadult MG males tend to associate with each other and to form all-male groups with one or two adult males (Robbins 1995; Yamagiwa 1987b). The MG formation of multi-male and all-male groups contrasts with ELG and WLG, which form predominantly single-male polygynous groups and have never been reported to form all-male groups (Jones & Sabater Pi 1971; Nishihara 1994; Remis 1994; Tutin 1996; Yamagiwa et al. 1993). The higher feeding competition costs caused by their frugivorous diets may limit the group size in the lowland tropical forest and the absence of infanticide may reduce the motivation of females to join multi-male groups in ELG and WLG (Yamagiwa & Kahekwa 2001).

Male bonobos tend to associate with each other in bisexual groups, but their associations and affiliative contacts are less frequent than those between females or those between males and females (Kano 1992). Unlike the other great apes, male bonobos do not form all-male groups or spend a solitary life but instead usually associate with females (Kuroda 1979). The loose association among males and males' frequent association with females are probably caused by males' indistinct dominance over females and females' prolonged sexual attractiveness (Furuichi 1997; Kano 1992; Parish 1994), made possible by relatively stable food availability year round and greater reliance on THV (White & Wrangham 1988). Although bonobos show the same degree of sexual dimorphism as chimpanzees, female bonobos occasionally dominate male bonobos and mothers' dominance ranks strongly influence their mature sons' social status (Kano 1992; Parish 1994). The length of females' maximal swelling is longer in bonobos than in chimpanzees (Table 12.4) and female bonobos resume estrus within one year after giving birth (Furuichi 1987; Kano 1992). Female bonobos' prolonged estrous may raise male bonobos' sexual motivation and decrease their mating competition (Furuichi 1992; Kano 1992). Among male bonobos G-G contact occurs frequently and may function as appeasement or reassurance (Kano 1989; Kitamura 1989). It also occurs between members of different social units during inter-unit encounters, and may contribute to peaceful relationships between units (Idani 1990; Kano 1992). Unlike chimpanzees, strong male bonding among males and male killing of conspecifics, including infanticide, have never been reported in bonobos in any habitat.

In summary, female and male great apes have evolved different social foraging strategies. Female orangutans and chimpanzees change the degree of fission–fusion grouping patterns based on individual foraging. The availability of fruits and the reproductive states of females may influence their decision to associate with adult conspecifics. By contrast, female gorillas and bonobos usually form bisexual groups while foraging. The higher folivorous content of gorilla and bonobo diets likely contributes to this pattern. Greater folivory may encourage female gorillas to form foraging groups and their vulnerability to large terrestrial predators and to infanticide may stimulate them to associate with protector males. The presence of THV combined with large fruit patches may reduce the cost of foraging groups for female bonobos, and their frequent sexual interactions and stronger female–female affiliation enable them to form large bisexual parties.

Male grouping patterns reflect mating strategies more than feeding strategies. Based on their great sexual dimorphism, male orangutans and gorillas experience stronger competition over access to females than male chimpanzees and bonobos, who may engage in sperm competition through promiscuous mating. Larger testes size and conspicuous swelling of female's sexual skin favor the latter system (Harcourt et al. 1981b; Short 1981). Accordingly, orangutan and gorilla males tend to range separately from other males to corral females for mating. Frugivorous diets based on small, dispersed fruit patches may not allow male orangutans to sustain prolonged access to females, while folivorous diets may facilitate male gorillas' maintaining small bisexual groups. With the same evolutionary trends in large sexual dimorphism with overt competition between males, frugivorous diets have permitted male orangutans to compete over priority of access to a female's range, while folivorous diets have permitted male gorillas to compete over priority of permanent access to females.

Male chimpanzees and bonobos, in contrast, associate with other males in their community within a dominance ranking system. This may be facilitated by their tendency to stay in their natal groups after maturity, so males who associate are commonly related to one another. Male chimpanzees also usually dominate female chimpanzees while male bonobos are occasionally dominated by female bonobos. These differences influence their mating strategies. In chimpanzees, the stronger competition among males over access to estrous females combined with their ability to dominate females may have generated three mating patterns (possessive, consort, and promiscuous) (Hasegawa & Hiraiwa-Hasegawa 1983; Tutin 1979); in bonobos, weaker competition among males and males' inability to dominate females may have promoted only promiscuous mating (Furuichi 1992; Kano 1992). The greater opportunity for mating in bonobos may reduce hostility between unrelated males living in neighboring communities. The generally stronger competition among male chimpanzees combined with associations among related males within communities may stimulate them to form male bonds to defend female ranges from neighboring communities of unrelated males. The weak competition among male bonobos usually enables them to form bisexual groups while keeping peaceful relationships with neighboring communities. The stable availability of large fruit patches year-round and their greater reliance on THV may also enable them to enjoy a lower level of feeding competition between communities.

Male mating strategies also affect female grouping patterns through infanticide or other forms of sexual aggression, like forced copulations in orangutans. The risk of infanticide or forced copulation may raise females' motivation to seek protector males and may promote females' prolonged association with males. The higher sociality of female chimpanzees than female orangutans as individual foragers, as well as the higher proximity of females to males in gorillas than in bonobos, may reflect such differences in the risk of infanticide.

Ecological factors reflecting female feeding strategies and social factors reflecting male mating strategies may form different combinations in the great apes. Such differences may have promoted different forms and perhaps levels of social and technical foraging abilities in each ape species. Such differences are seen in food sharing, hunting, and tool using.

SOCIAL FORAGING AND THE EVOLUTION OF HOMINOID FORAGING PATTERNS

Rigid hierarchies based on dominance rank systems within a group may have developed in group-living primate species to reduce overt competition over access to limited food resources and mating partners, by soliciting the subordinate's withdrawal or submissiveness. When food is strictly limited, the dominant individual always gains it with little or no dispute. Prolonged gaze is frequently used by dominants as a mild form of threat to subordinates (Redican 1975). Instead of returning a gaze, subordinates show submissive expressions or postures, which may possibly reduce social tension and mitigate risks of severe fights (van Hooff 1962, 1969).

By contrast, in all great apes, prolonged gazing or eye contact between conspecifics may fail to elicit recipients' submissiveness (Gómez 1996; Goodall 1968; Kano 1980; Nishida 1970; Yamagiwa 1992). Great apes' social relationships are not based on rigid ranking systems (see van Schaik et al., Chapter 11, this volume), even in multi-male and multi-female communities of chimpanzees and bonobos, and social staring has various functions such as initiation of play and copulation, invitation to reconciliation, greeting, and intervention in conflict (Bard 1990; de Waal & Yoshihara 1983; Idani 1995; van Schaik, van Deaner & Merrill 1999; Yamagiwa 1992). The most striking difference in prolonged gaze between great apes and other nonhuman primates is that in the great apes it is frequently subordinates that use it toward dominants.

Social staring accompanied by begging behavior is used to solicit food sharing, which is a unique foraging behavior of chimpanzees and bonobos (Idani 1995; Kano 1980, 1992; Kuroda 1980, 1984; Nishida 1970). It is also used as a begging gesture by orangutans, who occasionally share foods with conspecifics voluntarily (Bard 1990; van Schaik et al. 1999). Although food sharing does not occur among gorillas, gorillas sometimes use social staring to supplant other individuals from feeding spots (Yamagiwa 1992). Gorillas' requests for food sharing or withdrawal from feeding spots are made by subordinates to dominants more frequently than the reverse and they tend to be highly successful for acquiring food.

Food sharing patterns are different for meat than for plant foods. Most observations of meat sharing are in chimpanzees. Meat sharing usually followed hunting by adult males and was accompanied by excitement in all group members near the prey (Boesch & Boesch 1989; Goodall 1986; Nishida et al. 1992; Nishida, Uehara & Nyundo 1983; Stanford 1996). Meat was shared selectively with other individuals, and meat sharing was frequently used by the most dominant male as a coalition strategy (Mitani et al. 2002; Nishida et al. 1992). By contrast, plant food is the major resource shared by bonobos, who rarely hunt animals (Hohmann & Fruth 1996; Kuroda 1984). Females take the role of owner and frequently share foods with other females (Hohmann & Fruth 1996). Plant foods shared by bonobos are often available anywhere and beggars can easily access them without sharing, but nevertheless request the dominant to share (Kano 1992; Kuroda 1984). Plant food sharing might not be caused by strong nutritional needs but possibly by the need to reinforce social bonds between individuals (Kuroda 1984). Subordinate beggars may confirm their close relationship with food possessors by achieving food sharing, and possessors' desires to co-feed with beggars may underlie their great tolerance (Kuroda 1997). Unlike the rigid dominance rank system that inhibits subordinates' feeding in front of dominants, food sharing facilitates social foraging where multiple individuals feed on the same food resources together irrespective of dominance rank. In great apes' social foraging, food abundance does not strongly incite conflict between individuals because food is used as a social tool to reduce tension and maintain social relationships.

The relative rarity of food sharing in orangutans and gorillas may owe to orangutans' semi-solitary nature and gorillas' passive interactions. It is male chimpanzees that frequently hunt animals and their characteristically increased male association may facilitate hunting monkeys and meat sharing (Boesch & Boesch 1989; Stanford 1996; Uehara et al. 1992). Lack of male–male association may prevent orangutans and gorillas from both. The stronger solitary nature of female orangutans also hinders opportunities for food sharing. While female gorillas usually associate with unrelated females within groups, they rarely affiliate with them (Stewart & Harcourt 1987). Each female gorilla's proximity to the leading male is what produces female gregariousness (Harcourt 1978, 1979; Watts 1996). Gorillas' folivorous diets may mitigate feeding competition and reduce their needs for reinforcing social bonds.

Differences between chimpanzees and bonobos in the food resources they obtain for sharing may reflect their sex differences in association. Male chimpanzees tend to associate frequently and to form alliances to maintain their social status. Such male association is suitable for hunting and males may need meat to reinforce their male alliances and to obtain female compliance through food sharing (Boesch & Boesch 1989; McGrew 1992; Mitani et al. 2002; Nishida et al. 1992). Male bonobos may lack the motivation to seek meat for sharing. Male dominance rank may not profit male bonobos in obtaining mating success. Instead, it is female bonobos who need to share food, to facilitate their associations with unrelated females. They do not hunt animals but collect plant foods for sharing. Food sharing and G-G contacts may be efficient tools for females to ensure prolonged association with unrelated females (Furuichi 1989; Hohmann & Fruth 1996).

Differences in tool-using behavior among the great apes is almost the greater puzzle. Chimpanzees prepare and use various tools for fishing termites and ants, digging termite mounds and subterranean bee nests, drinking water, cracking nuts, and pounding oil-palm pith (Boesch & Boesch 1983; Goodall 1986; McGrew 1992; Yamakoshi 1998, Chapter 9, this volume). However, the other great apes almost completely lack tool using for feeding in the wild, although they exhibit a rich array of flexible tool use elsewhere (Boysen et al. 1999; Galdikas 1982; Jordan 1982; Lethmate 1982; Russon & Galdikas 1993, 1995; Wood 1984). Recent findings on orangutans may clarify the conditions that favor common tool use. Sumatran orangutans living in Suaq Balimbing manufacture tools for extracting insects or honey from tree holes or prying seeds from hard-husked fruits (Fox et al. 1999; van Schaik, Fox & Sitompul 1996). These findings provide hints to account for the differences in tool use among great apes. Both ecological and social factors influence the appearance of tool use. The exceptionally high density of orangutans in Suaq Balimbing coupled with their frugivorous diets may have produced severe scramble competition and spurred the invention of tool use to meet subsistence needs (Fox et al. 1999). Their frequent association may in turn have facilitated social learning of tool use and contributed to spreading these complex skills (van Schaik et al. 1999). Ecological conditions that increase feeding competition combined with the social tolerance that allows social learning may be necessary for creating and maintaining tool use. Social learning may be facilitated by the social staring common to all great apes. Although the ecological and social features of chimpanzees and some orangutan populations satisfy such conditions, gorillas and bonobos may lack them. Tool use may buffer the scarcity of high-quality foods by facilitating the extraction of embedded or hard-shelled foods for orangutans and chimpanzees (Fox et al. 1999; Yamakoshi 1998) – both of whom turn less to folivorous food sources than do gorillas and bonobos.

Monkey and ape brains are twice as large on average as those of mammals of equivalent body size. Their large brains are likely linked with evolutionary enhancements to both ecological and social intelligence (Byrne 1995). The need for mental maps of high-quality foods dispersed over a wide range and for memory of seasonal fluctuation in their distribution may have contributed to raising their ecological intelligence. The need for flexible social skills acquired by rapid learning of fluctuating social relations among others, such as kin relations, friendship and dominance rank, may have contributed to increasing their social intelligence. Great apes basically share these evolutionary trends with monkeys, but may have extended them to higher levels. However, great apes do not have more complex ecological niches or social groups than monkeys. Some monkey species live in wider home ranges in more seasonal habitats and form larger groups than great apes.

With their large body weight and unspecialized digestive systems, however, great apes experience different ecological constraints than monkeys. They have had to increase their dietary breadth and regulate their day range or foraging group size according to periodic fluctuations in food availability. Their capacity for sophisticated skills, as seen in tool using or hierarchically organized techniques, may have evolved to enable them to gain access to inaccessible foods, especially embedded ones (Byrne 2001; Gibson 1990; Parker & Gibson 1977). The greater seasonality in food availability and the particular types of social relationships that great apes experience may have stimulated such capacities. Food sharing may not be based on rigid dominance hierarchies; instead, it may be facilitated by relatively egalitarian associations among great apes who are motivated to gain high-quality foods from conspecifics as well as to reinforce and renegotiate social bonds with them. Hunting is enhanced by males' strong motivation to form alliances to protect their territories and female mates against other males. Great apes' tool-using skills may be promoted by frequent association with conspecifics and their motivation to obtain high-quality foods that are difficult to obtain. It seems likely that the seasonally severe shortages of high-quality foods and the fluid social relationships experienced by chimpanzees enabled them to develop various forms of tool manufacture.

Humans merely continue the trend from monkeys to great apes in increasing the complexity of their social relations and foraging strategies. Drier and or less predictable habitats in the Pliocene compared with the Miocene (Brain 1981; Kingston, Mrino & Hill 1994; Potts, Chapter 13, this volume) may be related to these changes. Enlargement in hominid brain size appeared in the early Pleistocene, when periodic swings between warmer and colder conditions occurred repeatedly (Potts, Chapter 13, this volume; Prentice & Denton 1988). It may reflect a rapid increase in hominid intelligence over a period of severe food conditions associated with highly fluctuating climate. This suggests that a wide variety of ecological and social problems co-occurred and interacted to raise both the social and the ecological cognitive abilities of Pleistocene hominids.

REFERENCES

Aveling, R. & Aveling, C. (1987). Report from the Zaire Gorilla Conservation Project. *Primate Conservation*, **8**, 162–4.

Badrian, N. & Malenky, R. (1984). Feeding ecology of *Pan paniscus* in the Lomako Forest, Zaire. In *The Pygmy Chimpanzee: Evolutionary Biology and Behavior*, ed. R. L. Susman, pp. 275–99. New York: Plenum Press.

Badrian, N., Badrian, A. & Susman, R. L. (1981). Preliminary observations on the feeding behavior of *Pan paniscus* in the Lomako Forest of central Zaire. *Primates*, **2**, 173–81.

Baldwin, P. J., McGrew, W. C. & Tutin, C. E. G. (1982). Wide-ranging chimpanzees at Mt. Assirik, Senegal. *International Journal of Primatology*, **3**, 367–85.

Bard, K. A. (1990). "Social tool use" by free-ranging orang-utans: a Piagetian and development perspective on the manipulation of an animate object. In *"Language" and Intelligence in Monkeys and Apes: Comparative Developmental Perspectives*, ed. S. T. Parker & K. R. Gibson, pp. 356–78. New York: Cambridge University Press.

Bauchop, T. & Martucci, R. W. (1968). Ruminant-like digestion of the langur monkey. *Science*, **161**, 698–9.

Bermejo, M. (1997). Study of western lowland gorillas in the Lossi Forest of North Congo and a pilot gorilla tourism plan. *Gorilla Conservation News*, **11**, 6–7.

Bermejo, M., Liera, G. & Sabater Pi, J. (1994). Animals and mushrooms consumed by bonobos (*Pan paniscus*): new record from Lilungu (Ikela), Zaire. *International Journal of Primatology*, **15**, 879–98.

Boekhorst, I. J. A. te, Schurmann, C. L. & Sugardjito, J. (1990). Residential status and seasonal movements of wild orangutans in the Gunung Leuser Reserve (Sumatra, Indonesia). *Animal Behaviour*, **39**, 1098–109.

Boesch, C. (1996). Social grouping in Taï chimpanzees. In *Great Ape Societies*, ed. W. C. McGrew, L. F. Marchant & T. Nishida, pp. 101–13. Cambridge, UK: Cambridge University Press.

Boesch, C. & Boesch, H. (1981). Sex differences in the use of natural hammers by wild chimpanzees: a preliminary report. *Journal of Human Evolution*, **10**, 585–93.

(1983). Optimization of nut-cracking with natural hammers by wild chimpanzees. *Behaviour*, **83**, 265–86.

(1989). Hunting behavior of wild chimpanzees in the Tai National Park. *American Journal of Physical Anthropology*, **78**, 547–73.

Boysen, S. T., Kuhlmeier, V. A., Halliday, P. & Halliday, Y. M. (1999). Tool use in captive gorillas. In *Mentalities of Gorillas and Orangutans: Comparative Perspectives*, ed. S. T. Parker, R. W. Mitchell & H. L. Miles, pp. 179–87. Cambridge, UK: Cambridge University Press.

Brain, C. (1981). Hominid evolution and climatic change. *South African Journal of Science*, **77**, 104–5.

Byrne, R. W. (1995). *The Thinking Ape: Evolutionary Origins of Intelligence*. Oxford: Oxford University Press.

(2001). Social and technical forms of primate intelligence. In *Tree of Origin: What Primate Behavior Can Tell us about Human Social Evolution*, ed, F. B. M. de Waal, pp. 147–72. Cambridge, MA: Harvard University Press.

Byrne, R. W. & Byrne, J. M. E. (1993). Complex leaf-gathering skills of mountain gorillas (*Gorilla g. beringei*): variability and standardization. *American Journal of Primatology*, **31**, 241–61.

Caro, T. M. (1976). Observations on the ranging behaviour and daily activity of lone silverback mountain gorillas (*Gorilla gorilla beringei*). *Animal Behaviour*, **24**, 889–97.

Casimir, M. J. (1975). Feeding ecology and nutrition of an eastern gorilla group in the Mt. Kahuzi region (République du Zaire). *Folia Primatologica*, **24**, 1–36.

Casimir, M. J. & Butenandt, E. (1973). Migration and core area shifting in relation to some ecological factors in a mountain gorilla group (*Gorilla gorilla beringei*) in the Mt. Kahuzi region (République du Zaire). *Zietschrift fur Tierpsychologie*, **33**, 514–22.

Chapman, C. A. & Wrangham, R. W. (1993). Range use of the forest chimpanzees of Kibale: implications for the understanding of chimpanzee social organization. *American Journal of Primatology*, **31**, 263–73.

Chapman, C. A., Wrangham, R. W. & Chapman, L. J. (1995). Ecological constraints on group size: an analysis of spider monkey and chimpanzee subgroups. *Behavioral Ecology and Sociobiology*, **36**, 59–70.

Chivers, D. J. & Hladik, C. M. (1980). Morphology of the gastrointestinal tract in primates: comparisons with other mammals in relation to diet. *Journal of Morphology*, **116**, 337–86.

(1984). Diet and morphology in primates. In *Food Acquisition and Processing*, ed. D. J. Chivers, B. A. Wood & A. Bilsborough, pp. 213–30. London: Plenum Press.

Chivers, D. J. & Langer, P. (1994). *The Digestive System in Mammals: Food, Form and Function*. Cambridge, UK: Cambridge University Press.

Clutton-Brock, T. H. & Harvey, P. H. (1977). Primate ecology and social organization. *Journal of the Zoological Society, London*, **183**, 1–39.

de Waal, F. B. M. & Yoshihara, D. (1983). Reconciliation and redirected affection in rhesus monkeys. *Behaviour*, **85**, 224–41.

Doran, D. & McNeilage, A. (1998). Gorilla ecology and behavior. *Evolutionary Anthropology*, **6**, 120–31.

(2001). Subspecific variation in gorilla behavior: the influence of ecological and social factors. In *Mountain Gorillas: Three Decades of Research at Karisoke*, ed. M. M. Robbins, P. Sicotte & K. J. Stewart, pp. 123–49. Cambridge, UK: Cambridge University Press.

Dunbar, R. I. M. (1988). *Primate Social Systems*. London: Croom Helm.

Elliott, R. C. (1976). Observations on a small group of mountain gorilla (*Gorilla gorilla beringei*). *Folia Primatologica*, **25**, 12–24.

Feeny, P. (1976). Plant apparency and chemical defence. *Recent Advances in Phytochemistry*, **10**, 1–40.

Fleming, T. H., Brettwisch, R. & Whitesides, G. H. (1987). Patterns of tropical vertebrate frugivore diversity. *Annual Review of Ecology and Systematics*, **18**, 91–109.

Fossey, D. (1974). Observations on the home range of one group of mountain gorilla (*Gorilla gorilla beringei*). *Animal Behaviour*, **22**, 568–81.

(1983). *Gorillas in the Mist*. Boston, MA: Houghton Mifflin.

(1984). Infanticide in mountain gorillas (*Gorilla gorilla beringei*) with comparative notes on chimpanzees. In *Infanticide: Comparative and Evolutionary Perspectives*, ed. G. Hausfater & S. Hrdy, pp. 217–36. Hawthorne, NY: Aldine.

Fossey, D. & Harcourt, A. H. (1977). Feeding ecology of free-ranging mountain gorilla (*Gorilla gorilla beringei*). In *Primate Ecology*, ed. T. H. Clutton-Brock, pp. 415–47. New York: Academic Press.

Fox, E. A., Sitompul, A. F. & van Schaik, C. P. (1999). Intelligent tool use in Sumatran orangutans. In *The Mentalities of Gorillas and Orangutans: Comparative*

Perspectives, ed. S. T. Parker, R. W. Mitchell & H. L. Miles, pp. 99–116. Cambridge, UK: Cambridge University Press.

Freeland, W. J. & Janzen, D. H. (1974). Strategies in herbivory by mammals: the role of plant secondary compounds. *American Naturalist*, **108**, 269–89.

Furuichi, T. (1987). Sexual swelling, receptivity and grouping of wild pygmy chimpanzee females at Wamba, Zaire. *Primates*, **20**, 309–18.

(1989). Social interactions and the life history of female *Pan paniscus* in Wamba, Zaire. *International Journal of Primatology*, **10**, 173–97.

(1992). The prolonged estrus of females and factors influencing mating in a wild group of bonobos (*Pan paniscus*) in Wamba, Zaire. In *Topics in Primatology. Vol. 2. Behavior, Ecology, and Conservation*, ed. N. Itoigawa, Y. Sugiyama, G. P. Sackett, & R. K. R. Thompson, pp. 170–90. Tokyo: University of Tokyo Press.

(1997). Agonistic interactions and matrifocal dominance rank of wild bonobos (*Pan paniscus*) at Wamba. *International Journal of Primatology*, **18**, 855–75.

Furuichi, T. & Hashimoto, C. (2002). Why female bonobos have a lower copulation rate during estrus than chimpanzees. In *Behavioral Diversity in Chimpanzees and Bonobos*, ed. C. Boesch & L. Marquardt, pp. 156–67. New York: Cambridge University Press.

Furuichi, T., Idani, G., Ihobe, H., Kuroda, S., Kitamura, K., Mori, A., Enomoto, T., Okayasu, N., Hashimoto, C. & Kano, T. (1998). Population dynamics of wild bonobos (*Pan paniscus*) at Wamba. *International Journal of Primatology*, **19**, 1029–43.

Galdikas, B. M. F. (1978). Orangutan Adaptation at Tanjung Puting Reserve, Central Borneo. Ph.D. dissertation, University of California, Los Angeles.

(1981). Orangutan reproduction in the wild. In *Reproductive Biology of the Great Apes*, ed. C. E. Graham, pp. 281–300. New York: Academic Press.

(1982). Orang-utan tool use at Tanjung Puting Reserve, Central Indonesian Borneo (Kalimantan Tengah). *Journal of Human Evolution*, **10**, 19–33.

(1984). Adult female sociality among wild orangutans at Tanjung Puting Reserve. In *Female Primates: Studies by Women Primatologists*, ed. M. F. Small, pp. 217–35. New York: Alan R. Liss.

(1985). Orangutan sociality at Tanjung Puting. *American Journal of Primatology*, **9**, 101–19.

(1988). Orangutan diet, range, and activity at Tanjung Puting, Central Borneo. *International Journal of Primatology*, **9**, 1–35.

Galdikas, B. M. F. & Teleki, G. (1981). Variations in subsistence activities of male and female pongids: new perspectives on the origins of human labor divisions. *Current Anthropology*, **22**, 241–56.

Galdikas, B. M. F. & Wood, J. (1990). Great ape and human birth intervals. *American Journal of Physical Anthropology*, **83**, 185–92.

Gautier-Hion, A., Duplantier, J. M., Quris, R., Feer, F., Sourd, C., Decoux, J.-P., Dubost, G., Emmons, L., Erard, C., Hecketsweller, P., Moungazi, A., Roussilhon, C. & Thiollay, J.-M. (1985). Fruit characters as a basis of fruit choice and seed dispersal in a tropical forest vertebrate community. *Oecologia*, **65**, 324–37.

Ghiglieri, M. P. (1984). *The Chimpanzees of Kibale Forest: A Field Study of Ecology and Social Structure*. New York: Columbia University Press.

Gibson, K. R. (1990). New perspectives on instincts and intelligence: brain size and the emergence of hierarchical mental construction skills. In *"Language" and Intelligence in Monkey and Apes*, ed. S. T. Parker & K. R. Gibson, pp. 97–128. New York: Cambridge University Press.

Goldsmith, M. L. (1999). Ecological constraints on the foraging effort of western gorillas (*Gorilla gorilla gorilla*) at Bai Hokou, Central African Republic. *International Journal of Primatology*, **20**, 1–23.

(1996). Ecological influences on the ranging and grouping behavior of western lowland gorillas at Bai Hokou in the Central African Republic. Ph.D. dissertation, State University of New York at Stony Brook.

Gómez, J. C. (1996). Ostensive behavior in great apes: the role of eye contact. In *Reaching into Thought. The Minds of the Great Apes*, ed. A. E. Russon, K. A. Bard & S. T. Parker, pp. 131–51. Cambridge, UK: Cambridge University Press.

Goodall, A. G. (1977). Feeding and ranging behaviour of a mountain gorilla group (*Gorilla gorilla beringei*) in the Tshibinda-Kahuzi region (Zaire). In *Primate Ecology*, ed. T. H. Clutton-Brock, pp. 450–79. New York: Academic Press.

Goodall, J. (1968). The behavior of free-living chimpanzees in the Gombe Stream Reserve. *Animal Behaviour Monographs*, **1**, 161–331.

(1986). *Chimpanzees of Gombe: Patterns of Behavior.* Cambridge, MA: The Belknap Press of Harvard University Press.

Goodall, J., Bandora, A., Bergmann, E., Busse, C., Matama, H., Mpongo, E., Pierce, A. & Riss, D. (1979). Intercommunity interactions in the chimpanzee population of the Gombe National Park. In *The Great Apes*, ed. D. A. Hamburg & E. R. McCown, pp. 13–53. Menlo Park, CA: Benjamin/Cummings.

Hall, J. S., White, L. J. T., Inogwabini, B-I., Omari, I., Morland, H. S., Williamson, E. A., Saltonstall, K., Walsh, P., Sikubwabo, C., Bonny, D., Kiswele, K. P., Vedder, A. & Freeman, K. (1998). Survey of Grauer's gorillas (*Gorilla gorilla graueri*) and Eastern chimpanzees (*Pan troglodytes schweinfurthii*) in the Kahuzi-Biega National Park Lowland Sector and adjacent forest in Eastern Democratic Republic of Congo. *International Journal of Primatology*, **19**, 207–35.

Harcourt, A. H. (1978). Strategies of emigration and transfer by primates, with particular reference to gorillas. *Zeitschrift für Tierpsychologie*, **48**, 401–20.

(1979). Social relationships among female mountain gorillas. *Animal Behaviour*, **27**, 251–64.

Harcourt, A. H., Fossey, D. & Sabater Pi, J. (1981a). Demography of *Gorilla gorilla. Journal of the Zoological Society, London*, **195**, 15–233.

Harcourt, A. H., Harvey, P. H., Larson, S. G. & Short, R. V. (1981b). Testes weight, body weight and breeding system in primates. *Nature*, **293**, 55–7.

Harcourt, A. H., Stewart, K. J. & Fossey, D. (1976). Male emigration and female transfer in wild mountain gorillas. *Nature*, **263**, 226–7.

Harcourt, A. H., Stewart, K. J. & Fossey, D. (1981c). Gorilla reproduction in the wild. In *Reproductive Biology of the Great Apes*, ed. C. Graham, pp. 265–79, New York: Academic Press.

Hasegawa, T. & Hiraiwa-Hasegawa, M. (1983). Opportunistic and restrictive mating among wild chimpanzees in the Mahale Mountains, Tanzania. *Journal of Ethology*, **1**, 75–85.

Hiraiwa-Hasegawa, M. (1987). Infanticide in primates and a possible case of male-biased infanticide in chimpanzees. In *Animal Societies: Theories and Facts*, ed. Y. Ito, J. L. Brown & J. Kikkawa, pp. 125–39, Tokyo.

Hladik, C. M. (1977). Chimpanzees of Gabon and chimpanzees of Gombe: some comparative data on diet. In *Primate Ecology*, ed. T. H. Clutton-Brock, pp. 481–501. New York: Academic Press.

Hohmann, G. & Fruth, B. (1993). Field observations on meat sharing among bonobos (*Pan paniscus*). *Folia Primatologica*, **60**, 225–9.

(1996). Nest building behavior in the great apes: the great leap forward? In *Great Ape Societies*, ed. W. C. McGrew, L. F. Marchant & T. Nishida, pp. 225–40. Cambridge, UK: Cambridge University Press.

Horr, D. A. (1975). The Borneo orang-utan: population structure and dynamics in relationship to ecology and reproductive strategy. In *Primate Behavior: Developments in the Field and Laboratory Research*, ed. L. A. Rosenblum, pp. 307–23. New York: Academic Press.

Idani, G. (1990). Relations between unit-groups of bonobos at Wamba, Zaire: encounters and temporary fusions. *African Studies Monographs*, **11**, 153–6.

(1991). Social relationships between immigrant and resident bonobo (*Pan paniscus*) females at Wamba. *Folia Primatologica*, **57**, 83–95.

(1995). Function of peering behavior among bonobos (*Pan paniscus*) at Wamba, Zaire. *Primates*, **36**, 377–83.

Ihobe, H. (1992). Observations on the meat-eating behavior of wild bonobos (*Pan paniscus*) at Wamba, Republic of Zaire. *Primates*, **33**, 247–50.

Izawa, K. (1970). Unit groups of chimpanzees and their nomadism in the savanna woodland. *Primates*, **11**, 1–45.

Janzen, D. H. (1974). Tropical blackwater rivers, animals, and mast fruiting by the Dipterocarpaceae. *Biotropica*, **6**, 69–103.

Jones, C. & Sabater Pi, J. (1971). Comparative ecology of *Gorilla gorilla* (Savage and Wyman) and *Pan troglodytes* (Blumenbach) in Rio Muni, West Africa. *Bibliotheca Primatologica*, **13**, 1–96.

Jordan, C. (1982). Object manipulation and tool-use in captive pygmy chimpanzees (*Pan paniscus*). *Journal of Human Evolution*, **11**, 35–9.

Kano, T. (1971). The chimpanzee of Filabanga, Western Tanzania. *Primates*, **12**, 229–46.

(1972). Distribution and adaptation of the chimpanzee on the eastern shore of Lake Tanganyika. *Kyoto University African Studies*, **7**, 37–129.

(1980). Social behavior of wild pygmy chimpanzees (*Pan paniscus*) of Wamba: a preliminary report. *Journal of Human Evolution*, **9**, 243–60.

(1982). The social group of pygmy chimpanzees (*Pan paniscus*) of Wamba. *Primates*, **23**, 171–88.

(1983). An ecological study of the pygmy chimpanzees (*Pan paniscus*) of Yalosidi, Republic of Zaire. *International Journal of Primatology*, **4**, 1–25.

(1989). The sexual behavior of pygmy chimpanzees. In *Understanding Chimpanzees*, ed. P. G. Heltne & L. A. Marcquardt, pp. 176–83. Cambridge, MA: Harvard University Press.

(1992). *The Last Ape: Pygmy Chimpanzee Behavior and Ecology*. Stanford, CA: Stanford University Press.

(1996). Male rank order and copulation rate in a unit-group of bonobos at Wamba, Zaire. In *Great Ape Societies*, ed. W. C. McGrew, L. F. Marchant & T. Nishida, pp. 135–45. Cambridge, UK: Cambridge University Press.

Kano, T. & Mulavwa, M. (1984). Feeding ecology of the pygmy chimpanzees (*Pan paniscus*) of Wamba. In *The Pygmy Chimpanzee: Evolutionary Biology and Behavior*, ed. R. L. Susman, pp. 233–324. New York: Plenum Press.

Kay, R. F. (1984). On the use of anatomical features to infer foraging behavior in extinct primates. In *Adaptation for Foraging in Nonhuman Primates: Contributions to an Organismal Biology of Prosimians, Monkeys and Apes*, ed. P. S. Rodman & J. G. H. Cant, pp. 21–53. New York: Columbia University Press.

Kay, R. N. B. & Davies, A. G. (1994). Digestive physiology. In *Colobine Monkeys: Their Ecology, Behavior and Evolution*, ed. A. G. Davies & J. F. Oates, pp. 229–59. Cambridge, UK: Cambridge University Press.

Kingston, J. D., Mrino, B. D. & Hill, A. (1994). Iso-tropic evidence for Neogene hominid paleo-environments in the Kenya Rift Valley. *Science*, **264**, 955–9.

Kitamura, K. (1989). Genito-genital contacts in the pygmy chimpanzee (*Pan paniscus*). *African Studies Monographs*, **10**, 46–67.

Knott, C. D. (1998a). Social system dynamics, ranging patterns and male and female strategies in wild Bornean orangutans (*Pongo pygmaeus*). *American Journal of Physical Anthropology, Supplement*, **140**.

(1998b). Changes in orangutan caloric intake, energy balance, and ketones in response to fluctuating fruit availability. *International Journal of Primatology*, **19**, 1061–79.

(1999). Reproductive, physiological and behavioral responses of orangutans in Borneo to fluctuations in food availability. Ph.D. dissertation, Harvard University.

Kuroda, S. (1979). Grouping of the pygmy chimpanzees. *Primates*, **20**, 161–83.

(1980). Social behavior of the pygmy chimpanzees. *Primates*, **21**, 181–97.

(1984). Interaction over food among pygmy chimpanzees. In *The Pygmy Chimpanzee: Evolutionary Biology and Behavior*, ed. R. L. Susman, pp. 301–24. New York: Plenum Press.

(1997). The sociological jump before hominization: a re-evaluation of food-sharing in *Pan* and its implications for social evolution. *Primate Research*, **13**, 149–59. (In Japanese with English summary.)

Kuroda, S., Nishihara, T., Suzuki, S. & Oko, R. A. (1996). Sympatric chimpanzees and gorillas in the Ndoki Forest, Congo. In *Great Ape Societies*, ed. W. C. McGrew, L. F. Marchant & T. Nishida, pp. 71–81. Cambridge, UK: Cambridge University Press.

Leigh, S. R. & Shea, B. T. (1995). Ontogeny and the evolution of adult body size dimorphism in apes. *American Journal of Primatology*, **36**, 37–60.

Leighton, M. (1993). Modeling dietary selectivity by Bornean orangutans: evidence for integration of multiple criteria in fruit selection. *International Journal of Primatology*, **14**, 257–313.

Lethmate, J. (1982). Tool-using skills of orangutans. *Journal of Human Evolution*, **11**, 49–64.

MacKinnon, J. R. (1974). The ecology and behavior of wild orangutans (*Pongo pygmaeus*). *Animal Behaviour*, **22**, 3–74.

(1979). Reproductive behavior in wild orangutan populations. In *The Great Apes*, ed. D. A. Hamburg & E. R. McCown, pp. 257–73. Menlo Park, CA: Benjamin/Cummings.

Malenky, R. W. & Stiles, E. W. (1991). Distribution of terrestrial herbaceous vegetation and its consumption by *Pan paniscus* in the Lomako Forest, Zaire. *American Journal of Primatology*, **23**, 153–69.

Malenky, R. K. & Wrangham, R. W. (1994). A quantitative comparison of terrestrial herbaceous food consumption by *Pan paniscus* in the Lomako Forest, Zaire, and *Pan troglodytes* in the Kibale Forest, Uganda. *American Journal of Primatology*, **32**, 1–12.

Matsumoto-Oda, A. (1999). Mahale chimpanzees: grouping patterns and cycling females. *American Journal of Primatology*, **47**, 197–207.

McGrew, W. C. (1979). Evolutionary implications of sex differences in chimpanzee predation and tool use. In

The Great Apes, ed. D. A. Hamburg & E. R. McCown, pp. 441–63. Menlo Park, CA: Benjamin/Cummings.

(1992). *Chimpanzee Material Culture: Implications for Human Evolution.* Cambridge, UK: Cambridge University Press.

McNeilage, A. (2001). Diet and habitat use of two mountain gorilla groups in contrasting habitats in the Virunga. In *Mountain Gorillas: Three Decades of Research at Karisoke*, ed. M. M. Robbins, P. Sicotte & K. J. Stewart, pp. 266–92. Cambridge, UK: Cambridge University Press.

Martin, R. D. (1990). *Primate Origins and Evolution: A Phylogenetic Reconstruction.* Princeton, NJ: Princeton University Press.

Milton, K. (1984). The role of food processing factors in primate food choice. In *Adaptations for Foraging in Nonhuman Primates: Contributions to an Organismal Biology of Prosimians, Monkeys and Apes*, ed. P. S. Rodman & J. G. H. Cant, pp. 249–79. New York: Columbia University Press.

(1986). Digestive physiology in primates. *News Physiological Science*, 1, 76–9.

Mitani, J. C. (1985). Mating behavior of male orangutans in the Kutai Reserve, East Kalimantan, Indonesia. *Animal Behaviour*, 33, 392–402.

Mitani, J. C. Watts, D. P. & Muller, M. N. (2002). Recent developments in the study of wild chimpanzee behavior. *Evolutionary Anthropology*, 11, 9–25.

Murnyak, D. F. (1981). Censusing the gorillas in Kahuzi-Biega National Park. *Biological Conservation*, 21, 163–76.

Nadler, R. D. (1977). Sexual behavior of captive orangutans. *Archives of Sexual Behavior*, 6, 457–75.

Nishida, T. (1968). The social group of wild chimpanzees in the Mahale Mountains. *Primates*, 9, 167–224.

(1970). Social behavior and relationship among wild chimpanzees of the Mahale mountains. *Primates*, 11, 47–87.

(1976). The bark-eating habits in primates, with special reference to their status in the diet of wild chimpanzees. *Folia Primatologica*, 25, 277–87.

(1979). The social structure of chimpanzees of the Mahale Mountains. In *The Great Apes*, ed. D. A. Hamburg and E. R. McCown, pp. 73–121. Menlo Park, CA: Benjamin/Cummings.

Nishida, T. & Hiraiwa-Hasegawa, M. (1987). Chimpanzees and bonobos: cooperative relationships among males.

In *Primate Societies*, ed. B. B. Smuts, D. L. Cheney, R. M. Seyfarth, R. W. Wrangham & T. T. Struhsaker, pp. 165–77. Chicago: The University of Chicago Press.

Nishida, T. & Kawanaka, K. (1972). Inter-unit-group relationships among wild chimpanzees of the Mahale Mountains. *Kyoto University African Studies*, 7, 131–69.

Nishida, T. & Uehara, S. (1983). Natural diet of chimpanzees (*Pan troglodytes schweinfurthii*): long-term record from the Mahale Mountains, Tanzania. *African Study Monographs*, 3, 109–30.

Nishida, T., Hasegawa, T., Hayaki, H., Takahata, Y. & Uehara, S. (1992). Meat-sharing as a coalition strategy by an alpha male chimpanzee? In *Topics in Primatology. Vol. 1. Human Origins*, ed. T. Nishida, W. C. McGrew, P. Marler, M. Pickford & F. B. M. de Waal, pp. 159–74. Tokyo: University of Tokyo Press.

Nishida, T., Hiraiwa-Hasegawa, M., Hasegawa, T. & Takahata, Y. (1985). Group extinction and female transfer in wild chimpanzees in the Mahale Mountains. *Zeitschrift fur Tierpsychologie*, 67, 284–301.

Nishida, T., Uehara, S. & Nyundo, R. (1983). Predatory behavior among wild chimpanzees of the Mahale Mountains. *Primates*, 20, 1–20.

Nishihara, T. (1994). Population density and group organization of gorillas (*Gorilla gorilla gorilla*) in the Nouabalé-Ndoki National Park, Congo. *Journal of African Studies*, 44, 29–45. (In Japanese.)

(1995). Feeding ecology of western lowland gorillas in the Nouabale-Ndoki National Park, Congo. *Primates*, 36, 151–68.

Parker, S. T. & Gibson, K. R. (1977). Object manipulation, tool use and sensorimotor intelligence as feeding adaptations in cebus monkeys and great apes. *Journal of Human Evolution*, 6, 623–41.

(1979). A development model for evolution of language and intelligence in early hominids. *Behavioral and Brain Sciences*, 2, 367–408.

Parra, R. (1978). Comparison of foregut and hindgut fermentation in herbivores. In *The Ecology of Arboreal Folivores*, ed. G. G. Montgomery, pp. 205–30. Washington, DC: Smithsonian Institution Press.

Parish, A. R. (1994). Sex and food control in the "uncommon chimpanzee": how bonobo females overcome a phylogenetic legacy of male dominance. *Ethology and Sociobiology*, 15, 157–79.

(1996). Female relationships in bonobos (*Pan paniscus*). *Human Nature*, **7**, 61–96.

Prentice, M. & Denton, G. (1988). The deep-sea oxygen isotope record, the global ice sheet system and hominid evolution. In *Evolutionary History of the "Robust" Australopithecines*, ed. F. E. Grine, pp. 383–403. New York: Aldine de Gruyter.

Redican, W. K. (1975). Facial expressions in nonhuman primates. In *Primate Behavior*, ed. L. A. Rosenblum, pp. 103–94. New York: Academic Press.

Remis, M. J. (1994). Feeding ecology and positional behavior of western lowland gorillas (*Gorilla gorilla gorilla*) in the Central African Republic. Ph.D. thesis, Yale University.

(1997a). Western lowland gorillas (*Gorilla gorilla gorilla*) as seasonal frugivores: use of variable resources. *American Journal of Primatology*, **43**, 87–109.

(1997b). Ranging and grouping patterns of a western lowland gorilla group at Bai Hokou, Central African Republic. *American Journal of Primatology*, **43**, 111–33.

Remis, M. J., Dierenfeld, E. S., Mowry, C. B. & Carroll, R. W. (2001). Nutritional aspects of western lowland gorilla (*Gorilla gorilla gorilla*) diet during seasons of fruit scarcity at Bai Hokou, Central African Republic. *International Journal of Primatology*, **22**, 807–36.

Reynolds, V. & Reynolds, F. (1965). Chimpanzees of the Budongo forest. In *Primate Behavior*, ed. I. DeVore, pp. 368–424. New York: Holt, Rinehart and Winston.

Rijksen, H. D. (1978). *A Field Study on Sumatran Orangutans (Pongo pygmaeus abelii, Lesson 1827): Ecology, Behaviour and Conservation*. Wageningen, The Netherlands: Veenan & Zonen.

Robbins, M. M. (1995). A demographic analysis of male life history and social structure of mountain gorillas. *Behaviour*, **132**, 21–48.

(2001). Variation in the social system of mountain gorillas: the male perspective. In *Mountain Gorillas: Three Decades of Research at Karisoke*, ed. M. M. Robbins, P. Sicotte & K. J. Stewart, pp. 29–58. Cambridge, UK: Cambridge University Press.

Rodman, P. S. (1973). Population composition and adaptive organization among orang-utans of the Kutai Nature Reserve. In *Comparative Ecology and Behaviour of Primates*, ed. R. P. Michael & J. H. Crook, pp. 171–209. New York: Academic Press.

(1977). Feeding behaviour of orangutans of the Kutai Nature Reserve, East Kalimantan. In *Primate Ecology*,

ed. T. H. Clutton-Brock, pp. 383–413. New York: Academic Press.

(1979). Individual activity and solitary nature of orangutans. In *The Great Apes*, ed. D. A. Hamburg & E. R. McCown, pp. 235–55. Menlo Park, CA: Benjamin/Cummings.

(2000). Great ape models for the evolution of human diet. www.cast.uark.edu/local/icaes/conferences/wburg/posters/psrodman/GAMHD.htm

Rodman, P. S. & Mitani, J. C. (1987). Orangutans: sexual dimorphism in a solitary species. In *Primate Societies*, ed. B. B. Smuts, D. L. Cheney, R. M. Seyfarth, R. W. Wrangham & T. T. Struksaker, pp. 146–54. Chicago, IL: University of Chicago Press.

Rogers, M. E., Maisels, F., Williamson, E. A., Fernandez, M. & Tutin, C. E. G. (1990). Gorilla diet in the Lope Reserve, Gabon: a nutritional analysis. *Oecologia*, **84**, 326–39.

Russon, A. E. & Galdikas, B. M. F. (1993). Imitation in free-ranging rehabilitant orangutans. *Journal of Comparative Psychology*, **107**, 147–61.

(1995). Constraints on great apes' imitation: model and action selectivity in rehabilitant orangutan (*Pongo pygmaeus*) imitation. *Journal of Comparative Psychology*, **109**, 5–17.

Sabater Pi, J. (1977). Contribution to the study of alimentation of lowland gorillas in the natural state, in Rio Muni, Republic of Equatorial Guinea (West Africa). *Primates*, **18**, 183–204.

Schaller, G. B. (1963). *The Mountain Gorilla: Ecology and Behavior*. Chicago, IL: University of Chicago Press.

Schurmann, C. L. (1982). Courtship and mating behavior of wild orangutans in Sumatra. In *Primate Behavior and Sociobiology*, ed. A. B. Chiarelli & R. S. Corruccini, pp. 129–35. Berlin: Springer.

Short, R. V. (1981). Sexual selection in man and the great apes. In *Reproductive Biology of the Great Apes*, ed. C. E. Graham, pp. 319–42. New York: Academic Press.

Singleton, I. S. & van Schaik, C. P. (2001). Orangutan home range size and its determinants in a Sumatran swamp forest. *International Journal of Primatology*, **22**, 877–911.

Stanford, C. B. (1996). The hunting ecology of wild chimpanzees: implications for the evolutionary ecology of Pliocene hominids. *American Anthropologist*, **98**, 96–113.

(1998). *Chimpanzee and Red Colobus: The Ecology of Predator and Prey*. Cambridge, MA: Harvard University Press.

Stanford, C. B., Wallis, J., Mpongo, E. & Goodall, J. (1994). Patterns of predation by chimpanzees on red colobus monkeys in Gombe National Park, Tanzania, 1982–1991. *American Journal of Physical Anthropology*, **94**, 213–28.

Stevens, C. E. & Hume, I. D. (1995). *Comparative Physiology of the Vertebrate Digestive System*, 2nd edn. Cambridge, UK: Cambridge University Press.

Stewart, K. J. & Harcourt, A. H., 1987. Variation in female relationships. In *Primate Societies*, ed. B. B. Smuts, D. L. Cheney, R. M. Seyfarth, R. W. Wrangham & T. T. Struksaker, pp. 155–64. Chicago, IL: University of Chicago Press.

Sugardjito, J. (1983). Selecting nest-sites of Sumatran orang-utans, *Pongo pygmaeus abelii* in the Gunung Leuser National Park, Indonesia. *Primates*, **24**, 467–74.

Sugardjito, J., Boekhorst, I. J. A. te & van Hooff, J. A. R. A. M. (1987). Ecological constraints on the grouping of wild orangutans (*Pongo pygmaeus*) in the Gunung Leuser National Park, Sumatra, Indonesia. *International Journal of Primatology*, **8**, 17–41.

Sugiyama, Y. & Koman, J. (1979). Tool-using and making behavior in wild chimpanzees at Bossou, Guinea. *Primates*, **20**, 513–24.

Suzuki, A. (1988). The socio-ecological study of orangutans and the forest conditions after the big forest fires and drought, 1983, in Kutai National Park, Indonesia. In *A Research on the Process of Earlier Recovery of Tropical Rain Forest After a Large Scale Fire in Kalimantan Timur, Indonesia*, ed. H. Tagawa & N. Wirawan, pp. 117–36. Kagoshima: Kagoshima University.

Symington, M. M. (1990). Fission–fusion social organization in *Ateles* and *Pan*. *International Journal of Primatology*, **11**, 47–61.

Takahata, Y. (1985). Adult male chimpanzees kill and eat a male newborn infant: newly observed intragroup infanticide and cannibalism in Mahale National Park, Tanzania. *Folia Primatologica*, **44**, 121–228.

Terborgh, J. (1986). Community aspects of frugivory in tropical forests. In *Frugivores and Seed Dispersals*, ed. A. Estrada & T. H. Fleming, pp. 371–84. Dordrecht: Dr. W. Junk Publishers.

Terborgh, J. & Janson, C. (1986). The socioecology of primate groups. *Annual Review of Ecological Systems*, **17**, 111–35.

Thompson-Handler, N. (1990). The pygmy chimpanzee: sociosexual behavior, reproductive biology and life history patterns. Ph.D. dissertation, Yale University.

Thompson-Handler, N., Malenky, R. K. & Badrian, N. (1984). Sexual behavior of *Pan paniscus* under natural conditions in the Lomako Forest, Equateur, Zaire. In *The Pygmy Chimpanzee: Evolution, Biology, and Behavior*, ed. J. R. Napier & P. H. Napier, pp. 347–68. New York: Plenum Press.

Trivers, R. L. (1972). Paternal investment and sexual selection. In *Sexual Selection and the Descent of Man, 1871–1971*, ed. B. Campbell, pp. 136–79. Chicago, IL: Aldine.

Tutin, C. E. G. (1979). Mating patterns and reproductive strategies in a community of wild chimpanzees (*Pan troglodytes schweinfurthii*). *Behavioral Ecology and Sociobiology*, **6**, 29–38.

(1996). Ranging and social structure of lowland gorillas in the Lopé Reserve, Gabon. In *Great Ape Societies*, ed. W. C. McGrew, L. F. Marchant & T. Nishida, pp. 58–70. Cambridge, UK: Cambridge University Press.

Tutin, C. E. G. & Fernandez, M. (1984). Nationwide census of gorilla (*Gorilla g. gorilla*) and chimpanzees (*Pan t. troglodytes*) populations in Gabon. *American Journal of Primatology*, **6**, 313–36.

(1992). Insect-eating by sympatric lowland gorillas (*Gorilla g. gorilla*) and chimpanzees (*Pan t. troglodytes*) in the Lope Reserve, Gabon. *American Journal of Primatology*, **28**, 29–40.

(1993). Composition of the diet of chimpanzees and comparisons with that of sympatric lowland gorillas in the Lope Reserve, Gabon. *American Journal of Primatology*, **30**, 195–211.

Tutin, C. E. G. & McGinnis, P. R. (1981). Chimpanzee reproduction in the wild. In *Reproductive Biology of the Great Apes*, ed. C. Graham, pp. 239–64. New York: Academic Press.

Uehara, S. (1984). Sex differences in feeding on Camponotus ants among wild chimpanzees in the Mahale Mountains, Tanzania. *International Journal of Primatology*, **5**, 389.

Uehara, S., Nishida, T., Hamai, M., Hasegawa, T., Hayaki, H., Huffman, M. A., Kawanaka, K., Kobayashi, S., Mitani, J. C., Takahata, Y., Takasaki, H. & Tsukahara, T. (1992). Characteristics of predation by the chimpanzees in the Mahale Mountains national Park, Tanzania. In *Topics in Primatology. Vol. 1. Human Origins*, ed. T. Nishida, W. C. McGrew, P. Marler, M. Pickford, M. & F. B. M. de Waal, pp. 143–58. Tokyo: University of Tokyo Press.

Utami, S. S. & van Hooff, J. A. R. A. M. (1997). Meat-eating by adult female Sumatran orangutans (*Pongo pygmaeus abelii*). *American Journal of Primatology*, **43**, 159–65.

Utami, S. S., Wich, S. A., Sterck, E. H. M. & van Hooff, J. A. R. A. M. (1997). Food competition between wild orangutans in large fig trees. *International Journal of Primatology*, **18**, 909–27.

van Hooff, J. A. R. A. M. (1962). Facial expressions in higher primates. *Symposia of the Zoological Society of London*, **8**, 97–125.

(1969). The facial displays of the catarrhine monkeys and apes. In *Primate Ethology*, ed. D. Morris, pp. 9–88. Garden City, NY: Anchor Doubleday.

van Schaik, C. P. (1983). Why are diurnal primates living in groups? *Behaviour*, **87**, 120–44.

(1986). Phenological changes in a Sumatran rain forest. *Journal of Tropical Ecology*, **2**, 327–47.

(1999). The socioecology of fission–fusion sociality in orangutans. *Primates*, **40**, 60–86.

van Schaik, C. P. & van Hooff, J. A. R. A. M. (1996). Toward an understanding of the orangutan's social system. In *Great Ape Societies*, ed. W. C. McGrew, L. F. Marchant & T. Nishida, pp. 3–15. Cambridge, UK: Cambridge University Press.

van Schaik, C. P., Fox, E. A. & Sitompul, A. F. (1996). Manufacture and use of tools in wild Sumatran orangutans. *Naturwissenschaften*, **83**, 186–8.

van Schaik, C. P. van Deaner, R. O. & Merrill, M. Y. (1999). The conditions for tool use in primates: implications for the evolution of material culture. *Journal of Human Evolution*, **36**, 719–41.

Wallis, J. (1997). A survey of reproductive parameters in the free-ranging chimpanzees of Gombe National Park. *Journal of Reproduction Fertility*, **109**, 297–307.

Watts, D. P. (1984). Composition and variability of mountain gorilla diets in the Central Virungas. *American Journal of Primatology*, **7**, 323–56.

(1989). Infanticide in mountain gorillas: new cases and a reconsideration of the evidence. *Ethology*, **81**, 1–18.

(1991). Mountain gorilla reproduction and sexual behavior. *American Journal of Primatology*, **24**, 211–25.

(1996). Comparative socio-ecology of gorillas. In *Great Ape Societies*, ed. W. C. McGrew, L. F. Marchant & T. Nishida, pp. 16–28. Cambridge, UK: Cambridge University Press.

(1998). Long-term habitat use by mountain gorillas (*Gorilla gorilla beringei*). I. Consistency, variation, and home range size and stability. *International Journal of Primatology*, **19**, 651–80.

Weber, A. W. & Vedder, A. (1983). Population dynamics of the Virunga gorillas: 1959–1978. *Biological Conservation*, **26**, 341–66.

White, F. J. (1988). Party composition and dynamics in *Pan paniscus*. *International Journal of Primatology*, **9**, 179–93.

(1992). Pygmy chimpanzee social organization: variation with party size and between study sites. *American Journal of Primatology*, **26**, 203–14.

(1994). Food sharing in wild pygmy chimpanzees, *Pan paniscus*. In *Current Primatology. Vol. 2. Social Development, Learning and Behavior*, ed. J. J. Roeder, B. Thierry, J. Anderson & N. Herrenschmidt, pp. 143–58. Strasbourg: Université Louis Pasteur.

(1996). Comparative socio-ecology of *Pan paniscus*. In *Great Ape Societies*, ed. W. C. McGrew, L. F. Marchant & T. Nishida, pp. 29–41. Cambridge, UK: Cambridge University Press.

White, F. J. & Wrangham, R. W. (1988). Feeding competition and patch size in the chimpanzee species *Pan paniscus* and *Pan troglodytes*. *Behaviour*, **105**, 148–63.

Whiten, A., Goodall, J., McGrew, W. C., Nishida, T., Reynolds, V., Sugiyama, Y., Tutin, C. E. G., Wrangham, R. W. & Boesch, C. (1999). Cultures in chimpanzees. *Nature*, **399**, 682–5.

Wich, S. A., Sterck, E. H. M. & Utami, S. S. (1999). Are orangutan females as solitary as chimpanzee females? *Folia Primatologica*, **70**, 23–8.

Williamson, E. A., Tutin, C. E. G., Rogers, M. E. & Fernandez, M. (1990). Composition of the diet of lowland gorillas at Lopé in Gabon. *American Journal of Primatology*, **21**, 265–77.

Wood, R. J. (1984). Spontaneous use of sticks by gorillas at Howletts Zoo Park, England. *International Zoo Yearbook*, **31**, 13–8.

Wrangham, R. W. (1977). Feeding behaviour of chimpanzees in Gombe National Park, Tanzania. In *Primate Ecology*, ed. T. H. Clutton-Brock, pp. 503–37. New York: Academic Press York.

(1979a). Sex differences in chimpanzee dispersion. In *The Great Apes*, ed. D. A. Hamburg & E. R. McCown, pp. 480–9. Menlo Park, CA: Benjamin/Cummings.

(1979b). On the evolution of ape social systems. *Social Sciences Information*, **18**, 334–68.

(1980). An ecological model of female-bonded primate groups. *Behaviour*, **75**, 262–300.

(1986). Ecology and social relationships in two species of chimpanzees. In *Ecological Aspects of Social Evolution: Birds and Mammals*, ed. D. I. Rubenstein & R. W. Wrangham, pp. 352–78. Princeton, NJ: Princeton University Press.

(1987). Evolution and social structure. In *Primate Societies*, ed. B.B. Smuts, D. L. Cheney, R. M. Seyfarth, R. W. Wrangham & T. T. Struhsaker, pp. 282–96. Chicago, IL: University of Chicago Press.

Wrangham, R. W. & Smuts, B. B. (1980). Sex differences in behavioral ecology of chimpanzees in Gombe National Park, Tanzania. *Journal of Reproduction and Fertility (Suppl.)*, **28**, 13–31.

Wrangham, R. W. & Bergmann-Riss, E. van Zinnicq (1990). Rates of predation on mammals by Gombe chimpanzees, 1972–1975. *Primates*, **31**, 157–70.

Wrangham, R. W., Chapman, C. A., Clark-Arcadi, A. P. & Isabyrye-Basuta, G. (1996). Social ecology of Kanyawara chimpanzees: implications for understanding the costs of great ape groups. In *Great Ape Societies*, ed. W. C. McGrew, L. F. Marchant & T. Nishida, pp. 45–57. Cambridge, UK: Cambridge University Press.

Wrangham, R. W., Clark, A. P. & Isabirye-Basuta, G. (1992). Female social relationships and social organization of Kibale Forest chimpanzees. In *Topics in Primatology. Vol. 1. Human Origin*, ed. T. Nishida, W. C. McGrew, P. Marler, M. Pickford & F. B. M. de Waal, pp. 81–98. Tokyo: University of Tokyo Press.

Yamagiwa, J. (1983). Diachronic changes in two eastern lowland gorilla groups (*Gorilla gorilla graueri*) in the Mt. Kahuzi region, Zaire. *Primates*, **24**, 174–83.

Yamagiwa, J. (1986). Activity rhythm and the ranging of a solitary male mountain gorilla (*Gorilla gorilla beringei*). *Primates*, **27**, 273–82.

(1987a). Male life history and social structure of wild mountain gorillas (*Gorilla gorilla beringei*). In *Evolution and Coadaptation in Biotic Communities*, ed. S. Kawano, J. H. Connell & T. Hidaka, pp. 31–51. Tokyo: University of Tokyo Press.

(1987b). Intra- and inter-group interactions of an all-male group of Virunga mountain gorillas (*Gorilla gorilla beringei*). *Primates*, **28**, 1–30.

(1992). Functional analysis of social staring behavior in an all-male group of mountain gorillas. *Primates*, **33**, 523–44.

(1999). Socioecological factors influencing population structure of gorillas and chimpanzees. *Primates*, **40**, 87–104.

(2001). Factors influencing the formation of ground nests by eastern lowland gorillas in Kahuzi-Biega National Park: some evolutionary implications of nesting behavior. *Journal of Human Evolution*, **40**, 99–109.

Yamagiwa, J. & Kahekwa, J. (2001). Dispersal patterns, group structure and reproductive parameters of eastern lowland gorillas at Kahuzi in the absence of infanticide. In *Mountain Gorilla: Three Decades of Research at Karisoke*, ed. M. M. Robbins, P. Sicotte & K. J. Stewart, pp. 90–122. New York: Cambridge University Press.

Yamagiwa, J. & Mwanza, N. (1994). Day-journey length and daily diet of solitary male gorillas in lowland and highland habitats. *International Journal of Primatology*, **15**, 207–24.

Yamagiwa, J., Kaleme, K., Milynganyo, M. & Basabose, K. (1996b). Food density and ranging patterns of gorillas and chimpanzees in the Kahuzi-Biega National Park, Zaire. *Tropics*, **6**, 65–77.

Yamagiwa, J., Maruhashi, T., Yumoto, T. & Mwanza, N. (1996a). Dietary and ranging overlap in sympatric gorillas and chimpanzees in Kahuzi-Biega National Park, Zaire. In *Great Ape Societies*, ed. W. C. McGrew, L. F. Marchant & T. Nishida, pp. 82–98. Cambridge, UK: Cambridge University Press.

Yamagiwa, J., Mwanza, N., Spangenberg, A., Maruhashi, T., Yumoto, T., Fischer, A. & Steinhauer, B. B. (1993). A census of the eastern lowland gorillas *Gorilla gorilla graueri* in Kahuzi-Biega National Park with reference to mountain gorillas *G. g. beringei* in the Virunga region, Zaire. *Biological Conservation*, **64**, 83–9.

Yamagiwa, J., Mwanza, N., Yumoto, T. & Maruhashi, T. (1991). Ant eating by eastern lowland gorillas. *Primates*, **32**, 247–53.

(1994). Seasonal change in the composition of the diet of eastern lowland gorillas. *Primates*, **35**, 1–14.

Yamakoshi, G. (1998). Dietary responses to fruit scarcity of wild chimpanzees at Bossou, Guinea: possible implications for ecological importance of tool use. *American Journal of Physical Anthropology*, **106**, 283–95.

Yamakoshi, G. & Sugiyama, Y. (1995). Pestle-pounding behavior of wild chimpanzees at Bossou, Guinea: a newly observed tool-using behavior. *Primates*, **36**, 489–500.

Part III
Fossil great ape adaptations

DAVID R. BEGUN

Department of Anthropology, University of Toronto

INTRODUCTION

In this part the contributors explore a variety of attributes of the paleobiology of fossil hominoids that may contribute to an understanding of the evolution of great ape intelligence. Our request to these experts, in each topic that is the focus of their contributions, was to focus attention on the way in which a single aspect of paleobiology may inform this issue. This is a two-edged sword. Many authors, most of whom had thought relatively little about the broader question of great ape intelligence (myself included), found new insights from the data they have long been contemplating from other perspectives. On the other hand, since we asked contributors to restrict themselves to a focused topic, many readers may feel that each contributor thinks their area of interest is the one most likely to "explain" great ape cognitive evolution. Nothing could be farther from the truth. All the contributors to this part of the volume recognize that their topic, whether it is environment and ecology, diet, locomotion, size, or life history, is one piece of the puzzle. When authors in this part of the book reach conclusions on the relevance of a specific aspect of fossil ape paleobiology to understanding great ape cognition, it is because the editors pushed them to do it, and all know that each attribute is but a facet of a very complex problem. In the end, all the contributors produced interesting new insights on the way in which the evolution of specific aspects of the biology of great apes could have contributed to the development of the great ape grade of cognition, rather than advocating on behalf of a single cause.

In Chapter 13, Potts provides an exhaustive summary of the ecological setting and dynamics of hominid evolution. Potts stresses the change from relatively predictable ecological conditions characteristic of early Miocene hominoid environments to the more variable and eventually extremely variable ecological conditions of middle and late Miocene and Pliocene hominids.

The latter are more likely to demand greater cognitive flexibility and complexity. He shows that ecological conditions and their predictability have changed sufficiently in the last 15 million years to account for at least one aspect of selection for increased intelligence in all hominids, variability selection, especially in the ancestors of humans.

Begun and Kordos, Chapter 14, summarize the cranial evidence for brain size in Miocene hominoids. From the relatively small brains of primitive catarrhines such as *Aegyptopithecus*, Begun and Kordos trace evidence of brain size increase through *Proconsul* to late Miocene hominids. The pattern that emerges is more complex and somewhat different from previous analyses. For example, *Proconsul* is argued to have had a papionin-sized brain rather than a hominoid-sized brain, modern hominid-sized brains appear in the late middle Miocene and do not change appreciably until the appearance of *Homo*, and brains became smaller over time in some hominoids (*Oreopithecus*, perhaps *Hylobates*). Relative brain mass is difficult to interpret in hominoids but appears to track changes in diet and life history more closely than other variables.

Kelley, Chapter 15, takes on the difficult task of relating life history variables to body mass, brain size, and intelligence in hominoid evolutionary history. He shows that changes in the duration and rate of several fundamental brain growth processes, themselves predictable products of altering life history along the fast–slow continuum, can have dramatic effects on brain size and complexity. Accordingly, he argues that such life history changes can account for the increases in brain size and complexity that enhance cognition, although parallel pressures on cognition are necessary to shape the form of cognitive enhancements. The timing and duration of one life history event that can be reconstructed from fossil evidence, M1 emergence, shows that a hominoid-like pattern of growth was already present in the early Miocene, leading to the conclusion that early Miocene hominoids had hominid-sized brains (note this is in some contrast to the conclusions of Begun and Kordos in Chapter 14, illustrating one of the many difficulties of interpreting fossil evidence).

Singleton, Chapter 16, reviews evidence of diet and foraging strategies in a number of fossil apes and finds evidence for increases in dietary challenges to cognition through time. Increasingly hominoids moved from relatively generalized, year round frugivory (with a few exceptions) to greater seasonal reliance on other foods, some probably embedded, which represent greater challenges to find and process. A few specialized early Miocene hominoids (*Afropithecus*) may have accomplished this with anatomical specializations, whereas late Miocene hominids may have relied more on cognitively mediated solutions.

Gebo, Chapter 17, explores the complex issue of reconstructing fossil hominoid positional behavior in light of a number of models of the evolution of great ape intelligence. He finds evidence, like Kelley, of relatively modern hominoid-like features (body form) in the early Miocene, again somewhat contrasting the views of Singleton, Chapter 16, and Begun and Kordos, Chapter 14, both of whom see greater discontinuity between early and late Miocene hominoids. Gebo concludes that terrestriality, to at least some degree, is more likely than arboreal clambering to have represented a challenge to great ape/human ancestors, to which enhanced cognition may have been one response.

Ward *et al.*, Chapter 18, focus their attention on body mass evolution in hominoids and its relationship to cognitive evolution. Many contributors discuss body mass in some detail, particularly Gebo, Chapter 17, and Begun and Kordos, Chapter 14, because it is essential to understanding nearly every aspect of a species' biology. Ward *et al.* identify increases in body mass in hominids as leading to ecological dominance and intra-specific arms races in cognitive abilities. Ecological dominance produces a number of changes that increase the overall complexity of both the social and ecological environment and may result, under the right circumstances, in an arms race within species for increasingly higher levels of intelligence in both of these domains.

The analysis of the paleobiology of fossil taxa is especially difficult when the taxa are poorly known anatomically, and this is the case for almost all fossil hominoids. Some aspects of paleobiology, such as positional behavior, diet and broad ecological preferences, are relatively straightforward, though not without difficulty. Others, such as brain and body mass, life history, and especially social organization and intelligence require so much anatomical and behavioral data that they are extremely difficult to reconstruct from fossil evidence. To paraphrase Darwin, we have only a few letters from some words from scattered pages of the book of hominoid evolution with which to reconstruct the entire text.

All of the authors of this part recognize the relatively low degree of certainty in their conclusions of behavior in extinct hominoids. As a consequence all attempt to bring as many aspects of the biology of extinct hominoids as possible to bear on the problem, accounting for the overlap in many chapters. The uncertainty caveat notwithstanding, the fossil record provides broad guidance for navigating hominoid evolutionary history. In many cases analysis of the fossil evidence cannot yield strong and confidence-inspiring bases for hypotheses of cognitive evolution in great apes. But all hypotheses of cognitive evolution, regardless of their origin, lead to predictions that can be tested by fossil evidence. If they are to remain viable hypotheses they must be consistent with the fossil record. We see several examples in this section where the fossil evidence tends to falsify existing hypotheses of cognitive evolution. This may be the most important contribution to a greater understanding of the evolution of intelligence in the great apes that we should expect from the fossil evidence.

13 · Paleoenvironments and the evolution of adaptability in great apes

RICHARD POTTS

Human Origins Program, National Museum of Natural History
Smithsonian Institution, Washington, DC

ENVIRONMENTS OF NATURAL SELECTION

Understanding the evolution of great ape cognition depends on identifying past adaptive settings and the factors that influenced early ape cognitive responses. Reconstructing past environments is not sufficient for developing and testing evolutionary arguments. It is the sequence of *selective environments* in which ancestral apes lived that is essential to determining how great ape mental abilities evolved. This requires us to assess the ways in which environmental settings (i.e., specific habitat reconstructions), trends (e.g., cooling, drying), and variability (e.g., seasonality and long-term oscillation) affected the resources and survival regimes of ancestral great apes – and thus posed adaptive problems in the places where they lived and the time periods when they evolved. This chapter investigates the evolutionary adaptability of great apes in light of local, regional, and global paleoenvironments; the geographic patterns of ape evolution; and the cognitive, social, environmental, and dietary characteristics of living great apes.

PRIOR HYPOTHESES OF GREAT APE COGNITIVE EVOLUTION

Although attempts to define unique aspects of great ape cognition have generated much debate (e.g., Tomasello & Call 1997), great apes appear to have achieved levels of cognitive sophistication and flexibility unknown in other nonhuman primates. Relative to cercopithecoid monkeys, unique achievements of great ape mentality are thought to include: self-recognition, some comprehension of others' mental states, intentional deception, causal and logical reasoning, planning, imitation, demonstration teaching, and the potential for using tools and symbols (Byrne 1995, 1997; Delgado &

van Schaik 2000; Parker 1996; Parker, Mitchell & Boccia 1994; see Byrne, Chapter 3, Parker, Chapter 4, Russon, Chapter 6, van Schaik *et al.*, Chapter 11, in this volume). To some, self-recognition and imitation suggest that great apes are cognitively capable of ascribing attributes (e.g., mental states, intentions) to other individuals (Byrne 1997; Frith & Frith 1999). Their accomplishments in social and technical problem solving further indicate that they comprehend cause–effect relations and depend on abstract problem representation (Byrne 1997). This list of mental functions ascribed to great apes serves as a starting point in determining what a coherent and sound hypothesis of great ape cognitive evolution needs to explain.

A variety of selective factors have been invoked in previous explanations of great ape cognitive evolution. A partial list includes arboreal travel of a large-bodied hominoid (Povinelli & Cant 1995), slow life history (Kelley 1997), extractive foraging (Parker & Gibson 1977), and processing technically difficult foods (Byrne 1997). Two other factors considered in general explanations of higher primate cognitive evolution are the complexity of social living (Byrne & Whiten 1988; Dunbar 1992, 1995) and temporo-spatially complex foraging (Garber 1989; Milton 1981).

Most hypotheses of great ape cognitive evolution give little consideration to the environmental conditions of great ape ancestry. Byrne's (1997) technical foraging hypothesis, for example, is based entirely on observations of living great apes and seeks to explain the unique dimensions of great ape mental functions in terms of observable aspects of their food acquisition. Although Povinelli and Cant's (1995) arboreal travel hypothesis and Parker's (1996) elaboration of the extractive foraging idea emphasize phylogenetic history, both hypotheses are largely devised to explain experimental findings and field observations of extant great apes. The

---

environments of ape ancestry, however, add to the suite of factors that may have affected great ape cognitive evolution.

Most hypotheses about the key ancestral mental functions, for example mental representation or self-conception, are essentially untestable because the fossil record is silent about when, where, and why they emerged. Virtually all hypotheses, however, stress the significance of *foraging* success. Foraging success is, in part, a function of an organism's response to the distribution and reliability of food sources in time and space. The abundance, patchiness, and availability of all food sources are sensitive to environmental variables. Information about environments encountered during great ape evolutionary history affords one of the very few means of testing the hypothetical foraging factors that shaped great ape cognitive evolution. The paleoenvironmental record also offers the advantage of a data set independent of the organisms themselves but that directly relates to the settings in which great apes evolved.

FOSSIL GREAT APES

Cladistic analysis of fossil ape taxa shows that homoplasy (parallel evolution) in the cranium and postcranium was abundant (Begun, Ward & Rose 1997b; C. V. Ward 1997; S. Ward 1997). This implies that Miocene populations were subject to multiple periods of interregional migration, vicariance, and independent adaptive evolution, all probably linked to an intricate environmental history. In order to study great ape adaptive history relative to cognitive evolution, it is necessary to determine the time, place, and environments in which great apes originated and diversified.

Early apes such as *Proconsul* lived in Africa prior to 18 Ma, but most researchers consider the morphology of these early Miocene apes too primitive to justify their inclusion as great apes – hominids *sensu lato*. To most Miocene ape specialists, the earliest definite fossil evidence of Hominidae is 17 to 12 Ma (Andrews *et al.* 1996; Begun *et al.* 1997b; but see Gebo, Chapter 17, this volume; MacLatchy *et al.* 2000; Pilbeam 1997). Most systematists working on Miocene apes place the better-known large-bodied hominoids that arose and lived after *Proconsul*, 12.5 to 7 Ma – *Dryopithecus*, *Sivapithecus*, and *Oreopithecus* – in Hominidae (see Begun, Ward & Rose 1997a). Some also include, but others dispute,

Afropithecus (18–17 Ma; East Africa and Arabian Peninsula), *Kenyapithecus* (15.5 to 14 Ma; East Africa), and *Griphopithecus* (16.5–14 Ma; the oldest known large ape of Eurasia) (Andrews *et al.* 1996; McCrossin & Benefit 1997).

Despite these taxonomic issues, a list of the known genera and species of Miocene large-bodied hominoids (Table 13.1) indicates that a diversity of great apes existed during the middle and late Miocene; following an initial radiation of ape-like catarrhines and stem (archaic) hominoids during the early Miocene, the main diversification of early great apes occurred between 12 and 9.5 Ma, and a drop in diversity followed.

ENVIRONMENTS AND BIOGEOGRAPHY OF MIOCENE APES

Local environments of Miocene apes

Table 13.2 summarizes paleoenvironmental interpretations of a sample of Miocene sites that preserve fossil great apes. One commonality in all is the presence of trees, ranging from relatively closed-canopy, subtropical and tropical forest to open and even dry woodland. Another commonality is fluctuation, some evidence of at least low-level climatic oscillation, from seasonal to longer-term cycles.

The radiation of late Miocene *Dryopithecus* is recorded primarily in seasonal, subtropical forest. Habitats associated with the last appearance of Miocene hominids in Eurasia involved moist, closed forests and swamps (e.g., *Oreopithecus* in southern Europe; *Lufengpithecus* in southeastern Asia), and grassy woodland associated with increasing monsoonal seasonality (e.g., *Sivapithecus* in South Asia). Reconstructions of middle to late Miocene ape habitats in Africa indicate, by contrast, overall drier, more open conditions – typically though not always a mosaic of seasonally dry forest and open woodland with grassy patches.

The Fort Ternan *Kenyapithecus* locality presents two habitat signals, a more closed woodland and a more open setting, suggestive of a savanna–woodland ecotone (Cerling *et al.* 1991, 1997b; Kappelman 1991; Pickford 1985, 1987; Retallack 1992; Retallack, Dugas & Bestland 1990; Shipman 1986; Shipman *et al.* 1981). Different authors stress different ends of the savanna–woodland spectrum based on analysis of vertebrates, invertebrates, and geochemistry. My own assessment agrees with Andrews, Begun & Zylstra (1997) and Shipman's

Table 13.1. *Current data on Miocene large-bodied apes (c. ≥10 kg): species, geographic distribution (representative sites), and time range (Early, Middle, Late Miocene; approximate ages)*

Species	Geographic distribution	Known time range (million years ago)
Africa and Arabia		
Morotopithecus bishopi[1]	East Africa (Moroto)	E Miocene (21.5–22 Ma) or M Miocene (15–17 Ma)
Proconsul major	East Africa (Songhor, Koru)	E Miocene (20–19 Ma)
Proconsul africanus	East Africa (Songhor, Koru)	E Miocene (20–19 Ma)
Proconsul heseloni	East Africa (Rusinga)	E Miocene (18 Ma)
Proconsul nyanzae	East Africa (Rusinga)	E Miocene (18 Ma)
Rangwapithecus gordoni	East Africa (Songhor)	E Miocene (20–19 Ma)
Ugandapithecus sp.	East Africa (Napak, Moroto)	E Miocene (20–19 Ma)
Afropithecus turkanensis	East Africa (Kalodirr, Buluk)	E Miocene (18–17 Ma)
Afropithecus leakeyi	Saudi Arabia (Ad Dabtiyah)	E Miocene (17 Ma)
Turkanapithecus kalakolensis	East Africa (Kalodirr)	E Miocene (18–17 Ma)
Nyanzapithecus vancouveringi	East Africa (Songhor, Rusinga)	E Miocene (20–18 Ma)
Kenyapithecus wickeri[1]	East Africa (Fort Ternan)	M Miocene (14 Ma)
Equatorius africanus[1] (also known as *K. africanus*)	East Africa (Tugen Hills, Maboko)	M Miocene (15.5–15 Ma)
Mabokopithecus clarki	East Africa (Maboko)	M Miocene (15 Ma)
Mabokopithecus pickfordi	East Africa (Maboko)	M Miocene (15 Ma)
Nacholapithecus sp.[1]	East Africa (Nachola)	M Miocene (15 Ma)
Otavipithecus namibiensis[1]	Southern Africa (Berg Aukas)	M Miocene (13 Ma)
Samburupithecus kiptalami[1]	East Africa (Samburu Hills)	L Miocene (9.5 Ma)
Eurasia		
cf. *Griphopithecus* sp.[1]	Central Europe (Engelswies)	M Miocene (16.5 Ma)
Griphopithecus alpani[1]	Western Asia (Paşalar, Çandır)	M Miocene (16.5–16 Ma)
Griphopithecus darwini[1]	C Europe (Neudorf-Sandberg)	M Miocene (15–14 Ma)
Ankarapithecus meteai[1]	Western Asia (Sinap)	M Miocene (10 Ma)
Dryopithecus fontani[2]	W Europe (St. Gaudens)	M–L Miocene (12–11 Ma)
Dryopithecus crusafonti[2]	W Europe (Can Ponsic)	L Miocene (10–9.5 Ma)
Dryopithecus brancoi[2]	Central Europe (Rudabánya)	L Miocene (10–9.5 Ma)
Dryopithecus laietanus[2]	W Europe (Can Llobateres)	L Miocene (9.5–9 Ma)
Sivapithecus sivalensis[2]	S Asia (Siwaliks)	M/L Miocene (12.7–8 Ma)
Sivapithecus indicus[2]	S Asia (Siwaliks)	L Miocene (?–7 Ma)
Sivapithecus parvada[2]	S Asia (Siwaliks)	L Miocene (10 Ma)
Gigantopithecus giganteus[2]	S Asia (Siwaliks)	L Miocene
Ouranopithecus macedoniensis[2] (= *Graecopithecus*)	S Europe (Ravin de la Pluie, Xirochori, Nikiti 1, Macedonia)	L Miocene (10–9.5 Ma)
Udabnopithecus garedziensis[1]	SE Europe (Udabno, Georgia)	L Miocene (9 Ma)
Oreopithecus bambolii[2]	S Europe (Baccinello, M. Bamboli)	L Miocene (8–7 Ma)
Lufengpithecus lufengensis[2]	SE Asia (Lufeng – S. China)	L Miocene (8–7 Ma)

Notes:

[1] Candidates.

[2] Strong candidates for inclusion in Hominidae (great apes).

Table 13.2. *Sample of Middle and Late Miocene hominoid sites and paleoenvironmental interpretation of each site.* (*Age range and type of paleoenvironmental evidence are also given*)

Site	Age (Ma)	Taxon	Paleoenvironmental interpretations	Source of interpretation	References
Eurasian sites					
Lufeng (Southeastern Asia)	8.0–7.0	*Lufengpithecus*	Moist tropical–subtropical forest with swamps and open patches	Fossil plants; sediments (lignites); fauna	Andrews *et al.* (1997) Badgeley *et al.* (1988) Begun & Kordos (1997)
Baccinello (Southern Europe)	8.5–7.0	*Oreopithecus*	Closed subtropical forest to warm temperate woodland with seasonal fluctuations; swamps/woodlands	Sediments (lignites); fossil pollen and animals	Andrews *et al.* (1997) Harrison & Harrison (1989) Harrison & Rook (1997)
Ravin de Pluie (Southern Europe)	9.0–10.0	*Ouranopithecus*	Seasonal forest to open setting	Mammalian fauna	Andrews *et al.* (1997) de Bonis & Koufos (1997)
Rudabánya (Central Europe)	10.0–9.5	*Dryopithecus*	Seasonal subtropical forest with swamp vegetation	Sediments, fossil mammals and plants	Andrews *et al.* (1997) and references therein
Can Ponsic (Western Europe)	10.0–9.5	*Dryopithecus*	Seasonal subtropical forest to tropical woodland	Ecological diversity spectra of mammalian fauna	Andrews *et al.* (1997)
Can Llobateres (Western Europe)	10.0–9.5	*Dryopithecus*	Seasonal subtropical forest to tropical woodland	Ecological diversity spectra of mammalian fauna	Andrews *et al.* (1997)
Paşalar (Western Asia)	16.0	*Griphopithecus*	Seasonal tropical to subtropical woodland and/or forest	Paleosols, stable isotopes of teeth and soils, fossil animals	Andrews *et al.* (1997) Bestland (1990)
Siwaliks (South Asia)	12.7–7.0	*Sivapithecus*	Warm forests and woodlands; change to more grassy (C_4-plant) wooded setting after 10 Ma; seasonality with monsoon development by 7 Ma	Sediments, stable C and O isotopes of soils and mammalian teeth, faunal analysis, dental microwear	e.g., Cering *et al.* (1993) Kappelman (1988) Morgan *et al.* (1994) Quade *et al.* (1989)

Site	Age (Ma)	Taxa	Environment	Evidence	References
African sites					
Tugen Hills, Lukeino Fm (East Africa)	6.2–5.6	*Orrorin*; Homininae indet.	Woodland to grassy woodland mosaic; lake margin setting	Sediments and fauna	Hill (1995) Hill *et al.* (1985) Kingston (1992) Pickford & Senut (2001)
Samburu Hills, Namurungule Fm (East Africa)	9.5	*Samburupithecus*	Woodland and open vegetation around a lake setting	Sediments and fauna (e.g., giraffids, bovids, equids)	Nakaya (1994) Nakaya *et al.* (1984)
Lothagam, Lower Apak Mbr (East Africa)		Homininae indet.	Grassy woodland mosaic; riverine forest along large river; seasonal drying and periodic soil formation	Fauna, stable isotopes on soils and teeth	Leakey *et al.* (1996)
Tugen Hills, Ngorora Fm (East Africa)	12.5	Hominoidea indet.	Fluctuating lacustrine period after a period of lowland rain forest	Sediments (alternating shales and laminated clays); fossil plants	Bishop & Pickford (1975) Hill (1995) Jacobs & Kabuye (1987)
Fort Ternan (East Africa)	14.0	*Kenyapithecus*	Seasonal closed woodland with forest and grassy patches	Fauna, stable isotopes on soils and teeth	Andrews *et al.* (1997) Cerling *et al.* (1992) Kappelman (1991) Shipman (1986)
Samburu Hulls, Aka Aiteputh Fm (East Africa)	15.0	*Kenyapithecus* (*Equatorius*)	Woodland associated with lake	Sediments and fauna	Nakaya (1994)
Maboko (East Africa)	15.0	*Kenyapithecus* (*Equatorius*)	Seasonally dry woodland	Ecological diversity spectra of mammalian fauna	Andrews *et al.* (1997)
Tugen Hills, Muruyur Fm (East Africa)	15.5–15.0	*Equatorius*	Widespread tropical forest with open habitat patches	Mammalian fauna	Hill (1995) Ward *et al.* (1999)

original interpretation, that a range of habitats is recorded in the fauna; seasonal closed woodland was dominant close to the site, while more open woodland and grassland patches occurred in close temporal and spatial proximity as the fossil bone assemblage was formed. Furthermore, isotopic evidence is clear about the overwhelming dominance of C_3 plants; while this suggests wooded or forested conditions, it also likely includes the presence of C_3 grasses.

In contrast with this East African site are two of the youngest Miocene hominid fossil localities, both in Eurasia – Lufeng (South China) and the *Oreopithecus* complex of sites in southern Tuscany (Italy). Both localities appear to represent insular areas of moist subtropical forest. At Lufeng, faunal remains associated with *Lufengpithecus* largely consist of forest and aquatic taxa. Pollen data indicate variation over time, yet arboreal pollen remains at a level of 60% to 90% throughout the strata in which hominid fossils are known (Badgley *et al.* 1988; Sun & Wu 1980). Based largely on the lignites, *Lufengpithecus* occurred in forested, freshwater swamps with forest vegetation on the immediately surrounding hillsides. The pollen flora is indicative of a moist tropical forest, open areas nearby, and moderate rainfall and humidity (Badgley *et al.* 1988).

The Baccinello and Monte Bamboli sites in southern Tuscany are comparable in age (late Miocene, *c.* 8 to 7 Ma) to Lufeng, and represent a southern refugium of the last recorded fossil great ape in Europe, *Oreopithecus bambolii*. At Baccinello, the lignites indicate a swampy setting associated with humid forest. Fossil pollen correlated with *Oreopithecus*-bearing sediments indicates mixed lowland mesophytic forest consisting of broad-leaved and coniferous species and a rich understory of bushes, small trees, and ferns. Overall, the landscape was largely forested with freshwater pools, swamps, and shallow lakes (Harrison & Harrison 1989; Harrison & Rook 1997). The bovids and rodents exhibit moderate hypsodonty (high-crowned dentition) consistent with a preference for drier, more open habitats, which may have occurred nearby at least periodically (Harrison & Rook 1997). The stratigraphic units above the *Oreopithecus*-bearing levels in the Baccinello basin indicate a change from warm/humid conditions to an inconsistent regime of irregularly alternating arid and moist phases (Benvenuti, Bertini & Rook 1995). This transition marks the final record of hominid apes in western Eurasia, *c.* 7 Ma, until the arrival of early humans

during the Pleistocene. The latest known Miocene ape sites in Eurasia thus occur in southern regions of the two continents and are characterized by lignites and pollen indicative of closed forest and swamp conditions. A similar setting is reconstructed for late Miocene *Dryopithecus* at Rudabánya, Hungary, and for extant *Pongo* in Southeast Asia.

Not all late Miocene ape localities of Eurasia were so moist and densely forested. The habitat of *Ouranopithecus* at Ravin de la Pluie (10 to 9 Ma), for example, is reconstructed by de Bonis and Koufos (1997) as "savanna-like," an open environment with relatively few trees. According to Andrews *et al.* (1997), however, none of the Ravin de la Pluie fossil faunas in de Bonis and Koufos' analysis corresponds to modern savanna or forest faunas; rather, the fossil data indicate little more than a seasonal climate and a vegetation that could range from seasonal forests to more open environment.

More promising evidence of a relatively open setting inhabited by a late Miocene ape may come from Çorakyerler in Central Anatolia, Turkey, tentatively dated 7.5 to 7.0 Ma (Sevim *et al.* 2001; Sevim & Begun pers. commun.). The rich fauna is dominated by grazing ungulates, such as *Gazella*. Initial finds from this site suggest that at least one lineage of late Miocene apes occupied a more open setting than did either *Oreopithecus* or *Lufengpithecus*.

Few Miocene ape localities provide long-term sequences of environmental change spanning several million years. The main ones that do are the Tugen Hills sequence in central Kenya and the Siwaliks sequence of Pakistan and India. The Siwaliks of Pakistan have been the target of long-term paleoenvironmental and fossil study (e.g., Badgley & Behrensmeyer 1980; Barry *et al.* 1985; Kappelman 1988; Morgan, Kingston & Marino 1994; Pilbeam *et al.* 1977; Quade *et al.* 1989; Raza *et al.* 1983; S. Ward 1997). Seasonality was strong throughout the 5 million years (*c.* 12.75 to 8.0 or 7.0 Ma) in which the fossil ape *Sivapithecus* is recorded in the Siwaliks. The overall environmental sequence of this hominid is considered to be forest and woodland, usually closed canopy woodland or forest with tropical to subtropical climate (Badgley & Behrensmeyer 1980; Kappelman 1988; Quade *et al.* 1989; S. Ward 1997). Ecological analysis of the fauna recovered from the long Siwalik sequence has suggested seasonal woodland to tropical deciduous forests (Andrews 1983), although Andrews *et al.* (1997) questioned the validity of interpreting such a mixed, time-averaged assemblage. One study has, however, addressed the habitat of *Sivapithecus* in a constrained time interval (locality Y311, Nagri Fm, northern Pakistan) (Scott, Kappelman & Kelley 1999). It showed that *S. parvada* occupied a continuous canopy forest approximately 10 Ma.

Sivapithecus disappears from the fossil record around 7.4 to 7.0 Ma, at the time of a major change in the plant community to C_4 vegetation – consistent with the transition from warm, humid forests and woodlands to drier, more open grasslands. This shift corresponded with intensification of the Asian monsoon (enhanced seasonality) and/or a significant drop in atmospheric CO_2 concentration (Cerling *et al.* 1997a; Quade *et al.* 1989). Some temporal and spatial variability characterized the environmental sequence in which *Sivapithecus* lived, although the details are not well documented (Morgan *et al.* 1994; Quade *et al.* 1995; S. Ward 1997).

In the Tugen Hills sequence, Miocene hominoids are recorded in the Muruyur and Ngorora Formations, between about 15.5 and 10 Ma, and in the Lukeino Formation, between 6.2 and 5.6 Ma (Hill 1995). The only definite records of large-bodied apes, however, are *Equatorius africanus* (originally *Kenyapithecus*), known between 15.5 and 15.0 Ma (Hill 1995; Ward *et al.* 1999) and the recently discovered hominin *Orrorin tugenensis* in the Lukeino Formation (Pickford & Senut 2001; Senut *et al.* 2001). Thus, despite the presence of fossiliferous sediments, a gap in the ape fossil record of about 9 million years occurs in the Tugen Hills. This gap appears to be consistent with the general dearth of hominid fossils from all of Africa between about 14 and 6 Ma (Begun 2001). *E. africanus* occurs in the Kipsaramon site complex, Muruyur Formation, which represented a widespread tropical forest, based on the presence of scaly-tailed flying squirrel, along with more open patches of vegetation, based on the presence of springhare (Hill 1995). A macrofossil plant locality dated 12.6 Ma in the Ngorora Formation preserves an extraordinary fossil leaf assemblage indicative of lowland rain forest with West African affinities (Jacobs & Kabuye 1987). Although there is no definite evidence of great ape fossils between the Muruyur and Lukeino Formations, it is often assumed that ancestors of extant African apes must have inhabited places like the Tugen Hills up until the split between hominins and *Pan*, by about 6 to 7 Ma (Brunet *et al.* 2002). Kingston, Marino and

Hill (1994) at first reconstructed a continuous open woodland mosaic throughout the Miocene and Pliocene from the Tugen Hills sequence, a remarkably uniform vegetational structure over a very long span. More recently, Kingston (1999) suggested a more variable environment through time, including the presence of widespread tropical forest (at *c.* 12.6, 7.0, and 6.3 Ma), seasonal woodland (*c.* 7 Ma), and arid, open woodland (*c.* 10 Ma). In addition, $\delta^{13}C$ of enamel apatite from a sample of fossil herbivores indicates the first evidence of C_4-plant-dominated diet in the Tugen Hill record at about 7 Ma (Kingston 1999: Figure 13.3), consistent with the results of Cerling *et al.* (1997a).

In summary, seasonal subtropical forest characterized the local environmental settings of Miocene great apes, particularly during the radiation of *Dryopithecus*, the most diverse great ape genus known. Two of the last recorded Miocene great apes, *Oreopithecus* and *Lufengpithecus*, were associated with moist, swampy forests in southern regions of Europe and East Asia. Evidence from Greece and recent finds from Turkey suggest that two or more lineages of late Miocene great apes ranged into relatively open habitat. The overwhelming evidence from Miocene ape sites implies, however, that early great apes were largely tied to heavily wooded habitats, ranging from moist forests to mosaics of forest and grassy woodland. In the two long stratigraphic sequences, in East Africa and South Asia, fossil great apes largely disappeared or were very rare in the former region over a 9-million-year period of fluctuating conditions ranging from tropical forest, seasonal woodland, and open woodland settings; in the latter region, *Sivapithecus* endured over at least 5 million years, mainly in seasonal forest and woodland habitats, and disappeared during a more-or-less permanent transition from warm, humid forest and woodland to drier open grassland.

Global and regional environments of Miocene apes

Figure 13.1 depicts $\delta^{18}O$ variation recorded in calcareous skeletons of the bottom-dwelling (benthic) foraminifer *Cibicidoides* recovered from a deep-sea core in the southwestern Pacific, covering the interval from 16 to 12 Ma. Measurement of $\delta^{18}O$ in benthic foraminifera provides a sequence of ^{18}O enrichment and depletion in the deep ocean, which partially reflects the worldwide pattern of temperature and ice variation. Enrichment is

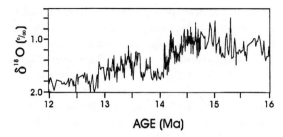

AGE (Ma)

Figure 13.1. Oxygen isotope record for the benthic foraminifer *Cibicidoides* from 16 to 12 million years ago at Deep Sea Drilling Project site 588A, southwest Pacific (Flower & Kennett 1993). An oxygen isotopic record (measured as $\delta^{18}O$ in parts per mil) for bottom-dwelling foraminifera is considered to reflect the overall effect of temperature and evaporation (water locked up as glacial ice) on oceans globally. This record shows the two major patterns of Miocene climate change – oxygen enrichment, which is indicated by the increase in $\delta^{18}O$; and an enlarged range of $\delta^{18}O$ oscillation, especially between 16 and 13 million years ago. Enrichment means a decrease in temperature and/or an increase in ocean water evaporation and global ice volume. Enlarged oscillatory amplitude implies variability in these climate parameters.

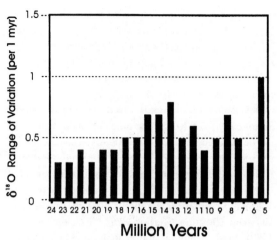

Million Years

Figure 13.2. $\delta^{18}O$ variability during the Miocene. The total range of variation in $\delta^{18}O$ is plotted in each 1-million-year interval from 24 to 5 million years ago. An increase in climate variability is indicated starting in the interval 18–17 Ma and reaching a height at 14–13 Ma. Miocene great apes diversified during the subsequent decrease in long-term $\delta^{18}O$ oscillation. Great ape diversity declined significantly in the context of increasing seasonality and rise in long-term variability beginning at 10–9 Ma. The oldest known hominins coincided with the large rise in $\delta^{18}O$ variability at the end of the Miocene. From Potts (1998a,b) using data from Woodruff, Savin and Douglas (1981), Wright and Miller (1992), and Miller and Mountain (1996).

caused by decreased temperature and increased water evaporation. Over most of the period of great ape evolutionary history, global temperature has been sufficiently low to periodically capture evaporated ocean water in ice caps (since the middle Miocene) and continental glaciers (since the late Pliocene), followed by its release back into the oceans. The periodicities at which these oscillations have occurred correlate with Milankovitch cycles, i.e., cyclical variations in Earth's orbit relative to the sun, which cause variation in incoming solar radiation (insolation). Although glacial oscillations have been particularly marked over the past 2.8 million years, ice volume fluctuation (and associated sea-level rise and fall) has been a feature of Earth's hydrological system over at least the past 15 million years, when the Antarctic ice cap became a permanent feature.

Two global climatic signals are evident in the oxygen isotope record of the middle Miocene: (1) an increase in the amplitude of $\delta^{18}O$ oscillation, indicative of wider environmental fluctuation, and (2) oxygen enrichment, indicative of cooler, more evaporative, and glacial climates (Figure 13.1). In the context of the entire Miocene, however, the period between 18 and 13 Ma exhibited particularly dramatic change in these two parameters (Figure 13.2). The first signal of

mid-Miocene climatic shift, involving increased fluctuation, coincides roughly with the appearance of the great ape clade. Between 24 and 18 Ma the recorded range of benthic $\delta^{18}O$ fluctuation is ≤0.3‰ (parts per mil) per million years. Between 18 and 17 Ma, variation in $\delta^{18}O$ rises to 0.5l per one million years. Between 16 and 14 Ma, the range continues to rise to 0.85‰ for the first time in the Cenozoic, followed by a decrease to around 0.3 to 0.7‰ in each million year interval between 13 and 6 Ma (Potts 1998a). The second signal, worldwide cooling, is registered primarily between 15 and 14 Ma; it coincided with the final closure of the circum-equatorial ocean current system (the Paratethys seaway) and the growth of the East Antarctic ice sheet at around 15 Ma (Flower & Kennett 1993; Kennett 1995).

The evolutionary histories of terrestrial animals were affected by global climate change moderated by regional tectonic events and physical geography, including the establishment of land bridges. The presence of the catarrhine *Dionysopithecus* in southern

Pakistan approximately 18–16 Ma (Bernor *et al.* 1988) and *Griphopithecus* at Engelswies, Germany, approximately 17–16.5 Ma (Heizmann & Begun 2001) indicates that a land bridge and suitable environmental conditions encouraged primate migration between Africa and Eurasia by 18–17 Ma. After this time, land bridges were established intermittently, partly due to sea-level fluctuation, and allowed faunal migration in waves until about 15 Ma when the African–Eurasian land bridge was more continuously established. Establishment of a continuous land bridge prevented circulation and heat exchange from the Indian Ocean to the western Mediterranean. In addition, uplift of the Tibetan Plateau, the Himalayas, and mountain ranges around the Mediterranean affected atmospheric circulation throughout Eurasia and northern Africa (Agustí, Rook & Andrews 1999a; Agustí *et al.* 1999b; Andrews *et al.* 1996; Jones 1999; O'Brien & Peters 1999; Rögl 1999).

Although paleotemperature analysis of European coral faunas broadly agrees with the evidence of global Miocene cooling (Rosen 1999), non-marine molluscs offer a more complex picture of European settings (Esu 1999). Cooling is evidenced around 15 Ma, but warm conditions were established shortly after, followed again by cooling in the early Vallesian (MN 9, in the system of Eurasian mammal biochronology), and then warming from MN 10 through MN 11 times (around 9 Ma). These data suggest a more fluctuating climatic regime than that registered in mammalian faunas.

Pollen assemblages studied by Suc *et al.* (1999) indicate a substantial shift in Miocene circum-Mediterranean vegetation, including the loss of tropical elements and a substantial decline in subtropical forest taxa in western Europe between 15 and 10 Ma. These changes corresponded to a temperature drop, a finding supported by paleobotanical evidence in central Europe (Kovar-Eder *et al.* 1996). According to Suc and colleagues, forests at the outset of this period would have been able to provide fruits all year long, whereas after 10 Ma, fruit production was reduced to several months per year. In accord with Andrews (1992), the pollen study suggests that this vegetational shift greatly affected European primates and may explain the extinction of hominoids in western Europe by about 9 Ma. These findings are consistent with the idea that Miocene great apes were largely, though not entirely, dependent on ripe fruit (see Singleton, Chapter 16, this volume) and that their geographic distribution would have thus

been affected by forest/woodland sources of such fruits (see below).

Dramatic biotic change occurred in western Eurasia and the circum-Mediterranean region between 10 and 9 Ma. Known as the mid-Vallesian crisis, this event involved diminishment or disappearance of warm and moist subtropical conditions, especially in western Europe. Numerous large mammals became extinct, including several groups of rodents and carnivores, suids, tapirs, rhinoceroses, and primates. Forest-dwelling species were most seriously affected, which suggests a climatic cause. Even though forests persisted in central Europe during this time, hominoids that had depended on such habitats in that region had their last recorded appearance in the fossil record by about 9 Ma (i.e., the boundary between Neogene Mammal zones MN 10 and 11) (Agustí *et al.* 1999b; Franzen & Storch 1999).

In an analysis of teeth from more than 500 equids and other hypsodont mammals, Cerling *et al.* (1997a) showed that carbon isotopic values shifted significantly in southern Asia, East Africa, North America, and South America between 8 and 6 Ma. This shift corresponded to a dietary change from predominantly cool, closed-canopy (C_3) plants to water- and/or heat-stressed (C_4) plants typical of open vegetation in latitudes below 37° N. Prior to 8 Ma, no mammals they tested showed evidence of any significant C_4-plant diet. Cerling *et al.* attribute this shift in diet and, by implication, in vegetation to a drop in atmospheric CO_2 concentration below an important threshold for C_3-photosynthesis. They thus conclude that a worldwide shift in vegetation occurred beginning around 8 Ma.

Other isotopic studies in East Africa (Kingston *et al.* 1994) and South Asia (Morgan *et al.* 1994; Quade *et al.* 1995), however, either do not detect a vegetational shift at this time or attribute the C_4-grass expansion to a gradual onset of monsoonal conditions rather than an abrupt change in global atmospheric pCO_2. Furthermore, the major faunal change in western Europe that reflects a transition from woodland–forest to open conditions took place closer to 9.5 Ma (Agustí *et al.* 1999b), well before the proposed global shift posited by Cerling *et al.* (1997a). No special faunal turnover is apparent in western Europe during the suggested critical span of 8 to 6 Ma, except toward the end of that period in association with the Messinian crisis, although this was a regional rather than a global event.

The Messinian "Salinity Crisis," dated 7.1 to 5.3 Ma, was associated with climatic cooling and drying and the temporary closure of the Gibraltar Strait, leading to the dessication of major portions of the Mediterranean basin (Benvenuti, Paplni & Testa 1999; Hsü *et al.* 1978; Jones 1999; Suc *et al.* 1999). During the Messinian, sea-level fluctuation evidently caused the Atlantic Ocean to breach the Mediterranean basin on numerous occasions. More than 60 cycles of Mediterranean filling and drying have been inferred, which are considered to reflect precessional cycles (approximately every 20 kyr) (Benvenuti *et al.* 1999).

Primate evolutionary responses to Miocene environments varied according to time and the range of climatic fluctuation. The first appearance and initial dispersal of great apes (e.g., *Griphopithecus*, Heizmann & Begun 2001) in Eurasia coincided broadly with a period of increased global climatic variability. The prominent radiation of great apes in the early part of the late Miocene (e.g., *Dryopithecus* and *Sivapithecus*) occurred, however, in a global context of relative environmental stability. Regional uplift and seaway closure exerted a powerful influence on terrestrial settings by buffering the impact of distant, major environmental events recorded in the marine record (e.g., growth of the Antarctic ice cap at 15 Ma). European and western Asian environmental data indicate, none the less, that Miocene great apes faced significant climatic transitions, particularly to drier conditions. Persistence of great apes in Eurasia from at least 17 to 7 Ma is thus evidence of a certain degree of ecological adaptability, possibly mediated by cognitive advances. By the end of the Miocene, when climate fluctuated widely during the Messinian crisis, great apes had already become extinct in Europe and southwest Asia (Andrews *et al.* 1996). That cercopithecoid monkeys persisted in these regions during the Messinian suggests their resilience to repeated climatic perturbation and aridity.

Biogeography and decline of Miocene great apes

On the basis of environmental reconstructions of ape fossil sites in western Eurasia, the middle and late Miocene radiation of great apes took place largely in settings of subtropical forest characterized by seasonal fluctuation (Table 13.2). Subsequent decline in species diversity, during the late Miocene, occurred as seasonality increased and temperature declined.

Fortelius and Hokkanen (2001) show that Miocene great apes disappeared in western Eurasia starting in the north (e.g., northwestern Europe) and proceeding to the south (Mediterranean and western Asia). This pattern, which occurred between 11 and 7 Ma, reflects the overall decline of great ape species diversity and a southward biogeographic shift of Eurasian apes (see also Begun 2001). The biogeographic pattern followed an environmental gradient in which great apes disappeared earliest from cooler and more seasonal settings and appear to have tracked warmer and less seasonal settings. The change to cooler and more seasonal environments took place later in the south and east (9 to 7 Ma), which is where great apes persisted longer.

Late Miocene apes were most commonly associated with a trophic structure typified by high numbers of animal-eaters and omnivores and lower numbers of hypsodont plant eaters, relative to fossil assemblages that lack apes. According to Fortelius and Hokkanen (2001), this ecological association reflects the low end of the seasonality spectrum. Their data indicate that since the Miocene apes of Eurasia persisted later in the south (e.g., *Oreopithecus*, 8 to 7 Ma in Italy), the temperature gradient (warmer to the south) may have been more important than seasonality in shaping both habitat preferences and the extinction pattern in Miocene great apes. Nevertheless, their data also suggest a gradient of increasing seasonality from west to east, which may help to explain why great apes died out earlier in the west during a period of declining temperature.

Oreopithecus' late appearance in southern Europe is paralleled by the last records of Miocene great apes in Asia, which also occur in the southerly Siwaliks and Lufeng regions. The combined geographic data for fossil and living apes strongly implies that, since the beginning of the late Miocene, hominoids experienced a diminishing range and south-eastward displacement toward the equator. Great apes became confined largely to wooded habitats, typically forests, in the tropical latitudes of Africa and Southeast Asia. Data amassed by Jablonski and co-workers on the spatial distribution of fossil apes in China offer a detailed picture of this range contraction during the Quaternary (Jablonski 1998; Jablonski *et al.* 2000). Their study shows a southerly displacement of *Pongo, Gigantopithecus*, and *Hylobates* associated with rising seasonality, increased climatic fluctuation, and a restriction of subtropical environments to the south.

In summary, the height of great ape species diversity occurred during the late Miocene, followed by decline under conditions of increasing seasonality, cooling, and forest retreat to southern refugia and continents. The later diversification of monkeys in Eurasia took place in the context of environmental trends and local habitat change that were apparently unfavorable to the persistence of great apes, despite possible differences in intelligence or cognitive adaptability between monkeys and apes. The last recorded great apes in Eurasia either occupied forested, swampy settings, perhaps similar to those described for extant populations of *Pongo* and lowland gorilla, or ventured into more open, arguably "savanna-like" conditions. Surviving great apes live in rain forest, seasonal forest, and closed and open woodland settings in tropical latitudes, consistent with the habitat preferences and biogeographic trend of late Miocene apes.

The dependence of Miocene apes on forested settings is reinforced by interpretations of the functional anatomy of their masticatory and postcranial systems. Most late Miocene great apes were frugivores, and most were arboreal, both strongly connected to forest ecologies (Andrews, *et al.* 1997; Begun & Kordos 1997; Harrison & Rook 1997; Kay 1977; Kay & Ungar 1997; King, Aiello & Andrews 1999; Leakey & Walker 1997; McCrossin & Benefit 1997; Rose 1997; Teaford & Walker 1984; Ungar 1996; Ungar & Kay 1995; C. V. Ward 1997; see also Gebo, Chapter 17, Singleton, Chapter 16 this volume).

EVOLUTIONARY ENVIRONMENTS OF EXTANT GREAT APES

Adaptive settings of the Pliocene and Pleistocene

A preference in great apes for relatively low-seasonality habitats is suggested by the local extinction of late Miocene apes from western Eurasia and of Plio-Pleistocene great apes from temperate-zone Eastern Asia as seasonality intensified. Although the habitats of living great apes tend to vary seasonally, they generally exhibit much smaller seasonal contrasts than those of monkeys such as baboons, vervets, macaques, and Asian colobines (e.g., Delgado & van Schaik 2000; Schoeninger, Moore & Sept 1999; Watts 1998; White 1998; Yamagiwa 1999).

While long-term stratigraphic records in forested and wooded regions currently inhabited by great apes

are lacking, a wealth of high-resolution climatic data indicate that tropical Africa and Southeast Asia sustained dramatic oscillation in monsoonal conditions over the past few million years (see below). Evidence of Pliocene and especially Pleistocene oscillation poses the question as to how great ape populations adjusted to large-scale habitat remodeling.

Based on a composite $\delta^{18}O$ marine record of benthic foraminifera over the past 6 million years, the period between 6 and 3 Ma entailed wider variability than typically occurred during the Miocene (Figure 13.3). The amplitude of oxygen isotopic oscillation again increased significantly between 3 and 2 Ma and also around 1 Ma (see Potts 1998b for details). During the latter third of the Quaternary, the last 500 000 years, the variability in global temperature, evaporation, ice volume, and sea level during periods of only 100 000 years typically exceeded the entire mean environmental change (cooling and drying) of the past 6 million years. Instability, then, is the pre-eminent feature of Quaternary climate and, therefore, of the paleoenvironments in which lineages of extant apes and other organisms evolved (Potts 1996).

Oscillation between aridity and heavy monsoon rainfall in Africa is indicated by organic-rich layers called sapropels deposited in the eastern Mediterranean Sea as a result of flooding peaks in the Nile drainage (Rossignol-Strick 1983; Rossignol-Strick *et al.* 1998). The sapropel record has been studied in detail back to 1.2 Ma, and it includes thick layers that mark intense shifts between relatively arid and heavy precipitation phases. Periods of intense African aridity and subsequent return to moist conditions over the past 900 000 years are further documented in airborne dust records recovered from deep-sea cores (deMenocal, Ruddiman & Pokras 1993; deMenocal & Bloemendal 1995).

These findings imply that for much of the Quaternary, African equatorial forests have sustained repeated contraction and expansion, and also fragmentation and coalescence. For most African primates, including great apes, the ultimate effect would have been profound variability in adaptive settings – shifts in species associations, diversity, competitors, predators, food abundances, and population densities. Populations reliant on specific habitats or resource types would have faced particularly severe adaptive problems related to the variable properties of their environments over time.

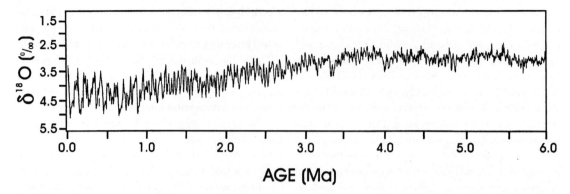

Figure 13.3. Oxygen isotope curve ($\delta^{18}O$) for the past 6 million years, based on composite data on benthic foraminifera from deep-sea cores (Shackleton 1995). The curve shows a trend during cooling and increased ice volume, especially since 2.8 Ma, and also many oscillations that increase in amplitude over time. Oscillation over the past 900 000 years represents the widest range of environmental fluctuation in the Cenozoic era.

Variability in great ape habitats was manifested not only in Africa but also in the East Asian settings of orangutans. While researchers have long believed that environments of southern China and Southeast Asia were remarkably stable throughout the Quaternary (e.g., Hutterer 1977; Teilhard de Chardin *et al.* 1935), recent research on the Loess Plateau, Yangtze River, the Bose basin, and the South China Sea has begun to furnish evidence of strong arid – moist oscillation and episodic disturbance to tropical and subtropical forests (An 2000; Guo *et al.* 2000; Hou *et al.* 2000; Wang 1990; Wang, Zhang & Jian 1991). Pronounced sea-level fluctuation, resulting from the growth and melting of glacial ice approximately every 100 000 years, was especially strong over the past 1 million years. Sea-level fall of more than 100 m (relative to the present) led to the expansion of terrestrial habitat onto the Sunda Shelf, including repeated increases in forest habitat by many thousands of square kilometers. Periods of expansion were followed by reduction in habitats favored by many primates, including orangutans.

In short, improving databases of global and regional paleoenvironments give good reason to assume that the lineages of living great apes confronted significantly heightened variability in their adaptive settings through much of the Quaternary. Due to the absence of Quaternary great ape fossil sites in Africa, and their relative rarity in Asia, there is no direct means of testing how great apes responded to this large range of habitat change. The question here is how the cognitive repertoires of living apes may have been shaped by these shifting conditions.

Great apes' adaptability to environmental variability

Primates have evolved diverse ways of buffering environmental variability, including physiological adaptations, dietary diversity, and cognitive solutions that are often related to acquiring food (Hladik 1981). During dry seasons, omnivorous primates often shift to a specific range of available "fallback" foods, while dietary specialists often maintain their distinctive feeding habits across seasonal variations by either tracking the specific foods they rely upon (e.g., *Presbytis*) or feeding on alternative food species within the same general food type (e.g., *Macaca*). Apes focus their frugivorous habits on ripe fruit, which is rarer, more patchily distributed, and more susceptible to seasonal and interannual fluctuation, so they have evolved an array of dietary, cognitive, and social behavioral responses (Hladik 1981).

In light of the Quaternary fluctuations, one of the most daunting adaptive problems facing frugivorous apes has been the shifts between continuous and patchy forest conditions. In order to survive in a region comprised of forest fragments, a primate group requires either a small home range (no larger than the fragment) or a very large range with the ability to move between fragments. In the Kibale Forest vicinity, chimpanzees and monkeys living in forest fragments vary their grouping, nesting, and dietary behavior from that of conspecific populations inhabiting nearby continuous forest (Onderdonk & Chapman 2000). Species' strategies for handling this problem also differ. In the Lopé Reserve, Gabon, chimpanzees and monkeys responded differently to fragmented and continuous forest habitats, which varied in their available food sources and predation risks (Tutin 1999). Relative to populations in neighboring continuous forest, each of four monkey species in the Lopé forest fragment spent less time feeding on fruit and tended to switch to other locally abundant foods like insects and leaves, whereas chimpanzees in the fragment increased their fruit feeding time by about 12%.

Chimpanzees appear to be able to depend on fruit sources that are dispersed and temporally variable by several means – memory; mental representation of the possible fruiting states of trees some distance away; calls that contribute information about the location of rich fruit sources; and fission–fusion formation of ephemeral parties (Goodall 1986; Newton-Fisher, Reynolds & Plumptre 2000; Sugiyama 1999; Wrangham 1980). Orangutans and gorillas have evolved similar cognitive and behavioral tactics.

The fluid nature of chimpanzees' social contacts plus their ability to vary their home range, from about 12.5 km^2 up to 400 km^2 in arid areas where woodland and forest are patchily distributed (Goodall 1986; Kingdon 1997), appear to offer means of maintaining frugivory under diverse environmental conditions and forest size (e.g., Schoninger *et al.* 1999; Wrangham, Conklin-Brittain & Hunt 1998; Yamagiwa 1999). Although fission–fusion is most clearly developed in *Pan troglodytes*, this fluid pattern of social grouping may characterize all great apes (Newton-Fisher *et al.* 2000; and see Yamagiwa, Chapter 12, van Schaik *et al.*, Chapter 11, this volume).

A dynamic tension between ripe-fruit frugivory and fallback on other food types is found in all great apes, although it may vary from one species to another and from one population to another within a species depending on environmental conditions (Goldsmith 1999; Sabater Pi 1977; Yamagiwa 1999, Chapter 12, this volume). In general, gorillas and orangutans adjust to environmental variation in different ways than chimpanzees (Yamagiwa, Chapter 12, this volume). Orangutans store fat during periods of fruit abundance and otherwise scrape by on lower-quality foods (allowing them to live in environments with irregular fruiting), alter ranging patterns to concentrate on areas with ripe fruit peaks, and change from nearly exclusive frugivory to more diverse diets, including bark (Delgado & van Schaik 2000). Gorillas shift to greater reliance on folivory, maintain relatively stable groups, and range over small distances. Their ability to track forest resources by eating terrestrial herbaceous vegetation stabilizes the food supply and buffers environmental alteration.

If great apes accommodate to environmental variation in a variety of ways, they approach the problem of adaptability differently from Old World monkeys. This is apparent in their dietary diversity: both colobine and cercopithecine monkeys are effective at processing a variety of foods and at foraging in a diversity of habitats that present difficulties for apes (Kingdon 1997; Yeager & Kirkpatrick 1998). The differences are also apparent in the life histories of the two clades (Harvey, Martin & Clutton-Brock 1987; Kelley 1997; Ross, Chapter 8, Kelley, Chapter 15, this volume). Monkeys' rapid reproductive rates are advantageous in strongly seasonal environments or settings prone to disturbance. Great apes' slow reproduction represents a different strategy, dependent on habitat tracking, which includes social and cognitive means of accessing resources, like ripe fruit, that are likely to manifest strong time–space variation.

A SYNTHETIC HYPOTHESIS OF GREAT APE COGNITIVE EVOLUTION

By combining the information presented here, it is possible to draw a coherent, if fragmentary, picture of the conditions under which great ape mentality evolved. One key difference in the evolutionary environments of apes and monkeys stands out from the start – apes' dietary bias toward ripe fruits. Since the Miocene, most hominoids have retained a primitive catarrhine molar form associated with frugivory. This constraint in overall molar design evidently led almost all later ape species to rely on fruit-producing environments, mainly forests, woodlands, or patches of such habitats – even apes like *Oreopithecus* and *Gorilla* that were, or are, more folivorous (Kay & Ungar 1997). The earliest apes are known from tropical African rain forests and the diversification of late Miocene apes (e.g., *Dryopithecus*) appears to have taken place mainly in the closed subtropical forests of Eurasia.

The biogeographic and extinction history of late Miocene hominids is also consistent with the idea that great apes have had a general bias toward frugivory and tree-dominated environments and have been constrained dentally and ecologically from severing this connection. The geographic pattern of late Miocene great ape extinction suggests a preference for warm, low seasonality, forested environments. Once seasonality began to increase in Europe, after about 9.6 Ma, apes first became extinct north of the Alps and persisted in the maritime south (Mediterranean coast) until 8.2 to 7.1 Ma (Begun 2001; Fortelius & Hokkanen 2001; Rook *et al.* 2000). During this period of decline, apes had to accommodate to more strongly seasonal environments, where experimentation with folivory and more open habitat foraging may have reached its peak. The late Miocene sequence of great ape diversification, followed by decline, and then the radiation of cercopithecids, suggests a non-competitive relationship between great ape and monkey evolution that resulted from these clades' differential responses to increasing seasonality, aridity, and the southern withdrawal of forest habitats. With further increases in seasonality and aridity after the late Miocene, great apes became extinct in the temperate latitudes of Eurasia for good (Fortelius & Hokkanen 2001; Jablonski *et al.* 2000).

Between the latest Miocene and mid-Pleistocene (7 to 1 Ma), great apes became restricted to equatorial regions. Their initial bias toward tree-dominated habitats and ripe-fruit frugivory created new adaptive problems as climatic oscillation increased in amplitude. Deep-sea isotopes, continental dust, sapropels, glaciations, and associated sea levels all point to recurrent and dramatic remodeling of the size, patchiness, and resource structure of forests and woodlands in equatorial regions where great ape populations lived. We may infer that strongly fluctuating Pleistocene settings posed stringent challenges to great apes with regard to maintaining their specific link to wooded habitats and consistently locating abundant sources of ripe fruits. This episodic revamping of adaptive environments, magnified over the past one million years, placed a premium on cognitive, social, and dietary means of coping with novel settings. Evolutionary change in these aspects of great ape life largely occurred, however, within the sphere dictated by their primal bias toward wooded habitats.

From these findings, a synthetic hypothesis of great ape cognitive evolution takes shape, which we may term the *fruit-habitat hypothesis*, which posits three main phases of selection pressure and ecological constraint. The *first phase* imposed an ecological constraint derived from the initial conditions of great ape evolutionary history – primarily a dental and metabolic bias toward high-energy, ripe fruit. This starting point predisposed great apes to wooded habitats where ripe fruits can be most easily located. High-quality fruit is, however, inherently dispersed in time and space. The bias toward ripe fruit suggests that ape foraging has always had a dimension of temporal–spatial complexity that did not affect cercopithecid monkeys so greatly. Accordingly, it is possible that a commitment to ripe-fruit frugivory from the outset of ape evolutionary history created a problem of food predictability. Enhanced memory and mental representation of the phenological properties of fruiting trees dispersed through the foraging range would have strongly assisted in solving the predictability problem. Since ripe fruit tends to occur in delimited patches *and* in delimited periods of time, competition for those patches would have been critical when compared against foods more evenly distributed in space or time. Dispersal of social group members across the foraging range as a means of locating fruit or information about fruit-tree properties seems to follow from the space–time predictability problem inherent in ripe-fruit frugivory. Commitment to a diet that relied, at least in part, on ripe fruit sources thus substantially extended the cognitive capacities required, compared with permanently available but spatially dispersed foods. Expansion of diets to include hard-object foods, herbivorous vegetation, and/or meat would have relaxed some of the cognitive demands related to the temporal–spatial predictability of ripe fruits, while exposing great apes to new mental challenges of technical food processing, memory, and selection of different foods in variable seasonal and interannual settings.

Great apes' affinity for ripe fruit sources was best satisfied where there were large, stable tracts of dense tree habitats. It is these environments, though, that were reduced in size soon after the height of great ape taxonomic diversity – around 10–9 Ma in Europe and South Asia, and 7–8 Ma in East Asia – and were thereafter confined to low latitudes. This *second phase* in great ape evolutionary history corresponded to the demise of great ape populations where they had once been the most abundant, the temperate latitudes of Eurasia. The geographic pattern of extinctions and displacement of surviving populations toward lower latitudes strongly suggest that great apes responded to late Miocene environmental change by tracking their favored habitat – warm forests and woodlands – in which their initial bias toward ripe-fruit frugivory could be maintained. During this second phase, then, habitat tracking would have helped sustain the conditions of natural selection in which Miocene great apes lived, even while populations had to deal with heightened seasonal contrasts.

Latest Miocene to Pleistocene restriction of great apes to low latitudes of Africa and Southeast Asia presented an especially critical time in great ape evolutionary history. During this *third phase*, heightened instability of ape habitats greatly exacerbated the predictability problem – i.e., the uncertainty of locating ripe fruits. Pleistocene forest contraction and expansion, fragmentation and coalescence, meant that whatever time–space template of fruiting existed at any given time would eventually be extensively revised.

For frugivorous apes, repeated episodes of forest contraction and expansion, created by long-term shifts in seasonality, placed a strong premium on the ability to ascribe or guess probabilistic qualities regarding distant (and therefore unseen) portions of one's foraging range – i.e., the presence or absence of ripe fruit. Selection pressure related to prediction and the ability to deal with uncertainty would have been strongest in populations with large foraging ranges, especially during times of forest fragmentation. The success of prediction would have largely been founded on memory and mental representation of temporally and spatially distant places (e.g., mental mapping, planning) – especially the density of trees, the likelihood of any of them being in fruit, the phenological properties of the fruit, and the likelihood that distant patches might attract mates and competitors. For groups occupying large areas, fluid social grouping maximized opportunities for locating fruit sources and buffering competition. In groups capable of eating a wide diversity of herbaceous vegetation, smaller ranging areas and more stable groups were possible.

In short, the *environments* of evolutionary adaptedness for great apes were highly dynamic, creating a challenge to adaptive versatility. The spatially and temporally dispersed quality of fruit ripening furnished the kind of adaptive complexity that favored improved problem solving. As the level of spatial and temporal complexity was amplified during the Quaternary, far greater demands were placed on forest-dwelling apes to respond to the problems of locating consistent sources of ripe fruit. Environmental inconsistency helped shape the cognitive, social, dietary, and other aspects of great ape adaptability – a phenomenon known as variability selection (Potts 1996, 1998a,b). Accordingly, the selective effects of environmental variability favored mental abilities to solve problems concerning resource unpredictability or uncertainty. Neuronal plasticity, increased memory, and mental imaging as a means of locating foods sensitive to changing surroundings, all augmented the adaptability of great apes to short- and long-term environmental dynamics.

If, as implied by the fruit-habitat hypothesis, environmental fluctuation was essential to the evolution of representational intelligence, planning ability, and self-conception, it would mean that these modern mental potentials were not as fully elaborated in Miocene as in living great apes. While *Dryopithecus* and other Miocene lineages (but not *Oreopithecus*) had relatively large, great-ape-sized brains (Begun & Kordos, Chapter 14, this volume), whatever enhanced level of intelligence this implied likely reflected the challenges of the first evolutionary phase described here. The reasons underlying great ape cognitive evolution, however, are not all to be found in the Miocene. Rather, this hypothesis suggests that the cognitive potentials manifested by chimpanzees, bonobos, gorillas, and orangutans arose more recently, possibly in parallel with the evolution of advanced mental abilities in humans, and can be expected to exhibit variation among the modern species.

From this perspective, the main issue underlying ape cognitive and social evolution involved food predictability and the changing distribution of favored habitat over time. Three different solutions are manifested in extant apes. Each of these solutions represents an idealized relationship between cognitive functions, on the one hand, and dietary, foraging, social, and ranging adaptations, on the other. Living great apes do, in fact, combine these strategies (see Yamagiwa, Chapter 12, this volume), and different populations of the same species may emphasize one or another strategy in different settings.

1. Maintain a small foraging range, which allows easy tracking of ripe fruit, and move when forest habitat contracts, expands, or breaks up into smaller fragments. This approach is evident in lesser apes

(hylobatids) and to some extent in orangutans (Leighton 1987; Rodman & Mitani 1987).

2. Maintain a large foraging range relative to day range, enhance the fluidity of social grouping (fission–fusion), develop the cognitive means for appraising ripe fruit availability in distant places, move with the shifting distribution of forest, and be able to move between forest fragments, including patches associated with relatively open habitat. This strategy seems to best characterize chimpanzees and some populations of orangutans (Newton-Fisher *et al.* 2000; van Schaik 1999).

3. Broaden the diet to include a sizeable component of terrestrial herbaceous vegetation (THV) or other fallback foods, which may present a diverse range of technical foraging problems applied to constantly visible foods within a small foraging range. As a result, group size and structure can be stabilized, and reliance on lower quality foods (in addition to ripe fruit where it is available) leads to larger guts and body size. This strategy is predominant in gorillas, especially eastern lowland and mountain populations (Stewart & Harcourt 1987). Fallback foods are typical also of orangutans and chimpanzees, and THV of bonobos (Badrian & Malenky 1984; Nishida & Hiraiwa-Hasegawa 1987; Rodman & Mitani 1987).

According to the hypothesis presented here, each of these three idealized strategies holds different implications in terms of cognitive function. In the first case, dependence on frugivory implies that evolving lineages have come to possess the cognitive and social means to cope with temporal variability in ripe fruit sources. However, small foraging ranges, as in hylobatids, imply that populations can be sustained without any need to deal with the problems of complex spatial variability in food sources, especially the uncertainty of distant sources that are faced by populations with larger home ranges, as in orangutans.

The second strategy would appear to maximize the opportunity for cognitive problem solving – i.e., responsiveness to both ecological variability and social complexity (caused by ever-changing group composition). In situations where this strategy predominates, enhanced memory and mental representation (e.g., of distant resources), self-concept, adaptability to novel social situations and social variability, and the capacity to ascribe emotional and mental qualities to other individuals, all would prove highly beneficial. These cognitive dimensions appear to have been at a premium in the evolutionary history of chimpanzees and, to some degree, orangutans. They also suggest a strong degree of interaction between the ecological and social domains in which mental problem solving occurs, possibly as a result of high-level cortical integration of information about ecological and social settings (Russon 1998). Expressions of adaptive flexibility in orangutans would seem to include a degree of fission–fusion grouping, prolonged learning, and the ability to make mental connections between ecological situations (e.g., mast fruiting) and the social realm (e.g., grouping behavior).

The third strategy, involving THV and other fallback foods, is the one that most reduces the impact of environmental variability on cognitive evolution. Reliable food sources occur within localized areas, usually visible throughout the foraging range. Dependence on a diversity of herbaceous foods requires, however, greater dexterity and solving of specific food processing problems. Under these conditions, a different type of mental acuity is favored probably based on solving the complexities of technical foraging (Byrne 1997). Cognitive mechanisms that ascribe characteristics to invisible food sources – e.g., seeds or other protected edible matter – would be crucial to the success of this strategy. It would be less dependent on cognitive mechanisms that enable one to ascribe characteristics to distant food sources, meanings to distant calls, or intentions to individuals encountered only after days of separation. By contrast, the second strategy depends entirely on such abilities.

It is thus possible to construe many aspects of cognitive and social evolution in humans from a similar perspective as in other great apes – i.e., a matter of response to resource uncertainty and habitat instability. The difference is that Pliocene human ancestors ultimately severed the connection with wooded habitats on which apes had largely relied up to that point. Human evolutionary history thus took place on a far more diverse and complex ecological stage, subject to more extensive revision, than that which occurred in the equatorial forests (Potts 1996). It is on this stage where tool dependence, purposeful resource sharing, extensive mental abstraction, complex social communication, and spatial/temporal mental maps of the world all became more elaborated in humans than in other great apes. All of these extreme cognitive and social expressions in humans make sense, however, within the prior context of great ape evolutionary history.

CONCLUSION

This survey of environmental and paleobiological findings suggests that a multiplicity of ecological factors contributed to the distinctive features of modern great ape cognition. The question arises that since environmental fluctuation was such a prevalent signal from the Miocene onward, why don't all primate species today exhibit similar types of advanced cognitive functions? This situation arises from the varied starting points in the evolutionary histories of different groups, from the variety of evolutionary responses that are possible to the same environmental history, and from parallel adaptations that evolved independently during the last several million years. All primates evolved one means or another of responding to environmental variability. The critical ingredient in the fruit–habitat hypothesis is the specific link that has existed between great apes and an ephemeral food resource, which tied them to tropical–subtropical forests and woodlands. Starting with this constraint, the great ape story and the factors responsible for their cognitive skills largely follow from the contingencies of late Cenozoic environmental history.

While certain aspects of diet and habitat preference have been conserved throughout their evolution, it is important to see that great apes are not living fossils from the Miocene. Like humans, they, too, had a Plio-Pleistocene evolutionary history. The central thesis of this chapter is that conserved elements from the Miocene combined with novel adaptive settings of the Pleistocene were responsible for the evolution of certain cognitive abilities distinctive to great apes. Evolutionary groups are defined by their initial adaptive tendencies, and these tendencies help define the evolutionary trajectories of those clades. Great apes and their cognitive evolutionary history are no exception to this rule.

ACKNOWLEDGMENTS

I am grateful for the opportunity offered by the volume editors to delve into the Miocene backdrop to ape evolution. Special thanks goes to colleagues who pointed me toward useful ideas, research directions, and published literature, including Mikael Fortelius, Peter Ungar, Andrew Hill, Ray Bernor, John Barry, Jay Kelley, Steve Leigh, David Begun, and David's student Louise Blundell, who kindly sent me a copy of her study of Miocene sites. John Finarelli carried out background research with typical efficiency, and I thank Jennifer Clark for her help in preparing the figures. David Begun and Anne Russon offered especially helpful reviews of the initial manuscript. This is a publication of the Smithsonian's Human Origins Program.

REFERENCES

Agustí, J., Rook, L. & Andrews, P. (ed.) (1999a). *The Evolution of Neogene Terrestrial Ecosystems in Europe*, Vol. 1. Cambridge, UK: Cambridge University Press.

Agustí, J., Cabrera, L., Garcés, M. & Llenas, M. (1999b). Mammal turnover and global climate change in the late Miocene terrestrial record of the Vallès-Penedès basin (NE Spain). In *The Evolution of Neogene Terrestrial Ecosystems in Europe*, Vol. 1, ed. J. Agustí, L. Rook & P. Andrews, pp. 397–412. Cambridge, UK: Cambridge University Press.

An, Z. (2000). The history and variability of the East Asian paleomonsoon climate. *Quaternary Science Reviews*, **19**, 171–87.

Andrews, P. (1983). The natural history of *Sivapithecus*. In *New Interpretations of Ape and Human Ancestry*, ed. R. L. Ciochon & R. Corruccini, pp. 441–63. New York: Plenum Press.

Andrews, P. (1992). Evolution and environment in the Hominoidea. *Nature*, **360**, 641–646.

Andrews, P., Begun, D. R. & Zylstra, M. (1997). Interrelationships between functional morphology and paleoenvironments in Miocene hominoids. In *Function, Phylogeny, and Fossils: Miocene Hominoid Evolution and Adaptations*, ed. D. R. Begun, C. V. Ward & M. D. Rose, pp. 29–58. New York: Plenum Press.

Andrews, P., Harrison, T., Delson, E., Bernor, R. L. & Martin, L. (1996). Distribution and biochronology of European and southwest Asian Miocene catarrhines. In *The Evolution of Western Eurasian Neogene Mammal Faunas*, ed. R. L. Bernor, V. Fahlbusch & H.-W. Mittmann, pp. 168–207. New York: Columbia University Press.

Badgley, C. & Behrensmeyer A. K. (1980). Paleoecology of Middle Siwalik sediments and faunas, northern Pakistan. *Palaeogeography, Palaeoclimatology, Palaeoecology*, **30**, 133–55.

Badgley, C., Qi, G., Chen, W. & Han, D. (1988). Paleoecology of a Miocene, tropical, upland fauna: Lufeng, China. *National Geographic Research*, **4**, 178–95.

Badrian, N. & Malenky, R. (1984). Feeding ecology of *Pan paniscus* in the Lomako Forest, Zaire. In *The Pygmy Chimpanzee: Evolutionary Biology and Behavior*, ed. R. L. Susman, pp. 275–99. New York: Plenum Press.

Barry, J. C., Johnson, N. M., Raza, S. M. & Jacobs, L. L. (1985). Neogene mammalian faunal change in southern Asia: correlations with climatic, tectonic, and eustatic events. *Geology*, **13**, 637–40.

Begun, D. R. (2001). African and Eurasian Miocene hominoids and the origins of the Hominidae. In *Hominoid Evolution and Environmental Change in the Neogene of Europe*, ed. P. Andrews, G. Koufos & L. de Bonis, pp. 231–53. Cambridge, UK: Cambridge University Press.

Begun, D. R. & Kordos, L. (1997). Phyletic affinities and functional convergence in *Dryopithecus* and other Miocene and living hominids. In *Function, Phylogeny, and Fossils: Miocene Hominoid Evolution and Adaptations*, ed. D. R. Begun, C. V. Ward & M. D. Rose, pp. 291–316. New York: Plenum Press.

Begun, D. R., Ward, C. V. & Rose, M. D. (ed.) (1997a). *Function, Phylogeny, and Fossils: Miocene Hominoid Evolution and Adaptations*. New York: Plenum Press.

(1997b). Events in hominoid evolution. In *Function, Phylogeny, and Fossils: Miocene Hominoid Evolution and Adaptations*, ed. D. R. Begun, C. V. Ward, & M. D. Rose, pp. 389–415. New York: Plenum Press.

Benvenuti, M., Bertini, A. & Rook, L. (1995). Facies analysis, vertebrate paleontology and palynology in the Late Miocene Baccinello-Cinigiano Basin (Southern Tuscany). *Memorie della Società Geologica Italiana*, **48**, 415–23.

Benvenuti, M., Papini, M. & Testa, G. (1999). Sedimentary facies analysis in palaeoclimatic reconstructions. Examples from the Upper Miocene–Pliocene successions of south-central Tuscany (Italy). In *The Evolution of Neogene Terrestrial Ecosystems in Europe*, Vol. 1, ed. J. Agustí, L. Rook & P. Andrews, pp. 355–77. Cambridge, UK: Cambridge University Press.

Bernor, R. L., Flynn, L., Harrison, T., Hussain, T. & Kelley, J. (1988). *Dionysopithecus* from southern Pakistan and the biochronology and biogeography of early Eurasian catarrhines. *Journal of Human Evolution*, **17**, 339–58.

Bestland, E. (1990). Sedimentology and paleopedology of Miocene alluvial deposits at the Paşalar hominoid site, western Turkey. *Journal of Human Evolution*, **19**, 363–78.

Bishop, W. W. & Pickford, M. (1975). Geology, fauna and palaeoenvironments of the Ngorora Formation, Kenya Rift Valley. *Nature*, **254**, 185–92.

Brunet, M., Franck, G., Pilbeam, D., *et al.* (2002). A new hominid from the Upper Miocene of Chad, Central Africa. *Nature*, **418**, 145–51.

Byrne, R. W. (1995). Primate cognition: comparing problems and skills. *American Journal of Primatology*, **37**, 127–41.

(1997). The technical intelligence hypothesis: an additional evolutionary stimulus to intelligence? In *Machiavellian Intelligence II: Extensions and Evaluations*, ed. A. Whiten & R. W. Byrne, pp. 290–311. Cambridge, UK: Cambridge University Press.

Byrne, R. W. & Whiten, A. (1988). *Machiavellian Intelligence*. Oxford: Clarendon Press.

Cerling, T. E., Harris, J. M., MacFadden, B. J., Leakey, M. G., Quade, J., Eisenmann, V. & Ehleringer, J. R. (1997a). Global vegetation change through the Miocene/Pliocene boundary. *Nature*, **389**, 153–8.

Cerling, T. E., Harris, J. M., Ambrose, S. H., Leakey, M. G. & Solounias, N. (1997b). Dietary and environmental reconstruction with stable isotope analyses of herbivore tooth enamel from the Miocene locality of Fort Ternan, Kenya. *Journal of Human Evolution*, **33**, 635–50.

Cerling, T. E., Kappelman, J., Quade, J., Ambrose, S. H., Sikes, N. E. & Andrews, P. (1992). Reply to comment on the paleoenvironment of *Kenyapithecus* at Fort Ternan. *Journal of Human Evolution*, **23**, 371–7.

Cerling, T. E., Quade, J., Ambrose, S. H. & Sikes, N. E. (1991). Fossil soils, grasses, and carbon isotopes from Fort Ternan, Kenya: grassland or woodland? *Journal of Human Evolution*, **21**, 295–306.

Cerling, T. E., Quade, J. & Wang, Y. (1993). Expansion of C4 ecosystems as an indicator of global ecological change in the Miocene. *Nature*, **361**, 344–5.

de Bonis, L. & Koufos, G. (1997). The phylogenetic and functional implications of *Ouranopithecus macedoniensis*. In *Function, Phylogeny, and Fossils: Miocene Hominoid Evolution and Adaptations*, ed. D. R. Begun, C. V. Ward & M. D. Rose, pp. 317–26. New York: Plenum Press.

Delgado, Jr., R. A. & van Schaik, C. P. (2000). The behavioral ecology and conservation of the orangutan (*Pongo pygmaeus*): a tale of two islands. *Evolutionary Anthropology*, **9**, 201–18.

deMenocal, P. B. & Bloemendal, J. (1995). Plio-Pleistocene climatic variability in subtropical Africa and the paleoenvironment of hominid evolution: a combined data-model approach. In *Paleoclimate and Evolution with Emphasis on Human Origins*, ed. E. S. Vrba, G. H. Denton, T. C. Partridge & L. H. Burckle, pp. 262–88. New Haven, CT: Yale University Press.

deMenocal, P. B., Ruddiman, W. F. & Pokras, E. M. (1993). Influences of high- and low-latitude processes on African climate: Pleistocene eolian records from equatorial Atlantic Ocean Drilling Program Site 663. *Paleoceanography*, **8**, 209–42.

Dunbar, R. I. M. (1992). Neocortex size as a constraint on group size in primates. *Journal of Human Evolution*, **20**, 469–93.

(1995). Neocortex size and group size in primates: a test of the hypothesis. *Journal of Human Evolution*, **28**, 287–96.

Esu, D. (1999). Contribution to the knowledge of Neogene climatic changes in western and central Europe by means of non-marine molluscs. In *The Evolution of Neogene Terrestrial Ecosystems in Europe*, Vol. 1, ed. J. Agustí, L. Rook & P. Andrews, pp. 328–54. Cambridge, UK: Cambridge University Press.

Flower, B. P. & Kennett, J. P. (1993). Middle Miocene ocean/climate transition: high-resolution oxygen and carbon isotopic records from DSDP Site 588A, southwest Pacific. *Paleoceanography*, **8**, 811–43.

Fortelius, M. & Hokkanen, A. (2001). The trophic context of hominoid occurrence in the later Miocene of western Eurasia: a primate-free view. In *Hominoid Evolution and Climatic Change in Europe*, ed. L. de Bonis, G. Koufos & P. Andrews, pp. 19–47. Cambridge, UK: Cambridge University Press.

Franzen, J. L. & Storch, G. (1999). Late Miocene mammals from Central Europe. In *The Evolution of Neogene Terrestrial Ecosystems in Europe*, Vol. 1, ed. J. Agustí, L. Rook & P. Andrews, pp. 165–90. Cambridge, UK: Cambridge University Press.

Frith, C. D. & U. Frith (1999). Interacting minds – a biological basis. *Science*, **286**, 1692–5.

Garber, P. A. (1989). Role of spatial memory in primate foraging patterns: *Saguinus mystax* and *Saguinus fuscicollis*. *American Journal of Primatology*, **19**, 203–16.

Gebo, D. L., MacLatchy, L., Kityo, R., Deino, A., Kingston, J. & Pilbeam, D. (1997). A hominoid genus from the early Miocene of Uganda. *Science*, **276**, 401–4.

Goldsmith, M. L. (1999). Ecological constraints on the foraging effort of western lowland gorillas (*Gorilla gorilla gorilla*) at Bai Hokou, Central African Republic. *International Journal of Primatology*, **20**, 1–23.

Goodall, J. (1986). *The Chimpanzees of Gombe*. Cambridge, MA: Harvard University Press.

Guo Z., Biscaye, P., Wei, L., Chen, X., Peng, S. & Liu, T. (2000). Summer monsoon variations over the last 1.2 Ma from the weathering of loess-soil sequences in China. *Geophysical Research Letters*, **27**, 1751–14.

Harrison, T. S. & Harrison, T. (1989). Palynology of the late Miocene *Oreopithecus*-bearing lignite from Baccinello, Italy. *Palaeogeography, Palaeoclimatology, Palaeoecology*, **76**, 45–65.

Harrison, T. & Rook, L. (1997). Enigmatic anthropoid or misunderstood ape? The phylogenetic status of *Oreopithecus bambolii* reconsidered. In *Function, Phylogeny, and Fossils: Miocene Hominoid Evolution and Adaptations*, ed. D. R. Begun, C. V. Ward & M. D. Rose, pp. 327–62. New York: Plenum Press.

Harvey, P. H., Martin, R. D. & Clutton-Brock, T. H. (1987). Life histories in comparative perspective. In *Primate Societies*, ed. B. B. Smuts, D. L. Cheney, R. M. Seyfarth, R. W. Wrangham & T. T. Struhsaker, pp. 181–96. Chicago, IL: University of Chicago Press.

Heizmann, E. & Begun, D.R. (2001). The oldest European Hominoid. *Journal of Human Evolution*, **41**, 465–81.

Hill, A. (1995). Faunal and environmental change in the Neogene of East Africa: evidence from the Tugen Hills Sequence, Baringo District, Kenya. In *Paleoclimate and Evolution with Emphasis on Human Origins*, ed. E. S. Vrba, G. H. Denton, T. C. Partridge & L. H. Burckle, pp. 178–93. New Haven, CT: Yale University Press.

Hill, A., Drake, R., Tauxe, L., Monaghan, M., Barry, J. C., Behrensmeyer, A. K., Curtis, G., Fine Jacobs, B., Jacobs, L., Johnson, N. & Pilbeam, D. (1985). Neogene palaeontology and geochronology of the Baringo Basin, Kenya. *Journal of Human Evolution*, **14**, 749–73.

Hladik, C.-M. (1981). Diet and the evolution of feeding strategies among forest primates. In *Omnivorous Primates*, ed. R. S. O. Harding & G. Teleki, pp. 215–54. New York: Columbia University Press.

Hou Y., Potts, R., Yuan B., Guo Z., Deino, A., Wang W., Clark, J., Xie G. & Huang W. (2000). Mid-Pleistocene Acheulean-like stone technology of the Bose basin, South China. *Science*, **287**, 1622–6.

Hsü, K. J., Montadert, L., Bernoulli, D., Cita, M. B., Erickson, A., Garrison, R. E., Kidd, R. B., Mélières, F., Müller, C. & Wright, R. (1978). History of the

Mediterranean salinity crisis. *Leg 42, Deep Sea, Drilling Project Initial Report*, **42**, 1053–78.

Hutterer, K. I. (1977). Reinterpreting the southeast Asian Paleolithic. In *Sunda and Sahul*, ed. J. Allen, J. Golson & R. Jones, pp. 31–72. London: Academic Press.

Jablonski, N. G. (1998). The response of catarrhine primates to Pleistocene environmental fluctuations in East Asia. *Primates*, **39**, 29–37.

Jablonski, N. G., Whitfort, M. J., Roberts-Smith, N. & Xu, Q. (2000). The influence of life history and diet on the distribution of catarrhine primates during the Pleistocene in eastern Asia. *Journal of Human Evolution*, **39**, 131–57.

Jacobs, B. F. & Kabuye, C. H. S. (1987). Environments of early hominoids: Evidence for Middle Miocene forest in East Africa. *Journal of Human Evolution*, **16**, 147–55.

Jones, R. W. (1999). Marine invertebrate (chiefly foraminiferal) evidence for the palaeogeography of the Oligocene–Miocene of western Eurasia, and consequences for terrestrial vertebrate migration. In *The Evolution of Neogene Terrestrial Ecosystems in Europe*, Vol. 1, ed. J. Agustí, L. Rook & P. Andrews, pp. 274–308. Cambridge, UK: Cambridge University Press.

Kappelman, J. (1988). Morphology and locomotor adaptations of the bovid femur in relation to habitat. *Journal of Morphology*, **198**, 119–30.

(1991). The paleoenvironment of *Kenyapithecus* at Fort Ternan. *Journal of Human Evolution*, **20**, 95–125.

Kay, R. F. (1977). Diets of early Miocene African hominoids. *Nature*, **268**, 628–30.

Kay, R. F. & Ungar, P. S. (1997). Dental evidence for diet in some Miocene catarrhines with comments on the effects of phylogeny on the interpretation of adaptation. In *Function, Phylogeny, and Fossils: Miocene Hominoid Evolution and Adaptations*, ed. D. R. Begun, C. V. Ward & M. D. Rose, pp. 131–52. New York: Plenum Press.

Kelley, J. (1997). Paleobiological and phylogenetic significance of life history in Miocene hominoids. In *Function, Phylogeny, and Fossils: Miocene Hominoid Evolution and Adaptations*, ed. D. R. Begun, C. V. Ward & M. D. Rose, pp. 173–208. New York: Plenum Press.

Kennett, J. P. (1995). A review of polar climatic evolution during the Neogene, based on the marine sediment record. In *Paleoclimate and Evolution with Emphasis on Human Origins*, ed. E. S. Vrba, G. H. Denton, T. C. Partridge & L. H. Burckle, pp. 49–64. New Haven, CT: Yale University Press.

King, T., Aiello, L. C. & Andrews, P. (1999). Dental microwear of *Griphopithecus alpani*. *Journal of Human Evolution*, **36**, 3–31.

Kingdon, J. (1997). *The Kingdon Guide to African Mammals*. San Diego, CA: Harcourt Brace.

Kingston, J. D. (1992). Stable isotopic evidence for hominid paleoenvironments in East Africa. Ph.D. dissertation, Harvard University.

(1999). Environmental determinants in early hominid evolution: issues and evidence from the Tugen Hills, Kenya. In *Late Cenozoic Environments and Hominid Evolution: A Tribute to Bill Bishop*, ed. P. Andrews & P. Banham, pp. 69–84. London: The Royal Geological Society.

Kingston, J. D., Marino, B. & Hill, A. (1994). Isotopic evidence for Neogene hominid paleoenvironments in the Kenya Rift Valley. *Science*, **264**, 955–9.

Kovar-Eder, J, Kvacek, Z., Zastawniak, E., Givulescu, R., Hably, L., Mihajlovic, D., Teslenko, J. & Walther, H. (1996). Floristic trends in the vegetation of the Paratethys surrounding areas during Neogene times. In *The Evolution of Western Eurasian Neogene Mammal Faunas*, ed. R. L. Bernor, V. Fahlbusch & H.-W. Mittmann, pp. 395–413. New York: Columbia University Press.

Leakey, M. & Walker, A. (1997). *Afropithecus*: function and phylogeny. In *Function, Phylogeny, and Fossils: Miocene Hominoid Evolution and Adaptations*, ed. D. R. Begun, C. V. Ward & M. D. Rose, pp. 225–40. New York: Plenum Press.

Leakey, M. G., Feibel, C. S., Bernor, R. L., Harris, J. M., Cerling, T. E., Stewart, K. M., Storrs, G. W., Walker, A., Werdelin, L. & Winkler, A. J. (1996). Lothagam: a record of faunal change in the late Miocene of East Africa. *Journal of Vertebrate Paleontology*, **16**, 556–70.

Leighton, D. R. (1987). Gibbons: territoriality and monogamy. In *Primate Societies*, ed. B. B. Smuts, D. L. Cheney, R. M. Seyfarth, R. W. Wrangham & T. T. Struhsaker, pp. 135–45. Chicago, IL: University of Chicago Press.

MacLatchy, L., Gebo, D., Kityo, R. & Pilbeam, D. (2000). Postcranial functional morphology of *Morotopithecus bishopi*, with implications for the evolution of modern ape locomotion. *Journal of Human Evolution*, **39**, 159–83.

McCrossin, M. L. & Benefit, B. R. (1997). On the relationships and adaptation of *Kenyapithecus*, a large-bodied hominoid from the middle Miocene of eastern Africa. In *Function, Phylogeny, and Fossils:*

Miocene Hominoid Evolution and Adaptations, ed. D. R. Begun, C. V. Ward & M. D. Rose, pp. 241–67. New York: Plenum Press.

Miller, K. G. & Mountain, G. S. (1996). Drilling and dating New Jersey Oligocene–Miocene sequences: ice volume, global sea level, and Exxon records. *Science*, **271**, 1092–5.

Milton, K. (1981). Distribution patterns of tropical plant foods as an evolutionary stimulus to primate mental development. *American Anthropologist*, **83**, 534–48.

Morgan M. E., Kingston, J. D. & Marino, B. D. (1994). Carbon isotopic evidence for the emergence of C4 plants in the Neogene from Pakistan and Kenya. *Nature*, **367**, 162–5.

Nakaya, H. (1994). Faunal change of late Miocene Africa and Eurasian mammalian fauna from the Namurungule Formation, Samburu Hills, northern Kenya. *African Study Monographs*, Supplement **20**, 1–112.

Nakaya, H., Pickford, M., Nakano, Y. & Ishida, H. (1984). The late Miocene large mammalian fauna from the Namurungule Formation, Samburu Hills, northern Kenya. *African Study Monographs*, Supplement Issue **5**.

Newton-Fisher, N. E., Reynolds, V. & Plumptre, A. J. (2000). Food supply and chimpanzee (*Pan troglodytes schweinfurthii*) party size in the Budongo Forest Reserve, Uganda. *International Journal of Primatology*, **21**, 613–28.

Nishida, T. & Hiraiwa-Hasegawa, M. (1987). Chimpanzees and bonobos: cooperative relationships among males. In *Primate Societies*, ed. B. B. Smuts, D. L. Cheney, R. M. Seyfarth, R. W. Wrangham & T. T. Struhsaker, pp. 165–77. Chicago, IL: University of Chicago Press.

O'Brien, E. M. & Peters, C. R. (1999). Climatic perspectives for Neogene environmental reconstructions. In *The Evolution of Neogene Terrestrial Ecosystems in Europe*, Vol. 1, ed. J. Agustí, L. Rook & P. Andrews, pp. 55–81. Cambridge, UK: Cambridge University Press.

Onderdonk, D. A. & Chapman, C. A. (2000). Coping with forest fragmentation: the primates of Kibale National Park, Uganda. *International Journal of Primatology*, **21**, 587–611.

Parker, S. T. (1996). Apprenticeship in tool-mediated extractive foraging: the origins of imitation, teaching, and self-awareness in great apes. In *Reaching Into Thought: The Minds of the Great Apes*, ed. A. E. Russon, K. A. Bard & S. T. Parker, pp. 348–70. Cambridge, UK: Cambridge University Press.

Parker, S. T. & Gibson, K. R. (1977). Object manipulation, tool use, and sensorimotor intelligence as feeding adaptations in cebus monkeys and great apes. *Journal of Human Evolution*, **6**, 623–41.

Parker, S. T., Mitchell, R. W. & Boccia, M. (ed.) (1994). *Self Awareness in Animals and Humans*. Cambridge, UK: Cambridge University Press.

Pickford, M. (1985). A new look at *Kenyapithecus* based on recent discoveries in western Kenya. *Journal of Human Evolution*, **14**, 113–43.

(1987). Fort Ternan (Kenya) paleoecology. *Journal of Human Evolution*, **16**, 305–9.

Pickford, M. & Senut, B. (2001). The geological and faunal context of Late Miocene hominid remains from Lukeino, Kenya. *Comptes Rendus de l'Académie des Sciences Paris*, **332**, 145–52.

Pilbeam, D. (1997). Research on Miocene hominoids and hominid origins: the last three decades. In *Function, Phylogeny, and Fossils: Miocene Hominoid Evolution and Adaptations*, ed. D. R. Begun, C. V. Ward & M. D. Rose, pp. 13–28. New York: Plenum.

Pilbeam, D., Barry, J., Meyer, G. E., Shah, S. M., Pickford, M. H. L., Bishop, W. W., Thomas, H. & Jacobs, L. L. (1977). Geology and palaeontology of Neogene strata of Pakistan. *Nature*, **270**, 684–9.

Potts, R. (1996). *Humanity's Descent: The Consequences of Ecological Instability*. New York: Avon.

(1998a). Variability selection in hominid evolution. *Evolutionary Anthropology*, **7**, 81–96.

(1998b). Environmental hypotheses of hominin evolution. *Yearbook of Physical Anthropology*, **41**, 93–136.

Povinelli, D. J. & Cant, J. G. H. (1995). Arboreal clambering and the evolution of self-conception. *The Quarterly Review of Biology*, **70**, 393–421.

Quade, J., Cater, J. M. L., Ojha, T. P., Adam, J. & Harrison, T. M. (1995). Late Miocene environmental change in Nepal and the northern Indian subcontinent: stable isotopic evidence from paleosols. *GSA Bulletin*, **107**, 1381–97.

Quade, J., Cerling, T. E., Bowman, J. R., *et al.* (1989). Development of Asian monsoon revealed by marked ecological shifts during the last Miocene in northern Pakistan. *Nature*, **342**, 163–5.

Raza, S. M., Barry, J. C., Pilbeam, D., Rose, M. D., Shah, S. M. I. & Ward, S. C. (1983). New hominoid primates from the middle Miocene Chinji Formations, Potwar, Plateau, Pakistan. *Nature*, **406**, 52–4.

Retallack, G. J. (1992). Comment on the paleoenvironment of *Kenyapithecus* at Fort Ternan. *Journal of Human Evolution*, **23**, 363–9.

Retallack, G. J., Dugas, D. P. & Bestland, E. A. (1990). Fossil soils and grasses of a middle Miocene East African grassland. *Science*, **247**, 1325–8.

Rodman, P. S. & Mitani, J. C. (1987). Orangutans: sexual dimorphism in a solitary species. In *Primate Societies*, ed. B. B. Smuts, D. L. Cheney, R. M. Seyfarth, R. W. Wrangham & T. T. Struhsaker, pp. 146–54. Chicago, IL: University of Chicago Press.

Rögl, F. (1999). Mediterranean and Paratethys palaeogeography during the Oligocene and Miocene. In *The Evolution of Neogene Terrestrial Ecosystems in Europe*, Vol. 1, ed. J. Agustí, L. Rook & P. Andrews, pp. 8–22. Cambridge, UK: Cambridge University Press.

Rook, L., Renne, P., Benvenuti, M. & Papini, M. (2000). Geochronology of *Oreopithecus*-bearing succession at Baccinello (Italy) and the extinction pattern of European Miocene hominoids. *Journal of Human Evolution*, **39**, 577–82.

Rose, M. D. (1997). Functional and phylogenetic features of the forelimb in Miocene hominoids. In *Function, Phylogeny, and Fossils: Miocene Hominoid Evolution and Adaptations*, ed. D. R. Begun, C. V. Ward & M. D. Rose, pp. 79–100. New York: Plenum.

Rosen, B. R. (1999). Palaeoclimatic implications of the energy hypothesis from Neogene corals of the Mediterranean region. In *The Evolution of Neogene Terrestrial Ecosystems in Europe*, Vol. 1, ed. J. Agustí, L. Rook & P. Andrews, pp. 309–27. Cambridge, UK: Cambridge University Press.

Rossignol-Strick, M. (1983). African monsoons, an immediate climate response to orbital insolation. *Nature*, **304**, 46–9.

Rossignol-Strick, M., Paterne, M., Bassinot, F. C., Emeis, K.-C. & De Lange, G. J. (1998). An unusual mid-Pleistocene monsoon period over Africa and Asia. *Nature*, **392**, 269–72.

Russon, A. E. (1998). The nature and evolution of intelligence in orangutans (*Pongo pygmaeus*). *Primates*, **39**, 485–503.

Sabater Pi, J. (1977). Contribution to the study of alimentation of lowland gorillas in the natural state, in Rio Muni, Republic of Equatorial Guinea (west Africa). *Primates*, **18**, 183–204.

Schoeninger, M. J., Moore, J. & Sept, J. M. (1999). Subsistence strategies of two "savanna" chimpanzee populations: the stable isotope evidence. *American Journal of Primatology*, **49**, 297–314.

Scott, R. S., Kappelman, J. & Kelley, J. (1999). The paleoenvironment of *Sivapithecus parvada*. *Journal of Human Evolution*, **36**, 245–74.

Senut, B., Pickford, M., Gommery, D., Mein, P., Cheboi, K. & Coppens, Y. (2001). First hominid from the Miocene (Lukeino Formation, Kenya). *Comptes Rendus de l'Académie des Sciences Paris*, **332**, 137–44.

Sevim, A., Begun, D. R., Gülec, E., Geraads, D. & Pehlevan, C. (2001). A new late Miocene hominid from Turkey. *American Journal of Physical Anthropology*, Supplement **32**, 134–5.

Shackleton, N. J. (1995). New data on the evolution of Pliocene climatic variability. In *Paleoclimate and Evolution with Emphasis on Human Origins*, ed. E. S. Vrba, G. H. Denton, T. C. Partridge & L. H. Burckle, pp. 242–8. New Haven, CT: Yale University Press.

Shipman, P. (1986). Paleoecology of Fort Ternan reconsidered. *Journal of Human Evolution*, **15**, 193–204.

Shipman, P., Walker, A., Van Couvering, J. A., Hooker, P. J. & Miller, J. A. (1981). The Fort Ternan hominoid site, Kenya: Geology, age, taphonomy, and paleoecology. *Journal of Human Evolution*, **10**, 49–72.

Stewart, K. J. & Harcourt, A. H. (1987). Gorillas: variation in female relationships. In *Primate Societies*, ed. B. B. Smuts, D. L. Cheney, R. M. Seyfarth, R. W. Wrangham & T. T. Struhsaker, pp. 155–64. Chicago, IL: Chicago University Press.

Suc, J. -P., Fauquette, S., Bessedik, M, Bertini, A., Zheng, Z., Clauzon, G., Suballyova, D., Diniz, F., Quézel, P., Feddi, N., Clet, M., Bessais, E., Taoufiq, N. B., Meon, H. & Combourieu-Nebout, N. (1999). Neogene vegetation changes in west European and west circum-Mediterranean areas. In *The Evolution of Neogene Terrestrial Ecosystems in Europe*, Vol. 1, ed. J. Agustí, L. Rook & P. Andrews, pp. 378–88. Cambridge, UK: Cambridge University Press.

Sugiyama, Y. (1999). Socioecological factors of male chimpanzee migration at Bossou, Guinea. *Primates*, **40**, 61–8.

Sun X. & Wu Y. (1980). Paleoenvironment during the time of *Ramapithecus lufengensis*. *Vertebrata PalAsiatica*, **18**, 247–55.

Teaford, M. F. & Walker, A. C. (1984). Quantitative differences in dental microwear between primate species with different diets and a comment on the presumed diet of *Sivapithecus*. *American Journal of Physical Anthropology*, **64**, 191–200.

Teilhard de Chardin, P., Young, C. C., Pei W. C. & Chang H. C. (1935). On the Cenozoic Formations of Kwangsi and Kwangtung. *Bulletin of Geological Society of China*, **14**, 179–205.

Tomasello, M. & Call, J. (1997). *Primate Cognition*. New York: Oxford University Press.

Tutin, C. E. G. (1999). Fragmented living: behavioural ecology of primates in a forest fragment in the Lopé Reserve, Gabon. *Primates*, **40**, 249–65.

Ungar, P. S. (1996). Dental microwear of European Miocene catarrhines: evidence for diets and tooth use. *Journal of Human Evolution*, **31**, 335–66.

Ungar, P. S. & Kay, R. F. (1995). The dietary adaptations of European Miocene catarrhines. *Proceedings of the National Academy of Sciences, USA*, **92**, 5479–81.

van Schaik, C. P. (1999). The socioecology of fission–fusion sociality in orangutans. *Primates*, **40**, 69–87.

Wang K.-F., Zhang Y.-L. & Jian H. (1991). Migration of paleovegetational zone and paleoclimatic changes of Quaternary in the coastal region of South China. *Chinese Science Bulletin*, **36**, 1721–4.

Wang, P. (1990). The ice-age China Sea – status and problems. *Quaternary Sciences*, **2**, 111–24.

Ward, C. V. (1997). Functional anatomy and phyletic implications of the hominoid trunk and hindlimb. In *Function, Phylogeny, and Fossils: Miocene Hominoid Evolution and Adaptations*, ed. D. R. Begun, C. V. Ward & M. D. Rose, pp. 101–30. New York: Plenum Press.

Ward, S. (1997). The taxonomy and phylogenetic relationships of *Sivapithecus* revisited. In *Function, Phylogeny, and Fossils: Miocene Hominoid Evolution and Adaptations*, ed. D. R. Begun, C. V. Ward & M. D. Rose, pp. 269–90. New York: Plenum.

Ward, S., Brown, B., Hill, A., Kelley, J. & Downs, W. (1999). *Equatorius*: a new hominoid genus from the middle Miocene of Kenya. *Science*, **285**, 1382–6.

Watts, D. P. (1998). Seasonality in the ecology and life histories of mountain gorillas (*Gorilla gorilla beringei*). *International Journal of Primatology*, **19**, 929–48.

White, F. J. (1998). Seasonality and socioecology: the importance of variation in fruit abundance to bonobo sociality. *International Journal of Primatology*, **19**, 1013–27.

Woodruff, F., Savin, S. M. & Douglas, R. G. (1981). Miocene stable isotope record: a detailed deep Pacific Ocean study and its paleoclimatic implications. *Science*, **212**, 665–8.

Wrangham, R. W. (1980). An ecological model of female-bonded primate groups. *Behaviour*, **75**, 262–99.

Wrangham, R. W., Conklin-Brittain, N. L. & Hunt, K. D. (1998). Dietary response of chimpanzees to seasonal variation in fruit abundance. I. Antifeedants. *International Journal of Primatology*, **19**, 949–70.

Wright, J. D. & Miller, K. G. (1992). Miocene stable isotope stratigraphy, Site 747, Kerguelen Plateau. *Proceedings of the Ocean Drilling Program, Scientific Results*, **120**, 855–66.

Yamagiwa, J. (1999). Socioecological factors influencing population structure of gorillas and chimpanzees. *Primates*, **40**, 87–104.

Yeager, C. P. & Kirkpatrick, R. C. (1998). Asian colobine social structure: ecological and evolutionary constraints. *Primates*, **39**, 147–55.

14 · Cranial evidence of the evolution of intelligence in fossil apes

DAVID R. BEGUN[1] AND LÁSZLÓ KORDOS[2]

[1]Department of Anthropology, University of Toronto, Toronto
[2]The Geological Institute of Hungary, Budapest

INTRODUCTION

Fossil endocasts, natural or artificial casts of the inside of a cranial vault, provide the most direct evidence of the evolution of the brain. Among fossil hominoids, the vast majority of endocasts come from Plio-Pleistocene hominids, and these have been described in detail, (Conroy, Vannier & Tobias 1990; Dart 1925; Falk 1980a,b, 1983a,b, 1987, 1990; Falk & Conroy 1983; Holloway 1974a, 1982, 1983a, 1984, 1995; Holloway & De la Coste-Lareymondie 1982; Martin 1983, 1990; Martin & Harvey 1985; Schepers 1946, 1950; Tobias 1967, 1971a,b, 1975, 1978, 1983, 1991, 1995). Fossil great ape endocasts are extremely rare and are thus far undescribed. Therefore, beyond extrapolation from an outgroup, little is known of the primitive condition from which modern great ape and human brains could have evolved.

Six specimens of the primitive Oligocene catarrhine *Aegyptopithecus zeuxis* from about 33–33.5 Ma are described (Radinsky 1973, 1974, 1977; Rasmussen 2002; Simons 1993). Among hominoids, only four specimens are sufficiently complete to estimate brain size: one for *Proconsul nyanzae*, an early Miocene (*c.* 18 Ma) primitive or stem[1] hominoid that predates the emergence of the great ape and human clade, and three for the great apes *Dryopithecus brancoi* and *Oreopithecus bambolii* from between about 10 to 6 Ma (Begun 2002; Falk 1983a; Harrison 1989; Kordos 1990; Kordos & Begun 1997, 1998, 2001a; Walker *et al.* 1983).

The only fossil hominoid for which the endocast has yet been described is *Proconsul. Proconsul* is said to be more encephalized than monkeys of similar size, and close to living great apes (Walker *et al.* 1983), though this conclusion is revisited here. Most authorities have also concluded that the endocast of *Proconsul* is morphologically more primitive than that of any living hominoid

(Falk 1983a; Radinsky 1974). Between the primitive endocast of the early catarrhine *Aegyptopithecus* and the stem hominoid *Proconsul* there is about a 15 Ma gap.

There is another 8 Ma gap from *Proconsul* to the late Miocene great ape *Dryopithecus brancoi* (Kordos & Begun 1997, 2001a). *Oreopithecus* and *Sahelanthropus*, a newly described hominid from Chad, both between 6 and 7 Ma in age, fill the gap between *Dryopithecus* and the earliest australopithecine for which brain and body size data are available, *Australopithecus afarensis* (Brunet *et al.* 2002; Harrison & Rook 1997). *Oreopithecus* appears unique in brain size (see below) while *Sahelanthropus*, like *Dryopithecus*, appears to have a great-ape-sized brain relative to its body mass (see below and Brunet *et al.* 2002). *A. afarensis*, from 3.6–2.9 Ma, shows a level of encephalization comparable to or slightly above that seen in living great apes and *Dryopithecus* and clearly above that seen in *Proconsul, Oreopithecus*, and most other anthropoids (see below and Jerison 1973, 1975; Kappelman 1996; Martin 1983, 1990; Pilbeam & Gould 1974; White 2002). In this chapter we review the available fossil evidence and assess its relevance to a reconstruction of the evolution of the brain in great apes.

APE ANCESTORS

Aegyptopithecus is a propliopithecoid, a primitive catarrhine (Harrison 1987; Rasmussen 2002), and its endocasts are informative as a precursor of the brain in hominoids. *Aegyptopithecus* lived before the divergence of the Old World monkeys and apes, and is known primarily from Oligocene deposits in the Fayum depression of Egypt (Fleagle 1983; Fleagle *et al.* 1986; Fleagle & Kay 1983; Harrison 1987; Kappelman, Simons & Swisher 1992; Kay, Fleagle & Simons 1981; Rasmussen 2002; Simons 1965, 1968, 1987, 1993).

Table 14.1. *Body mass and endocranial volume estimates in some fossil catarrhines*

Body mass (kg)[1]	*Proconsul* RU-7290[2]	*D. brancoi* RUD 77	*D. brancoi* RUD 200	*Aegyptopithecus*[3]
Log BM = 4.718 (log OHT) − 2.56	13.5	31.0	22.7	
Log BM = 4.445 (log OHT) − 2.155	14.3	31.2	23.2	5.3–6.0
Log BM = 4.420 (log OHT) − 2.12	14.0	30.4	22.6	
Log BM = 5.22 (log OB) − 3.35	18.5	28.7	20.3	
Ln BM = 1.62 (ln M^{1}S.A.) + 2.72	16.2	22.8	21.6	
Ln BM = 1.37 (ln M^{2}S.A.) + 3.49	17.9	19.7	19.0	
Endocranial volume[2] EV = 2.5 (CL) + 55.3	167 cc (155–181)	330 cc (302–350)	305 cc (280–330)	27–33 cc

Notes:

Abbreviations: BM = body mass; OHT = orbital height; OB = orbital breadth; M$^\#$S.A. = molar surface area; EV = endocranial volume; CL = cranial length. Ranges of endocranial volume estimates are 95% confidence intervals except for *Aegyptopithecus*. Ranges of body mass estimates for *Dryopithecus* are 95% confidence intervals. For *Proconsul*, see text.

[1] Body mass formulae from Aiello and Wood (1994), Kappelman (1996), Gingerich, Smith & Rosenberg (1982)

[2] Endocranial volume estimates formulae from Walker *et al.* (1983). Body mass estimates from this study (see text).

[3] Body mass endocranial volume estimates for *Aegyptopithecus* from Simons (1993).

Radinsky originally interpreted the brain of *Aegyptopithecus* to be large compared with living prosimians, but later suggested that it is probably most similar in relative size to prosimians (Radinsky 1977; see also Jerison 1979 and Table 14.1). Simons (1993) estimated volumes of the most complete of the *Aegyptopithecus* endocasts at between about 27 and 33 cc and the body mass of *Aegyptopithecus* at between about 5300 to 6000 g, suggesting a very small brain size compared with living anthropoids. Radinsky (1973, 1974) stressed the more modern anthropoid-like qualities of the *Aegyptopithecus* endocast, including evidence of a larger visual cortex, reduced olfactory lobes, and a well-defined central sulcus between the primary somatic and motor cortices. He also listed a number of possibly primitive, more prosimian-like qualities, including smaller frontal lobes with fewer sulci and more rostral (anterior) olfactory lobes. He interpreted these primitive traits either as indications of the primitive nature of the brain of *Aegyptopithecus*, or due to allometric effects (Radinsky 1973, 1974). Radinsky noted that "primitive" aspects of the endocasts of *Aegyptopithecus* and *Alouatta*, the largest members of their respective clades, may be related to large body mass. Most recent analyses of *Aegyptopithecus*, however, have concluded that the primitive aspects of the endocranium are just that, primitive and prosimian-like (e.g., Simons 1993).

In summary, *Aegyptopithecus* was about as encephalized as many living prosimians and shares primitive characters with prosimians and derived characters with living anthropoids. However, in overall size, the brain of *Aegyptopithecus* was primitive anthropoid-like (see below). The exact behavioral implications of the brain of *Aegyptopithecus* are not clear. It is possible that increasing the amount of neurological tissue devoted to processing visual stimuli, decreasing that devoted to olfaction, and an emerging distinction between primary sensory and motor cortices reflect a greater dependence on vision over olfaction and more refined neurological control over movement (Radinsky, 1973). These are conceivably precursors to more complex forms of cognitive processing that distinguish anthropoids from prosimians, and which are most elaborately developed in great apes and humans.

FOSSIL EARLY APES

The earliest direct evidence of hominoid brain evolution comes from the endocast of KNM RU-7290, a well-preserved skull of *Proconsul nyanzae*[1] from Rusinga Island, Kenya. Le Gros Clark & Leakey (1951) concluded that it had a relatively small frontal lobe and a simple, relatively primitive sulcal pattern. Radinsky (1974) concluded that it most closely resembles *Hylobates*. He noted that the only apparently primitive character of this endocast is the absence of a frontal sulcus, which is otherwise usually present in modern cercopithecoids and hominoids.

Falk (1983a) noted that the sulcal pattern of KNM RU-7290 (BMNH 32363)[1] is more complicated. She supported Radinsky's conclusion that the endocast is anthropoid-like, but did not see the hominoid affinities that Radinsky stressed. She noted that many of the sulci are also present in the brains of Old and New World Monkeys, suggesting that they are primitive for the anthropoids and not derived similarities shared with hominoids. However, the absence of sulci typical of apes in the *Proconsul* endocast is more ambiguous than, for example, the absence of skeletal attributes of the elbow joint or the face. Endocasts are trace fossils of brain surfaces. Even under the best circumstances, endocasts often do not reveal the sulci that are present on the surface of the cerebrum, due to the fact that the meninges, meningeal blood vessels, and cerebrospinal fluid intervene between the endocranial and brain surface. Sulci are notoriously difficult to identify on partial endocasts. The absence of specific sulci on an endocast is not especially strong evidence that they were not present on a cerebrum. In addition, at least one sulcus present on the *Proconsul* endocast is present only in great apes (Falk 1983a). The absence in hylobatids of various sulci present in *Proconsul*, great apes, and some cercopithecoids may simply reflect the absolutely small size of hylobatid brains, which is known to be correlated to sulcal complexity (Jerison 1973; MacLeod, Chapter 7, this volume). *Proconsul* endocast morphology represents a likely starting point for the evolution of all subsequent hominoid brains.

Radinsky (1974) estimated the volume of the *Proconsul* endocast at about 150 cc, but he later decided that the specimen was too damaged to estimate its volume accurately (Radinsky 1979). Falk (1983a) agreed with Radinsky's "ball park" estimate of the brain size being close to that of *Papio*. Walker *et al.* (1983) used a regression analysis to predict the brain size of KNM RU-7290 from a new reconstruction of the skull with additional conjoining fragments. They estimated the cranial capacity of this *Proconsul* individual to be 167 cc (95% CI=155–181). With this and an estimate of body mass, they calculated an encephalization quotient or EQ for *Proconsul*, which they characterized as larger than in monkeys of similar size. They suggest that this may be a great ape trait, but express uncertainty given the fact that all living great apes are much larger than this individual of *Proconsul*. Manser and Harrison (1999) estimated the cranial capacity of the same specimen at 130.3 cc based on foramen magnum size. Based on the overall size of the cranium and damage to the foramen magnum, we place more confidence in the estimates of Walker *et al.* (1983), which are used here.

Walker *et al.*'s (1983) estimate of body mass (11 kg) is based on a second individual, R114 or KNM RU-2036, a partial skeleton of a subadult *Proconsul* and the type specimen of *Proconsul heseloni* (Walker *et al.* 1993). Only small fragments are known of the cranium of KNM RU-2036. Comparable portions are smaller overall than in KNM RU-7290, and there are a number of morphological differences as well (Walker *et al.* 1993; Begun & Kordos pers. obs.). These specimens may not represent the same species of *Proconsul*, and even if they do, the evidence suggests that KNM RU-7290 represents a larger individual than KNM RU-2036. While KNM RU-2036 is about the skeletal size of an average adult male *Colobus polykomos*, KNM RU-7290 is larger than this monkey and *Hylobates symphalangus*, as well as larger than male *Macaca fuscata* and *Alouatta pigra* (11–13 kg) in almost all cranial measurements (Delson *et al.* 2000; Smith & Jungers 1997). It is closer to male *Semnopithecus entellus* and female *Papio anubis* and *Mandrillus sphinx* (Begun pers. obs.), which are all about 13–18 kg in mean body mass.

KNM RU-7290 preserves a complete dentition and orbits, which have been shown to have a close relationship to body mass (Aiello & Wood 1994; Conroy 1987; Dagasto & Terranova 1992; Gingerich 1979; Kappelman 1996). Following methods described by these authors we estimated the body mass of this individual at between 13.5 and 18.5 kg (Table 14.1). At this body mass, the EQ of the species of *Proconsul* represented by KNM RU-7290 is within the range of values for similarly sized Old World monkeys. There is no strong evidence that *Proconsul* shares any degree of increased encephalization with living great apes. The significance of EQ measures of relative brain size is discussed in more detail below.

FOSSIL GREAT APES

Following Walker *et al.*'s lead, we focus in this section on the bases for estimating body size, brain size, and EQ in fossil great apes. Comparative evolutionary implications of these EQ estimates are discussed in the following section.

Dryopithecus

Two cranial specimens of *Dryopithecus* provide the earliest direct evidence of brain size and morphology in a fossil great ape. RUD 77 is a partial cranium from the late Miocene locality of Rudabánya attributable to *Dryopithecus brancoi* (Begun & Kordos 1993; Kordos & Begun 1997). It preserves much of the cranial vault of an adult female, which allows for a relatively confident estimate of cranial capacity, using techniques similar to those described above for *Proconsul* (Kordos & Begun 1997, 1998, 2001a; Walker *et al.* 1983). RUD 200 is a more recently discovered and more completely preserved cranium, which makes the estimate of cranial capacity in this specimen even more certain than in RUD 77 (Kordos & Begun 2001a). Both specimens of *Dryopithecus* preserve a few details of endocranial surface morphology and general features of relative cerebral lobe size.

Both endocasts of *Dryopithecus* preserve portions of the frontal and parietal lobes, but very little of the temporal and occipital lobes, and none of the cerebellum, olfactory lobes, or any structure of the ventral surface of the brain. The frontal lobes are preserved anteriorly and superiorly but not inferiorly. They are broader relative to length than in *Proconsul* and *Hylobates* but narrower rostrally compared with living great apes. The parietal lobes are also broad transversely compared with *Proconsul* and *Hylobates*. The endocast is asymmetric, with subtle right frontal and more pronounced left occipital petalia (for comparison, see MacLeod, Chapter 7, this volume).

The sulci of the frontal lobes are more discernable than on the parietal and occipital lobes. They are clearly more complex than in *Proconsul* or the typical pattern in *Hylobates*. The rectus sulcus is short and immediately superior to the superior orbital surface, and it is surrounded caudally by a clear arcuate sulcus, which does not occur in *Proconsul* or *Hylobates* (Falk 1983a). Between the arcuate and central sulci on both endocasts, two additional sulci are apparent, which probably correspond to the precentral and the superior frontal sulcus. Only hominids have such complexity to their lateral frontal endocasts, reinforcing the view that *Dryopithecus* shares brain morphology with living hominids.

We took three neurocranial measurements on *Dryopithecus* and a sample of great ape specimens of known cranial capacity to estimate brain size in the fossils. Based on these measurements we calculated least squares (LS) and reduced major axis (RMA) regressions, and used the resulting regression formulae (shown in Table 14.1) to estimate cranial capacity. The six formulae produce consistent results; one is reproduced in Table 14.1 and Figure 14.1. Our best estimates of brain size in these two specimens are 305 and 330 cc.

The mean percentage predictor errors (MPE) for all equations were well under 10%, which is quite low (Dagasto & Terranova 1992). The frequency with which predicted endocranial volumes were within 20% of the observed cranial capacities (% ± 20%) was over 99%. The MPE and % ± 20% analysis of these regressions suggest that the predictions are reasonable, despite relatively modest correlation coefficients.

A separate sample of four bonobo (*Pan paniscus*) crania of similar size to *Dryopithecus* was also used to assess the reliability of the predictions. For all these bonobos, the regression predicts a cranial capacity within 10% of the known values for each cranium. Finally, the predicted size of the endocranial volumes of RUD 77 and RUD 200 was compared to actual volumes of great ape endocasts of similar linear dimensions. The overall sizes of the endocasts are close to small endocasts of *Pan* and *Pongo*, in the range of 300 to 350 cc.

RUD 77 and RUD 200 have the orbits and dentition sufficiently well preserved to make reasonable estimates of body mass. Based on orbital dimensions, the estimates range from 28.7 to 31.2 kg for RUD 77 and 20.3 to 23.2 kg for RUD 200 (Table 14.1). These estimates are consistent with overall cranial and postcranial dimensions in extant catarrhines of known body mass. RUD 77 and RUD 200 are larger than monkeys in most

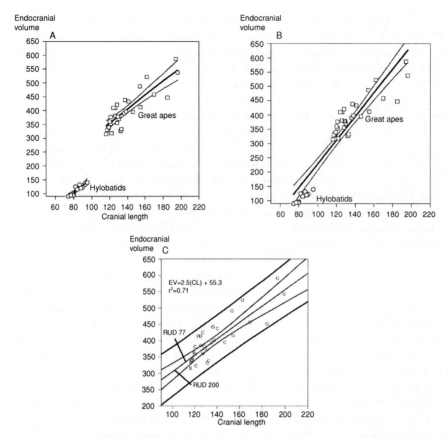

Figure 14.1. (A) Least squares regressions and 95% confidence intervals of endocranial volume against cranial length in hylobatids and great apes. Note the downward displacement of the hylobatid line, which has statistically the same slope but a different *y*-intercept from the great ape line. (B) Combining these data sets produces a regression with a tighter fit, but this is an artifact of the large size range and results in a slope that is not meaningful for either group. (C) Although the *Dryopithecus* specimens are slightly smaller than the smallest specimens of the comparative sample, the morphological similarities and close evolutionary relationship with these taxa (extant great apes) make this regression more informative for predicting brain mass in *Dryopithecus*. See text for discussion.

dimensions and smaller overall than the smallest great apes (*Pan troglodytes schweinfurthii* and *Pan paniscus*) (Smith & Jungers 1997). Female *Dryopithecus* postcrania are much smaller than their homologues in *Pan* and larger than in most monkeys (Begun 1992b, 1993, 1994; Jungers 1982; Kordos & Begun 2001b; Morbeck 1983). Most monkeys are smaller than 20 kg and great apes are larger than 27 kg (Smith & Jungers 1997).

Oreopithecus

The only other fossil ape for which an estimate of cranial capacity has been made directly from the cranial evidence is *Oreopithecus*. Straus (1963) and Straus and Schön (1960) estimated the cranial capacity of a very severely crushed adult male specimen of *Oreopithecus*

to have been between about 276–529 cc, with a best guess estimate of about 400 cc, which they said compares favorably with australopithecines as well as great apes. This estimate is based on external dimensions of a reconstruction and is much too high. Harrison (1989) estimated cranial capacity from another individual of unknown sex with a well-preserved foramen magnum, using a regression of foramen magnum area on brain size in a sample of modern anthropoids. His estimate was 128 cc, with a range between 83 and 173 cc. Using an estimated species mean body mass of 22.5 kg, Harrison (1989) calculated that the relative brain size of *Oreopithecus* as quite low by modern anthropoid standards. A male *Oreopithecus* has a body mass estimated at about 30 kg (Jungers 1987) and females, which are thought

Table 14.2. *Brain mass, body mass and relative brain size in fossil (bold) and extant primates. Taxa are listed in increasing order of brain mass*

Taxon	Brain mass (g)	Body mass (kg)	Encephalization EQ
Callitrichids	9.5–17.6	0.35–0.67	1.43–1.92
Aegyptopithecus **female**	**29**	**6**	**0.78**
Cebids	24.8–118.4	0.63–8.89	1.38–4.79
Old World Monkeys[1]	41.1–119.4	1.38–21.32	1.05–2.76
Hylobatids	87.5–133	5.70–12.74	1.93–2.74
Gibbons	87.5–105	5.70–7.37	1.93–2.74
Oreopithecus **female**	**112**	**15**	**1.49**
Most papionins[2]	116–179	8.68–32	1.48–2.76
Siamang	133	12.74	2.03
Proconsul **female**	**146**	**15**	**1.94**
Papio	179–222	16–35	1.73–2.35
Pongo female	288	44.45	1.63
Dryopithecus **female**	**289**	**31**	**2.35**
Pan troglodytes female	325	43.90	2.17
Pan paniscus female	314	38	2.24
Pan paniscus male	334	61	1.73
Pongo male	395	90.72	1.91
Gorilla female	426	90.72	1.76
Pan troglodytes male	440	56.69	2.48
Gorilla male	570	172.37	1.53

Notes:

Data on most extant primates are from Jerison (1973). They are the largest brain and body mass data, taken from Bauchot and Stephan (1969) from the same individuals. These body masses should not be considered accurate species means, as provided more reliably by Smith and Jungers (1997). Values for fossil catarrhines are from Harrison (1989), Radinsky (1977), Walker *et al.* (1983) and this chapter. Values for cranial capacity were divided by 1.14 to convert brain volume into brain mass in grams (Hartwig-Scherer 1993; Kappleman 1996). EQ for fossil taxa and *Pan paniscus* were calculated using formulae from Jerison (1973) for comparability. It is noteworthy that data from Jungers and Susman (1984) and Tobias (1971b) are generally higher than those from Bauchot and Stephan (1969) that are reproduced in Jerison (1973).

[1] Includes only *Cercopithecus*, *Miopithecus*, *Macaca*, and *Semnopithecus*.

[2] Includes *Cercocebus*, *Lophocebus*, and *Mandrillus* and excludes *Papio*.

to have been about half the size of males, have an estimated mean body mass of about 15 kg (hence Harrison's 22.5 kg species mean). Using any of these body mass estimates, the low degree of encephalization in *Oreopithecus* found by Harrison is confirmed (see Table 14.2). Our examination of the specimen used by Straus and Schön supports Harrison's conclusions. The cranium appears large due to the presence of massive ectocranial crests, but the neurocranial cavity itself was clearly short and quite small transversely. If *Oreopithecus* is a great ape, which is likely based on the preponderance of

fossil evidence (Begun 2002; Begun, Ward & Rose 1997; Harrison 1986; Harrison & Rook 1997; Hürzeler 1949, 1951, 1958, 1960; Straus 1961, 1963), then it represents a relatively unusual case of "de-encephalization," which is discussed briefly below. Nothing has been published to date on the morphology of the brain of *Oreopithecus*, for which no endocast is currently described.

Other fossil great apes

Kelley (1997, Chapter 15, this volume) carried out a detailed analysis of the pattern and timing of dental maturation in *Sivapithecus parvada*, a fossil great ape from South Asia widely believed to be closely related to *Pongo* (Andrews 1992; Andrews & Cronin 1982; Andrews & Martin 1987; Begun & Güleç 1998; Begun *et al.* 1997; Kelley 2002; Kelley & Pilbeam 1986; McCollum & Ward 1997; Pilbeam 1982; Ward 1997b; Ward & Brown 1986; Ward & Kimbel 1983; Ward & Pilbeam 1983; but see Pilbeam 1997; Pilbeam & Young 2001). Kelley's analysis indicates that *Sivapithecus* matured dentally in a manner essentially identical to living great apes. He used a well-known correlation between the rate of dental maturation in primates, particularly the age at which the first molar M1 erupts, and brain size, to estimate an older-than-expected (i.e., hominoid-like) age of M1 emergence for *Proconsul* (Smith 1989, 1991). Applying the same logic to the finding of a great-ape-like age of M1 emergence for *Sivapithecus* suggests that this taxon had a brain size in the modern great ape range (Kelley, Chapter 15, this volume). Unfortunately, the neurocranium of *Sivapithecus* is not known, so this prediction cannot be tested directly at present.

A male and female cranium of the Chinese fossil great ape *Lufengpithecus lufengensis* are described in Kordos (1988), Schwartz (1984a, b, 1990), and Wu, Qinghua & Quingwu 1983. They are very badly crushed, but, as is the case with *Oreopithecus*, careful scrutiny can reveal some important anatomical details. It is clear from our examination of these specimens that the crania are very close in overall size to those of small- to medium-sized living great apes (female *Pongo* to male/female *Pan*) and that they lacked the ectocranial cresting of *Oreopithecus*. The neurocrania, though crushed to the thickness of a thick pancake, were large in relation to the face, and the brains were probably in the range of modern great apes. No numerical estimate of cranial capacity is possible, but the conclusion that *Lufengpithecus* probably had a great-ape-sized brain is consistent with its phylogenetic position as closely related to *Sivapithecus* and *Pongo* (Kordos 1988; Schwartz 1984a, b, 1990; Wu *et al.* 1983). As with *Sivapithecus*, the morphology of the brain of *Lufengpithecus* is not currently known.

RELATIVE BRAIN SIZE

Encephalization quotient

With body mass and cranial capacity estimates from the same individuals of *Dryopithecus* it is possible for the first time to quantify relative brain size in a fossil great ape. New data also allow for a proposed revision of the relative brain size calculation of *Proconsul*. There are many methods of normalizing brain size, most of which give equivalent results (Bauchot & Stephan 1969; Begun & Walker 1993; Hartwig-Scherer 1993; Harvey 1988; Jerison 1973, 1979; Kappelman 1996; Martin 1983, 1990; Martin & Harvey 1985; Pagel & Harvey 1988; Radinsky 1974, 1977, 1979, 1982; Tobias 1971a, 1975). The most widely used techniques employ regression analysis to compare predicted brain sizes at a given body mass with observed brain sizes in animals of known or estimated (in the case of fossils) body mass. Primates with brains that are larger than expected for mammals of the same body mass are considered "encephalized," which is generally the case for hominoids. The techniques basically vary in the assumptions made with regard to the expected relationship of brain mass to body mass and depend in large part on the animals included in the comparison. Here we calculate EQ, probably the most widely used brain size normalizing statistic, using the formula from Jerison (1973) to facilitate comparisons across the large number of primates included in his analysis and widely reproduced elsewhere. However, EQ is not without its problems (see below).

Estimates of EQ are shown in Table 14.2. With a revised estimate of the body mass of *Proconsul* (KNM RU-7290), a revised EQ is reported. The EQ estimate for *Dryopithecus* is based on the maximum body mass estimate and the larger of endocranial volume estimates, following the methods outlined by Jerison (1973). The EQ estimates for *Aegyptopithecus* are also based on maxima and for *Oreopithecus*, on the only available values.

The EQ of *Aegyptopithecus* is low by anthropoid standards, which is consistent with many previous assessments of encephalization in this taxon. In *Oreopithecus*, EQ is also quite low, toward the low end of the range of variation in monkeys and below all hominoid ranges. The revised EQ for *Proconsul* is not especially hominoid-like, which is consistent with the analysis of Falk (1983a) concerning endocranial morphology. The EQ estimate for *Dryopithecus* is among the highest values for living apes. However, the significance of these EQ values with regard to an understanding of intelligence in these taxa is not immediately clear. For example, *Gorilla* and male *Pongo* EQ values are equal to or lower than EQs for hylobatids and many monkeys, though most agree that they are cognitively superior to hylobatids and

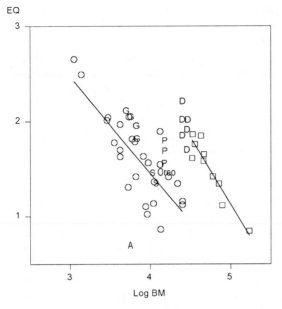

EQ

Log BM

Figure 14.2. The relationship between EQ and body mass (BM). Open circles represent living cercopithecoids and open squares represent living great apes. G = gibbons, S = siamangs, A = *Aegyptopithecus*, P = *Proconsul*, Oreo = *Oreopithecus*, D = *Dryopithecus*. Multiple points for *Proconsul* and *Dryopithecus* represent different possible brain mass and body mass estimates. Note that *Dryopithecus* clusters with great apes while the other hominoids cluster with cercopithecoids. *Aegyptopithecus* has a much smaller EQ relative to body mass. Note also that the largest cercopithecoids that approach hominids in body mass have much lower EQ values. See text.

monkeys and certainly equivalent to other great apes of smaller body mass (*Pan* and female *Pongo*) (see other contributions in this volume).

The EQ allows for a comparison of brain size among animals of differing body sizes but retains a body size artifact. Jerison (1973) recognized this uncertainty and noted that the EQ is most useful in comparisons at higher taxonomic levels. Many statistical or sampling factors have been proposed to account for the residual effects of body mass on EQ (Harvey 1988; Martin 1983, 1990; Pagel & Harvey 1988). Other biological or behavioral causes of EQ diversity have also been suggested, which are indirectly related or even unrelated to intelligence (Barton & Dunbar 1997; Clutton-Brock & Harvey 1980; Gibson, Rumbaugh & Beran 2001; Harvey 1988; Jerison 1973; Kappelman 1996; Martin 1983, 1990; Milton 1988; Radinsky 1977). While reviewing these is beyond

the scope of this chapter, Figure 14.2 illustrates that EQ declines with increasing body mass at similar rates in Old World monkeys and great apes, but along displaced trajectories.

Despite the effects of body mass, a few facts about EQ diversity remain clear. Figure 14.2 shows that no non great ape of body mass close to that of any great ape approaches EQ values for great apes. Monkeys that overlap in body mass with the smallest great apes, the largest papionins, have much lower EQ values than the smallest great apes, even though they are the largest-brained cercopithecoids. In Table 14.2, the papionins that have EQ values exceeding those of great apes are all at the low end of the range of variation in body mass in this group, much smaller than any great ape (Figure 14.2). Hylobatids follow the trend line for monkeys and have EQ values that are consistent with Old World monkeys of similar body mass. *Dryopithecus* follows the trend set by living non-*Homo* hominids, clustering around living great apes with the smallest body masses. *Proconsul* is intermediate though somewhat more monkey/hylobatid-like than great-ape-like. *Oreopithecus* is more clearly cercopithecoid-like. *Aegyptopithecus* is well below both trends, with a much lower EQ than other catarrhines of similar body mass.

In sum, although issues of body mass and analytical artifacts make EQ difficult to interpret, the analysis presented here addresses some of the body mass issues and suggests that the EQ of *Dryopithecus* indicates a level of encephalization equivalent to that of living great apes. This level of encephalization in a fossil great ape that is both closely related to living great apes, and of similar body mass, is most probably a shared derived trait of the great ape clade.

Absolute brain size

Jerison (1973) noted that while EQ effectively measures relative brain size and intelligence across broad taxonomic levels, a second measurement that quantifies the amount of brain mass beyond that determined exclusively by body mass was needed at finer taxonomic levels. His "theory of brain size" or "extra neurons" attempts to calculate the number of neurons required for normal metabolism and basic or "primitive" patterns of behavior at a given body mass in mammals, and the number of "extra neurons" represented by larger than expected brain masses. This idea is dependent on a number of definitions and assumptions, most of which are highly debatable (Holloway, 1969; 1974a). However, Jerison's theory of brain size resembles current theories of intelligence that emphasize absolute size and is consistent with observations of intelligence differences among primates (Dunbar 1993; Gibson *et al.* 2001).

While Jerison's calculations do attempt to account for differences in body mass in assessing the significance of brain mass, too many uncertainties remain to be confident in the accuracy of his neuron counts (Holloway 1969; 1974a). Although his measurements of extra neurons (N_c) have recently been used in an analysis of brain mass and intelligence in primates (Gibson *et al.* 2001), we agree in part with Holloway that the precise numbers are controversial and so they are not reproduced here. However, in Table 14.2 we list values for brain mass, body mass and EQ in a diversity of primates, mainly from data taken from Jerison (1973). Taxa in Table 14.2 are listed in order of brain mass, exactly the same as the order in which they would have been if listed in order of N_c (Jerison, 1973).

Jerison (1973) reported EQ, brain, and body mass values for a large number of primates from the largest specimens in each taxon. To make comparisons to fossil taxa more directly comparable to the values for extant taxa, the largest reasonable estimates of body and brain mass in *Aegyptopithecus*, *Proconsul*, *Oreopithecus*, and *Dryopithecus* are also used here. Table 14.2 updates and reinforces the conclusions reached by Gibson *et al.* (2001) and Jerison (1973) that absolute brain size appears to track broadly accepted categories of cognitive capabilities better than EQ. Brain mass is lowest in the most primitive anthropoids, higher in cebids and Old World monkeys excluding papionins, and highest in great apes, with no overlap among these groups. Hylobatids have great-ape-like EQ values, but Old-World-monkey-like brain sizes, with gibbons clustering with non-papionins and siamangs with papionins. This is consistent with the conclusions presented earlier regarding the effects of body mass on EQ. Interestingly, siamangs fall within the range of papionins other than *Papio*, i.e., *Cercocebus*, *Lophocebus*, and *Mandrillus*. *Papio* has a larger brain that does not overlap with the ranges in other catarrhines. These results are generally similar to those obtained by Gibson *et al.* (2001), with finer categories discriminated here. It is beyond the scope of this chapter to interpret the significance of these differences, though tempting to suggest that it may be related to the unique aspects of baboon adaptation (social, dietary, ecological, or all of the above) (Parker, Chapter 4, this volume).

Not only are the brain mass values for great apes above those of all other living nonhuman primates, the range of great ape values is essentially the same as within the papionins, the minimum value being about 50% of the maximum in each set. When male gorillas and *Papio* are excluded, the minimums of papionin and great ape values both climb to about 65% of the maximum, which is about the same as the minimum/maximum ratio in hylobatids (Table 14.2). This pattern provides a context to interpret the significance of differences in brain size among great apes. They appear to be no more important or extensive than are brain size differences among papionins or even within the single genus *Hylobates* when outliers are removed. Finally, it is noteworthy that the order in which the taxa in Table 14.2 are listed would be nearly the same if they were listed in increasing order of body mass. While this may be taken to imply that body mass alone is sufficient to estimate relative brain size, the interesting exceptions represented by *Papio* and the positions of some of the fossil taxa would be difficult to interpret using body mass alone (see below).

Brain mass for *Aegyptopithecus* is above the range of variation for the anthropoids with the smallest brains, callitrichids, while its EQ is unusually low. This is consistent with Radinsky (1973), who noted that the appearance of a low relative brain size in *Aegyptopithecus*, reflected here in its low EQ, may be an artifact of its large body mass compared with other paleogene primates. In contrast, brain mass in *Aegyptopithecus* is at the low end of the range of variation for cebids, while its body mass is toward the upper end of the range of variation in cebids. *Aegyptopithecus* is probably anthropoid-like in brain mass, i.e., intermediate between callitrichids and cebids, and represents a reasonable ancestral morphotype for catarrhines.

Proconsul has a brain mass in the range of papionins other than *Papio*, but above those of other Old World monkeys and hylobatids. *Proconsul* EQ is low in comparison with hylobatids, which are considerably smaller in body mass, and within the range of all Old World monkeys (papionins are not distinguished from other Old World monkeys by EQ). This pattern is difficult to interpret in isolation. One explanation suggests itself, given observations of behavioral complexity in papionins (Parker, Chapter 4, this volume and references therein), and the consensus view is that *Proconsul* is a basal hominoid. It may be that *Proconsul* shows the ancestral brain mass pattern for hominoids. Hylobatids have artificially high EQ values, in part due to phyletic dwarfing resulting in their unusually low body mass (Begun, Chapter 2, this volume). They also have smaller brains, probably mostly the direct effect of body mass decrease, which may imply a lowering of cognitive capabilities if smaller body masses led to some reduction in selection for or ability to support large brains in hominoids. Papionins, especially *Papio*, have converged on the relative brain mass increases shown by hominids, though not to the same degree. This last observation has intriguing implications for interpreting the significance of brain size increase in the separate lineages of hominins ("robust australopithecines" and *Homo*), but this too is beyond the scope of this chapter (Elton, Bishop & Wood 2001; Falk *et al.* 2000).

Dryopithecus has EQ, body mass, and brain mass values within the range of variation of living great apes. This is consistent with the view that *Dryopithecus* is phyletically a great ape (Begun 1992a, 1994; Begun, Ward & Rose 1997; Kordos 1990; Kordos & Begun 1997, 2001a). It is also consistent with the observation of probable great ape levels of encephalization in other fossil great apes of similar age, *Sivapithecus* and *Lufengpithecus* (see above). These three fossil great apes belong to the two main clades of living great apes, pongines (*Sivapithecus*, *Lufengpithecus*, and *Pongo*) and hominines (*Dryopithecus* and the African apes and humans) (Andrews & Cronin 1982; Andrews & Martin 1987; Begun 1994; Begun & Kordos 1997; Kelley 2002; Kelley & Pilbeam 1986; Pilbeam 1982; Schwartz 1990, 1997; Ward, 1997b). That levels of encephalization are indistinguishable in the ancestors of both clades of living hominids suggests that this level of encephalization was probably inherited from the common ancestor of all hominids (Begun, Chapter 2, this volume; but see Potts, Chapter 13, this volume for suggestions of parallelism). Brain size increase beyond that seen both in more primitive hominoids such as *Proconsul* and *Hylobates*, and in the most encephalized monkey, *Papio*, may be part of a suite of characters that define the Hominidae and distinguish them from all other primates.

What about *Oreopithecus*? Typically for this taxon, its body mass–brain mass relationship does not follow the same pattern in other anthropoids. The *Oreopithecus* female brain mass reported in Table 14.2 is slightly larger

than the largest gibbon brain cited by Jerison (1973), but its body mass is over twice that of the same gibbon individual. Its brain mass is also below that in siamangs of somewhat smaller body mass. In body mass this *Oreopithecus* female falls in the middle of the range of variation in papionins excluding *Papio*, while its brain mass is lower than in papionins. The smallest *Papio* is very close in body mass to this *Oreopithecus* female but has a 60% larger brain (Table 14.2). Finally, the *Oreopithecus* female has a considerably smaller brain than a similarly sized *Proconsul* female (Table 14.2). *Oreopithecus* appears to cluster more closely with non-papionin Old World monkeys than with hominoids (Table 14.2).

Oreopithecus is generally considered to be a basal great ape (Begun *et al.* 1997; Harrison & Rook 1997) and as such probably has experienced a reduction in relative brain mass given its considerably smaller brain compared with the basal hominoid *Proconsul*. This may well be convergent on brain mass reduction in hylobatids since it is not accompanied by (or caused by) body mass reduction, as appears to be the case in hylobatids. Both of these cases reveal a surprising diversity in hominoid brain evolution, with lineages appearing to be as likely to lose brain mass as to gain it. However, hominids have maintained relatively stable levels of encephalization. Early humans ("australopithecines") are marginally encephalized, if at all, compared with living great apes, *Dryopithecus*, and probably other fossil great apes (Hartwig-Scherer 1993; Kappelman 1996). The first clear evidence of substantial increases in absolute and relative brain mass in hominids comes with the origin of the genus *Homo* (Begun & Walker 1993; Falk 1980a, 1987; Falk *et al.* 2000; Kappelman 1996; Martin 1983; Tobias 1971a). The brain size–body mass relations among the taxa reviewed here are summarized in Figure 14.3.

Reorganization

One constant feature in the evolution of catarrhine brains is the partial de-coupling of size and morphology. *Aegyptopithecus*, *Proconsul*, and *Dryopithecus* endocasts all have more primitive features of cerebral morphology than living catarrhines of similar brain size. *Aegyptopithecus* appears to retain smaller frontal lobes, fewer sulci, and more rostral olfactory lobes compared with most living catarrhines. *Proconsul* brain size is small for a hominoid of its size, with possibly fewer sulci.

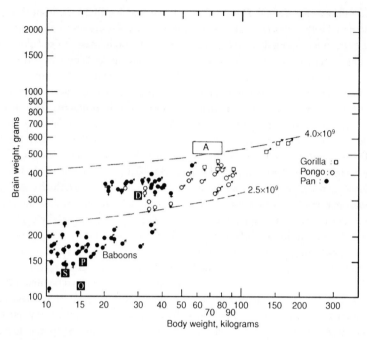

Figure 14.3. Brain weight and body mass in selected catarrhines. D = *Dryopithecus*, P = *Proconsul*, S = Siamang, O = *Oreopithecus*. A = *Australopithecus afarensis*. The position of fossil apes is based on the largest values (see text). *Australopithecus afarensis* is also based on the largest published specimens for this species, the AL 444-2 skull (*c.* 550 cc) and the AL 333-3 femur (67 kg, the mean of eight estimates ranging from 50 to 91 kg) (Lockwood, Kimbel & Johanson 2000; Jungers 1988; McHenry 1988, 1992; McHenry & Berger 1998). Modified from Jerison (1973:398).

Dryopithecus has the brain size and sulcal complexity (at least on the frontal lobes) of a great ape of its size, but may retain comparatively narrower frontal lobes rostrally. This result supports recent research challenging the idea of Finlay and Darlington (1995) that most if not all cerebral evolution is a direct result of overall size increase (Barton & Harvey 2000; Rilling & Insel 1998; Winter & Oxnard 2001; MacLeod, Chapter 7, this volume). It is also consistent with many analyses of human brain evolution that document morphological changes in relative lobe size and sulcal patterning in spite of little size difference compared with living great apes (Falk *et al.* 2000; Holloway 1974a, 1983a,b, 1984, 1995; Holloway & De la Coste-Lareymondie 1982; Tobias 1971a, 1983, 1991, 1995). Others have stressed the overall importance of brain size change in accounting for gross brain morphological evolution among hominids, so at least within this group the relative contributions of size and organization must be considered unresolved (Falk 1980a, 1987; Gibson *et al.* 2001; Preuss, Qi & Kaas 1999). Semendeferi and Damasio (2000) and Semendeferi *et al.* (1997) have shown that living

hominids including *Homo* differ little in the relative size of the frontal lobe, but that hylobatids have smaller frontal lobes, and gorillas may have unique cerebral proportions (see also MacLeod, Chapter 7, this volume). Visually, *Dryopithecus* appears to have comparatively small frontal lobes, but this is impossible to confirm without more complete material. If true, it would suggest independent expansion of this portion of the brain in living hominines and pongines.

CONCLUSIONS

A new cerebral size rubicon?

The idea of a critical brain mass defining a certain adaptive grade was common in interpreting the evidence of brain evolution in *Homo* (Falk 1980a, 1987; Holloway 1995; Jerison 1973; Tobias 1971a, 1995). Nonhominid catarrhine brain size evolution is labile, and in this way it is similar to the evolution of other biological attributes and their anatomical correlates (body mass, positional behavior, diet, etc.). Excluding *Homo*, brain size has been surprisingly stable in hominid evolution, despite dramatic changes and diversity in body mass, diet, and positional behavior. It may be that the typical nonhuman hominid level of encephalization (a brain of at least 270g) represents a rubicon that allows for the production of great-ape-like levels of behavioral complexity. This is very close to the limits proposed by Jerison (1973) (Figure 14.3).

Which came first?

Brain size is correlated to many other biological variables (life history, ecological and social pressures) and it is likely that significant changes in brain size cannot occur without affecting other biologically critical variables (Aiello & Dunbar 1993; Aiello & Wheeler 1995; Clutton-Brock & Harvey 1980; Dunbar 1992; Falk 1987, 1990; Gibson *et al.* 2001; Holloway 1995; Kelley 1997; Martin 1983, 1990, 1996; Parker 1996; Parker & Gibson 1979; Smith 1991). The converse is probably also true, i.e., significant changes, particularly in life history variables, may very strongly affect brain mass evolution. Many other authors stress one or a few variables (diet, foraging, social relations, group size, body size, positional behavior, etc.) as critical to the evolution of higher levels of intelligence in hominids, but this intelligence is made possible by the presence of a large brain, whether brain mass is the direct result of selection or not.

The earliest, albeit suggestive, evidence of hominoid-like brain mass is in *Proconsul*, which also appears to exhibit a hominoid-like life history (Kelley 1997; Chapter 15, this volume). Selection may have operated on one or more life history variables (rate of maturation, length of infancy, number of offspring, age of first birth, etc.) or on brain mass directly. The *Proconsul* individual on which our brain and body mass calculations were based was the size of a large monkey, most of which have smaller brains. This suggests that body mass selection was not the prime mover for brain mass increase, at least in *Proconsul*. The ecology and diet of *Proconsul* were not remarkable, as far as we can tell (Andrews, Begun & Zylstra 1997; Kay & Ungar 1997; Teaford & Walker 1984; Ungar & Kay 1995; Walker 1997; Walker & Teaford 1989; Singleton, Chapter 16, this volume). One unusual aspect of *Proconsul* is the combination of incipiently hominoid-like capabilities in the hip joint, wrist joints, and phalanges along with the absence of an external tail (Beard *et al.* 1993; Begun *et al.* 1994; Kelley 1997; Ward 1993, 1997a; Ward, Walker & Teaford 1991; Ward *et al.* 1993). It is possible that

Proconsul, which is for the most part larger than hylobatids, responded to the challenges of negotiating an arboreal setting with incipiently hominoid-like encephalization and postcranial anatomy. However, *Proconsul* was clearly not a suspensory hominoid, so that arboreal challenges, while reminiscent of those on which Povinelli & Cant (1995; see Gebo, Chapter 17, Hunt, Chapter 10, this volume) focus, would have been qualitatively different. At any rate, the initial phase of hominoid brain evolution is represented by the evidence of *Proconsul*.

Dryopithecus was a suspensory great ape and had a large brain but *Oreopithecus*, which was at least as suspensory but somewhat smaller in size, did not. The positional behavior of *Sivapithecus* is not completely clear, though most indications point to some degree of arboreality with suspensory postures in most species (Rose 1983, 1984, 1986, 1989, 1997; Spoor, Sondaar & Hussain 1991). *Lufengpithecus* postcrania are very poorly known and almost undescribed, but indications are that it was as suspensory as *Dryopithecus* and *Oreopithecus* (Begun pers. obs.). Both Asian fossil great apes were large and both probably had large brains. All four fossil great apes have distinctive dental and gnathic morphologies indicative of diverse dietary preferences (Singleton, Chapter 16, this volume). However, while *Oreopithecus* was a specialized folivore the other three taxa were all frugivores of one sort or another.

Taken as a whole, large body mass, suspensory positional behavior, and brain size in fossil great apes do not offer unambiguous support for a clambering hypothesis of the evolution of a distinctive great ape intelligence (Povinelli & Cant 1995). *Proconsul* was relatively large for an arboreal primate but non-suspensory with a small brain compared with hominids. *Dryopithecus* and probably *Lufengpithecus* and *Sivapithecus* overlap with *Proconsul* in body mass but are more suspensory and larger brained. *Oreopithecus* is in the same size range and is highly suspensory but had an even smaller brain than *Proconsul*. This is not to say that the Povinelli and Cant hypothesis is falsified by the paleontological evidence, since we do not know whether or not any fossil great ape clambered in the manner they propose. It could be that most fossil great apes broadly fit the predictions of the Povinelli and Cant hypothesis, but *Oreopithecus*, the fossil great ape they proposed as a good fit, does not. Why?

Oreopithecus has a highly specialized dentition and postcranial morphology suggestive of uniquely specialized folivory and exclusive, highly suspensory arboreality (Harrison & Rook 1997; Kay & Ungar 1997).

Some have suggested that *Oreopithecus* was bipedal and terrestrial, but this is based on a questionable reconstruction of the foot, and a very poorly preserved innominate and set of vertebrae (Köhler & Moyà-Solà 1997; Moyà-Solà, Kohler & Rook 1999; Rook *et al.* 1999). In contrast, there are many clear-cut suspensory arboreal characters of the *Oreopithecus* postcranium (Harrison & Rook 1997; Jungers 1987; E. E. Sarmiento & Marcus 2000; S. Sarmiento 1987; Straus 1963; Szalay & Langdon 1986). Whatever the positional behavior of *Oreopithecus*, the diversity of opinions probably reflects its lack of close modern analogues. It is possible that *Oreopithecus*, while suspensory, was unlike any living hominoid in the details of its positional behavior. It has been likened by some to sloths (Wunderlich, Walker & Jungers 1999), and may have been not a clamberer but a slowly (?stereotypically) moving suspensory quadruped, which may explain its departure from the prediction of Povinelli and Cant (see Hunt, Chapter 10, Gebo, Chapter 17, this volume).

The pattern of brain size diversity in fossil great apes more closely matches broad patterns of diet. Hominid-like craniodental characters of *Dryopithecus*, *Sivapithecus*, and *Lufengpithecus* are associated with specialized hominid-like frugivory (large incisors, robust, elongated anterior palates, large postcanine dentitions, large brains), all absent from the highly folivorous *Oreopithecus* (Singleton, Chapter 16, this volume). A dietary shift may instead be associated with brain size increase in early great apes. Or, it may be that all of these factors (life history, diet, and positional behavior) are necessary to account for the evolution of the early hominid brain.

Fossil great apes tell us much about the timing of the origin of hominoid and hominid-like characters of the brain, and set some broad parameters for understanding the causes of these changes and their relationship to the evolution of intelligence. The biggest difficulty in interpreting this evidence is the bias intrinsic in the fossil record that turns the attention of researchers to behavior very closely related to or constrained by morphology. The fossil evidence of great apes is not suitable for testing hypotheses of social cognition, communication, group size, technical abilities, or foraging strategies. They do tell us, however, that many of the anatomical correlates of large brain mass (and by extension, intelligence) in living hominids, whether they are prime movers in great ape intelligence evolution or not, were already present in the fossil great apes of the late Miocene.

ACKNOWLEDGMENTS

We thank Bruce Latimer, Cleveland Museum of Natural History, and Wim Van Neer of the Musée Royal d'Afrique Centrale, Tervuren, for access to the collections in their care, R. D. Martin for constructive criticism of a previous version of this paper, and Ralph Holloway, Anne Russon and an anonymous reviewer for many additional useful insights. This work is supported by grants from NSERC, Wenner-Gren, Leakey Foundation and OTA.

ENDNOTE

1 Stem catarrhine, or stem hominoid, etc., refers to a taxon that cannot be attributed to any living taxon but that is essentially more closely related to the taxon to which it is a stem than to anything else. A stem catarrhine is more closely related to Old World monkeys and apes than to other primates, but is not more closely related to either Old World monkeys or apes. A stem hominoid is a hominoid but is no more closely related to hylobatids than to hominids. On a related nomenclatural issue, there is some debate on the species designation of the specimen of *Proconsul* that provides evidence of the brain (KNM RU-7290). In my view it is a female *Proconsul nyanzae*, while Walker *et al.* (1993) regard it as *Proconsul heseloni*. Radinsky (1974) analyzed this same specimen when it was known as *Dryopithecus (Proconsul) africanus*, following Simons and Pilbeam (1965). It was at that time accessioned in the collections of the British Museum, with the catalogue number BMNH 32363. Falk (1983a) uses the old catalogue number but assigns the specimen to *Proconsul africanus*, following Le Gros Clark and Leakey (1951). The specimen was returned to Kenya and given the new catalogue number used here. The genus *Proconsul* is probably in need of revision, which accounts for the differences of opinion between Walker *et al.* and myself. For the purposes of this chapter it does not matter, because the body and brain mass estimates used to assess relative brain size come from the same specimen.

REFERENCES

Aiello, L. C. & Dunbar, R. I. M. (1993). Neocortex size, group size and the evolution of language. *Current Anthropology*, **34**, 184–93.

Aiello, L. C. & Wheeler, P. (1995). The expensive tissue hypothesis: the brain and the digestive system in human and primate evolution. *Current Anthropology*, **36**, 199–221.

Aiello, L. C. & Wood, B. A. (1994). Cranial predictors of hominine body mass. *American Journal of Physical Anthropology*, **95**, 409–26.

Andrews, P. (1992). Evolution and environment in the Hominoidea. *Nature*, **360**, 641–6.

Andrews, P. & Cronin, J. (1982). The relationships of *Sivapithecus* and *Ramapithecus* and the evolution of the orang-utan. *Nature*, **297**, 541–6.

Andrews, P. & Martin, L. (1987). Cladistic relationships of extant and fossil hominoids. *Journal of Human Evolution*, **16**, 101–18.

Andrews, P., Begun, D. R. & Zylstra, M. (1997). Interrelationships between functional morphology and paleoenvironments in Miocene hominoids. In *Function, Phylogeny, and Fossils: Miocene Hominoid Evolution and Adaptations*, ed. D. R. Begun, C. V. Ward & M. D. Rose, pp. 29–58. New York: Plenum Press.

Barton, R. A. & Dunbar, R. I. M. (1997). Evolution of the social brain. In *Machiavellian Intelligence II; Extensions and Evaluations*, ed. A. Whiten & R. W. Bryne, pp. 240–63. Cambridge, UK: Cambridge University Press.

Barton, R. A. & Harvey, P. H. (2000). Mosaic evolution of brain structure in mammals. *Nature*, **405**, 1055–8.

Bauchot, R. & Stephan, H. (1969). Encéphalisation et niveau évolutif chez les simiens. *Mammalia*, **33**, 225–75.

Beard, K. C., Teaford, M. F. & Walker, A. (1993). New hand bones of the early Miocene hominoid *Proconsul* and their implications for the evolution of the hominoid wrist. In *Hands of Primates*, ed. H. Preuschoft & D. J. Chivers, pp. 387–403. New York: Springer-Verlag.

Begun, D. R. (1992a). Miocene fossil hominids and the chimp–human clade. *Science*, **257**, 1929–33.

(1992b). Phyletic diversity and locomotion in primitive European hominids. *American Journal of Physical Anthropology*, **87**, 311–40.

(1993). New catarrhine phalanges from Rudabánya (Northeastern Hungary) and the problem of parallelism and convergence in hominoid postcranial morphology. *Journal of Human Evolution*, **24**, 373–402.

(1994). Relations among the great apes and humans: new interpretations based on the fossil great ape *Dryopithecus*. *Yearbook of Physical Anthropology* **37**, 11–63.

(2002). European hominoids. In *The Primate Fossil Record*, ed. W. Hartwig, pp. 339–68. Cambridge, UK: Cambridge University Press.

Begun, D. R. & Güleç, E. (1998). Restoration of the type and palate of *Ankarapithecus meteai*: taxonomic, phylogenetic, and functional implications. *American Journal of Physical Anthropology*, **105**, 279–314.

Begun, D. R. & Kordos, L. (1993). Revision of *Dryopithecus brancoi* SCHLOSSER 1901 based on the fossil hominoid material from Rudabánya. *Journal of Human Evolution*, **25**, 271–86.

(1997). Phyletic affinities and functional convergence in *Dryopithecus* and other Miocene and living hominids. In *Function, Phylogeny, and Fossils: Miocene Hominoid Evolution and Adaptations*, ed. D. R. Begun, C. V. Ward & M. D. Rose, pp. 291–316. New York: Plenum Press.

Begun, D. R. & Walker, A. (1993). The endocast. In *The Nariokotome* Homo erectus *Skeleton*, ed. A. Walker & R. Leakey, pp. 326–58. Cambridge, MA: Harvard University Press.

Begun, D. R., Teaford, M. F. & Walker, A. (1994). Comparative and functional anatomy of *Proconsul* phalanges form the Kaswanga primate site, Rusinga Island, Kenya. *Journal of Human Evolution*, **26**, 89–165.

Begun, D. R., Ward, C. V. & Rose, M. D. (1997). Events in hominoid evolution. In *Function, Phylogeny, and Fossils: Miocene Hominoid Origins and Adaptations*, ed. D. R. Begun, C. V. Ward & M. D. Rose, pp. 389–415. New York: Plenum Press.

Brunet, M., Guy, F., Pilbeam, D., *et al.* (2002). A new hominid from the Upper Miocene of Chad, Central Africa. *Nature*, **418**, 145–51.

Clutton-Brock, T. H. & Harvey, P. H. (1980). Primates, brains and ecology. *Journal of the Zoological Society of London*, **190**, 309–23.

Conroy, G. C. (1987). Problems of body-weight estimation in fossil primates. *International Journal of Primatology*, **8**, 115–37.

Conroy, G. C., Vannier, M. W. & Tobias, P. V. (1990). Endocranial features of *Australopithecus africanus* revealed by 2- and 3-D computed tomography. *Science*, **247**, 838–41.

Dagasto, M. & Terranova, C. J. (1992). Estimating the body size of Eocene primates: a comparison of results from dental and postcranial variables. *International Journal of Primatology*, **13**, 307–44.

Dart, R. A. (1925). *Australopithecus africanus*: the man-ape of South Africa. *Nature*, **115**, 195–9.

Delson, E., Terranova, C. J., Jungers, W. L., Sargis, E. J., Jablonski, N. G. & Dechow, P. C. (2000). Body mass in Cercopithecidae (Primates, Mammalia): estimation and scaling in extinct and extant taxa. *American Museum of Natural History, Anthropological Papers*, **83**, 1–159.

Dunbar, R. I. M. (1992). Neocortex size as a constraint on group size in primates. *Journal of Human Evolution*, **20**, 469–93.

(1993). Coevolution of neocortical size, group size and language in humans. *Behavioral and Brain Sciences*, **16**, 681–735.

Elton, S., Bishop, L. C. & Wood, B. (2001). Comparative context of Plio-Pleistocene hominin brain evolution. *Journal of Human Evolution*, **41**, 1–27.

Falk, D. (1980a). Hominid brain evolution: the approach from paleoneurology. *Yearbook of Physical Anthropology*, **23**, 93–107.

(1980b). A reanalysis of the South African Australopithecine natural endocasts. *American Journal of Physical Anthropology*, **53**, 525–39.

(1983a). A reconsideration of the endocast of *Proconsul africanus*: implications for primate brain evolution. In *New Interpretations of Ape and Human Ancestry*, ed. R. L. Ciochon & R. S. Corruccini, pp. 239–48. New York: Plenum Press.

(1983b). The Taung endocast: a reply to Holloway. *American Journal of Physical Anthropology*, **60**, 479–89.

(1987). Hominid paleoneurology. *Annual Review of Anthropology*, **16**, 13–30.

(1990). Brain evolution in *Homo*: the "radiator theory." *Behavioral and Brain Sciences*, **13**, 333–81.

Falk, D. & Conroy, G. C. (1983). The cranial venous sinus system in *Australopithecus afarensis*. *Nature*, **306**, 779–81.

Falk, D., Redmond, J. C., Guyer, J., Conroy, G. C., Recheis, W., Weber, G. W. & Seidler, H. (2000). Early hominid brain evolution: a new look at old endocasts. *Journal of Human Evolution*, **38**, 695–717.

Finlay, B. L. & Darlington, R. B. (1995). Linked regularities in the development and evolution of mammalian brains. *Science*, **268**, 1578–84.

Fleagle, J. G. (1983). Locomotor adaptations of Oligocene and Miocene hominoids and their phyletic implications. In *New Interpretations of Ape and Human Ancestry*, ed. R. L. Ciochon & R. S. Corruccini, pp. 301–24. New York: Plenum Press.

Fleagle, J. G. & Kay, R. F. (1983). New interpretations of the phyletic position of Oligocene hominoids. In *New Interpretations of Ape and Human Ancestry*, ed. R. L. Ciochon & R. S. Corruccini, pp. 181–210. New York: Plenum Press.

Fleagle, J. G., Brown, T. M., Obradovitch, J. D. & Simons, E. L. (1986). Age of the earliest African anthropoids. *Science*, **234**, 1247–1249.

Gibson, K. R., Rumbaugh, D. & Beran, M. (2001). Bigger is better: primate brain size in relationship to cognition. In *Evolutionary Anatomy of the Primate Cerebral Cortex*, ed. D. Falk & K. R. Gibson, pp. 79–97. Cambridge, UK: Cambridge University Press.

Gingerich, P. D. (1979). Patterns of tooth size variability in the dentition of primates. *American Journal of Physical Anthropology*, **51**, 457–66.

Gingerich, P. D., Smith, B. H. & Rosenberg, K. (1982). Scaling in the dentition of primates and prediction of body weight from tooth size in fossils. *American Journal of Physical Anthropology*, **58**, 81–100.

Harrison, T. (1986). A reassessment of the phylogenetic relationship of *Oreopithecus bambolii*. *Journal of Human Evolution*, **15**, 541–83.

(1987). The phylogenetic relationships of the early catarrhine primates: a review of the current evidence. *Journal of Human Evolution*, **16**, 41–80.

(1989). New estimates of cranial capacity, body size, and encephalization in *Oreopithecus bambolii*. *American Journal of Physical Anthropology*, **78**, 237.

Harrison, T. & Rook, L. (1997). Enigmatic anthropoid or misunderstood ape: the phylogenetic status of *Oreopithecus bambolii* reconsidered. In *Function, Phylogeny, and Fossils: Miocene Hominoid Origins and Adaptations*, ed. D. R. Begun, C. V. Ward & M. D. Rose, pp. 327–62. New York: Plenum Press.

Hartwig-Scherer, S. (1993). Body weight prediction in early fossil hominids: towards a taxon "independent" approach. *American Journal of Physical Anthropology*, **92**, 17–36.

Harvey, P. H. (1988). Allometric analysis and brain size. In *Intelligence and Evolutionary Biology*, ed. H. J. Jerison & I. Jerison, pp. 199–210. Berlin: Springer-Verlag.

Holloway, R. L. (1969). Cranial capacity and neuron number: a critique and proposal *American Journal of Physical Anthropology*, **25**, 305–14.

(1974a). On the meaning of brain size. *Science*, **184**, 677–9.

(1974b). The casts of fossil hominid brains. *Scientific American*, **23**, 106–15.

(1982). *Homo erectus* brain endocasts: volumetric and morphological observations, with some comments on cerebral asymmetries. *Congrés Internationale de Paleontologie Humaine, Premier Congrés*, pp. 355–69.

(1983a). Cerebral brain endocast pattern of *Australopithecus afarensis* hominid. *Nature*, **303**, 420–2.

(1983b). Human paleontological evidence relevant to language behaviour. *Human Neurobiology*, **2**, 105–14.

(1984). The Taung endocast and the lunate sulcus: a rejection of the hypothesis of its anterior position. *American Journal of Physical Anthropology*, **64**, 285–7.

(1995). Toward a synthetic theory of human brain evolution. In *Origins of the Human Brain*, ed. J.-P. Changeux & J. Chavaillon, pp. 42–60. Oxford: Clarendon Press.

Holloway, R. L. & De la Coste-Lareymondie, M. C. (1982). Brain endocast asymmetry in pongids and hominids: some preliminary findings on the paleontology of cerebral dominance. *American Journal of Physical Anthropology*, **58**, 101–10.

Hürzeler, J. (1949). Neubeschreibung von *Oreopithecus bambolii* Gervais. *Schweizerische Paläontologische Abhandlungen*, **66**, 3–20.

(1951). Contribution à l'étude de la dentition de lait d'*Oreopithecus bambolii* Gervais. *Ecologae Geologica Helvetica*, **44**, 404–11.

(1958). *Oreopithecus bambolii* Gervais: a preliminary report. *Verhandlungen Naturforschende Gesellschaft Basel*, **69**, 1–47.

(1960). The significance of *Oreopithecus* in the genealogy of man. *Triangle*, **4**, 164–74.

Jerison, H. J. (1973). *The Evolution of Brain and Intelligence*. New York: Academic Press.

(1975). Fossil evidence of the evolution of the human brain. *Annual Review of Anthropology*, **4**, 27–58.

(1979). Brain, body and encephalization in early primates. *Journal of Human Evolution*, **8**, 615–35.

Jungers, W. L. (1982). Lucy's limbs: skeletal allometry and locomotion in *Australopithecus afarensis*. *Nature*, **297**, 676–8.

(1987). Body size and morphometric affinities of the appendicular skeleton in *Oreopithecus bambolii* (IGF 11778). *Journal of Human Evolution*, **16**, 445–56.

(1988). New estimates of body size in Australopithecines. In *Evolutionary History of the "Robust" Australopithecines*, ed. F. E. Grine, pp. 115–125. New York: Aldine de Gruyter.

Jungers, W. L. & Susman, R. L. (1984). Body size and skeletal allometry in African apes. In *The Pygmy Chimpanzee: Evolutionary Biology and Behavior*, ed. R. L. Susman, pp. 131–77. New York: Plenum Press.

Kappelman, J. (1996). The evolution of body mass and relative brain size in fossil hominids. *Journal of Human Evolution*, **30**, 243–76.

Kappelman, J., Simons, E. L. & Swisher, C. (1992). New age determinations for the Eocene–Oligocene boundary sediments in the Fayum Depression, Northern Egypt. *Journal of Geology*, **100**, 647–68.

Kay, R. F. & Ungar, P. S. (1997). Dental evidence for diet in some Miocene catarrhines with comments on the effects of phylogeny on the interpretation of adaptation. In *Function, Phylogeny, and Fossils: Miocene Hominoid Evolution and Adaptations*, ed. D. R. Begun, C. V. Ward & M. D. Rose, pp. 131–51. New York: Plenum Press.

Kay, R. F., Fleagle, J. G. & Simons, E. L. (1981). A revision of Oligocene apes of the Fayum Province, Egypt. *American Journal of Physical Anthropology*, **55**, 293–322.

Kelley, J. (1997). Paleobiological and phylogenetic significance of life history in Miocene hominoids. In *Function, Phylogeny, and Fossils: Miocene Hominoid Evolution and Adaptations*, ed. D. R. Begun, C. V. Ward & M. D. Rose, pp. 173–208. New York: Plenum Press.

(2002). The hominoid radiation in Asia. In *The Primate Fossil Record*, ed. W. Hartwig, pp. 369–84. Cambridge, UK: Cambridge University Press.

Kelley, J. & Pilbeam, D. R. (1986). The Dryopithecines: taxonomy, comparative anatomy, and phylogeny of Miocene large hominoid. In *Comparative Primate Biology. Vol. 1. Systematics, Evolution and Anatomy*, ed. D. R. Swindler & J. Erwin, pp. 361–411. New York: Alan R. Liss.

Köhler, M. & Moyà-Solà, S. (1997). Ape like or hominid-like? The positional behavior of *Oreopithecus bambolii* reconsidered. *Proceedings of the National Academy of Sciences USA*, **94**, 11 747–50.

Kordos, L. (1988). Comparison of early primate skulls from Rudabánya and China. *Anthropologia Hungarica*, **20**, 9–22.

(1990). Descriptions and reconstruction of the skull of *Rudapithecus hungaricus* KRETZOI (Mammalia). *Annals of the National Museum of Hungary*, **21**, 11–24.

Kordos, L. & Begun, D. R. (1997). A new reconstruction of RUD 77, a partial cranium of *Dryopithecus brancoi* from Rudábanya, Hungary. *American Journal of Physical Anthropology*, **103**, 277–94.

(1998). Encephalization and endocranial morphology in *Dryopithecus brancoi*: implications for brain evolution in early hominids. *American Journal of Physical Anthropology*, **Suppl. 26**, 141–2.

(2001a). A new cranium of *Dryopithecus* from Rudabánya, Hungary. *Journal of Human Evolution*, **41**, 689–700.

(2001b). Fossil catarrhines from the late Miocene of Rudabánya. *Journal of Human Evolution*, **40**, 17–39.

Le Gros Clark, W. E. & Leakey, L. S. B. (1951). The Miocene Hominoidea of East Africa. *British Museum of Natural History, Fossil Mammals of Africa*, **1**, 1–117.

Lockwood, C. A., Kimbel, W. H. & Johanson, D. C. (2000). Temporal trends and metric variation in the mandibles and dentition of *Australopithecus afarensis*. *Journal of Human Evolution*, **39**, 23–55.

Manser, J. & Harrison, T. (1999). Estimates of cranial capacity and encephalization in *Proconsul* and *Turkanapithecus*. *American Journal of Physical Anthropology*, **28**, 189.

Martin, R. D. (1983). *Human Brain Evolution in an Ecological Context: Fifty-Second James Arthur Lecture on the Evolution of the Human Brain*. New York: American Museum of Natural History.

(1990). *Primate Origins and Evolution*. Princeton, NJ: Princeton University Press.

(1996). Scaling of the mammalian brain: the maternal energy hypothesis. *Newsletter of Physiological Sciences*, **11**, 149–56.

Martin, R. D. & Harvey, P. H. (1985). Brain size allometry: ontogeny and phylogeny. In *Size and Scaling in Primate biology*, ed. W. L. Jungers, pp. 147–74. New York: Plenum Press.

McCollum, M. A. & Ward, S. C. (1997). Subnasoalveolar anatomy and hominoid phylogeny evidence from comparative ontogeny. *American Journal of Physical Anthropology*, **102**, 377–405.

McHenry, H. M. (1988). New estimates of body weight in early hominids and their significance to encephalization and megadontia in robust australopithecines. In *The Evolution of the "Robust" Australopithecines*, ed. F. E. Grine, pp. 133–48. New York: Aldine de Gruyter.

(1992). Body size and proportions in early hominids. *American Journal of Physical Anthropology*, **87**, 407–30.

McHenry, H. M. & Berger, L. R. (1998). Body proportions in *Australopithecus afarensis* and *A. africanus* and the origin of the genus *Homo*. *Journal of Human Evolution*, **35**, 1–22.

Milton, K. (1988). Foraging behaviour and the evolution of primate intelligence. In *Machiavellian Intelligence*, ed. R. W. Bryne & A. Whiten, pp. 285–305. Oxford: Clarendon Press.

Morbeck, M. E. (1983). Miocene hominoid discoveries from Rudabánya: implications from the postcranial skeleton. In *New Interpretations of Ape and Human Ancestry*, ed. R. L. Ciochon & R. S. Corruccini, pp. 369–404. New York: Plenum Press.

Moyà-Solà, S., Köhler, M. & Rook, L. (1999). Evidence of hominid-like precision grip capability in the hand of the Miocene ape *Oreopithecus*. *Proceedings of the National Academy of Sciences USA*, **96**, 313–7.

Pagel, M. D. & Harvey, P. H. (1988). The taxon-level problem in the evolution of mammalian brain size: facts and artefacts. *American Naturalist*, **132**, 344–59.

Parker, S. T. (1996). Apprenticeship in tool-mediated extractive foraging: the origins of imitation, teaching, and self-awareness in great apes. In *Reaching into Thought: The Minds of the Great Apes*, ed. A. E. Russon, K. A. Bard & S. T. Parker, pp. 348–370. Cambridge, UK: Cambridge University Press.

Parker, S. T. & Gibson, K. R. (1979). A developmental model for the evolution of language and intelligence in early hominids. *Behavioral and Brain Sciences*, **2**, 367–408.

Pilbeam, D. R. (1982). New hominoid skull material from the Miocene of Pakistan. *Nature*, **295**, 232–4.

(1997). Research on Miocene hominoids and hominid origins: the last three decades. In *Function, Phylogeny, and Fossils: Miocene Hominoid Evolution and Adaptations*, ed. D. R. Begun, C. V. Ward & M. D. Rose, pp. 13–28. New York: Plenum Press.

Pilbeam, D. R. & Gould, S. J. (1974). Size and scaling in human evolution. *Science*, **186**, 892–901.

Pilbeam, D. R. & Young, N. M. (2001). *Sivapithecus* and hominoid evolution: some brief comments. In *Hominoid Evolution and Environmental Change in the Neogene of Europe*, ed. L. de Bonis, G. Koufos & P. Andrews, pp. 349–64. Cambridge, UK: Cambridge University Press.

Povinelli, D. J. & Cant, J. G. H. (1995). Arboreal clambering and the evolution of self-cognition. *Quarterly Review of Biology*, **70**, 393–421.

Preuss, T. M., Qi, H. & Kaas, J. H. (1999). Distinctive compartmental organization of the human primary visual cortex. *Proceedings of the National Academy of Sciences USA*, **96**, 11 601–6.

Radinsky, L. (1973). *Aegyptopithecus* endocasts: oldest record of a pongid brain. *American Journal of Physical Anthropology*, **39**, 239–48.

(1974). The fossil evidence of anthropoid brain evolution. *American Journal of Physical Anthropology*, **41**, 15–28.

(1977). Early primate brains: facts and fiction. *Journal of Human Evolution*, **6**, 79–86.

(1979). *The Fossil Record of Primate Brain Evolution. 49th James Arthur Lecture, 1979*. New York: American Museum of Natural History.

(1982). Some cautionary notes on making inferences about relative brain size. In *Primate Brain Evolution: Methods and Concepts*, ed. E. Armstrong & D. Falk, pp. 29–38. New York: Academic Press.

Rasmussen, D. T. (2002). Early catarrhines of the African Eocene and Oligocene. In *The Primate Fossil Record*, ed. W. Hartwig, pp. 203–20. Cambridge, UK: Cambridge University Press.

Rilling, J. K. & Insel, T. R. (1998). Evolution of the cerebellum in primates: differences in relative volume among monkeys, apes and humans. *Brain and Behavioral Evolution*, **52**, 308–14.

Rook, L., Bondioli, L., Köhler, M., Moyà-Solà, S. & Macchiarelli, R. (1999). *Oreopithecus* was a bipedal ape after all: evidence from the iliac cancellous architecture. *Proceedings of the National Academy of Sciences*, **96**, 8795–9.

Rose, M. D. (1983). Miocene hominoid postcranial morphology: monkey-like, ape-like, neither, or both? In *New Interpretations of Ape and Human Ancestry*, ed. R. L. Ciochon & R. S. Corruccini, pp. 405–17. New York: Plenum Press.

(1984). Hominoid postcranial specimens from the middle Miocene Chinji Formation, Pakistan. *Journal of Human Evolution*, **13**, 503–16.

(1986). Further hominoid postcranial specimens from the late Miocene Nagri Formation of Pakistan. *Journal of Human Evolution*, **15**, 333–67.

(1989). New postcranial specimens of catarrhines from the middle Miocene Chinji Formation, Pakistan: description and a discussion of proximal humeral functional morphology in anthropoids. *Journal of Human Evolution*, **18**, 131–62.

(1997). Functional and phylogenetic features of the forelimb in Miocene hominoids. In *Function, Phylogeny, and Fossils: Miocene Hominoid Evolution and Adaptations*, ed. D. R. Begun, C. V. Ward & M. D. Rose, pp. 79–100. New York: Plenum Press.

Sarmiento, E. E. & Marcus, L. F. (2000). The Os Navicular of humans, great apes, OH 8, Hadar, and *Oreopithecus*: function, phylogeny, and multivariate analyses. *American Museum Novitates*, **3288**, 1–38.

Sarmiento, S. (1987). The phylogenetic position of *Oreopithecus* and its significance in the origin of the Hominoidea. *American Museum Novitates*, **2881**, 1–44.

Schepers, G. W. H. (1946). The endocranial casts of the South African ape-men. In *The South African Fossil Ape-Men: The Australopithecinae*, ed. R. Broom & G. W. H. Schepers, pp. 153–72. Pretoria: Transvaal Museum Memoirs.

(1950). The brain casts of the recently discovered *Plesianthropus* skulls. In *Sterkfontein Ape-Man, Plesianthropus*, ed. R. Broom, J. T. Robinson & G. W. H. Schepers, pp. 85–117. Pretoria: Transvaal Museum Memoirs.

Schwartz, J. H. (1984a). The evolutionary relationships of man and orang-utans. *Nature*, **308**, 501–5.

(1984b). Hominoid evolution: a review and a reassessment. *Current Anthropology*, **25**, 655–72.

(1990). *Lufengpithecus* and its potential relationship to an orang-utan clade. *Journal of Human Evolution*, **19**, 591–605.

(1997). *Lufengpithecus* and Hominoid phylogeny: problems in delineating and evaluating phylogenetically relevant characters. In *Function, Phylogeny, and Fossils: Miocene Hominoid Evolution and Adaptations*, ed. D. R. Begun, C. V. Ward & M. D. Rose, pp. 363–88. New York: Plenum Press.

Semendeferi, K. & Damasio, H. (2000). The brain and its main anatomical subdivisions in living hominoids using magnetic resonance imaging. *Journal of Human Evolution*, **38**, 317–32.

Semendeferi, K., Damasio, H., Frank, R. & Van Hoesen, G. W. (1997). The evolution of the frontal lobes: a volumetric analysis based on three-dimensional reconstructions of magnetic resonance scans of human and ape brains. *Journal of Human Evolution*, **32**, 375–88.

Simons, E. L. (1965). New fossil apes from Egypt and the initial differentiation of Hominoidea. *Nature*, **205**, 135–9.

(1968). Assessment of a fossil hominid. *Science*, **160**, 672–5.

(1987). New faces of *Aegyptopithecus* from the Oligocene of Egypt. *Journal of Human Evolution*, **16**, 273–89.

(1993). New endocasts of *Aegyptopithecus*: oldest well-preserved record of the brain in anthropoidea. *American Journal of Science*, **293**, 383–90.

Simons, E. L. & Pilbeam, D. R. (1965). Preliminary revision of the Dryopithecinae (Pongidae, Anthropoidea). *Folia Primatologica*, **3**, 81–152.

Smith, B. H. (1989). Dental development as a measure of life history in primates. *Evolution*, **43**, 683–8.

(1991). Dental development and the evolution of life history in the Hominidae. *American Journal of Physical Anthropology*, **86**, 157–74.

Smith, R. J. & Jungers, W. L. (1997). Body mass in comparative primatology. *Journal of Human Evolution*, **32**, 523–59.

Spoor, C. F., Sondaar, P. Y. & Hussain, S. T. (1991). A new hominoid hamate and first metacarpal from the late Miocene Nagri Formation of Pakistan. *Journal of Human Evolution*, **21**, 413–24.

Straus, W. L. (1961). The phylogenetic status of *Oreopithecus bambolii*. *Philadelphia Anthropological Society Bulletin*, **14**, 12–13.

(1963). The classification of *Oreopithecus*. In *Classification and Human Evolution*, ed. S. L. Washburn, pp. 146–77. Chicago, IL: Aldine.

Straus, W. L. & Schön, M. A. (1960). Cranial capacity of *Oreopithecus bambolii*. *Science*, **132**, 670–2.

Szalay, F. S. & Langdon, J. H. (1986). The foot of *Oreopithecus*: an evolutionary assessment. *Journal of Human Evolution*, **15**, 585–621.

Teaford, M. F. & Walker, A. C. (1984). Quantitative differences in the dental microwear between primates with different diets and a comment on the presumed diet of *Sivapithecus*. *American Journal of Physical Anthropology*, **64**, 191–200.

Tobias, P. V. (1967). *Olduvai Gorge. Vol. 2. The Cranium and Maxillary Dentition of Australopithecus (Zinjanthropus) boisei*. Cambridge, UK: Cambridge University Press.

(1971a). *The Brain in Hominid Evolution*. New York: Columbia University Press.

(1971b). The distribution of cranial capacity values among living hominoids. *Proceedings of the 3rd International Congress of Primatology, Zurich 1970*, Vol. 1, pp. 18–35. Basel: Karger.

(1975). Brain evolution in Hominoidea. In *Primate Functional Morphology and Evolution*, ed. R. H. Tuttle, pp. 353–92. The Hague: Mouton.

(1978). The place of *Australopithecus africanus* in hominid evolution. In *Recent Advances in Primatology*, ed. D. J. Chivers & K. A. Joysey, pp. 373–94. New York: Academic Press.

(1983). Recent advances in the evolution of the hominids with special reference to brain and speech. In *Recent Advances in the Evolution of Primates*, ed. C. Chagas,

pp. 85–140. Citta del Vaticano: Pontificia Academia Scientiarum.

(1991). *The Skulls, Endocasts and Teeth of Homo habilis*. Cambridge, UK: Cambridge University Press.

(1995). The brain of the first hominids. In *Origins of the Human Brain*, ed. J.-P. Changeux & J. Chavaillon, pp. 61–83. Oxford: Clarendon Press.

Ungar, P. S. & Kay, R. F. (1995). The dietary adaptations of European Miocene catarrhines. *Proceedings of the National Academy of Sciences USA*, **92**, 5479–81.

Walker, A. (1997). *Proconsul*: function and phylogeny. In *Function, Phylogeny, and Fossils: Miocene Hominoid Evolution and Adaptations*, ed. D. R. Begun, C. V. Ward & M. D. Rose, pp. 209–24. New York: Plenum Press.

Walker, A. & Teaford, M. F. (1989). The hunt for *Proconsul*. *Scientific American*, **260**, 76–82.

Walker, A., Teaford, M. F., Martin, L. B. & Andrews, P. (1993). A new species of *Proconsul* from the early Miocene of Rusinga/Mfango Island, Kenya. *Journal of Human Evolution*, **25**, 43–56.

Walker, A. C., Falk, D., Smith, R. & Pickford, M. F. (1983). The skull of *Proconsul africanus*: reconstruction and cranial capacity. *Nature*, **305**, 525–7.

Ward, C. V. (1993). Torso morphology and locomotion in *Proconsul nyanzae*. *American Journal of Physical Anthropology*, **92**, 291–328.

(1997a). Functional anatomy and phyletic implications of the hominoid trunk and hindlimb. In *Function, Phylogeny, and Fossils: Miocene Hominoid Evolution and Adaptations.*, ed. D. R. Begun, C. V. Ward & M. D. Rose, pp. 101–30. New York: Plenum Press.

Ward, C. V., Walker, A. C. & Teaford, M. F. (1991). *Proconsul* did not have a tail. *Journal of Human Evolution*, **21**, 215–20.

Ward, C. V., Walker, A., Teaford, M. F. & Odhiambo, I. (1993). A partial skeleton of *Proconsul nyanzae* from Mfango Island, Kenya. *American Journal of Physical Anthropology*, **90**, 77–111.

Ward, S. (1997). The taxonomy and phylogenetic relationships of *Sivapithecus* revisited. In *Function, Phylogeny, and Fossils: Miocene Hominid Origins and Adaptations*, ed. D. R. Begun, C. V. Ward & M. D. Rose, pp. 269–90. New York: Plenum Press.

Ward, S. C. & Brown, B. (1986). The facial skeleton of *Sivapithecus indicus*. In *Comparative Primate Biology*, ed. D. R. Swindler & J. Erwin, pp. 413–52. New York: Alan R. Liss.

Ward, S. C. & Kimbel, W. H. (1983). Subnasal alveolar morphology and the systemic position of *Sivapithecus*. *American Journal of Physical Anthropology*, **61**, 157–71.

Ward, S. C. & Pilbeam, D. R. (1983). Maxillofacial morphology of Miocene hominoids from Africa and Indo-Pakistan. In *New Interpretations of Ape and Human Ancestry*, ed. R. L. Corruccini & R. S. Ciochon, pp. 211–38. New York: Plenum Press.

White, T. D. (2002). Earliest hominids. In *The Primate Fossil Record*, ed. W. Hartwig, pp. 407–17. Cambridge, UK: Cambridge University Press.

Winter, W. de. & Oxnard, C. E. (2001). Evolutionary radiations and convergences in the structural organization of mammalian brains. *Nature*, **409**, 710–14.

Wu, R., Qinghua, X. & Qingwu, L. (1983). Morphological features of *Ramapithecus* and *Sivapithecus* and their phylogenetic relationships: morphology and comparison of the crania. *Acta Anthropologica Sinica*, **2**, 1–10.

Wunderlich, R. E., Walker, A. & Jungers, W. L. (1999). Rethinking the positional behavior of *Oreopithecus*. *American Journal of Physical Anthropology*, **Suppl. 28**, 282.

15 · Life history and cognitive evolution in the apes

JAY KELLEY

Department of Oral Biology, University of Illinois, Chicago

INTRODUCTION

It has become almost axiomatic in discussions of brain size increase within primates that causation lies with cognition. However, it is also worth exploring other possible causative factors. For instance, it has long been known that within mammals, brain size is broadly correlated with the pace of life history (see Deaner, Barton & van Schaik 2003; van Schaik & Deaner 2003, and references therein). It is not surprising therefore that large-brained great apes have greatly prolonged life histories compared with smaller-brained monkeys (Harvey & Clutton-Brock 1985; Harvey, Martin & Clutton-Brock 1987; Kelley 1997). In discussions of the life-history/brain size relationship, particularly concerning human evolution, the arrow of causation is almost universally suggested to point from brain size to life history. In this view, life-history changes are passive consequences of selection for brain size and by implication, cognitive capacity. However, this is contradicted by a substantial body of theory and empirical evidence pointing to species demographics in the shaping of life history. The relationship between brain size and life history could also be due to correlations to another variable, such as body size, without any direct cause-and-effect relationship between the two, but this seems unlikely (van Schaik & Deaner 2003; and below).

There are further difficulties with the proposition that selection for enhanced cognitive capacity leads to increasing brain size. In spite of the broad correlation between cognitive capacity and brain size across major primate higher taxa (Byrne 1997; Gibson, Rumbaugh & Beran 2001, Hart & Karmel 1996), there is no evidence of correlations between cognitive capacity and brain size within species. Thus, it is not obvious why selection for increased cognitive capacity should lead to increases in brain size across species. For these and other reasons,

it might be useful to explore an alternative ordering of cause and effect in the relationships between brain size, cognition, and life history. Might brain size increase be a largely passive consequence of life-history prolongation, and enhanced cognitive capacity an emergent property of this developmental process?

In the following discussion, I first describe life history and briefly review the body of theory relating to the evolution of mammalian life histories. I then describe the relationship between brain size and life history and its causation, focusing on the processes and patterns of brain growth and development. Finally, I review what is known about the evolution of life history and brain size within the ape clade based on the fossil record, using both indirect evidence from dental development and direct neurocranial evidence where available.

MAMMALIAN LIFE HISTORY AND LIFE-HISTORY EVOLUTION

Most simply, life history is about the pace of life and the progression through life stages, including both prenatal and postnatal growth and maturation, as well as the reproductive and, for humans, post-reproductive phases of adulthood. It is most frequently characterized in terms of key developmental, maturational, and reproductive milestones such as gestation period, age at weaning, age at sexual maturity, age at first breeding, interbirth interval, reproductive span, and longevity.

The durations and ages of occurrence of life-history variables tend to be highly correlated within species (Figure 15.1). For example, a species with a short gestation period will also tend to have an early age at weaning, an early age at first reproduction, etc. Consequently, life history as a whole is expressed as distinct suites or syndromes, and organisms can be arrayed along a

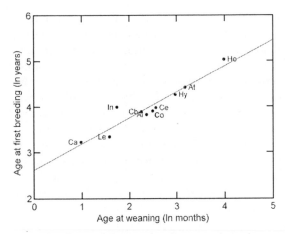

Figure 15.1. Least squares regression of age at weaning against age at first breeding, both log-transformed, in several primate higher taxa. Values are averages of the included species. Symbols (numbers of included species in parentheses): Al, Alouattini (2); At, Atelini (2); Ca, Callitrichinae (3); Cb, Cebinae (3); Ce, Cercopithecinae (6); Co, Colobinae (8); Ho, Hominidae (3); Hy, Hylobatidae (2); In, Indriidae (3); Le, Lemuridae (7). Data from Godfrey *et al.* (2001) and from K. Strier, personal communication, for *Brachyteles arachnoides* (Atelini). Indriids are exceptional among extant primates in their early age at weaning and their precocious dental development in relation to other life-history traits (Godfrey *et al.* 2002).

continuum of general life-history schedules (Harvey, Promislow & Read 1989a; Harvey, Read & Promislow 1989b; Kowalewski, Blomquist & Leigh 2003; Promislow & Harvey 1990; Read & Harvey 1989). This has been referred to as the fast–slow life-history continuum. When the pace of life history changes within a species, the durations of individual life-history variables tend to be either compressed or extended in concordant and roughly proportionate fashion, although individual taxon-specific circumstances or constraints can impact the process to produce disproportionate change in one or more variables (e.g., Godfrey, Petto & Sutherland 2002; Leigh & Bernstein, 2003; Lycett & Barrett, 2003).

The positions of different taxa along the fast–slow life-history continuum are generally highly correlated with body size. This allometric reality has frequently been offered as the "explanation" for observed life-history variation. However, in the absence of hypotheses about causation, the allometric relationship between life history and body size merely restates the facts it is supposed to explain (Boyce 1988; Harvey *et al.* 1989b).

Further, across mammals, there is variation in life history that does not correlate with body size (Read & Harvey 1989). This suggests that, (1) factors other than body size influence life history, and (2) the correlation between life history and body size might owe in part to the correlation of each with one or more of these other factors.

A substantial body of theory now characterizes life history primarily in terms of the scheduling of reproduction, particularly the age at first reproduction (e.g., Caswell 1982; Charlesworth 1980; Charnov 1991, 1993; Stearns 1992). Selection for the scheduling of reproduction clearly impacts the entire ontogenetic process, reflected in the high degree of correlation among life-history traits. Changes in life history are brought about primarily through the cascading effects on ontogeny resulting from selection to alter the timing and frequency of reproduction. Since fecundity must be balanced by mortality, in life-history theory mortality rate is the principal selective agent responsible for observed life-history variation.

Harvey and co-workers have examined the relationship between mortality and life-history variation in mammals empirically (Harvey 1991; Harvey & Read 1988; Harvey & Zammuto 1985; Harvey *et al.* 1989a,b; Promislow 1991; Promislow & Harvey 1990, 1991; Read & Harvey 1989; Sutherland, Grafen & Harvey 1986). They found that with the effects of body size removed, there are significant correlations between mortality, especially age-specific mortality, and the residual variation in a number of life-history variables. While the pattern of correlations between life-history traits and age-specific mortality is complex (Promislow & Harvey 1990), in general taxa that suffer relatively high adult mortality in relation to infant/juvenile mortality tend to have relatively fast life histories. Those that suffer relatively low rates of adult mortality tend to have slower life histories (see also Horn 1978; Stearns 2000). The explanation for this phenomenon lies with the "tradeoffs" intrinsic to life history. If the probability of death as an adult is relatively low and that of infants and juveniles is relatively high, greater lifetime reproductive output will result from delaying reproduction and minimizing the many deleterious consequences of early reproduction (Stearns 1992, 2000). If, on the other hand, the probability of death as an adult is relatively high, greater lifetime reproductive output will result from early reproduction despite its costs. As expected, given the influence of body size on predator–prey relations, mortality is also significantly correlated with body size.

Considering the catarrhine primates, great apes have greatly prolonged life histories compared with Old World monkeys, with longer gestation, slower maturation and much later ages at first reproduction (Harvey & Clutton-Brock 1985; Godfrey *et al.* 2001; Harvey *et al.* 1987; Kelley 1997, 2002). Limited data on age-specific mortality in the two groups are consistent with their relative positions on the primate fast–slow life-history continuum (Kelley 1997).

Sequential hypermorphosis (Progenesis) in life-history change

As noted above, change in the overall pace of life history is expressed as more-or-less concordant changes in every life-history stage. That is, the duration or age of onset of every life-history stage is changed to some degree in the same direction. In the terminology of heterochrony, this is known either as sequential hypermorphosis (prolongation) or sequential progenesis (acceleration). These are often linked with changes in the termination of growth (terminal hypermorphosis or progenesis) (McNamara 2002; McKinney, 2002). Since life history impacts all developmental phenomena, life-history change will clearly affect the timing and patterning of growth, for both the organism as a whole and for individual organs and tissues. In the absence of significant changes in growth rates or counteracting selection pressures affecting specific organs, the end result of sequential hypermorphosis with respect to growth will be peramorphosis, or the development of traits or organs beyond the condition present in the ancestral adult.

MAMMALIAN LIFE HISTORY, BRAIN SIZE, AND COGNITION

Among mammals, there is a broad correlation between the overall pace of life history and brain size, reflected in significant correlations between brain size and a variety of life-history traits (e.g., Eisenberg 1981; Promislow 1991; Sacher 1959, 1975, 1978; Sacher & Staffeldt 1974). This relationship can be demonstrated within the primate order as well (Deaner *et al.* 2003; Godfrey *et al.* 2001; Smith 1989; Smith, Gannon & Smith 1995). Nevertheless, there has been debate about the degree to which life-history traits and brain size co-vary, and

whether the correlations reflect an actual causal relationship. For example, Harvey (Harvey *et al.* 1989a; Read & Harvey 1989) has suggested that the relationships are nothing more than statistical artifacts of the correlation between brain size and body size on the one hand, and life history and body size on the other. However, more recent analyses incorporating better data and more appropriate statistical methods have greatly strengthened the argument for correlated evolution of brain size and life history independent of body size in both primates and mammals as a whole (Deaner *et al.* 2003; Godfrey *et al.* 2001; Promislow 1991; van Schaik & Deaner 2003).

Linking life history, brain size, and cognition: brain growth

If great ape cognitive capacities owe to absolutely large brains (Gibson 1990; Gibson & Jessee 1999; Gibson *et al.* 2001), then causation in the life-history/brain size relationship is of paramount importance for understanding their evolution. Before examining the issue of causation, however, it is useful briefly to explore brain growth, which is relevant to brain size increase and to cognitive development and the evolution of cognitive capacity as well. Neocortex growth is most critical because variation in brain size and differences in cognitive capacity among primates are largely attributable to differences in the size of the neocortex (Gibson *et al.* 2001; Northcutt & Kaas 1995; Rakic 1988).

With respect to cognitive potential, growth and development of the neocortex can be divided into two basic phases, an initial phase of neuron production and a subsequent phase of neural network formation (Caviness, Takahashi & Nowakowski 1995; Kornack & Rakic 1998; Rakic 1988, 1995; Rakic & Kornack 2001). In mammals, neuron production takes place during the first half of gestation and is divided into two phases. During the initial proliferative phase, each mitotic event produces two daughter progenitor cells (symmetrical division), resulting in a doubling of progenitor cell numbers with each mitotic cell cycle. Since cortical area is proportional to neuronal number, species differences in cortical surface area, and thus in overall brain size, are largely due to differences in the duration of the proliferative phase and to a lesser degree, to the rate of cell cycling during this phase. Even small changes in the duration of the proliferative phase can have profound effects on the founder cell population size. As Rakic (1995: 386) has noted, "Conceivably, a slight prolongation of this phase . . . of proliferation could be responsible indirectly for a significant surface enlargement of the cerebral cortex."

The proliferative phase is followed by the period of neuron formation, or neuronogenesis. During this phase symmetrical cell division progressively shifts to asymmetrical division, wherein each mitotic event produces one progenitor cell and one postmitotic cell (cell differentiation). The postmitotic cell then develops into a neuron as it migrates outward toward the cortical plate. The neuronogenetic period ends when all cells have either undergone terminal differentiation, that is, the production of two postmitotic cells, or succumbed to apoptosis.

The consequences for both brain size and cognitive capacity of relatively modest changes in these two phases of brain growth can be dramatic. A 100-fold increase in cortical area from mouse to macaque results from a fourfold increase in the duration of the proliferative phase (10 days versus 40 days) combined with a five- to tenfold increase in the duration of the neurogenetic phase (6 days versus 30–60 days depending on the cortical region). The result is nearly three times as many cell cycles in the neurogenetic phase alone, despite slower cell cycling in the macaque than in the mouse. Assuming approximately equal cell cycle length, it takes only a few additional days of the proliferative phase to produce the difference in order of magnitude in cortical expansion between monkey and human (Rakic, 1995).

Neural network formation, the development of neural organization, begins during late gestation and continues through infancy and the juvenile period. It results from both genetically and epigenetically or experientially determined processes (e.g., Gibson 1990, 1991; McKinney 2002; McNamara 2002; Rakic & Kornack 2001). Synaptogenesis, for instance, including dendritic growth, is a strongly time-dependent and partly experientially mediated process involving competitive, selective reinforcement and elimination of synaptic pathways During this period, accumulated experience and learning contribute to shaping the increasingly complex connectivity in the neurally enriched substrate of larger brains. The degree to which the final topology of neuronal connectivity owes to genetically based, species-specific patterns of connectivity, versus afferent stimulation acting to regulate gene expression and

cellular interactions is only now being explored in depth (Rakic & Kornack 2001). Nevertheless, in species with larger neocortices and larger numbers of neurons, there is greater and more complex connectivity, which forms the anatomical basis for greater cognitive capacity. In primates, the increase in neuronal connectivity with increasing brain and neocortical size is reflected in a pronounced relative increase in the mass of neocortical white matter, or axonal mass (Hofman 2001). Lengthening the period of neural network formation also leads ultimately to more and longer, qualitatively distinct stages of cognitive development. This is most evident in humans; significantly, it appears to be true of great apes as well in comparison with monkeys (Byrne 1997; Langer 1996; Parker 2002; Parker & McKinney 1999).

Sequential hypermorphosis and brain growth

The observed differences between species in brain size are most simply explained as the products of sequential delay in the stages of brain growth in larger-brained species, resulting in peramorphosis. Given the correlation between brain size and the pace of life history across mammals, this sequential hypermorphosis in brain growth is most reasonably explained as just another of the many outcomes of sequential hypermorphosis in life history more generally. In a model based on sequential hypermorphosis, it is also reasonable to presume that temporal changes in any particular life stage, such as gestation or infancy, are likely to be reflected in each of its component phases as well. Thus, in the case of gestation, for example, it would be expected that, in the absence of counteracting pressures to alter brain growth trajectory, general growth prolongation would result in concomitant and concordant prolongation in both the proliferative and neuronogenetic phases of neuron production, as well as in the initial, prenatal phase of neural network formation. What is known about the durations of brain developmental stages in various species is consistent with this model. Therefore, since the two principal phases of brain development, neuronal production and neural network formation, are prolonged in concert, brain enlargement through sequential hypermorphosis also explains why larger brains are also more complex.

Given that life history change through sequential hypermorphosis affects all growth and maturational stages, it is difficult to see how correlated response in brain size and complexity would not result from general life-history prolongation, again, unless there are counteracting selection pressures limiting brain enlargement. In fact, Deaner *et al.* (2003) and van Schaik and Deaner (2003) have identified one mammalian group, the Chiroptera (bats), in which life-history prolongation has not led to the expected degree of brain size increase. These authors demonstrated that bats have very small brains compared with their size-adjusted life histories, but declined to speculate on explanations for the dissociation. One plausible explanation is the need to reduce weight in volant species. Minimizing brain size increase during life-history prolongation in this group might have been strongly favored, as one of the many adaptations to limit weight in bats.

Regarding catarrhine primates, there is sufficient knowledge of the ontogeny of brain development at the cellular level in monkeys (macaques) and humans to describe the developmental basis for differences in brain size and cognitive capacity between the two. Almost nothing is known about brain development in great apes at this level, although it can be inferred that they occupy an approximately intermediate position in terms of the durations of developmental phases. More is known in all three groups about rates of ontogenetic size increase of the brain, in which great apes more closely resemble the multi-phasic pattern of humans (Rice 2002; Vrba 1998). Vrba has shown that, (1) in humans and chimpanzees, the initiation, duration, and/or termination of the different rate phases of brain size increase generally correspond to identified growth phases of the brain, and (2) that differences in the durations of the different phases in the two species are explainable by proportional prolongation of each phase in humans (sequential hypermorphosis).

The foregoing account of brain growth and its relationship to life history in primates, and particularly humans, is not novel (e.g., McKinney 1998, 2002; McNamara 2002, Rakic & Kornack 2001; Vrba 1998). What has not been adequately addressed, however, is causation in the life-history/brain size relationship.

CAUSATION IN THE LIFE-HISTORY/ BRAIN SIZE CORRELATION

For those primarily concerned with primate evolution, or human evolution specifically, hypotheses about evolutionary increase in brain size almost invariably invoke selection for enhanced cognitive ability of one sort or another as the driving force for brain size increase (see many contributions in this volume). The same is true

concerning the arrow of causation in the relationship between life history and brain size. The earliest descriptions of the life-history/brain size connection posited that the brain must be the pacemaker of life history, due to the high energetic requirements of neural tissue (Hofman 1983; Sacher 1959, 1975; Sacher & Staffeldt 1974). In this view, the costs of growing and maintaining a large brain are so high that other selective factors relating to growth, development, and reproduction are secondary. Similar, albeit more sophisticated, arguments continue to be made (e.g., Foley & Lee 1991; Lee 1999; Martin 1996). A number of other specific hypotheses have been proposed as to why large brains and slow life histories go hand in hand (see van Schaik & Deaner 2003). These are broadly of two types, those in which selection for increasing brain size directly and necessarily leads to slower life history (e.g., maturational constraints and cognitive buffer hypotheses), and those in which selection for brain size increase is permitted by a coincident slowing of life history (e.g., brain malnutrition risks and delayed benefits hypotheses). What all of these hypotheses share is the notion that increasing brain size is being selected, presumably because of the adaptive advantages of enhanced cognition. In this view, large brains are invariably adaptive.

One problem with all such hypotheses is that the relationship between life history and brain size extends to all mammals. It is not clear that hypotheses formulated to explain brain size increase in primates, or only one group of primates, and based on selection for cognitive capacity, apply equally well to mammals as a whole (van Schaik & Deaner 2003). There is no general theory of cognitive evolution in mammals comparable to the very robust and taxonomically encompassing theory of life-history evolution based on mortality rates and the selective advantages of altering reproductive schedules. Moreover, if selection for cognitive capacity underlies the life-history/brain size relationship, then life-history prolongation becomes an essentially passive consequence of selection for brain size. This is the view adopted by most primatologists who have investigated this relationship, particularly those concerned primarily with brain expansion in humans. However, adopting such explanations renders meaningless the substantial body of life-history theory oriented around reproductive scheduling and based on demographics, or it at least presumes that causation in the relationship is different in primates than in other mammals. In light of these concerns, and given that abandonment of general theories in favor of *ad hoc* explanations should be avoided, it is worth exploring a reversal of cause and effect in the life-history/brain size relationship.

If life-history prolongation cascades through ontogeny, altering the durations of all life stages and growth phases to some degree in the same direction, then some degree of peramorphic brain enlargement will inevitably result, again assuming the absence of counteracting pressures to selectively alter rates and/or the duration of brain growth. This ordering of cause and effect eliminates the conundrum noted earlier, wherein selection for cognitive capacity leads to increased brain size despite the lack of evidence for correlations between cognitive capacity and brain size within species. While cognitive abilities can in principle influence mortality (cognitive buffer hypothesis, e.g., Rakic 1995; Rakic & Kornack 2001), empirical studies reveal that, overwhelmingly, body size and habitat (including substrate preference) are the major influences on mortality (Deaner *et al.* 2003; Partridge & Harvey 1988; Ross 1992, 1998; Southwood 1988; van Schaik & Deaner 2003; Wootton 1987) and life-history evolution.

Thus, in a way that hypotheses of brain size increase based on cognitive selection cannot do, a hypothesis based on life-history change adequately explains the relationship between life history and brain size in all mammals. Further, in a life-history driven process, brain enlargement is inextricably linked to increasing complexity in neural organization. Since extension of the infant and juvenile stages will also occur with life-history prolongation, this process provides a developmental link between an augmented neural substrate and the epigenetic components of enhanced cognitive capacity that are instrumental in shaping the organization of that substrate. The extension of the infant/juvenile learning period has often been suggested to be the principal target of selection for enhanced cognitive capacity and, therefore, increased brain size, particularly in great ape and human evolution. However, in a model of life-history evolution based on sequential hypermorphosis, prolongation of this period is, like the other developmental phases, an expected outcome of life-history prolongation as a whole.

Finally, life-history prolongation through sequential hypermorphosis can also account for differential enlargement of specific brain areas through the application of a uniform time extension to areas with different growth allometries (see also MacLeod, Chapter 7, this volume). Sequential hypermorphosis in life history can therefore explain the entirety of the enhanced neural substrate associated with larger brains, as well as differential enlargement of specific brain areas, without recourse to selection for enhanced cognitive capacity as a whole or specific cognitive abilities.

EVOLUTION OF ENHANCED COGNITIVE CAPACITY

Importantly, the life-history based hypothesis of brain size increase proposed here does not negate or alter hypotheses of cognitive evolution *per se*; the former is not a substitute for the latter. What it does is eliminate causation from the relationship between cognitive evolution and evolutionary increase in brain size, with both ultimately dependent upon life history. Hypotheses of cognitive evolution on the one hand and brain size increase through life-history change on the other are, in fact, complementary rather than mutually exclusive. Brain size increase through life-history prolongation is in itself an insufficient explanation for enhanced cognitive capacity. Selection pressures favoring enhanced cognition are necessary for shaping and modifying the basic, heritable portion of neural organization from which the potential for greater cognitive capacity emerges. At the same time, without the enriched neural substrate provided by increased brain size, there are limits to the qualitatively different cognitive responses to truly novel selection pressures that might lead to adaptive increases in cognitive capacity; the organism would effectively be adaptively "blind" to the existence of such pressures. In the model proposed here, life-history evolution and cognitive evolution are coincident, but are responses to different sets of selection pressures. They are linked through brain size, with life-history prolongation providing a more compelling explanation for brain size increase for mammals as a whole than does selection for enhanced cognition.

This framework for cognitive evolution is similar to, but subtly different from, that proposed by van Schaik & Deaner (2003), which is also based on coincident selection for life-history prolongation and increased cognitive capacity. Here, brain enlargement is seen as a largely pleiotropic, developmental phenomenon related to selection for prolonged life history. On the other hand, van Schaik and Deaner see brain enlargement being driven by selection for cognitive development, but occurring only when there is coincident slowing of life history to provide a sufficiently long growth period.

Another important element of both models is taxon specificity in the degree of brain size increase and increases in cognitive capacity. The potential for brain size increase will depend in part on species-specific factors influencing brain development and function (Preuss 2001; Rakic & Kornack 2001). Critical factors include the size of the founder population of neuron progenitor cells and rates of cell cycling (Rice 2002), both of which are strongly phylogenetically dependent. Equally important is the capacity of the organism to support the energy requirements for the development and maintenance of an enlarging brain (Aiello & Wheeler 1995; Aiello et al. 2001; Martin 1996; Parker 1990). Animals whose basic trophic adaptations preclude them from a high-quality, energy-rich diet will face limits to brain enlargement, regardless of the progress of life-history prolongation. There may well be other selection pressures, largely or wholly unrelated to life history, that might act to limit evolutionary increase in brain size, as the selection pressures suggested that have limited brain size increase in bats. The nature of the selection pressures for increasing cognitive capacity will also vary with the habitat and habitus of the organism. In the case of great apes, the stimuli favoring enhanced cognition might have included any of the many cognitive challenges explored in this book.

FOSSIL EVIDENCE FOR THE EVOLUTION OF LIFE HISTORY AND BRAIN SIZE IN APES

Life-history inference from dental development

Because life history and brain size have undergone correlated evolution, measures of brain size in fossil species are reliable proxies for inferring the overall pace of life history. Unfortunately, there are very few fossil apes for which there are sufficiently well preserved neurocrania to provide reliable estimates of cranial capacity (Begun & Kordos, Chapter 14, this volume). The principal means, therefore, for inferring the life histories of fossil species has been through the chronology of dental development.

The timing of dental development in all mammals is highly correlated with ontogeny as a whole, yielding significant correlations between specific events in dental development and individual life-history variables (Smith 1989, 1991, 1992). Dental development is in a sense simply another life-history trait (Smith &

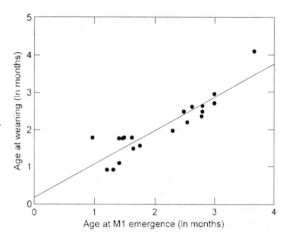

Figure 15.2. Least squares regression of age at weaning against age at M1 emergence, both log-transformed, in 20 extant nonhuman primate species. Included species are those from Table 15.1, with the following exclusions because of a lack of weaning age data: *Cheirogaleus medius*, *Galago senegalensis*, *Macaca fuscata*, and *Homo sapiens*. Age at weaning from Godfrey et al. (2001); age at M1 emergence from Smith et al. (1994). Other life-history variables are similarly correlated with age at M1 emergence (Smith, 1989).

Tompkins 1995), but one that is preserved in the fossil record. There are variations in the relationships between dental development and life-history attributes that are systematic and primarily associated with differences in diet (Godfrey et al. 2001), but, within a broad framework, the pace of dental development serves as a reliable proxy for the pace of life history as a whole. Smith (1989, 1991) has demonstrated that, among living primates, age at first molar (M1) emergence is a particularly good correlate of various life-history traits (Smith 1989, 1991), emergence being defined as the initial penetration of the oral gingiva by the molar cusps (Figure 15.2). Thus, if the average age at M1 emergence can be established for a fossil species, then its general life-history profile can be characterized as well.

There are two approaches to estimating age at M1 emergence in fossil species. The most straightforward is to determine the age at death for individuals that died while in the process of erupting their M1s, making necessary adjustments if the stage of eruption differs from that associated with gingival emergence. The second is to determine the crown formation time of M1 and to add to this the time taken to form the amount of root that would have been present at the time of emergence. The underlying developmental basis of this approach is that the M1 begins to form just prior to birth in nearly all primates (Beyon, Dean & Read 1991; Dean 1989). This method still requires a tooth from an individual that died not too long after the M1 erupted, because crown formation time can only be reliably determined from teeth that are unworn or that show minimal wear. Presently, it is also less precise than the first method because there is only a limited amount of information from living primates on the extent of root development at the time of first M1 emergence and how this varies across species (Kelley & Smith 2003). Nevertheless, with some reasonable assumptions, this approach can still be used to establish approximate minimum values for this key event in dental development (see below).

Both approaches to estimating age at M1 emergence rely on the record of incremental growth lines that is preserved in all teeth (Boyde 1963). Regular short-period and long-period incremental features record daily secretions of the enamel and dentine-forming cells and periodic disruptions in secretion across the developing enamel and dentine fronts (Bromage & Dean 1985; Dean 1987, 1989; FitzGerald 1998). These incremental features are preserved in all teeth, including fossilized teeth, and permit calculation of the periods of crown and root formation, from which ages at death and M1 emergence can be derived. They are analogous to the growth rings in trees, but with a daily rather than annual period of resolution. Describing how the incremental lines of teeth are used to determine crown and root formation times, and how these are then used to determine age at M1 emergence, is beyond the scope of this chapter. The details of methodology can be found in any of the following sources: Beynon *et al.* (1991), Bromage & Dean (1985), Dean (1987, 1989), Dean *et al.* (1986, 1993), Dirks (1998), Kelley (1997, 2002), Kelley, Dean & Reid (2001), Kelley & Smith (2003), Macho & Wood (1995).

To date, age at M1 emergence has been directly calculated for only two fossil apes (Figures 15.3 and 15.4). The first is an individual of *Sivapithecus parvada* from a 10 Ma locality in the Siwaliks of Pakistan (Kelley 1997, 2002). *Sivapithecus* is widely regarded to be a member of the orangutan lineage (Andrews & Cronin 1982; Begun *et al.* 1997; Pilbeam 1982; Ward 1997; Ward & Brown 1986). The second is an individual of *Afropithecus turkanensis* from the 17 Ma site of Moruorot in Kenya (Kelley 2002; Kelley & Smith 2003). *Afropithecus* is generally regarded to be a stem ape, outside the great

ape and human clade (Andrews 1992; Begun, Ward & Rose, 1997; Leakey & Walker 1997).

For both individuals, some of the dental growth parameters necessary for calculating age at M1 emergence were obtainable from the specimens themselves. Others had to be estimated from growth data for extant great apes and humans. Since there is both intra- and interspecific variation in these growth parameters, a range of estimates of age at first molar emergence was calculated. These estimates also included slight adjustments to account for the fact that neither individual died precisely at the stage of M1 eruption corresponding to gingival emergence. As determined by, respectively, the position of the M1 within the mandible and the degree of M1 root development, the *Afropithecus* individual died just prior to gingival emergence (Kelley & Smith 2003), while the *Sivapithecus* individual died soon afterward (Kelley 1997).

Even the minimum estimates for both *Afropithecus* (28.2 months) and *Sivapithecus* (39.0 months), which incorporate the minimum known values for the various estimated growth parameters, are well within the range of values for chimpanzees (25.7–48.0 months). The minimum estimate for *S. parvada* reported here differs from that in Kelley (1997, 2002) based on new, unpublished data on tooth growth in this species. Since it is unlikely that any one individual would express the minimum known values for each of the growth parameters, it is more probable that the actual ages at first molar emergence for the two individuals are closer to the mean estimates of approximately 36 and 43 months, respectively. Since age at M1 emergence is broadly correlated with body mass (Kelley & Smith 2003), these estimates are close to what would be expected for apes of this size based on the relationship between age at M1 emergence and body mass in living great apes (*A. turkanensis* was roughly the size of small chimpanzees while the body size range for the highly sexually dimorphic *S. parvada* was between that of average-sized chimpanzees and female gorillas).

A minimum value only for age at M1 emergence has been calculated for another Miocene hominoid species, *Dryopithecus laietanus*, from the 9.5-million-year-old site of Can Llobateres in Spain (Kelley *et al.* 2001, 2002). *Dryopithecus* is generally considered to belong to the great ape and human clade, either as a primitive member or as a member specifically of the African ape and human clade (Begun *et al.* 1997) or the orangutan clade (Moyà Solà & Köhler 1995). The minimum value for age at

(a)

(b)

Figure 15.3. Infant mandible of *Sivapithecus parvada* from the late Miocene Siwalik sediments of Pakistan: (a) buccal, (b) occlusal. The specimen preserves the deciduous premolars and the permanent central incisor exposed in its crypt. The erupted first molar fell out of the jaw prior to fossilization, as revealed by the matrix filled alveolus distal to the last deciduous premolar. Several pieces of evidence suggest that the M1 was no more than about six months past gingival emergence when the individual died (see Kelley, 1997).

M1 emergence was determined from the incremental growth lines in the enamel, and root dentine, of an isolated lower M1. It is a minimum estimate because only the initial two millimeters of root are sufficiently well preserved for analysis, which is almost surely less than would have been present at tooth emergence (Kelley *et al.* 2001, 2002). The calculated minimum value, at 31.7 months, is again well within the range of M1 emergence ages of chimpanzees. Since *D. laietanus* was approximately the size of small chimpanzees, the actual age at M1 emergence in this individual would probably have been within the expected range for a great ape of this size.

Assuming the values for the three fossil individuals were representative of their respective species, they suggest life-history profiles that were broadly like those of extant great apes. This in turn suggests that prolonged life histories evolved early in the hominoid lineage. Whether this occurred as adaptive prolongation of life history from an ancestral state in which size-adjusted life histories were faster, or as a simple extension with increasing body mass of the ancestral condition,

(a)

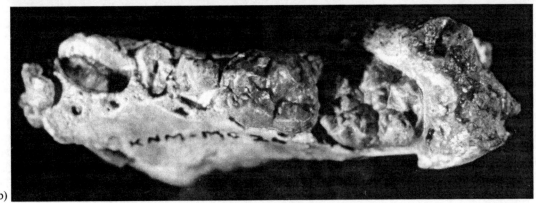

(b)

Figure 15.4. Infant mandible of *Afropithecus turkanensis* from the early Miocene site of Moruorot, Kenya showing the erupting M1 and the lateral incisor germ within its crypt; (a) lingual, (b) occlusal. Based on a longitudinal radiographic study of M1 development and eruption in extant baboons, and similar but more limited data from chimpanzees, it was determined that this individual died approximately 2 to 4 months before the M1 would have undergone gingival emergence (see Kelley & Smith, 2003).

cannot be determined from available data (Kelley & Smith 2003). However, with respect to the correlated evolution between life history and brain size, it makes little difference.

Dental development and brain size

Given the correlation between brain size and life history on the one hand, and age at M1 emergence and life history on the other, it is not surprising that age at M1 emergence and brain size are also strongly correlated (Smith 1989; Smith, Crummet & Brandt 1994) (Figure 15.5; Table 15.1). In fact, the correlation between age at M1 emergence and brain size is stronger than the majority of correlations between age at M1 emergence and life history variables relating to life stages and reproduction. This probably has to do with greater intra-specific variability in certain life history parameters than in dental development (Kelley & Smith 2003; Smith 1989). Using this correlation, brain size in the

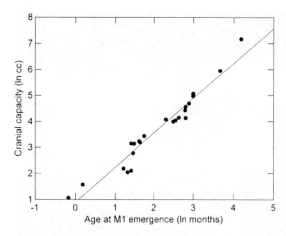

Figure 15.5. Least squares regression of age at M1 emergence against adult cranial capacity, both log-transformed, in 23 of the 24 species of extant primates in Table 15.1. *Propithecus* was excluded because of the anomalous dental development in this genus compared to all other extant primates (see Godfrey *et al.* 2002). Regression equation: ln brain size = 1.33 (ln M1 age) + 0.90.

fossil species can be estimated from the calculated M1 emergence ages. Based on the single estimates of age at M1 emergence in *S. parvada* and *A. turkanensis*, cranial capacity in the two species is estimated at approximately 365 and 290 cc, respectively. By comparison, the average cranial capacity of chimpanzees is 383 cc (Smith *et al.* 1995) with an approximate range of 270–470 cc. However, chimpanzees and especially humans have fairly large positive residuals from the age at M1 emergence/brain size regression line (Figure 15.5), and the estimated average cranial capacity of chimpanzees based on the regression is only 322 cc. Therefore, the estimates for *Sivapithecus* and *Afropithecus* might also be somewhat low. Nevertheless, even these estimates are within the chimpanzee range. Obviously, these figures are tentative but it is intriguing that the cranial capacity estimate of even the 17 Ma *Afropithecus* individual is within the chimpanzee range.

Measures of brain size in fossil apes

Actual cranial capacity estimates are available from two other late Miocene great ape species, *Dryopithecus brancoi* from Hungary and *Oreopithecus bambolii* from Italy (see Begun & Kordos, Chapter 14, this volume). Kordos and Begun (2001) estimated cranial capacity for a partial

calvaria of a female *D. brancoi* individual at 305 cc, with a 95% confidence interval of 280 to 332 cc. Cranial capacity in a second, less complete female calvaria was estimated at 320 cc, with a confidence interval of 305 to 329 cc (Kordos & Begun 1998). These values are also well within the chimpanzee range, but in the lower end of that range as would be expected for a species that was, like *Afropithecus*, the size of small female chimpanzees on average.

Oreopithecus is generally regarded as a stem great ape, either lacking clear relationships to other taxa (Begun *et al.* 1997) or perhaps closely related to *Dryo-pithecus* (Harrison & Rook 1997). Moyà Solà and Köhler (1997) also consider *Oreopithecus* to be closely related to *Dryopithecus*, but they see both as belonging to a broad orangutan clade. Estimating the cranial capacity of *Oreopithecus bambolii* has proved challenging as the one complete skull of the species was crushed flat during fossilization and has only recently been reconstructed (Clarke 1997). Estimates of the cranial capacity of this individual have varied greatly (Harrison 1989), but it is now generally agreed that it would have been quite small (see Begun & Kordos, Chapter 14, this volume; Clarke 1997), perhaps less than 200 cc (Szalay & Berzi 1973). Harrison (1989) was able to estimate the cranial capacity in another, subadult individual using foramen magnum size. This estimate was approximately 130 cc, with a con-fidence interval of 85 to 175 cc. Like *Dryopithecus* and *Afropithecus*, *Oreopithecus* was smaller than extant great apes, with males estimated to have again been about the size of small female chimpanzees and females the size of large macaques (Harrison 1989; Jungers 1987). Nevertheless, this would still leave *Oreopithecus* with a relatively small brain for an ape this size, and a very small, monkey-sized brain in absolute terms.

If large brains are common to great apes, including very early great apes as implied above, then how is the surprisingly small brain of *Oreopithecus* to be explained? Given the general correlation between brain size and life history, it is reasonable to conclude that *Oreopithecus* had an unusually fast life history for a great ape. It is also reasonable to ask if there is any supporting evidence for this supposition. One such piece of evidence is found in dental morphology. *Oreopithecus* was the most foliv-orous of all fossil apes for which dietary inference has been attempted (Ungar & Kay 1995). Folivorous pri-mates have relatively much smaller brains and faster life histories than similarly sized frugivores of the same clade

Table 15.1. *Age at M1 (lower molar) emergence and brain size in extant primates*

Species	Age at M1 emergence (months)	Average brain size (cc)
Cheirogaleus medius[1]	0.84	2.9
Varecia variegata	5.76	31.2
Lemur catta	4.08	23.4
Eulemur fulvus	5.04	25.6
E. macaco	5.16	24.6
Propithecus verreauxi	2.64	29.7
Galago senegalensis[1]	1.20	4.8
Callithrix jacchus	3.72	7.7
Saguinus fuscicollis	4.10	8.2
S. nigricollis	3.35	8.9
Cebus albifrons	12.72	56.8
C. apella	13.80	63.1
Saimiri sciureus	4.44	23.2
Aotus trivirgaus	4.32	16.1
Cercopithecus aethiops	9.96	59.2
Macaca fascicularis	16.44	62.5
M. fuscata[1]	18.00	109.1
M. mulatta	16.20	81.3
M. nemestrina	16.44	96.2
Papio anubis	20.04	158.9
P. cynocephalus	20.04	145.5
Trachypithecus cristata[2]	12.00	54.5
Pan troglodytes[3]	39.12	383.4
Homo sapiens[3,4]	66.03	1292.5

Notes:

M1 emergence data from Smith *et al.* (1994) and cranial capacities from Godfrey *et al.* (2001), with the following exceptions:

[1] Cranial capacity from Harvey *et al.* (1987).

[2] M1 emergence data from Wolf (1984).

[3] Cranial capacity from Smith *et al.* (1995).

[4] M1 emergence data from Smith *et al.* (1995).

(e.g., Blomquist, Kowalewski & Leigh 2003; Clutton-Brock & Harvey 1980; Godfrey *et al.* 2002; Harvey *et al.* 1987; Martin, 1984). Additionally, remains of *Oreopithecus* are found only from regions that, at that time, were islands in the northern Tethys Sea (Harrison & Rook 1997 and references therein). Insular environments tend to produce life-history convergence among large and small species, most likely a consequence of altered patterns of mortality having mainly to do with predation (Boyce 1988). In non-insular environments, larger species suffer relatively high rates of infant/juvenile predation in comparison with predation on adults, whereas predation on small mammals is largely independent of age (therefore, relatively high rates of predation on adults compared with large species). With the relaxation of predation pressure that typically occurs in insular environments, infant/juvenile survival is disproportionately increased in large species. This favors earlier and more frequent reproduction and, therefore, accelerating life history. The converse is true for small mammals because mortality from intrinsic causes and extrinsic causes other than predation is still higher in infants and juveniles, and adult mortality is therefore relatively diminished.

Both direct and indirect evidence therefore support accelerated life history to explain small brain size in *Oreopithecus bambolii*. This hypothesis can be tested by determining ages at M1 emergence in individuals of the species. What the consequences of such small brain size might have been for cognition can only be speculated upon. With respect to the thesis that the cognitive abilities that differentiate great apes from monkeys ultimately depend upon the absolutely larger brains of the great apes, that speculation would be that cognitive capacity in *Oreopithecus* was no more advanced than in extant monkeys.

Oreopithecus notwithstanding, the weight of the current evidence from the fossil record suggests, first, that a large extant-great-ape-sized brain was present in the last common ancestor of modern great apes and humans. This is not particularly surprising. What is more interesting is the dental evidence from early Miocene *Afropithecus* for a life-history profile, and brain size, which were broadly like those of extant great apes as well. I have suggested elsewhere that a shift toward prolonged life history in apes might have been a key adaptation in the divergence of apes and monkeys in the earliest Miocene (Kelley 1997). Given the correlated evolution between life history and brain size, it may therefore be that the enhanced cognitive capacities seen in the living great apes began to emerge early in the evolutionary history of the ape clade.

CONCLUSIONS

In the foregoing discussion, it has been assumed that cognitive capacity and brain size are strongly linked. Traditionally, evolutionary increase in brain size, particularly among primates, has been viewed as being driven by selection for cognitive abilities. Selection for cognitive ability and increasing brain size have also been viewed as the principal determinants of the pace of life history, either acting directly through limitations on general growth and development imposed by the energy requirements of the developing brain, or indirectly through selection for lengthening the infant/juvenile learning period. There are reasons, however, to consider a reversal of cause and effect in the life-history/brain size relationship, that evolutionary increase in brain size is primarily a developmental outcome of selection for

prolonged life history, not only in primates but, importantly, in mammals as a whole.

Life-history prolongation proceeds through the heterochronic process of sequential hypermorphosis, wherein all developmental and maturational stages are extended to some degree in the same direction. Extending the developmental periods of the various stages of brain growth results not only in a larger neocortex with a larger complement of neurons, but also in more numerous and more complex neuronal connections both within and between different regions of the brain. There is a considerable body of theory, backed by empirical tests, relating life-history variation among mammals to patterns of mortality and consequent selection for either delayed or accelerated reproduction. Thus, variation in brain size, and brain size increase in particular, is more reasonably viewed as an almost inevitable effect of selection for the timing and scheduling of reproduction, which govern the overall pace of life history.

In this model, great apes (and humans) have absolutely large brains in comparison with monkeys because they have much slower life histories. Their greater cognitive capacities relative to monkeys can be viewed as emergent properties, made possible by their enlarged brains and promoted by selection pressures that favored these capacities. In this model, neither enlarged brains nor selection pressures that might lead to increased cognitive capacity will alone result in cognitive evolution in the absence of the other. Cognitive evolution occurs with the coincidence of the two, with brain size increase resulting from selection for life-history prolongation and enhanced cognitive capacity resulting from taxon-specific pressures that promote the reorganization of the now enhanced neural substrate. The presence of taxon-specific selection provides a plausible explanation for why the cognitive capacities of the various mammalian species with absolutely large brains, while uniformly impressive, differ in fundamental ways (see also van Schaik & Deaner 2003, and contributions to this volume). Life-history prolongation plays a further critical role in cognitive evolution by stretching out the time period during which neural connections are being established and modified, a partly epigenetic process by which the organizational "blueprint" of the brain is elaborated and reshaped so that cognitive potential is fully realized (Rakic & Kornack 2001). Thus, the lengthened learning period, often viewed as the target of selection in cognitive hypotheses of brain size increase, is also more reasonably viewed as an inevitable consequence of selection for life-history prolongation.

Direct evidence from the fossil record of brain evolution in apes is very limited (Begun & Kordos, Chapter 14, this volume). However, the correlated evolution of brain size and life history provides an indirect means for inferring brain size in fossil ape species. The chronology of dental development can be viewed as simply another life-history variable, and one that is particularly strongly correlated with brain size in primates (Smith 1989; Smith *et al.* 1994). Accumulating evidence from the timing of dental development and eruption in several fossil ape taxa, including *Sivapithecus* and *Dryopithecus* from the late Miocene and, importantly, *Afropithecus* from the early Miocene, suggests that life-history prolongation and brain size increase began early in the evolutionary history of the group.

When examining the causes of evolutionary increase in brain size, it is at least prudent to look at possible causative agents other than those having to do with cognition. In particular, it is worthwhile to examine developmental processes that impact brain growth and development, and their causation. In this respect, the ideas developed here are similar to those of Gibson (1990), Vrba (1998), McNamara (2002), and McKinney (2002) among others. The major difference is the focus here on life history as the target of selection leading to evolutionary increase in brain size. From this perspective, cognitive capacities, in particular the enhanced cognitive capacities of great apes, are as much due to changes in life history as they are to the selective pressures that favored the emergence of those capacities.

ACKNOWLEDGMENTS

It is not enough merely to thank the editors of this volume, Anne Russon and David Begun, for inviting my participation. I have in fact never been more indebted to the editors of a volume for their patience in my entire career. I would also like to thank all those with whom I have collaborated over the last few years in the area of life history and dental development, as well as those who have helped shape my thoughts through discussions or through their work. Notable among these individuals are Chris Dean, Wendy Dirks, Paul Harvey, Don Reid, Gary Schwartz, Holly Smith, Tanya Smith, Karen Strier, and Carel van Schaik. Lastly, I wish to thank the Geological Survey of Pakistan and the National Museums of Kenya, and their staffs, for support of fieldwork and for access to specimens in their care. The research on dental development in fossil apes was supported by grants from the National Science Foundation.

This contribution is dedicated to the late James Fuller, an exceptional neurophysiologist and an even better colleague. Jim's untimely death cut short our very stimulating and lively discussions on brain evolution just as I began to work on this chapter in earnest.

REFERENCES

Aiello, L. C. & Wheeler, P. (1995). The expensive tissue hypothesis: the brain and the digestive-system in human and primate evolution. *Current Anthropology*, **36**, 199–221.

Aiello, L. C., Bates, N. & Joffe, T. (2001). In defense of the expensive tissue hypothesis. In *Evolutionary Anatomy of the Primate Cerebral Cortex*, ed. D. Falk & K. R. Gibson, pp. 57–78. Cambridge, UK: Cambridge University Press.

Andrews, P. (1992). Evolution and environment in the Hominoidea. *Nature*, **360**, 641–6.

Andrews, P. & Cronin, J. E. (1982). The relationship of *Sivapithecus* and *Ramapithecus* and the evolution of the orang-utan. *Nature*, **297**, 541–6.

Begun, D. R., Ward, C. V. & Rose, M. D. (1997). Events in hominoid evolution. In *Function, Phylogeny and Fossils: Miocene Hominoid Evolution and Adaptation*, ed. D. R. Begun, C. V. Ward & M. D. Rose, pp. 389–415. New York: Plenum Press.

Beynon, A. D., Dean, M. C. & Reid, D. J. (1991). Histological study on the chronology of the developing dentition in gorilla and orangutan. *American Journal of Physical Anthropolology*, **86**, 189–204.

Blomquist, G. E., Kowalewski, M. M. & Leigh, S. R. (2003). A phylogenetic approach to quantifying the relationship between age of first reproduction and maximum lifespan. *American Journal of Physical Anthropology, Supplement*, **36**, 68.

Boyce, M. S. (1988). Evolution of life histories: theory and patterns from mammals. In *Evolution of Life Histories of Mammals*, ed. M. S. Boyce, pp. 3–30. New Haven, CT: Yale University Press.

Boyde, A. (1963). Estimation of age at death of young human skeletal remains from incremental lines in the dental enamel. *Proceedings, Third International Meeting in Forensic Immunology, Medical Pathology and Toxicology*,

London, Excerpta Medica International Congress Series, **80,** 36–46.

Bromage, T. G. & Dean, M. C. (1985). Re-evaluation of the age at death of immature fossil hominids. *Nature,* **317,** 525–7.

Byrne, R. W. (1997). The technical intelligence hypothesis: an additional evolutionary stimulus to intelligence. In *Machiavellian Intelligence II: Extensions and Evaluations,* ed. A. Whiten & R. W. Byrne, pp. 289–311. Cambridge, UK: Cambridge University Press.

Caswell, H. (1982). Optimal life histories and the age specific costs of reproduction. *Journal of Theoretical Biology,* **98,** 519–29.

Caviness, V. S. Jr., Takahashi, T. & Nowakowski, R. S. (1995). Numbers, time and neocortical neurogenesis: a general developmental and evolutionary model. *Trends in Neuroscience,* **18,** 379–83.

Charlesworth, B. (1980). *Evolution in Age-structured Populations.* Cambridge, UK: Cambridge University Press.

Charnov, E. L. (1991). Evolution of life history variation among female mammals. *Proceedings of the National Academy of Sciences, USA,* **88,** 1134–7.

(1993). *Life History Invariants: Some Explorations of Symmetry in Evolutionary Ecology.* Oxford: Oxford University Press.

Clarke, R. J. (1997). First complete restoration of the *Oreopithecus* skull. *Human Evolution,* **12,** 221–32.

Clutton-Brock, T. H. & Harvey, P. H. (1980). Primates, brains and ecology. *Journal of Zoology, London,* **190,** 309–23.

Dean, M. C. (1987). Growth layers and incremental markings in hard tissues: a review of the literature and some preliminary observations about enamel structure in *Paranthropus boisei. Journal of Human Evolution,* **16,** 157–72.

(1989). The developing dentition and tooth structure in hominoids. *Folia Primatologica,* **53,** 160–77.

Dean, M. C., Beynon, A. D., Thackeray, J. F. & Macho, G. A. (1993). Histological reconstruction of dental development and age at death of a juvenile *Paranthropus robustus* specimen, SK 63, from Swartkrans, South Africa. *American Journal of Physical Anthropology,* **91,** 401–19.

Dean, M. C., Stringer, C. B. & Bromage, T. G. (1986). A new age at death for the Neanderthal child from the Devil's Tower, Gibralter and the implications for studies of general growth and development in Neanderthals.

American Journal of Physical Anthropology, **70,** 301–9.

Deaner, R. O., Barton, R. A. & van Schaik, C. P. (2003). Primate brains and life histories: renewing the connection. In *Primate Life Histories and Socioecology,* ed. P. M. Kappeler & M. E. Pereira, pp. 233–65. Cambridge, UK: Cambridge University Press.

Dirks, W. (1998). Histological reconstruction of dental development and age at death in a juvenile gibbon (*Hylobates lar*). *Journal of Human Evolution,* **35,** 411–25.

Eisenberg, J. F. (1981). *The Mammalian Radiations.* Chicago, IL: University of Chicago Press.

FitzGerald, C. M. (1998). Do enamel microstructures have regular time dependency? Conclusions from the literature and a large-scale study. *Journal of Human Evolution,* **35,** 371–86.

Foley, R. A. & Lee, P. C. (1991). Ecology and energetics of encephalization in hominid evolution. *Philosophical Transactions of the Royal Society of London B,* **334,** 223–32.

Gibson, K. R. (1990). New perspectives on instincts and intelligence: brain size and the emergence of hierarchical mental construction skills. In *"Language" and Intelligence in Monkeys and Apes: Comparative Developmental Perspectives,* ed. S. T. Parker & K. R. Gibson, pp. 197–228. Cambridge, UK: Cambridge University Press.

(1991). Myelination and behavioural development: a comparative perspective on questions of neotany, altriciality and intelligence. In *Brain Maturation and Cognitive Development,* ed. K. R. Gibson & A. C. Petersen, pp. 29–64. New York: De Gruyter.

Gibson, K. R. & Jessee, S. (1999). Language evolution and the expansion of multiple neurological processing areas. In *The Origins of Language: What Nonhuman Primates Can Tell Us,* ed. B. J. King, pp. 189–227. Santa Fe, CA: SAR Press.

Gibson, K. R., Rumbaugh, D. & Beran, M. (2001). Bigger is better: primate brain size in relation to cognition. In *Evolutionary Anatomy of the Primate Cerebral Cortex,* ed. D. Falk & K. R. Gibson, pp. 79–97. Cambridge, UK: Cambridge University Press.

Godfrey, L. R., Petto, A. J. & Sutherland, M. R. (2002). Dental ontogeny and life-history strategies: the case of the giant extinct indrioids of Madagascar. In *Reconstructing Behavior in the Primate Fossil Record,* ed. J. M. Plavcan, R. F. Kay, C. P. van Schaik & W. L. Jungers, pp. 113–57. New York: Kluwer.

Godfrey, L. R., Samonds, K. E., Jungers, W. L. & Sutherland, M. R. (2001). Teeth, brains, and primate life histories. *American Journal of Physical Anthropology*, **114**, 192–214.

Harrison, T. (1989). New estimates of cranial capacity, body size and encephalization in *Oreopithecus bambolii*. *American Journal of Physical Anthropology*, **78**, 237.

Harrison, T. & Rook, L. (1997). Enigmatic anthropoid or misunderstood ape? The phylogenetic status of *Oreopithecus bambolii* reconsidered. In *Function, Phylogeny, and Fossils: Miocene Hominoid Evolution and Adaptations*, ed. D. R. Begun, C. V. Ward & M. D. Rose, pp. 327–62. New York: Plenum Press.

Hart, D. & Karmel, M. P. (1996). Self-awareness and self-knowledge in humans, apes and monkeys. In *Reaching Into Thought: The Minds of the Great Apes*, ed. A. E. Russon, K. A. Bard & S. T. Parker, pp. 325–47. Cambridge, UK: Cambridge University Press.

Harvey, P. H. (1991). Comparing life histories. In *Evolution of Life*, ed. S. Osawa & T. Honjo, pp. 215–28. Berlin: Springer-Verlag.

Harvey, P. H. & Clutton-Brock, T. H. (1985). Life history variation in primates. *Evolution* **39**, 559–81.

Harvey, P. H. & Read, A. F. (1988). How and why do mammalian life histories vary? In *Evolution of Life Histories of Mammals: Theory and Pattern*, ed. M. S. Boyce, pp. 213–32. New Haven, CT: Yale University Press.

Harvey, P. H. & Zammuto, R. M. (1985). Patterns of mortality and age at first reproduction in natural populations of mammals. *Nature*, **315**, 319–20.

Harvey, P. H., Martin, R. D. & Clutton-Brock, T. H. (1987). Life histories in comparative perspective. In *Primate Societies*, ed. B. B. Smuts, D. L. Cheney, R. M. Seyfarth, R. W. Wrangham, and T. T. Struhsaker, pp. 181–96. Chicago: Chicago University Press.

Harvey, P. H., Promislow, D. E. L. & Read, A. F. (1989a). Causes and correlates of life history differences among mammals. In *Comparative Socioecology: The Behavioural Ecology of Humans and Other Mammals*, ed. R. Foley & V. Standen, pp. 305–18. Oxford: Blackwell.

Harvey, P. H., Read, A. F. & Promislow, D. E. L. (1989b). Life history variation in placental mammals: unifying the data with the theory. *Oxford Surveys in Evolutionary Biology*, **6**, 13–31.

Hofman, M. A. (1983). Energy metabolism, brain size and longevity in mammals. *Quarterly Review of Biology*, **58**, 495–512.

(2001). Brain evolution in hominids: are we at the end of the road? In *Evolutionary Anatomy of the Primate Cerebral Cortex*, ed. D. Falk & K. R. Gibson, pp. 113–27. Cambridge, UK: Cambridge University Press.

Horn, H. S. (1978). Optimal tactics of reproduction and life history. In *Behavioural Ecology: An Evolutionary Approach*, ed. J. R. Krebs & N. B. Davies, pp. 272–94. Oxford: Blackwell.

Jungers, W. L. (1987). Body size and morphometric affinities of the appendicular skeleton in *Oreopithecus bambolii* (IGF 11778). *Journal of Human Evolution*, **16**, 445–56.

Kelley, J. (1997). Paleobiological and phylogenetic significance of life history in Miocene hominoids. In *Function, Phylogeny, and Fossils: Miocene Hominoid Evolution and Adaptations*, ed. D. R. Begun, C. V. Ward & M. D. Rose, pp. 173–208. New York: Plenum Press.

(2002). Life-history evolution in Miocene and extant apes. In *Human Evolution Through Developmental Change*, ed. N. Minugh-Purvis & K. J. McNamara, pp. 223–48. Baltimore, MD: Johns Hopkins University Press.

Kelley, J. & Smith, T. (2003). Age at first molar emergence in early Miocene *Afropithecus turkanensis* and the evolution of life history in the Hominoidea. *Journal of Human Evolution*, **44**, 307–29.

Kelley, J., Dean, M. C. & Reid, D. J. (2001). Molar growth in the late Miocene hominoid, *Dryopithecus laietanus*. In *Dental Morphology 2001*, ed. A. Brook, pp. 123–34. Sheffield, UK: Sheffield Academic Press.

(2002). Molar growth in the late Miocene hominoid, *Dryopithecus laietanus*. *American Journal of Physical Anthropology*, *Supplement*, **34**, 93.

Kordos, L. & Begun, D. R. (1998). Encephalization and endocranial morphology in *Dryopithecus brancoi*: implications for brain evolution in early hominids. *American Journal of Physical Anthropology*, *Supplement* **26**, 141–2.

(2001). A new cranium of *Dryopithecus* from Rudabánya, Hungary. *Journal of Human Evolution*, **41**, 689–700.

Kornack, D. R. & Rakic, P. (1998). Changes in cell-cycle kinetics during the development and evolution of primate neocortex. *Proceedings of the National Academy of Sciences USA*, **95**, 1242–6.

Kowalewski, M. M., Blomquist, G. E. & Leigh, S. R. (2003). Life history and folivory in primate species. *American Journal of Physical Anthropology*, *Supplement*, **36**, 132–3.

Langer, J. (1996). Heterochrony and the evolution of primate cognitive development. In *Reaching Into Thought: The Minds of the Great Apes*, ed. A. E. Russon, K. A. Bard &

S. T. Parker, pp. 257–77. Cambridge, UK: Cambridge University Press.

Leakey, M. & Walker, A. (1997). *Afropithecus* function and phylogeny. In *Function, Phylogeny, and Fossils: Miocene Hominoid Evolution and Adaptations*, ed. D. R. Begun, C. V. Ward & M. D. Rose, pp. 225–39. New York: Plenum Press.

Lee, P. C. (1999). Comparative ecology of postnatal growth and weaning among haplorhine primates. In *Comparative Primate Socioecology*, ed. P. C. Lee, pp. 111–39. Cambridge, UK: Cambridge University Press.

Leigh, S. R. & Bernstein, R. M. (2003). Ontogeny, life history, and maternal reproductive strategies in baboons. *American Journal of Physical Anthropology, Supplement*, **36**, 138.

Lycett, J. E. & Barrett, L. (2003). Whose life is it anyway? Maternal investment and life history strategies in baboons. *American Journal of Physical Anthropology, Supplement*, **36**, 143.

Macho, G. A. & Wood, B. A. (1995). The role of time and timing in hominid dental evolution. *Evolutionary Anthropology*, **4**, 17–31.

Martin, R. D. (1984). Body size, brain size and feeding strategies. In *Food Acquisition and Processing in Primates*, ed. D. J. Chivers, B. A. Wood & A. Bilsborough, pp. 73–103. New York: Plenum.

(1996). Scaling of the mammalian brain: the maternal energy hypothesis. *News in the Physiological Sciences*, **11**, 149–56.

McKinney, M. L. (1998). The juvenilized ape myth: our overdeveloped brain. *Bioscience*, **48**, 109–16.

(2002). Brain evolution by stretching the global mitotic clock of development. In *Human Evolution Through Developmental Change*, ed. N. Minugh-Purvis & K. J. McNamara, pp. 173–88. Baltimore, MD: The Johns Hopkins University Press.

McNamara, K. J. (2002). Sequential hypermorphosis: stretching ontogeny to the limit. In *Human Evolution Through Developmental Change*, ed. N. Minugh-Purvis & K. J. McNamara, pp. 102–21. Baltimore, MD: The Johns Hopkins University Press.

Moyà Solà, S. & Köhler, M. (1995). New partial cranium of *Dryopithecus* Lartet, 1863 (Hominoidea, Primates) from the upper Miocene of Can Llobateres (Barcelona, Spain). *Journal of Human Evolution*, **29**, 101–39.

(1997). The phylogenetic relationships of *Oreopithecus bambolii* Gervais, 1872. *Comptes Rendu Académie des Sciences, Paléontologie, Série Iia*, **324**, 141–8.

Northcutt, R. G. & Kaas, J. H. (1995). The emergence and evolution of the mammalian neocortex. *Trends in Neurosciences*, **18**, 373–9.

Parker, S. T. (1990). Why big brains are so rare: energy costs of intelligence and brain size in anthropoid primates. In *"Language" and Intelligence in Monkeys and Apes: Comparative Developmental Perspectives*, ed. S. T. Parker & K. R. Gibson, pp. 129–54. Cambridge, UK: Cambridge University Press.

(2002). Evolutionary relationships between molar eruption and cognitive development in anthropoid primates. In *Human Evolution Through Developmental Change*, ed. N. Minugh-Purvis & K. J. McNamara, pp. 305–16. Baltimore, MD: The Johns Hopkins University Press.

Parker, S. T. & McKinney, M. L. (1999). *Origins of Intelligence: The Origin of Cognitive Development in Monkeys, Apes and Humans*. Baltimore, MD: The Johns Hopkins University Press.

Partridge, L. & Harvey, P. H. (1988). The ecological context of life history evolution. *Science*, **241**, 1449–55.

Pilbeam, D. (1982). New hominoid skull material from the Miocene of Pakistan. *Nature*, **295**, 232–4.

Preuss, T. M. (2001). The discovery of cerebral diversity: an unwelcome scientific revolution. In *Evolutionary Anatomy of the Primate Cerebral Cortex*, ed. D. Falk & K. R. Gibson, pp. 138–64. Cambridge, UK: Cambridge University Press.

Promislow, D. E. L. (1991). Senescence in natural populations of mammals: a comparative study. *Evolution*, **45**, 1869–87.

Promislow, D. E. L. & Harvey, P. H. (1990). Living fast and dying young: a comparative analysis of life-history variation among mammals. *Journal of Zoology, London*, **220**, 417–37.

(1991). Mortality rates and the evolution of mammal life histories. *Acta Oecologia*, **12**, 119–37.

Rakic, P. (1988). Specification of cerebral cortical areas. *Science*, **241**, 170–6.

(1995). A small step for the cell, a giant leap for mankind: a hypothesis of neocortical expansion during evolution. *Trends in Neurosciences*, **18**, 383–8.

Rakic, P. & Kornack, D. R. (2001). Neocortical expansion and elaboration during primate evolution: a view from neuroembryology. In *Evolutionary Anatomy of the Primate Cerebral Cortex*, ed. D. Falk & K. R. Gibson, pp. 30–56. Cambridge, UK: Cambridge University Press.

Read, A. F. & Harvey, P. H. (1989). Life history differences among the eutherian radiations. *Journal of Zoology, London*, **219**, 329–53.

Rice, S. H. (2002). The role of heterochrony in primate brain evolution. In *Human Evolution Through Developmental Change*, ed. N. Minugh-Purvis & K. J. McNamara, pp. 154–70. Baltimore, MD: The Johns Hopkins University Press.

Ross, C. R. (1992). Environmental correlates of the intrinsic rate of natural increase in primates. *Oecologia*, **90**, 383–90.

(1998). Primate life histories. *Evolutionary Anthropology*, **6**, 54–63.

Sacher, G. A. (1959). Relation of life span to brain weight and body weight in mammals. In *CIBA Foundation Colloquia on Aging. Vol. 5. The Lifespan of Animals*, ed. G. E. W. Wolstenholme & M. O'Conner, pp. 115–33. London: Churchill.

(1975). Maturation and longevity in relation to cranial capacity in hominid evolution. In *Primate Functional Morphology and Evolution*, ed. R. H. Tuttle, pp. 417–41. The Hague: Mouton.

(1978). Longevity, aging and death: an evolutionary perspective. *Gerontologist*, **18**, 112–9.

Sacher, G. A. & Staffeldt, E. F. (1974). Relation of gestation time to brain weight for placental mammals: implications for the theory of vertebrate growth. *American Naturalist*, **108**, 593–616.

Smith, B. H. (1989). Dental development as a measure of life history in primates. *Evolution* **43**, 683–8.

(1991). Dental development and the evolution of life history in Hominidae. *American Journal of Physical Anthropology*, **86**, 157–74.

(1992). Life history and the evolution of human maturation. *Evolutionary Anthropology*, **1**, 134–42.

Smith, B. H. & Tompkins, R. L. (1995). Toward a life history of the Hominidae. *Annual Reviews of Anthropology*, **24**, 257–79.

Smith, B. H., Crummet, T. L. & Brandt, K. L. (1994). Age of eruption of primate teeth: a compendium for aging individuals and comparing life histories. *Yearbook of Physical Anthropology*, **37**, 177–231.

Smith, R. J., Gannon, P. J. & Smith, B. H. (1995). Ontogeny of australopithecines and early *Homo*: evidence from cranial capacity and dental eruption. *Journal of Human Evolution*, **2**, 155–68.

Southwood, T. R. E. (1988). Tactics, strategies and templates. *Oikos*, **52**, 3–18.

Stearns, S. C. (1992). *The Evolution of Life Histories*. Oxford: Oxford University Press.

(2000). Life history evolution: successes, limitations, prospects. *Naturwissenschaften*, **87**, 476–86.

Sutherland, W. J., Grafen, A. & Harvey, P. H. (1986). Life history correlations and demography. *Nature*, **320**, 88.

Szalay, F. S. & Berzi, A. (1973). Cranial anatomy of *Oreopithecus*. *Science*, **180**, 183–5.

Ungar, P. S. & Kay, R. F. (1995). The dietary adaptations of European Miocene catarrhines. *Proceedings of the National Academy of Sciences USA*, **92**, 5479–81.

van Schaik, C. & Deaner, R. (2003). Life history and cognitive evolution in primates. In *Animal Social Complexity: Intelligence, Culture and Individualized Societies*, ed. F. de Waal & P. Tyack, pp. 5–25. Cambridge, MA: Harvard University Press.

Vrba, E. S. (1998). Multiphasic growth models and the evolution of prolonged growth exemplified by human brain evolution. *Journal of Theoretical Biology*, **190**, 227–39.

Ward, S. C. (1997). The taxonomy and phylogenetic relationships of *Sivapithecus* revisited. In *Function, Phylogeny, and Fossils: Miocene Hominoid Evolution and Adaptations*, ed. D. R. Begun, C. V. Ward & M. D. Rose, pp. 269–90. New York: Plenum Press.

Ward, S. C. & Brown, B. (1986). Facial anatomy of Miocene hominoids. In *Comparative Primate Biology. Vol. 1. Systematics, Evolution and Anatomy*, ed. D. R. Swindler & J. Erwin, pp. 413–52. New York: Liss.

Wolf, K. (1984). Reproductive competition among co-resident male silvered leaf monkeys (*Presbytis cristata*). Ph.D. Dissertation. Yale University.

Wootton, J. T. (1987). The effects of body mass, phylogeny, habitat and trophic level on mammalian age at first reproduction. *Evolution*, **41**, 732–49.

16 · Fossil hominoid diets, extractive foraging, and the origins of great ape intelligence

MICHELLE SINGLETON

Department of Anatomy, Midwestern University, Downers Grove

> The History of every major Galactic Civilization
> tends to pass through three distinct and
> recognizable phases . . . the first phase is
> characterized by the question How can we eat? the
> second by the question Why do we eat? and the
> third by the question Where shall we have lunch?
>
> Douglas Adams, *The Hitchhiker's Guide to the Galaxy*

INTRODUCTION

Ecological hypotheses for the evolution of great ape intelligence relate selective pressures for increased intelligence to biological and environmental parameters such as body size, metabolic rate, life history, diet, home range size, habitat stratification, and predation risk (Clutton-Brock & Harvey 1980; Dunbar 1992; Gibson 1986; Milton 1981, 1988; Sawaguchi 1989, 1992). Of these, diet is the ecological selective pressure most frequently invoked to explain the emergence of great ape cognitive abilities. A correlation between diet and relative brain size in primates has long been established; frugivorous primates tend to have relatively larger brains than closely related folivorous taxa (Clutton-Brock & Harvey 1980; Milton 1981, 1988; Sawaguchi 1992). This pattern was most often explained in terms of the differing nutritional properties of fruits and leaves. A high-energy, fruit-based diet, it was thought, released energetic and metabolic constraints, allowing accelerated neonatal brain growth and maintenance of relatively greater adult brain mass (Jolly 1988; Martin 1981). However, the expansion of energy-hungry brain tissue will occur only where it confers an immediate adaptive advantage (Dunbar 1992). In other words, adequate energy supply is a necessary precondition for, but not in itself a sufficient stimulus to, increased encephalization.

Researchers seeking such a stimulus have tended to focus upon the adaptive role of intelligence in solving the unique foraging problems posed by primate diets. Cognitive mapping hypotheses (Clutton-Brock & Harvey 1980; Milton 1981, 1988) posit that primates' reliance on foods that are clumped, spatially dispersed, and temporally ephemeral necessitates maintenance of complex mental maps spurring evolution of increased mental capacity (Milton 1981, 1988). The extractive foraging hypothesis (Gibson 1986; Parker & Gibson 1977) emphasizes the importance of "embedded" food resources such as nuts, tubers, social insects, and pith that require skilled manipulation. This hypothesis and its variants stress reliance upon tool-mediated extractive foraging and complex food preparation techniques as key to differences in cognitive capacity between great apes and other anthropoids (Byrne 1996, 1997; Byrne & Byrne 1993; Parker 1996; and see Byrne, Chapter 3, Yamagiwa, Chapter 12, Yamakoshi, Chapter 9, this volume).

Dietary hypotheses for the origins of great ape intelligence posit specific selective pressures favoring the evolution of this unique suite of cognitive and technical capacities. Such adaptationist scenarios are notoriously difficult to test (Byrne 1997; Gould & Lewontin 1984), but their assumptions and predictions may be evaluated via the comparative method. Unfortunately, this avenue of inquiry is severely limited by the evolutionary history of the hominoids. The extant apes represent geographically restricted relict populations, the last survivors of a taxonomically diverse and geographically dispersed radiation with its roots in the early Miocene. The divergence of Asian and African great apes is dated to a minimum of 10 Ma; the separation of the gorilla and chimpanzee lineages to approximately 7 Ma; and the split between the two chimpanzee species, to as recently as 2 Ma (Begun 1999). Thus, modern great apes are products of several million years of independent evolution, and each exhibits distinct and highly specialized ecological adaptations. This combination of ecological diversity and taxonomic poverty precludes statistical testing of ecological correlates of ape intelligence and makes even qualitative comparisons difficult. Efforts to reconstruct ancestral great ape dietary patterns on the basis of extant great ape characteristics are similarly fraught. This leads to the paradoxical situation in which ecological hypotheses for the evolution of great ape intelligence may be inspired by extant ape adaptations but are unlikely to be strongly corroborated by them.

Fortunately, comparisons are not restricted to modern forms. Hominoid paleoecology is well studied (Andrews 1981, 1992; Andrews *et al.* 1996; Andrews & Martin 1991; Benefit 2000; Fleagle & Kay 1985; Temerin & Cant 1983), and the fossil record is sufficiently speciose to document a more representative range of hominoid ecological adaptations. Adaptations of the immediate predecessors to and earliest members of the great ape clade furnish evidence of the dietary adaptations of the last common ancestor of modern great apes. This fossil-based approach is more than a convenient means to reconstructing the ancestral great ape ecotype. A paleontological perspective is absolutely necessary to understand the origins of great ape cognition. Discussions of "great ape intelligence" assume, explicitly or implicitly, that the enhanced cognitive capacities of extant great apes are homologous. If these unique mental faculties are, in fact, shared, derived features inherited from a common ancestor (Parker & Mitchell 1999; Russon & Bard 1996), the selective pressures to which this ancestor was subject formed the adaptive milieu in which great ape intelligence arose. Logically, hypotheses for the origins of great ape intelligence must address the ecological adaptations of the earliest great apes. Accordingly, this paper reviews current evidence for fossil hominoid diets with the goals of tracing major trends in hominoid dietary evolution, reconstructing the ancestral great ape dietary adaptation, and evaluating dietary hypotheses for the evolution of great ape intelligence.

RECONSTRUCTION OF FOSSIL PRIMATE DIETS

Primates are traditionally classified into three major dietary groups: folivores, frugivores, and insectivores (Martin 1990). While all anthropoid primates are omnivorous to varying degrees, the term "omnivory" is generally reserved for primates such as chimpanzees,

which have particularly catholic dietary preferences (Martin 1990). Because all known catarrhines exceed the metabolically determined maximum body mass for insect specialization (Kay 1975), insect consumption occurs primarily as a supplement to plant-based diets. Folivorous primates are those that consume substantial quantities of leaves or herbaceous matter such as grasses, stems, and piths, supplemented with varying amounts of fruit and animal protein. Frugivores consume a fruit-based diet supplemented with higher protein foods such as leaves, nuts, insects, and small vertebrates. Frugivores may be further categorized based on preferences for small versus large fruit; ripe versus unripe fruit; or soft, pulpy fruits versus those with hard skins or fibrous flesh. Hard-object feeding, usually treated as a subclass of frugivory, encompasses a variety of resistant food items, including nuts, seeds, tubers, rhizomes, and bark, usually as a substantial component of a fruit-based diet.

There are two principal forms of dental evidence for fossil primate diets and foraging behavior (Kay 1984): comparative dental morphology – the study of tooth size, shape, and tissue composition (see Table 16.1 for terminology) – and dental wear analysis. A third line of evidence, stable isotope analysis of dental tissues, is routinely employed in the reconstructions of primate paleoenvironments (Behrensmeyer *et al.* 2002; Cerling *et al.* 1997; Quade *et al.* 1995), but has not been widely applied to nonhominin primate fossils (but see Quade *et al.* 1995). All reconstructions of fossil primate diets are drawn within a classic comparative framework and are limited by the availability of suitable extant comparative models (Kay 1984). While this limitation is particularly salient to the reconstruction of fossil hominoid diets, dental evidence remains our most reliable source of paleo-dietary information. Combined with information concerning body mass, cranial anatomy, locomotor behavior, and paleoenvironment, it allows us to reconstruct the dietary patterns of fossil primates with reasonable accuracy. The literature pertaining to fossil primate diets is both extensive and extensively reviewed (cf. Butler 2000; Kay 1977a, 1984; Kay & Covert 1984; Rose & Ungar 1998; Teaford 1994, 2000; Ungar 1998); readers are referred to these papers and citations therein.

Functional dental morphology

Comparative functional analysis of primate dental morphology focuses primarily upon molars and incisors, the

Table 16.1. *Glossary of morphological terminology*

Apical – of or towards the biting surface, especially the cusp tips (ant. Cervical).
Buccal – the tooth surface oriented towards the cheek (ant. Lingual).
Cervical – of or towards the tooth root (ant. Apical).
Cingulum (pl. cingula) – an elevated band of enamel encircling a tooth crown.
Corpus – the bony body of the lower jaw (mandible).
Dentognathic – relating to the anatomy of the teeth and jaws.
Diastema – a space between adjacent teeth, usually to accommodate a projecting canine.
Distal – a tooth or tooth surface farther from the anterior midline of the jaw (ant. Mesial).
Labial – the tooth surface oriented toward the lips (ant. Lingual).
Lingual – the tooth surface oriented towards the tongue (ant. Buccal or Labial).
Mesial – a tooth or tooth surface closer to the anterior midline of the jaw (ant. Distal).
Occlusal – relating to the biting or grinding surface of a tooth.
Symphysis – the bony union between the right and left halves of the lower jaw (mandible).
Transverse torus – a bony shelf projecting lingually from the mandibular symphysis.
Zygomatic – a bone of the check region to which a principal masticatory muscle attaches.

principal agents of mastication and ingestion, respectively. In comparison with other mammalian orders, primates possess relatively generalized molars. Still, certain features of molar morphology are known to be strongly correlated with diet across extant primates (Kay 1975, 1978, 1984; Kay & Hiiemae 1974; Rosenberger & Kinzey 1976). In qualitative terms, frugivorous primates possess relatively short, broad molars, with low crowns, minimal cusp relief, expanded occlusal basins and poorly developed shearing crests (Figure 16.1). By contrast, folivorous taxa possess relatively long, narrow molars with tall crowns, high cusp relief, and increased shearing capacity (Kay 1978, 1984). Efforts to quantify these features have been variably successful. Kay's "Shearing Quotient" (SQ), a quantitative measure of relative molar shearing capacity, is strongly functionally correlated with diet (Kay 1975; Kay & Ungar 1997) and has been widely applied to paleodietary studies. Indices of molar crown shape and cusp relief are less reliable but have some value as general dietary indicators (Benefit 2000; Singleton 2001).

Dental enamel, the mineralized surface layer that gives teeth their hardness, is one of the most intensely studied features of primate molar morphology (Beynon *et al.* 1998). There is disagreement regarding the most appropriate quantification of relative enamel thickness and definitions of thickness categories vary among authors (Martin 1985; Shellis *et al.* 1998). However, it is generally accepted that thicker enamel is associated with the mastication of resistant, abrasive, or brittle food

items, while thinner enamel is more efficient for the processing of soft or pliant items (Kay 1981; Kinzey 1992; Teaford 2000). Thus, relatively thin enamel is found both in folivores, where it appears to encourage the formation and maintenance of shearing crests, and in soft fruit feeders, whose molars develop lacunar enamel deficits that may increase retention of soft, juicy food items between the teeth (Teaford 2000). Conversely, primates specializing on hard or abrasive foods have thick or hyper-thick enamel. Increased enamel thickness alters external crown geometry, maximizing crushing efficiency and increasing force dissipation while decreasing shearing capacity (Kay 1981; Macho & Spears 1999; Shellis *et al.* 1998; Ungar 1998). Under abrasive dietary regimes, it extends the functional life of the tooth simply by increasing the volume of enamel available to be worn away before the softer, underlying dentine is exposed (Shellis *et al.* 1998). For similar reasons, increased relative molar size is thought to be an adaptation to abrasive or fibrous diets (Lucas Corlett & Luke 1986; Shellis *et al.* 1998). Attempts to correlate tooth size and diet have been largely unsuccessful (Ungar 1998), but postcanine megadonty is frequently associated with nut cracking and hard seed consumption (Kay 1981). Enamel crenulation (wrinkling) is likewise associated with hard diets and may serve to increase grinding efficiency by trapping particles between opposed crushing surfaces (Lucas & Luke 1984).

Unlike molars, whose function is solely masticatory, the incisors and the canine–premolar complex are

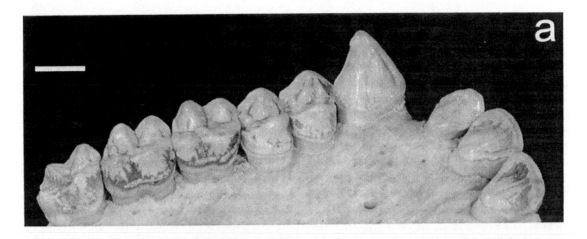

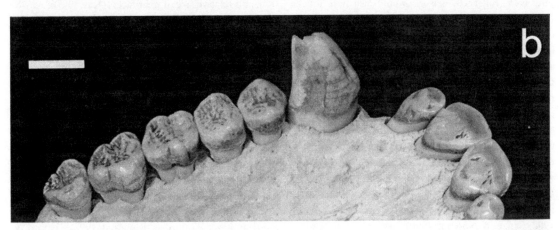

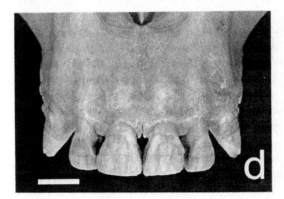

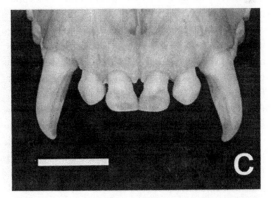

Figure 16.1. Hominoid functional dental morphology. Specimens represent the extremes of extant hominoid dental adaptation (scale bar = 1 cm). (a) Maxillary molars of gorillas (*Gorilla gorilla gorilla*) exhibit the tall crowns, high cusp relief, and well-developed shearing crests associated with diets dominated by leaves or herbaceous matter. (b) Maxillary molars of orangutans (*Pongo pygmaeus pygmaeus*) show the low crowns, minimal cusp relief, expanded occlusal basins and densely crenulated (wrinkled) enamel characteristic of frugivores that also consume hard or abrasive food items. (c) The incisors of gibbons (*Hylobates concolor gabriellae*) – which primarily consume smaller fruits and, in some taxa, leaves – are relatively smaller and narrower than those of (d) chimpanzees (*Pan troglodytes troglodytes*), which possess the enlarged, spatulate incisors associated with consumption of large fruits requiring incisal preparation.

subject to the competing selective demands of dietary and non-dietary functions. Because of their role in grooming, defense, and social display, the morphology of these teeth is considered a less reliable indicator of dietary patterns (Kay 1981; Teaford 2000). However, correlations between anterior tooth form and diet have been noted (Figure 16.1). Anthropoid primates that feed on leaves or small fruits have proportionately smaller and narrower incisors relative to body size than those specializing on large, tough-skinned fruits or other objects requiring incisal preparation (Eaglen 1984; Hylander 1975). Presumably, large, spatulate incisors provide greater working surface area, increasing their efficiency for tasks such as opening thick-skinned fruits and stripping bark. Enlarged incisors should also have longer functional lives (Ungar 1998), and thus are thought to be selectively advantageous to omnivores and large-object frugivores whose incisors are subject to heavy attrition (Eaglen 1984; Ungar 1998). Dietary adaptations of canine morphology are less common and more idiosyncratic. In particular, South American saki and uakari monkeys (tribe Pitheciini) possess robust, laterally splayed canines in combination with bilaterally compressed and procumbent lower incisors. This functional complex supports a specialized mode of seed predation in which the anterior dentition is employed to husk tough-skinned (sclerocarp) fruits to gain access to their nutrient-rich seeds (Anapol & Lee 1994; Kinzey 1992; Kinzey & Norconk 1990).

Dietary inferences based on comparative morphology must be drawn with caution (Kay 1984). Primates entering new niches will exploit novel food resources whether their teeth are well-adapted to them or not, and natural selection for improved dental function is expected to lag somewhat behind major dietary shifts (Teaford 1994). Because it is under close genetic control, dental morphology may not track intra-specific dietary variation and is frequently subject to phylogenetic effects (Teaford 1994). For example, incisor size (Eaglen 1984) and enamel thickness (Dumont 1995) both vary systematically across major primate groups and these differences must be factored into dietary analyses. Paleodietary studies must also account for changes in functional dental morphology through time (Kay & Ungar 1997). Average molar shearing capacity increases in Miocene catarrhines through time (Kay & Ungar 1997), and Singleton (2001) has documented similar temporal trends in molar flare, another feature associated

with diet (Benefit 2000). Clearly, it is important to maintain appropriate phylogenetic and temporal controls when drawing morphologically based dietary inferences (Ungar 1998).

Dental wear analysis

Dental wear includes macrowear, gross features such as dentine exposures and honing facets, and microwear, the microscopic scratches and pits created in dental enamel by tooth on tooth contact (attrition) and by contact with food items or exogenous materials such as grit (abrasion) (Rose & Ungar 1998; Teaford 1994). Interpretations of dental macrowear are based upon the location, orientation, and relative size of wear facets (Kay 1977b; Kay & Hiiemae 1974; Teaford 1994). High molar wear gradients are considered indicative of abrasive diets (Ungar 1998), and distinctive patterns of incisor wear, for example heavy labial attrition, signal specific premasticatory behaviors such as stripping of vegetation (Kilgore 1989). These assessments are largely qualitative, and more rigorous functional interpretation of gross wear features has only recently been undertaken (Teaford 2000; Ungar & Williamson 2000). By contrast, dental microwear analysis is a well-established and widely accepted method of reconstructing fossil primate diets (Gordon 1982, 1984; Kay & Ungar 1997; King 2001; Rose & Ungar 1998; Teaford 1985, 1994, 2000; Teaford & Oyen 1989; Teaford & Runestad 1992; Teaford & Walker 1984; Ungar 1990, 1995, 1996; Ungar & Kay 1995). Microwear studies are premised on the fact that food items of varying chemical composition and hardness create characteristic patterns of microscopic defects in dental enamel. Traditionally, enamel defects are classified as either scratches or pits, and microwear patterns are characterized by the number of pits expressed as a percentage of total microwear features (Teaford & Oyen 1989). Comparative studies of extant primates have established that the molars of highly folivorous primates show low pit percentages (Teaford 1985; Teaford & Runestad 1992; Teaford & Walker 1984), soft fruit eaters show a high percentage of pits, and hard-object feeders exhibit the highest pit percentages (Teaford & Walker 1984). It has also been suggested that pit width is indicative of diet, with hard-object feeders showing relatively wider pits than soft-object feeders (Teaford & Oyen 1989; Teaford & Runestad 1992). Incisor microwear has been studied in the context of ingestive behavior as well as premasticatory behaviors including fruit husking and leaf stripping (Teaford 1994; Rose & Ungar 1998). The incisors of frugivorous primates show a higher density of microwear features than those of folivores (Ungar 1990), and characteristic incisor wear patterns have been associated with incisal preparation of tough-skinned fruits, stripping of leaves and pith, and consumption of terrestrial resources such as rhizomes and tubers (Ryan 1981). Microwear of the canine–premolar complex is poorly studied (but see Ryan 1981), and patterns associated with behaviors such as the canine-assisted fruit-husking characteristic of pitheciin seed predators are largely uninvestigated (Anapol & Lee 1994; Kinzey 1992; Kinzey & Norconk 1990).

Dental microwear analysis is subject to several potential confounding factors. Individual microwear features are quickly obliterated by subsequent feeding bouts (Teaford & Oyen 1989); thus microwear preserves only the signal of food items consumed in the last several days preceding death, the so-called "Last Supper" effect (Grine 1986). Taken alone, incisor microwear can be an unreliable indicator of dietary patterns (Kelley 1990). Microwear patterns differ along the molar row as well as between shearing and crushing facets (Gordon 1982, 1984; Rose & Ungar 1998), and can be subtly influenced by seasonal and environmental variation, sex, age, and even reproductive status (Teaford 2000). While such patterns hold out the possibility of discerning fine-grained dietary variation in the fossil record, they also mandate the analysis of large samples to avoid erroneous inferences based on sampling artifacts (Rose & Ungar 1998).

REVIEW OF MIOCENE HOMINOID DIETS

Early Miocene (23–17 Ma)

The early Miocene East African primate radiation encompasses numerous basal catarrhines of uncertain phylogenetic affinities (Harrison 1988) as well as the earliest stem hominoids – species more closely related to modern apes than to any other group. The best-known stem hominoid, *Proconsul*, retains a primitive catarrhine locomotor pattern while sharing numerous similarities with later Miocene and extant apes (Rose 1994, 1997; Walker 1997). *Proconsul* incisors are relatively narrow, but the I^1 is slightly enlarged relative to M^1 (Andrews

& Martin 1991). The lower incisors are extremely high-crowned and narrow and frequently show heavy lingual wear (Andrews 1978). *Proconsul* molars are low-crowned with crenulated enamel, large cusps, moderate cusp relief, strong cingula, and poorly developed shearing crests. SQ values are most similar to those of *Pan troglodytes* (Kay & Ungar 1997), a soft-fruit frugivore, and analyses of molar microwear are likewise consistent with frugivory (Walker, Teaford & Ungar 1994; Walker 1997). *Proconsul nyanzae* shows both lower SQ values and higher microwear pit percentages than either *P. major* or *P. heseloni*, suggesting it may have consumed relatively harder food items (Kay & Ungar 1997; Walker et al. 1994). This is consistent with Andrews, (1978) observation that *P. nyanzae* shows a stronger molar wear gradient than other *Proconsul* species. Relative enamel thickness also varies among taxa, with the Rusinga Island species (*P. nyanzae* and *P. heseloni*) showing thicker enamel than either *P. africanus* or *P. major* (Andrews & Martin 1991; Beynon et al. 1998). Songhor and Koru, the sites from which the latter species are known, are reconstructed as wet tropical forests, while Rusinga Island represents a drier, more seasonal woodland habitat (Andrews, Begun & Zylstra 1997). Thus, *Proconsul* encompasses a cohort of medium- to large-bodied arboreal frugivores whose dietary differences track local environmental variation (Beynon et al. 1998).

Afropithecus turkanensis, another stem hominoid, displays a unique suite of dentognathic features, clearly derived relative to the primitive catarrhine condition (Leakey & Walker 1997; Leakey & Leakey 1986; Leakey, Leakey & Walker 1988). Its upper central incisors are large, mesiodistally broad, and strongly procumbent. Mandibular incisors are elongate, bilaterally compressed, and also strongly procumbent. Canines are stout, low-crowned, and laterally splayed. *Afropithecus* molars are low-crowned, with marked basal flare, little cuspal relief, and densely crenulated enamel. Dental enamel is described as "extremely thick" (Leakey & Walker 1997). Molar crowns exhibit heavy occlusal wear with significant loss of crown height and extensive dentine exposure (Leakey et al. 1988). In contrast with *Proconsul*, the mandible of *Afropithecus* is characterized by a deep corpus and elongated symphysis with a distinct inferior transverse torus (Brown 1997), and the facial skeleton exhibits features consistent with powerful mastication (Leakey & Walker 1997).

Leakey and Walker (1997) likened the anterior dentition of *Afropithecus* to that of pitheciin seed predators (Anapol & Lee 1994; Kinzey 1992; Kinzey & Norconk 1990). They point to numerous similarities of the facial skeleton and the unusual pattern of apical canine wear as indicative of pitheciin-like ingestive behaviors; however, *Afropithecus* differs from pitheciins in its molar morphology. Sakis and uakaris have thin molar enamel and show relatively little occlusal wear, features related to the physical properties – tough but neither brittle nor abrasive – of the seeds they consume (Kinzey 1992). By contrast, the thick enamel and heavy occlusal wear of *Afropithecus* molars indicate consumption of food items that were hard, abrasive, or both. *Afropithecus* faunas are consistent with wooded settings and *Afropithecus* has been reconstructed as an arboreal quadruped, in most respects indistinguishable from *Proconsul* (Andrews *et al.* 1997; Leakey & Walker 1997). This suggests *Afropithecus* foraged arboreally, consuming large, hard-skinned fruits with resistant mesocarps or hard seeds.

Middle Miocene (16–13 Ma)

The middle Miocene was a period of significant environmental change characterized by decreased mean annual temperatures, increased seasonality, and, in Africa, aridification and expansion of open woodland and grassland habitats (Andrews *et al.* 1997; Potts, Chapter 13, this volume; Wynn & Retallack 2001). In response, middle Miocene hominoids evolved new locomotor and dietary adaptations (McCrossin & Benefit 1997; McCrossin *et al.* 1998; Nakatsukasa *et al.* 1998), the true diversity of which has only recently been recognized with the naming of two new hominoid genera (Ishida *et al.* 1999; Ward *et al.* 1999). Relationships among these taxa remain unresolved, but they are generally acknowledged to be derived in the direction of the modern ape clade with which they share key postcranial features (Andrews 1992; Begun 2001; Ishida *et al.* 1999; McCrossin & Benefit 1997; McCrossin *et al.* 1998; Nakatsukasa *et al.* 1998; Ward *et al.* 1999).

With the exception of *Otavipithecus namibiensis*, a southern African hominoid with idiosyncratic molar morphology and poorly understood dietary adaptations (Singleton 2000), middle Miocene hominoids share a suite of dental features associated with hard-object frugivory. The most broadly distributed middle Miocene taxon, *Griphopithecus*, is represented at several localities in Germany and the Vienna Basin (Andrews *et al.* 1996; Heizmann & Begun 2001) but is best known from the Anatolian localities of Çandır and Paşalar. The Paşalar sample is believed to comprise two species, *Griphopithecus alpani* and a second unnamed taxon (Alpagut, Andrews & Martin 1990). Upper central incisors assigned to *G. alpani* are mesiodistally narrow but robust, with poorly developed lingual cingula and strong lingual pillars that are frequently obliterated by heavy lingual wear (Alpagut *et al.* 1990). Small, asymmetrical lateral incisors wear quickly to horizontal dentine exposures (Alpagut *et al.* 1990). Lower incisors are tall but not bilaterally compressed and show moderate lingual wear extending from the incisal edge toward the cervix (Alpagut *et al.* 1990). Canines referred to *G. alpani* are robust and low crowned with massive roots; mandibular canines show distinctive apical wear facets (Alpagut *et al.* 1990). *Griphopithecus* possesses low-crowned molars with low, rounded cusps, poorly developed shearing crests and thick, densely crenulated enamel (Alpagut *et al.* 1990; King *et al.* 1998). Consistent with this pattern, dentine exposures are not observed until a crown has worn almost flat. The molar crowns are quite broad relative to length and show variable expression of a shelf-like cingulum which falls relatively higher on the crown than in early Miocene forms such as *Proconsul* (Alpagut *et al.* 1990). With moderate wear, the cingulum is incorporated into the occlusal surface, possibly a secondary adaptation to extend functional tooth life (Alpagut *et al.* 1990). King *et al.* (1998) found that *Griphopithecus* microwear is similar to that of *Pongo*, suggesting a frugivorous diet. However, it consistently shows higher pit percentages than either *Pan* or *Pongo*, indicating consumption of harder foods. Further evidence for hard-object consumption is found in mandibles attributed to *G. alpani* that are characterized by robust corpora with massive muscle insertions and strongly developed transverse tori (Alpagut *et al.* 1990; Andrews & Tekkaya 1976; Güleç & Begun 2003). Paleodietary reconstructions for the Paşalar fauna are consistent with a closed, forested environment (Andrews *et al.* 1997; Geraads *et al.* 2003; Quade *et al.* 1995), making hard fruits or nuts the most likely candidates for the hard-object component of the *Griphopithecus* diet.

Equatorius (formerly *Kenyapithecus*, see Ward *et al.* 1999), like *Afropithecus*, is thought to have been a sclerocarp specialist convergent in many features on pitheciins (McCrossin & Benefit 1997). Among the traits cited in support of this interpretation are externally rotated, robust and tusk-like canines; high-crowned, bilaterally compressed, and strongly procumbent mandibular incisors; enlarged upper premolars; and low-crowned molars with crenulated enamel. Cranial features indicative of forceful incision and powerful mastication include anteriorly positioned zygomatic roots; strong maxillary canine pillars; and a robust mandible with pronounced symphyseal buttressing (McCrossin & Benefit 1997). Also like *Afropithecus*, *Equatorius* diverges from the pitheciin model in its possession of thick molar enamel caps and heavy molar occlusal wear. Dental microwear analysis of *Equatorius* molars from Maboko Island showed large pit widths and high pit percentages, both indicative of hard-object feeding (McCrossin *et al.* 1998; Palmer *et al.* 1998; Teaford & Oyen 1989). The Maboko Island habitat has been reconstructed as a seasonal open woodland, and postcranial remains indicate that *Equatorius* was at least semi-terrestrial (McCrossin & Benefit 1997; McCrossin *et al.* 1998; Sherwood *et al.* 2002). Thus, *Equatorius* would have had access to terrestrial resources such as tubers and rhizomes as well as dry forest foods such as sclerocarp fruits, seed pods, and nuts.

Initially attributed to *Kenyapithecus* (Ishida *et al.* 1984) and subsequently transferred to *Equatorius* (Ward *et al.* 1999), the hominoid material from Nachola, Kenya, is now recognized as a distinct genus, *Nacholapithecus* (Ishida *et al.* 1999). The Nachola fauna is provisionally interpreted as a forest or woodland community (Tsujikawa & Nakaya 1998), and *Nacholapithecus* is distinguished from *Equatorius* on the basis of its postcranial morphology, which shows adaptations to forelimb-dominated orthograde climbing and clambering (Nakatsukasa *et al.* 1998; Rose, Nakano & Ishida 1996). The *Nacholapithecus* dental sample remains largely undescribed, but the molars are thickly-enameled with low crown relief and reduced cingula (Ishida *et al.* 1984). Ishida *et al.* (1984) described a symphyseal fragment with a strong inferior transverse torus but no appreciable superior torus, and Kunimatsu *et al.* (1998) report mandibular proportions similar to *Proconsul*. This morphology is unlike the robust and strongly buttressed mandibles of *Equatorius*, thus *Nacholapithecus* may have eaten somewhat less-resistant food items than its more terrestrial contemporary.

Kenyapithecus sensu stricto exhibits an anterior dental pattern distinct from that of *Equatorius* (Ward *et al.* 1999). Maxillary incisors are more symmetrical, with well-developed enamel features; the canine is high

crowned and bilaterally compressed (Kelley *et al.* 2002; Ward *et al.* 1999). The molar morphology and robust mandibular architecture of *Kenyapithecus* are indicative of hard-object feeding, but the high-crowned, relatively narrow canines preclude paramasticatory use as hypothesized for the tusk-like canines of *Equatorius* and *Afropithecus* (Leakey & Walker 1997; McCrossin & Benefit 1997). A humerus from Fort Ternan attributed to *Kenyapithecus wickeri* is said to lack key features indicative of terrestriality (McCrossin 1997; Sherwood *et al.* 2002), and the Fort Ternan environment has been reconstructed as both less open and wetter than Maboko Island (Andrews *et al.* 1997). This suggests arboreal foraging as the dominant dietary pattern.

Late Miocene (12–5 Ma)

The late Miocene radiation of hominoids in Western Europe and Asia Major coincides with the emergence of the great ape clade and the evolution of modern hominoid suspensory adaptations. While the precise phylogenetic relationships of the late Miocene hominoids are a source of ongoing debate (Begun 2001; Begun, Ward & Rose 1997; de Bonis & Koufos 1997, 2001; Harrison & Rook 1997; Köhler, Moyà-Solà & Alba 2001; Moyà-Solà & Köhler 1996), their dietary adaptations are among the most thoroughly studied and are largely uncontroversial (Kay & Ungar 1997; Teaford & Walker 1984; Ungar 1996; Ungar & Kay 1995; Ward, Beecher & Kelley 1991).

The most cosmopolitan of the late Miocene hominoid genera, *Dryopithecus*, is known from localities in Austria, France, Germany, Hungary, Spain and possibly Georgia (Begun 1994; Gabunia *et al.* 2001). *Dryopithecus* species are nevertheless fairly uniform in their dental and dietary adaptations (Begun 1994; Ungar & Teaford 1996). In contrast with other Eurasian hominoids and extant great apes, *Dryopithecus* maxillary incisors are moderately tall and narrow but reduced relative to molar area, a feature in which they resemble hylobatids and gorillas (Begun 1994). Incisor microwear is consistent with labiolingual and apicocervical stripping of moderately abrasive food items, perhaps young leaves (Ungar 1996). Canines are bilaterally compressed and mesiodistally elongated but small relative to molar size. *Dryopithecus* molars are characterized by high crowns with moderate cusp relief; buccolingually restricted cusps with peripheral apices; and broad, shallow occlusal basins (Begun 1994). With the exception of *D. fontani*, molar cingula are absent. Molar enamel is thin, and worn cusps exhibit discrete apical dentine exposures (Begun 1994). Shearing quotients, most similar to those of *Pan paniscus* and the more frugivorous gibbons, indicate a soft fruit diet (Kay & Ungar 1997; Ungar 1996), as does molar microwear (Kay & Ungar 1997; Ungar 1996). *Dryopithecus* habitats range from wet subtropical evergreen forest conditions (Andrews & Bernor 1999) to more seasonal tropical or subtropical forest environments (Andrews *et al.* 1997). Postcranial remains indicate that *Dryopithecus* shared modern great ape adaptations for orthograde body posture and below-branch suspension (Begun 1993; Morbeck 1983; Moyà-Solà & Köhler 1996; Rose 1994). Thus, *Dryopithecus* appears to have been an arboreal specialist similar in many respects to the orangutan. However, its narrow incisors and microwear patterns indicate a diet emphasizing young leaves and smaller, softer fruits, more similar to that of extant gibbons.

The dental characters of *Oreopithecus bambolii* leave little doubt as to its dietary adaptations; its dental apparatus is unequivocally designed for a highly folivorous diet (Harrison & Rook 1997). The incisors are small, vertically implanted, and robust (Harrison & Rook 1997; Hürzeler 1958). The canines are ovoid in cross-section and, in males, projecting. The pattern and extent of incisor and canine wear are consistent with nipping of leaves. The molars are elongate and high crowned with voluminous conical cusps, high cusp relief, well-developed shearing crests, and restricted occlusal basins (Harrison & Rook 1997). The cheek teeth exhibit a steep wear gradient (Harrison & Rook 1997). Cranial features including a relatively short face, anteriorly positioned zygomatic root, deep and heavily buttressed mandibular corpus, and tall, vertical mandibular ramus are likewise consistent with a folivorous adaptation (Harrison & Rook 1997). Both shearing quotients and microwear analyses place *Oreopithecus* among the most highly folivorous anthropoids (Ungar 1996; Ungar & Kay 1995). Paleoenvironmental reconstructions of *Oreopithecus* localities suggest an insular environment with subtropical swampy forest conditions (Andrews *et al.* 1997). *Oreopithecus* has features consistent with orthograde body postures (Harrison & Rook 1997), and it has been suggested that it engaged in a novel form of bipedal locomotion (Köhler & Moyà-Solà 1997; Rook *et al.* 1999). However, its post-cranial morphology is more plausibly interpreted as adapted for quadrupedal clambering, vertical climbing and below-branch

suspensory behavior (Harrison & Rook 1997; Jungers 1987; Rose 1997; Sarmiento 1995), all consistent with an arboreal, folivorous ecological niche.

Ouranopithecus is a monospecific genus known almost exclusively from craniodental remains. Body mass estimates vary widely (de Bonis & Koufos 2001; Kelley 2001), but it is clearly among the largest of the Eurasian hominoids. The anterior dental complex is consistent with ingestion of foods requiring significant premasticatory preparation. The premaxilla is projecting and the incisors are strongly procumbent (de Bonis & Koufos 1993). The maxillary central incisor is spatulate, and most specimens exhibit heavy wear with significant loss of crown height and large labial dentine exposures (de Bonis & Koufos 1993; de Bonis & Melentis 1978). The asymmetrical lateral incisors are smaller but equally heavily worn. Mandibular incisors are narrow (de Bonis & Melentis 1978), only slightly procumbent, and show heavy wear characterized by continuous dentine exposures from the incisal edge onto the lingual surface (personal observation). Incisor microwear is characterized by high feature density and a relatively high incidence of mesiodistally oriented striations, suggesting lateral stripping of vegetation (Ungar 1996). In contrast with other late Miocene hominoids, maxillary canines are stout rather than bilaterally compressed and exhibit heavy apical wear with significant loss of crown height, indicating heavy paramasticatory use (de Bonis & Melentis 1978). *Ouranopithecus* molars are large relative to estimates of body mass (Kelley 2001), with hyperthick dental enamel, inflated cusps, and low occlusal relief (de Bonis & Koufos 1993). As in other thickly enameled forms, molar wear is heavy and characterized by loss of crown relief and rapidly expanding dentine exposures. Shearing quotients are extremely low, suggesting a frugivorous diet with a significant hard-object component (Ungar 1995), an inference supported by microwear feature density and pit percentages (Ungar 1996). The mandible is characterized by deep corpora, heavily buttressed symphyses, strongly defined muscle markings, and condylar proportions consistent with forceful mastication (de Bonis & Koufos 1993, 1997, 2001). Paleoenvironmental reconstructions of Macedonian hominoid localities indicate a dry, seasonal and possibly open environment (Andrews *et al.* 1997; de Bonis & Koufos 2001). Incisor microwear patterns are consistent with near-ground or terrestrial feeding (Ungar 1996), but the locomotor adaptations of *Ouranopithecus* are currently unknown. Taken in total, the dental evidence suggests a diet incorporating highly abrasive food items requiring significant incisal preparation and possibly including terrestrial resources (Ungar 1996).

Paleoecological interpretation of late Miocene Asian hominoids has been influenced both by their purported hominid (*sensus usus*) affinities (Simons 1976) and morphological similarities to the orangutan (Ward 1997). Like those of the orangutan, *Sivapithecus* maxillary incisors are heteromorphic and strongly procumbent. The I^1 is large and spatulate, with moderate lingual cingula and a distinct lingual pillar. The lateral incisor is both smaller and less symmetrical, and is set well posterior to I^1 (Pilbeam & Smith 1981). The mandibular incisors are homomorphic, parallel-sided teeth with moderately developed basal tubercles. Incisor wear is heavy, producing significant loss of crown height (Pilbeam & Smith 1981). The canines are robust and moderately high crowned and usually heavily worn; mandibular canines exhibit apical facets, presumably from occlusion with the I^2 (Pilbeam 1982). The molars are high-crowned with thick, coarsely crenulated enamel, low occlusal relief, and peripheral cusp apices. In comparison with those of *Pongo*, *Sivapithecus* molars show relatively greater cusp relief and more restricted basins (Ward *et al.* 1991). Molar occlusal wear is heavy, with a strong buccolingual wear gradient and extensive dentine exposure (Pilbeam & Smith 1981). While thick enamel is generally associated with hard-object feeding, dental microwear analysis of several *Sivapithecus indicus* specimens shows pit percentage values similar to those of *Pan troglodtyes*, a soft fruit eater (Teaford & Walker 1984). However, the observed pattern of molar wear indicates consumption of relatively resistant food items, and the maxillary incisor morphology and heavy anterior dental attrition suggest consumption of food items requiring extensive premasticatory preparation. Mandibular proportions, particularly corpus depth, vary among species, but all *Sivapithecus* mandibles are robust, with massive medial and lateral buttresses and well-developed symphyseal tori, features also indicative of powerful incision and forceful mastication (Brown 1997). Altogether, the dentognathic morphology of *Sivapithecus* points toward consumption of large fruits with tough skins and fibrous or otherwise resistant flesh. The locomotor adaptation of *Sivapithecus* is still debated (Moyà-Solà & Köhler 1996), but most analysis support arboreal quadrupedalism as the dominant pattern of locomotion (Pilbeam *et al.* 1990; Richmond & Whalen 2001; Rose 1997). The paleoenvironment of the Siwaliks region has been reconstructed as seasonally dry tropical deciduous forest (Andrews *et al.* 1997; Retallack 1991), thus, it seems likely *Sivapithecus* foraged arboreally. Differences in microwear and positional behavior notwithstanding, *Pongo* – which consumes significant quantities of large, hard-husked fruits as well as relatively high proportions of unripe fruit – remains the most appropriate extant dietary analog for this taxon (Ungar 1995).

In comparison with *Sivapithecus* and *Pongo*, the Chinese pongine *Lufengpithecus* exhibits narrower incisors, higher-crowned and slenderer canines, and relatively gracile mandibles with little buttressing (Brown 1997; Kelley & Pilbeam 1986; Schwartz 1997; Wu & Xu 1985). Paleoecological reconstructions suggest *Lufengpithecus* was an arborealist, perhaps with some suspensory capabilities, living in a moist, tropical forest environment (Andrews *et al.* 1997). It probably had an orangutan-like diet, primarily fruvigorous with a hard-object component. The basal pongine *Ankarapithecus* also shares many dentognathic features with *Sivapithecus* and *Pongo*. However, enlargement of the postcanine dentition, heavy dental attrition, and a robust facial skeleton point to greater emphasis on forceful mastication of hard or abrasive foods requiring extensive incisal preparation (Alpagut *et al.* 1996; Andrews & Alpagut 2001; Andrews & Tekkaya 1980; Begun & Güleç 1998). *Ankarapithecus* is associated with a high-diversity open woodland fauna (Lunkka *et al.* 1999) but its locomotor patterns are unknown.

TRENDS IN HOMINOID DIETARY EVOLUTION

Table 16.2 summarizes inferred paleoecological and dietary patterns for the major large-bodied Miocene hominoid taxa. Reconstructing the ancestral great ape dietary adaptation requires placing these patterns in an explicit phylogenetic context, yet the recent literature attests to the diversity of opinions concerning the phylogenetic relationships and taxonomic status of extant apes and their fossil relatives (Begun 2000, 2001; Begun *et al.* 1997; de Bonis & Koufos 1997, 2001; Harrison & Rook 1997; Kelley 2001; Köhler *et al.* 2001; McCrossin & Benefit 2000; Moyà-Solà & Köhler 1996; Sherwood *et al.* 2002; Ward *et al.* 1999). For present purposes,

Table 16.2. *Summary of evidence for large-bodied Miocene hominoid diet and foraging*

	Age (Ma)	Habitat	Locomotion	Enamel thickness	Diet		
					Microwear	Shearing quotient[f]	Dental morphology
Proconsul	18–20	Tropical forest	Arboreal quadruped	Thick[a] to average[b]	Frugivory	Frugivory	Soft fruit
Afropithecus	17	Woodland?	Arboreal quadruped	"Extremely thick"[c]	?	?	Sclerocarp specialist
Equatorius	15	Open woodland	Semi-terrestrial	"Thick"	Hard object	Hard object	Sclerocarp specialist
Nacholapithecus	15	Woodland?	Forelimb–dominated	"Thick"	?	?	Hard object?
Griphopithecus	15–16	Forest	?	Thick[a]	Hard Object	?	Hard object
Kenyapithecus	14	Woodland	?	"Thick"	?	?	Hard object
Oreopithecus	8	Swamp forest	Vertical climbing & suspensory	Intermediate thick[b]	Folivory	Folivory	Folivory
Ankarapithecus	10	Seasonal forest	?	Thick[d]	?	?	Hard fruit
Lufengpithecus	9–10	Moist forest	Suspensory?	"Thick"[e]	?	?	Hard fruit
Sivapithecus	7–12	Seasonal forest	Arboreal quadruped	Thick[a]	Frugivory	?	Hard fruit
Ouranopithecus	9	Open country	?	Thick/hyper–thick[a]	Hard object	Hard object	Hard object
Dryopithecus	10–12	Subtropical forest	Suspensory	Thin[a]	Frugivory	Frugivory	Soft fruit

Notes:

[a] Andrews & Martin (1991)
[b] Beynon et al. (1998)
[c] Leakey & Walker (1997)
[d] Andrews & Alpagut (2001)
[e] Wu & Xu (1985)
[f] Kay & Ungar (1997)

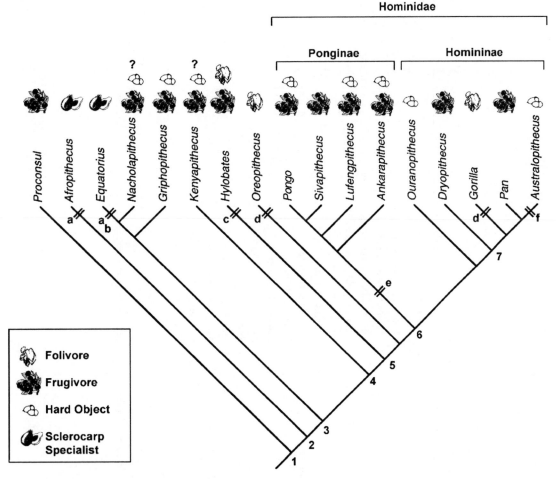

Figure 16.2. Cladogram based on Begun *et al.* (1997, figure 2c). *Equatorius*, *Nacholapithecus*, and *Griphopithecus* are grouped to indicate morphological and probable phylogenetic affinities. The "*Equatorius* clade" is rooted to indicate its postcranial affinities with later hominoids; branching order within the clade is arbitrary and does not signify specific cladistic relationships. The position of *Ankarapithecus* follows Begun & Güleç (1998). Icons indicate major dietary categories; see Table 16.3 for explanation of other symbols.

Begun *et al.*'s (1997) cladistic analysis was taken as the starting point from which to develop a working hypothesis of hominoid phylogenetic relationships (see Figure 16.2 and Table 16.3). Mapping key morphological and ecological characters onto the resulting tree makes it possible to trace trends in hominoid dietary evolution and infer the ecological adaptation of the hypothetical great ape ancestor.

Hominoids of archaic aspect

The primitive ecological pattern for large-bodied Miocene hominoids (Figure 16.2, Node 1) is exemplified by *Proconsul*, a frugivorous, above-branch arboreal quadruped restricted to forested environments (Walker 1997). While *P. nyanzae* appears to have consumed harder food items (Beynon *et al.* 1998; Kay & Ungar 1997), no proconsulid exhibits a true hard-object feeding adaptation. A dietary shift toward hard-object consumption is established in the late early Miocene (Node 2). Features linked to hard-object frugivory, including enlarged incisors, thickly-enameled molars, and development of an inferior transverse torus, are first expressed in *Afropithecus* and persist for the remainder of the Miocene. Beginning in the middle Miocene, hominoids move into a range of woodland and open country

Table 16.3. *Reconstruction of hominoid dietary evolution (see Figure 16.2)*

Hypothetical ancestral condition	Major adaptations
Node 1 Average enamel thickness Frugivorous diet Above-branch arboreal quadrupedalism Forest environment	a Pithecin-like anterior dentition Maxillary incisors enlarged & procumbent Mandibular incisors bilaterally compressed & procumbent Canines tusklike & splayed
Node 2 Increased enamel thickness Reduced molar occlusal relief Inferior transverse torus development	b Semi-terrestrial adaptation
Node 3 Derived elbow morphology Increased locomotor diversity Forest woodland habitats	c Hylobatid radiation Soft fruit frugivory Incisor reduction Molars thin enameled & low crowned d Autapomorphic folivorous adaptation
Node 4 Symmetrical[1] Increased canine height	e Hard fruit frugivory f Hominin dental morphology Anterior dental reduction Postcanine megadonty Hyper-thick enamel
Node 5 LCA of extant ape clade	
Node 6 Enlarged, spatulate I[1] Enlarged inferior transverse torus	
Node 7 Thin Dental Enamel Reduced Maxillary Incisors	

(a)–(f), see Figure 16.2.

habitats (Node 3), thus gaining access to dry forest and terrestrial food resources. This shift is accompanied by a marked increase in locomotor and ecological diversity (Sherwood *et al.* 2002), but all members of the *Equatorius* clade retain dental features indicative of hard-object feeding. *Kenyapithecus sensu stricto* also retains the characteristic thickly enameled molar morphology, even as its incisor and canine morphologies (Node 4) anticipate the crown great ape condition (Kelley *et al.* 2002; Ward *et al.* 1999). Both *Afropithecus* and *Equatorius* possess a derived anterior dental complex consistent with pitheciin-like ingestive behaviors (Leakey & Walker 1997; McCrossin & Benefit 1997) interpreted here as functional convergences related to sclerocarp feeding.

Hominoids of modern aspect

Reconstructing the ancestral dietary pattern of the extant ape clade (Figure 16.2, Node 5) is problematic. The origins of the hylobatid radiation are obscure, and its position relative to middle Miocene stem hominoids is disputed. Current opinion rejects an early Miocene catarrhine ancestry for gibbons and siamangs (Begun *et al.* 1997), and Figure 16.2 reconstructs hylobatids as descended from a thickly enameled middle Miocene ancestor. Under this scenario, hylobatid features such as reduced incisor height and low-crowned, thin-enameled molars (Figure 16.2, Table 16.3 c) arise as secondary functional adaptations to softer-consistency, small-object diets. The specialized dental morphology of *Oreopithecus* – reconstructed here as descended from a thickly enameled ancestral form (Figure 16.2, Table 16.3d) – is uniquely derived and therefore immaterial to the present argument.

Dietary inferences for the hypothetical common ancestor of the great ape clade are more straightforward. The ancestral great ape morphology (Node 5) is reconstructed here as characterized by enlarged, spatulate, and moderately procumbent central incisors, enlarged premolars, low-crowned molars with thick, crenulated enamel, and robust mandibles with deep symphyses and well-developed inferior transverse tori. These features are present in the pongines and are largely retained by *Ouranopithecus*. The pongines vary in habitat preference and locomotor pattern, but all are characterized by morphological features associated with hard-object frugivory and, where known, arboreal foraging. Differences in tooth proportions and dental microwear indicate

varying levels of hard-object consumption (Teaford & Walker 1984; Ward *et al.* 1991), but macrowear patterns, incisor form, and mandibular morphology are clearly indicative of diets dominated by resistant food items requiring incisal manipulation and powerful mastication (Andrews & Alpagut 2001). Only *Ouranopithecus* appears to have been a committed hard-object specialist, as indicated by derived features such as postcanine megadonty and incisor reduction. *Dryopithecus* and the African apes (Node 7), subtropical forest dwellers with suspensory capabilities, evolved adaptations to soft-object feeding, most notably decreased molar enamel thickness. This trend is subsequently reversed in the hominin lineage (Figure 16.2, Table 16.3f).

Ancestral great ape dietary adaptations

As reconstructed here, hominoid dietary evolution is characterized by increasingly efficient exploitation of open country and seasonal forest resources, culminating in a primitive great ape functional complex adapted for the consumption of sclerocarp fruits and hard objects. This picture of hominoid dietary evolution is largely congruent with previous analyses (Andrews *et al.* 1997; Andrews & Martin 1991; Benefit 2000) that accept soft fruit frugivory as the primitive hominoid dietary adaptation, and link trends in middle Miocene dental evolution, particularly the increase in enamel thickness, to a shift toward more varied diets incorporating hard fruits. Andrews & Martin (1991) also considered thick molar enamel to be the primitive great ape condition, but suggested that its presence might be due to phylogenetic inertia or developmental factors and thus not indicative of ancestral great ape dietary adaptations. However, the apparently independent evolution of thin-enameled molar morphologies in *Hylobates* and the *Dryopithecus*–African ape clade suggests that this trait is relatively labile and is not maintained under soft feeding regimes.

Microwear analysis has yet to confirm hard-object feeding in any late Miocene hominoid other than *Ouranopithecus*, and late Miocene Asian hominoids do not seem to have been hard-object specialists *per se*. Instead, they appear to resemble *Pongo* in their ability to exploit hard fruits and their capacity for opportunistic hard-object consumption. By analogy with modern orangutans, basal great apes almost certainly consumed (and possibly preferred) soft fruits (Nowak 1999; Ungar

1995), but the ability to process hard objects, especially nutrient-rich nuts and seeds, would have conferred a significant selective advantage in the seasonal forest environments of late Miocene Eurasia. The last common ancestor of the great ape clade is thus inferred to have consumed a frugivorous diet based on large, resistant fruits supplemented by a range of softer foods, with opportunistic, perhaps seasonal, hard-object consumption playing a significant role in its dietary repertoire.

EXTRACTIVE FORAGING AND THE ORIGINS OF GREAT APE INTELLIGENCE

The picture of hominoid dietary evolution presented here is consistent with the major assumptions of hypotheses emphasizing the exploitation of technically challenging food resources as a major selective force favoring the evolution of increased intelligence. The shift from primitive, soft fruit diets to frugivorous hard-object feeding can be seen as initiating a pattern of reliance on "embedded" food resources that persists and is refined by selection for increased foraging efficiency and dietary flexibility. Thus, the earliest hard-object feeders, *Afropithecus* and *Equatorius*, exhibit specializations of the anterior dentition indicative of highly specific ingestive behaviors. This strategy resembles that of "anatomical extractors," such as the aye-aye and pitheciin monkeys, whose morphologies are adapted for intense exploitation of a narrow spectrum of key resources (Gibson 1986). By contrast, later middle Miocene hominoids and basal great apes possess "multipurpose" dentitions combining somewhat more generalized anterior teeth with powerful masticatory systems. These animals had access to a broad array of forest and open country resources, and all non-folivorous Eurasian hominoids show dietary adaptations at least consistent with more omnivorous feeding regimes. The most recent common ancestor of the great ape clade is reconstructed as a frugivore with hard-object feeding capacities living in a seasonal tropical forest environment, a niche likely to encourage dietary ecumenicism and reward exploitation of embedded resources such as nuts and seed. On the basis of the present evidence, a role for extractive foraging in the evolution of great ape intelligence is highly plausible.

Because extractive foraging behaviors are present in primates other than great apes, most notably *Cebus*

monkeys (Parker & Gibson 1977), technological and behavioral innovations unique to hominoid foraging are key to dietary explanations for great ape intelligence. Under Byrne's (1997) technical intelligence hypothesis, primitive hominoid adaptations are expected to give way to more varied diets secured by increasingly complex and technically sophisticated foraging behaviors (Byrne 1997). Selection for the ability to organize and plan such behaviors would then drive the evolution of increased cognitive capacity. The apparent transition from "anatomical extraction" in the early and middle Miocene to omnivorous hard-object frugivory in the late middle and late Miocene is consistent with a scenario whereby behavioral flexibility and technical innovation supplant anatomical specialization as the dominant hominoid foraging strategy. The cognitive capacities enabling tool-assisted foraging and hierarchical food processing behaviors in extant great apes might then be viewed as the product of primarily ecological factors (Byrne 1997; McGrew 1992; van Schaik & Knott 2001).

Consistency is not confirmation, and positive evidence in support of hypotheses linking foraging behavior to the origins of great ape intelligence is largely lacking. While early great apes possess features appropriate to exploit a broad range of technically challenging foods, the actual complexity of fossil hominoid feeding behaviors is unknown and probably unknowable. The tools used by orangutans and chimpanzees – and likely to have been employed by fossil great apes – leave little paleontological record (McGrew 1992; Mercader, Panger & Boesch 2002; van Schaik & Knott 2001). Recent reports of true tool use in semifree-ranging *Cebus* (Ottoni & Mannu 2001), if accurate, cast further doubt on our ability to draw strong causal links between tool-assisted extractive foraging and the emergence of great ape cognitive capacities. This failure suggests that dietary models are, if not incompatible, then certainly incomplete, and the likelihood that a multifaceted capacity such as intelligence may be attributed to any single factor seems remote. While undoubtedly important, hominoid dietary adaptations and foraging strategies are most prudently viewed as but one element in a nexus of social and ecological factors leading to the evolution of great ape intelligence.

SUMMARY AND CONCLUSIONS

Dietary hypotheses for the origins of great ape intelligence link specific characteristics of extant hominoid diet and foraging behavior to the evolution of great ape cognitive capacities. Seasonal reliance on embedded food resources, complex, hierarchical processing techniques, and tool-mediated resource extraction have all been seen as favoring the evolution of true imitation, enhanced learning capabilities, and technical insight (Byrne 1997; Gibson 1986; Parker 1996). If the complex of cognitive capabilities shared by extant great apes is assumed to be homologous, and thus present in the most recent common ancestor of the great ape clade, the dietary patterns of this ancestor are key to evaluating the extractive foraging hypothesis and its variants. Hominoid dietary evolution is inferred to be characterized by a shift from generalized frugivory to increasingly efficient exploitation of open country and dry forest resources. The most recent common ancestor of the great ape clade is reconstructed as an arboreal hard-fruit frugivore with hard-object feeding capabilities living in a seasonal tropical forest environment. This pattern is broadly consistent with the predictions of extractive foraging theory, but does not provide strong support for its role in the emergence of great ape cognition.

ACKNOWLEDGMENTS

I wish to thank D. R. Begun and A. E. Russon for inviting me to contribute to this volume. Their editorial guidance, humor, and patience over the course of this project have been much appreciated. Thanks also to the Field Museum of Natural History and W. Stanley, Division of Mammals, for access to specimens and to S. Inouye for the use of photographic equipment. Finally, I thank J. M. Plavcan and J. P. Hunter for stimulating and fruitful (pun, regrettably, intended) conversations on this subject and many others.

REFERENCES

Alpagut, B., Andrews, P., Fortelius, M., Kappelman, J., Temizsoy, I., Çelebi, H. & Lindsay, W. (1996). A new specimen of *Ankarapithecus meteai* from the Sinap formation of central Anatolia. *Nature*, **382**, 349–51.

Alpagut, B., Andrews, P. & Martin, L. (1990). New hominoid specimens from the Middle Miocene site at Paşalar, Turkey. *Journal of Human Evolution*, **19**, 397–422.

Anapol, F. & Lee, S. (1994). Morphological adaptation to diet in platyrrhine primates. *American Journal of Physical Anthropology*, **94**, 239–61.

Andrews, P. (1978). A revision of the Miocene Hominoidea of East Africa. *Bulletin of the British Museum of Natural History (Geology)*, **30**, 85–224.

(1981). Species diversity and diet in monkeys and apes during the Miocene. In *Aspects of Human Evolution*, ed. C. B. Stringer, pp. 25–61. London: Taylor & Francis.

(1992). Evolution and environment in the Hominoidea. *Nature*, **360**, 641–6.

Andrews, P. & Alpagut, B. (2001). Functional morphology of *Ankarapithecus meteai*. In *Hominoid Evolution and Climatic Change in Europe. Vol. 2. Phylogeny of the Neogene Hominoid Primates of Eurasia*, ed. L. de Bonis, G. D. Koufos & P. Andrews, pp. 454–87. New York: Cambridge University Press.

Andrews, P. & Bernor, R. L. (1999). Vicariance biogeography and paleoecology of Eurasian Miocene hominoid primates. In *Evolution and Climatic Change in Europe. Vol. 1. The Evolution of Neogene Terrestrial Ecosystems in Europe*, ed. J. Agustí, L. Rook & P. Andrews, pp. 454–87. New York: Cambridge University Press.

Andrews, P. & Martin, L. (1991). Hominoid dietary evolution. *Philosophical Transactions of the Royal Society of London B*, **334**, 199–209.

Andrews, P. & Tekkaya, I. (1976). *Ramapithecus* from Kenya and Turkey. In *Les Plus Anciens Hominidés*, ed. P. V. Tobias & Y. Coppens, pp. 7–25. Nice: International Congress of Prehistoric and Protohistoric Sciences.

(1980). A revision of the Turkish Miocene hominoid *Sivapithecus meteai*. *Paleontology*, **23**, 85–95.

Andrews, P., Begun, D. R. & Zylstra, M. (1997). Interrelationships between functional morphology and paleoenvironments in Miocene hominoids. In *Function, Phylogeny and Fossils: Miocene Hominoid Evolution and Adaptation*, ed. D. R. Begun, C. V. Ward & M. D. Rose, pp. 29–58. New York: Plenum Press.

Andrews, P., Harrison, T., Delson, E., Bernor, R. L. & Martin, L. (1996). Distribution and biochronology of European and Southwest Asian Miocene catarrhines. In *The Evolution of Western Eurasian Neogene Mammal Faunas*, ed. R. L. Bernor, V. Fahlbusch & H.-W. Mittmann, pp. 29–58. New York: Columbia University Press.

Begun, D. R. (1993). New catarrhine phalanges from Rudabánya (Northeastern Hungary) and the problem of parallelism and convergence in hominoid postcranial morphology. *Journal of Human Evolution*, **24**, 373–402.

(1994). Relations among the great apes and humans: new interpretations based on the fossil great ape

Dryopithecus. Yearbook of Physical Anthropology, **37**, 11–63.

(1999). Hominid family values: morphological and molecular data on relations among the great apes and humans. In *The Mentalities of Gorillas and Orangutans: Comparative Perspectives*, ed. S. T. Parker, R. W. Mitchell & H. L. Miles, pp. 3–42. Cambridge, UK: Cambridge University Press.

(2000). Technical comments. *Science*, **287**, 2375a.

(2001). African and Eurasian Miocene hominoids and the origins of the Hominidae. In *Hominoid Evolution and Climatic Change in Europe. Vol. 2. Phylogeny of the Neogene Hominoid Primates of Eurasia*, ed. L. de Bonis, G. D. Koufos & P. Andrews, pp. 231–53. New York: Cambridge University Press.

Begun, D. R. & Güleç, E. (1998). Restoration of the type and palate of *Ankarapithecus meteai*: taxonomic and phylogenetic implications. *American Journal of Physical Anthropology*, **105**, 279–314.

Begun, D. R., Ward, C. V. & Rose, M. D. (1997). Events in hominoid evolution. In *Function, Phylogeny and Fossils: Miocene Hominoid Evolution and Adaptation*, ed. D. R. Begun, C. V. Ward & M. D. Rose, pp. 389–415. New York: Plenum Press.

Behrensmeyer, A. K., Deino, A. L., Hill, A., Kingston, J. D. & Saunders, J. J. (2002). Geology and geochronology of the middle Miocene Kipsaramon site complex, Muruyur Beds, Tugen Hills, Kenya. *Journal of Human Evolution*, **42**, 11–38.

Benefit, B. R. (2000). Old World monkey origins and diversification: an evolutionary study of diet and dentition. In *Old World Monkeys*, ed. P. F. Whitehead & C. J. Jolly, pp. 133–79. New York: Cambridge University Press.

Beynon, A. D., Dean, M. C., Leakey, M. G., Reid, D. J. & Walker, A. (1998). Comparative dental development and microstructure of *Proconsul* teeth from Rusinga Island, Kenya. *Journal of Human Evolution*, **24**, 163–209.

Brown, B. (1997). Miocene hominoid mandibles – functional and phylogenetic perspectives. In *Function, Phylogeny and Fossils: Miocene Hominoid Evolution and Adaptation*, ed. D. R. Begun, C. V. Ward & M. D. Rose, pp. 153–71. New York: Plenum.

Butler, P. M. (2000). The evolution of tooth shape and tooth function in primates. In *Development, Function and Evolution of Teeth*, ed. M. F. Teaford, M. M. Smith & M. W. J. Ferguson, pp. 201–11. New York: Cambridge University Press.

Byrne, R. W. (1996). The misunderstood ape: cognitive skills of the gorilla. In *Reaching into Thought: The Minds of the Great Apes*, ed. A. E. Russon, K. A. Bard & S. T. Parker, pp. 111–30. Cambridge, UK: Cambridge University Press.

(1997). The technical intelligence hypothesis: an additional evolutionary stimulus to intelligence. In *Machiavellian Intelligence II: Extensions and Evaluations*, ed. A. Whiten & R. W. Byrne, pp. 289–311. Cambridge, UK: Cambridge University Press.

Byrne, R. W. & Byrne, J. M. E. (1993). Complex leaf-gathering skills of mountain gorillas (*Gorilla g. beringei*): variability and standardization. *American Journal of Primatology*, **31**, 241–61.

Cerling, T. E., Harris, J. M., Ambrose, S. H., Leakey, M. G. & Solounias, N. (1997). Dietary and environmental reconstruction with stable isotope analyses of herbivore tooth enamel from the Miocene locality of Fort Ternan, Kenya. *Journal of Human Evolution*, **33**, 635–50.

Clutton-Brock, T. H. & Harvey, P. H. (1980). Primates, brains and ecology. *Journal of Zoology, London*, **190**, 309–23.

de Bonis, L. & Koufos, G. D. (1993). The face and the mandible of *Ouranopithecus macedoniensis*: description of new specimens and comparisons. *Journal of Human Evolution*, **24**, 469–91.

(1997). The phylogenetic and functional implications of *Ouranopithecus macedoniensis*. In *Function, Phylogeny and Fossils: Miocene Hominoid Evolution and Adaptation*, ed. D. R. Begun, C. V. Ward & M. D. Rose, pp. 317–26. New York: Plenum.

(2001). Phylogenetic relationships of *Ouranopithecus macedoniensis* (Mammalia, Primates, Hominoidea, Hominidae) of the late Miocene deposits of Central Macedonia (Greece). In *Hominoid Evolution and Climatic Change in Europe. Vol. 2. Phylogeny of the Neogene Hominoid Primates of Eurasia*, ed. L. de Bonis, G. D. Koufos & P. Andrews, pp. 254–68. New York: Cambridge University Press.

de Bonis, L. & Melentis, J. (1978). Les primates hominoïdes du Miocène superiéur de Macédoine. *Annales Paléontogie (Vertébrés)*, **64**, 185–202.

Dumont, E. R. (1995). Enamel thickness and dietary adaptation among extant primates and chiropterans. *Journal of Mammalogy*, **76**, 1127–36.

Dunbar, R. I. M. (1992). Neocortex size as a constraint on group size in primates. *Journal of Human Evolution*, **20**, 469–93.

Eaglen, R. H. (1984). Incisor size and diet revisited: the view from a platyrrhine perspective. *American Journal of Physical Anthropology*, **69**, 262–75.

Fleagle, J. G. & Kay, R. F. (1985). The paleobiology of catarrhines. In *Ancestors: The Hard Evidence*, ed. E. Delson, pp. 23–36. New York: Alan R. Liss.

Gabunia, L., Gabashvili, E., Vekua, A. & Lordkipanidze, D. (2001). The late Miocene hominoid from Georgia. In *Hominoid Evolution and Climatic Change in Europe. Vol. 2. Phylogeny of the Neogene Hominoid Primates of Eurasia*, ed. L. de Bonis, G. D. Koufos & P. Andrews, pp. 316–25. New York: Cambridge University Press.

Geraads, D., Begun, D. R. & Güleç, E. (2003). The middle Miocene hominoid site of Çandır, Turkey: general paleoecological conclusions from the mammalian fauna. *Courier Forschungsinstitut Senckenburg*, **240**, 251–265.

Gibson, K. R. (1986). Cognition, brain size and the extraction of embedded food resources. In *Primate Ontogeny, Cognition and Social Behaviour*, ed. J. G. Else & P. C. Lee, pp. 93–103. New York: Cambridge University Press.

Gordon, K. D. (1982). A study of microwear on chimpanzee molars: implications of dental microwear analysis. *American Journal of Physical Anthropology*, **59**, 195–215.

(1984). Hominoid dental microwear: complications in the use of microwear analysis to detect diet. *Journal of Dental Research*, **63**, 1043–6.

Gould, S. J. & Lewontin, R. C. (1984). The spandrels of San Marco and the Panglossian paradigm: a critique of the adaptationist programme. In *Conceptual Issues in Evolutionary Biology: An Anthology*, ed. E. Sober, pp. 252–70. Cambridge, MA: MIT Press.

Grine, F. E. (1986). Dental evidence for dietary differences in *Australopithecus* and *Paranthropus*. *Journal of Human Evolution*, **15**, 783–822.

Güleç E., & Begun, D. R. (2003). Functional morphology and affinities of the hominoid mandible from Çandır. *Courier Forschungsinstitut Senckenburg*, **240**, 89–112.

Harrison, T. (1988). A taxonomic revision of the small catarrhine primates from the early Miocene of East Africa. *Folia Primatologica*, **50**, 59–108.

Harrison, T. & Rook, L. (1997). Enigmatic anthropoid or misunderstood ape? The phylogenetic status of *Oreopithecus bambolii* reconsidered. In *Function, Phylogeny, and Fossils: Miocene Hominoid Evolution and Adaptations*, ed. D. R. Begun, C. V. Ward & M. D. Rose, pp. 327–62. New York: Plenum Press.

Heizmann, E. & Begun, D. R. (2001). The oldest European hominoid. *Journal of Human Evolution*, **41**, 465–81.

Hürzeler, J. (1958). *Oreopithecus bambolii* Gervais: a preliminary report. *Verhandlungen Naturforschungs Gesellschaft Basel*, **69**, 1–48.

Hylander, W. L. (1975). Incisor size and diet in anthropoids with special reference to the Cercopithecoidea. *Science*, **189**, 1095–8.

Ishida, H., Kunimatsu, Y., Nakatsukasa, M. & Nakano, Y. (1999). New hominoid genus from the Middle Miocene of Nachola, Kenya. *Anthropological Sciences*, **107**, 189–91.

Ishida, H., Pickford, M., Nakaya, H. & Nakano, Y. (1984). Fossil anthropoids from Nachola and Samburu Hills, Samburu District, Kenya. *African Studies Monographs*, Supplement **2**, 73–85.

Jolly, A. (1988). The evolution of purpose. In *Machiavellian Intelligence: Social Expertise and the Evolution of Intellect in Monkeys, Apes and Humans*, ed. R. W. Byrne & A. Whiten, pp. 363–78. New York: Oxford University Press.

Jungers, W. L. (1987). Body size and morphometric affinities of the appendicular skeleton in *Oreopithecus bambolii* (IGF 11778). *Journal of Human Evolution*, **16**, 445–56.

Kay, R. F. (1975). The functional adaptations of primate molar teeth. *American Journal of Physical Anthropology*, **43**, 195–216.

(1977a). Diets of early Miocene African hominoids. *Nature*, **268**, 628–30.

(1977b). The evolution of molar occlusion in the cercopithecidae and early catarrhines. *American Journal of Physical Anthropology*, **46**, 327–52.

(1978). Molar structure and diet in extant Cercopithecidae. In *Development, Function and Evolution of Teeth*, ed. P. M. Butler & K. A. Joysey, pp. 319–39. New York: Academic Press.

(1981). The nut-crackers – a new theory of the adaptations of the Ramapithecinae. *American Journal of Physical Anthropology*, **55**, 141–51.

(1984). On the uses of anatomical features to infer foraging behavior in extinct primates. In *Adaptations for Foraging in Nonhuman Primates – Contributions to an Organismal Biology of Prosimians, Monkeys and Apes*, ed. P. S. Rodman & J. G. H. Cant, pp. 21–53. New York: Columbia University Press.

Kay, R. F. & Covert, H. H. (1984). Anatomy and behaviour of extinct primates. In *Food Acquisition and Processing in Primates*, ed. D. J. Chivers, B. A. Wood & A. Bilsborough, pp. 467–508. New York: Plenum Press.

Kay, R. F. & Hiiemae, K. M. (1974). Jaw movements and tooth use in recent and fossil primates. *American Journal of Physical Anthropology*, **40**, 227–56.

Kay, R. F. & Ungar, P. S. (1997). Dental evidence for diet in some Miocene catarrhines with comments on the effects of phylogeny on the interpretation of adaptation. In *Function, Phylogeny and Fossils: Miocene Hominoid Evolution and Adaptation*, ed. D. R. Begun, C. V. Ward & M. D. Rose, pp. 131–51. New York: Plenum Press.

Kelley, J. (1990). Incisor microwear and diet in three species of *Colobus*. *Folia Primatologica*, **55**, 73–84.

(2001). Phylogeny and sexually dimorphic characters: canine reduction in *Ouranopithecus*. In *Hominoid Evolution and Climatic Change in Europe. Vol. 2. Phylogeny of the Neogene Hominoid Primates of Eurasia*, ed. L. de Bonis, G. D. Koufos & P. Andrews, pp. 269–83. New York: Cambridge University Press.

Kelley, J. & Pilbeam, D. R. (1986). The Dryopithecines: taxonomy, comparative anatomy, and phylogeny of Miocene large hominoids. In *Comparative Primate Biology. Vol. 1. Systematics, Evolution, and Anatomy*, ed. D. R. Swindler & J. Erwin, pp. 362–409. New York: A. R. Liss.

Kelley, J., Ward, S., Brown, B., Hill, A. & Duren, D. L. (2002). Dental remains of *Equatorius africanus* from Kipsaramon, Tugen Hills, Baringo District, Kenya. *Journal of Human Evolution*, **42**, 39–62.

Kilgore, L. (1989). Dental pathologies in ten free-ranging chimpanzees from Gombe National Park, Tanzania. *American Journal of Physical Anthropology*, **80**, 219–27.

King, T. (2001). Dental microwear and diet in Miocene Eurasian catarrhines. In *Hominoid Evolution and Climatic Change in Europe. Vol. 2. Phylogeny of the Neogene Hominoid Primates of Eurasia*, ed. L. de Bonis, G. D. Koufos & P. Andrews, pp. 102–17. New York: Cambridge University Press.

King, T., Aiello, L. C. & Andrews, P. (1998). Dental microwear of *Griphopithecus alpani*. *Journal of Human Evolution*, **36**, 3–31.

Kinzey, W. G. (1992). Dietary and dental adaptations in the Pitheciinae. *American Journal of Physical Anthropology*, **88**, 499–514.

Kinzey, W. G. & Norconk, M. A. (1990). Hardness as a basis of fruit choice in two sympatric primates. *American Journal of Physical Anthropology*, **81**, 5–15.

Köhler, M. & Moyà-Solà, S. (1997). Ape-like or hominid-like? The positional behavior of *Oreopithecus bambolii* reconsidered. *Proceedings of the National Academy of Sciences USA*, **94**, 11 747–50.

Köhler, M., Moyà-Solà, S. & Alba, D. M. (2001). Eurasian hominoid evolution in the light of recent *Dryopithecus* findings. In *Hominoid Evolution and Climatic Change in Europe. Vol. 2. Phylogeny of the Neogene Hominoid Primates of Eurasia*, ed. L. de Bonis, G. D. Koufos & P. Andrews, pp. 192–212. New York: Cambridge University Press.

Kunimatsu, Y., Nakatsukasa, M., Yamanaka, A., Shimizu, D. & Ishida, H. (1998). The newly discovered maxillomandibular specimen of *Kenyapithecus* from the site BG-K in Nachola. *Anthropological Sciences*, **106**, 137–8.

Leakey, M. G. & Walker, A. (1997). *Afropithecus* – function and phylogeny. In *Function, Phylogeny and Fossils: Miocene Hominoid Evolution and Adaptation*, ed. D. R. Begun, C. V. Ward & M. D. Rose, pp. 225–39. New York: Plenum.

Leakey, R. E. & Leakey, M. G. (1986). A new Miocene hominoid from Kenya. *Nature*, **324**, 143–8.

Leakey, R. E., Leakey, M. G. & Walker, A. (1988). Morphology of *Afropithecus turkanensis* from Kenya. *American Journal of Physical Anthropology*, **76**, 289–307.

Lucas, P. W. & Luke, D. A. (1984). Chewing it over: basic principles of food breakdown. In *Food Acquisition and Processing in Primates*, ed. D. J. Chivers, B. A. Wood & A. Bilsborough, pp. 283–301. New York: Plenum Press.

Lucas, P. W., Corlett, R. T. & Luke, D. A. (1986). Postcanine tooth size and diet in anthropoid primates. *Zeitschrift für Morphologie und Anthropologie*, **76**, 253–76.

Lunkka, J. P., Fortelius, M., Kappelman, J. & Sen, S. (1999). Chronology and mammal faunas of the Miocene Sinap Formation, Turkey. In *Hominoid Evolution and Climatic Change in Europe. Vol. 1. The Evolution of Neogene Terrestrial Ecosystems in Europe*, ed. J. Agustí, L. Rook & P. Andrews, pp. 238–64. New York: Cambridge University Press.

Macho, G. A. & Spears, I. R. (1999). Effects of loading on the biochemical behavior of molars of *Homo*, *Pan*, and *Pongo*. *American Journal of Physical Anthropology*, **109**, 211–27.

Martin, L. (1981). New specimens of *Proconsul* from Koru, Kenya. *Journal of Human Evolution*, **10**, 139–50.

(1985). Significance of enamel thickness in hominoid evolution. *Nature*, **314**, 260–3.

Martin, R. D. (1990). *Primate Origins and Evolution: A Phylogenetic Reconstruction*. Princeton, NJ: Princeton University Press.

McCrossin, M. L. (1997). New postcranial remains of *Kenyapithecus* and their implications for understanding the origins of hominoid terrestriality. *American Journal of Physical Anthropology*, Supplement **24**, 164.

McCrossin, M. L. & Benefit, B. R. (1997). On the relationships and adaptations of *Kenyapithecus*, a large-bodied hominoid from the middle Miocene of eastern Africa. In *Function, Phylogeny, and Fossils: Miocene Hominoid Evolution and Adaptations*, ed. D. R. Begun, C. V. Ward & M. D. Rose, pp. 241–67. New York: Plenum.

(2000). Technical comments. *Science*, **287**, 2375a.

McCrossin, M. L., Benefit, B. R., Gitau, S. N., Palmer, A. K. & Blue, K. T. (1998). Fossil evidence for the origins of terrestriality among Old World higher primates. In *Primate Locomotion: Recent Advances*, ed. E. Strasser, J. G. Fleagle, A. Rosenberger & H. McHenry, pp. 253–396. New York: Plenum.

McGrew, W. C. (1992). *Chimpanzee Material Culture: Implications for Human Evolution*. Cambridge, UK: Cambridge University Press.

Mercader, J., Panger, M. & Boesch, C. (2002). Chimpanzee-produced stone assemblages from the tropical forests of Taï, Côte d'Ivoire. *Journal of Human Evolution*, **42**, A23–4.

Milton, K. (1981). Distribution patterns of tropical plant foods as an evolutionary stimulus to primate mental development. *American Anthropologist*, **83**, 534–48.

(1988). Foraging behaviour and the evolution of primate intelligence. In *Machiavellian Intelligence: Social Expertise and the Evolution of Intellect in Monkeys, Apes and Humans*, ed. R. W. Byrne & A. Whitten, pp. 283–305. New York: Oxford University Press.

Morbeck, M. E. (1983). Miocene hominoid discoveries from Rudabánya – implications from the postcranial skeleton. In *New Interpretations of the Ape and Human Ancestry*, ed. R. L. Ciochon & R. S. Corruccini, pp. 369–404. New York: Plenum.

Moyà-Solà, S. & Köhler, M. (1996). A *Dryopithecus* skeleton and the origins of great-ape locomotion. *Nature*, **376**, 156–9.

Nakatsukasa, M., Yamanaka, A., Shimizu, D., Ishida, H. & Kunimatsu, Y. (1998). A new discovery of postcranial skeleton of *Kenyapithecus* from BG-K fossil site. *Anthropological Sciences*, **106**, 138.

Nowak, R. M. (1999). *Walker's Primates of the World*. Baltimore, MD: Johns Hopkins University Press.

Ottoni, E. B. & Mannu, M. (2001). Semifree-ranging tufted capuchins (*Cebus apella*) spontaneously use tools to crack open nuts. *International Journal of Primatology*, **22**, 347–58.

Palmer, A. K., Benefit, B. R., McCrossin, M. L. & Gitau, S. N. (1998). Paleoecological implications of dental microwear analysis for the middle Miocene primate fauna from Maboko Island, Kenya. *American Journal of Physical Anthropology*, Supplement **26**, 175.

Parker, S. T. (1996). Apprenticeship in tool-mediated extractive foraging: the origins of imitation, teaching, and self-awareness in great apes. In *Reaching Into Thought: The Minds of the Great Apes*, ed. A. E. Russon, K. A. Bard & S. T. Parker, pp. 348–70. Cambridge, UK: Cambridge University Press.

Parker, S. T. & Gibson, K. R. (1977). Object manipulation, tool use and sensorimotor intelligence as feeding adaptations in *Cebus* monkeys and great apes. *Journal of Human Evolution*, **6**, 623–41.

Parker, S. T. & Mitchell, R. W. (1999). The mentalities of gorillas and orangutans in phylogenetic perspective. In *The Mentalities of Gorillas and Orangutans: Comparative Perspectives*, ed. S. T. Parker, R. W. Mitchell & H. L. Miles, pp. 397–411. Cambridge, UK: Cambridge University Press.

Pilbeam, D. (1982). New hominoid skull material from the Miocene of Pakistan. *Nature*, **295**, 232–4.

Pilbeam, D. & Smith, R. J. (1981). New skull remains of *Sivapithecus* from Pakistan. *Memoirs of the Geological Survey of Pakistan*, **2**, 1–13.

Pilbeam, D., Rose, M. D., Barry, J. C. & Ibrahim Shah, S. M. (1990). New *Sivapithecus* humeri from Pakistan and the relationship of *Sivapithecus* and *Pongo*. *Nature*, **348**, 237–9.

Quade, J., Cerling, T. C., Andrews, P. & Alpagut, B. (1995). Paleodietary reconstruction of Miocene faunas from Paşalar, Turkey using stable carbon and oxygen isotopes of fossil tooth enamel. *Journal of Human Evolution*, **28**, 373–84.

Retallack, G. J. (1991). *Miocene Paleosols and Ape Habitats of Pakistan and Kenya*. New York: Oxford University Press.

Richmond, B. G. & Whalen, M. (2001). Forelimb function, bone curvature and phylogeny of *Sivapithecus*. In *Hominoid Evolution and Climatic Change in Europe. Vol. 2. Phylogeny of the Neogene Hominoid Primates of Eurasia*, ed. L. de Bonis, G. D. Koufos & P. Andrews, pp. 269–83. New York: Cambridge University Press.

Rook, L., Bondioli, L., Köhler, M., Moyà-Solà, S. & Macchiarelli, R. (1999). *Oreopithecus* was a bipedal ape after all: evidence from the iliac cancellous architecture. *Proceedings of the National Academy of Sciences USA*, **96**, 8795–9.

Rose, J. C. & Ungar, P. S. (1998). Gross dental wear and dental microwear in historical perspective. In *Dental Anthropology: Fundamentals, Limits, and Prospects*, ed. K. W. Alt, F. W. Rösing & M. Teschler-Nicola, pp. 349–84. New York: Springer.

Rose, M. D. (1994). Quadrupedalism in some Miocene catarrhines. *Journal of Human Evolution*, 26, 387–411.

(1997). Functional and phylogenetic features of the forelimb in Miocene hominoids. In *Function, Phylogeny and Fossils: Miocene Hominoid Evolution and Adaptation*, ed. D. R. Begun, C. V. Ward & M. D. Rose, pp. 79–100. New York: Plenum.

Rose, M. D., Nakano, Y. & Ishida, H. (1996). *Kenyapithecus* postcranial specimens from Nachola, Kenya. *African Studies Monographs*, Supplement 24, 3–56.

Rosenberger, A. W. & Kinzey, W. G. (1976). Functional patterns of molar occlusion in platyrrhine primates. *American Journal of Physical Anthropology*, 45, 281–98.

Russon, A. E. & Bard, K. A. (1996). Exploring the minds of the great apes: issues and controversies. In *Reaching into Thought: The Minds of the Great Apes*, ed. S. T. Parker, pp. 1–22. Cambridge, UK: Cambridge University Press.

Ryan, A. S. (1981). Anterior dental microwear and its relationship to diet and feeding behavior in three African primates (*Pan troglodytes troglodytes*, *Gorilla gorilla gorilla*, and *Papio Hamadryas*). *Primates*, 22, 533–50.

Sarmiento, E. (1995). Cautious climbing and folivory: a model of hominoid differentiation. *Journal of Human Evolution*, 10, 289–321.

Sawaguchi, T. (1989). Relationships between cerebral indices for "extra" cortical parts and ecological cateories in anthropoids. *Brain, Behavior and Ecology*, 34, 131–45.

(1992). The size of the neocortex in relation to ecology and social structure in monkeys and apes. *Folia Primatologica*, 5, 131–45.

Schwartz, J. H. (1997). *Lufengpithecus* and hominoid phylogeny – problems in delineating and evaluating phylogenetically revelant characters. In *Function, Phylogeny and Fossils: Miocene Hominoid Evolution and Adaptation*, ed. D. R. Begun, C. V. Ward & M. D. Rose, pp. 363–88. New York: Plenum.

Shellis, R. P., Beynon, A. D., Reid, D. J. & Hiiemae, K. M. (1998). Variations in molar enamel thickness among primates. *Journal of Human Evolution*, 35, 507–22.

Sherwood, R. J., Ward, S., Duren, D. L., Brown, B. & Downs, W. (2002). Preliminary description of the *Equatorius africanus* partial skeleton (KNM-TH 28860) from Kipsaramon, Tugen Hills, Baringo District, Kenya. *Journal of Human Evolution*, 42, 63–73.

Simons, E. L. (1976). *Ramapithecus. Scientific American*, 236, 28–35.

Singleton, M. (2000). The phylogenetic affinities of *Otavipithecus namibiensis. Journal of Human Evolution*, 38, 537–73.

(2001). Molar flare in Miocene hominoids – function or phylogeny? *American Journal of Physical Anthropology*, Supplement 32, 137.

Teaford, M. F. (1985). Molar microwear and diet in the genus *Cebus. American Journal of Physical Anthropology*, 66, 363–70.

(1994). Dental microwear and dental function. *Evolutionary Anthropology*, 3, 17–30.

(2000). Primate dental functional morphology revisited. In *Development, Function and Evolution of Teeth*, ed. M. F. Teaford, M. M. Smith & M. W. J. Ferguson, pp. 290–304. New York: Cambridge University Press.

Teaford, M. F. & Oyen, O. J. (1989). *In vivo* and *in vitro* turnover in dental microwear. *American Journal of Physical Anthropology*, 66, 363–70.

Teaford, M. F. & Runestad, J. A. (1992). Dental microwear and diet in Venezuelan primates. *American Journal of Physical Anthropology*, 88, 347–64.

Teaford, M. F. & Walker, A. C. (1984). Quantitative differences in dental microwear between primate species with different diets and a comment on the presumed diet of *Sivapithecus. American Journal of Physical Anthropology*, 64, 191–200.

Temerin, L. A. & Cant, J. G. H. (1983). The evolutionary divergence of Old World monkeys and apes. *American Naturalist*, 122, 335–51.

Tsujikawa, H. & Nakaya, H. (1998). Faunal analysis of palaeoenvironment of *Kenyapithecus. Anthropological Sciences*, 106, 139.

Ungar, P. S. (1990). Incisor microwear and feeding behavior in *Aloutta seniculus* and *Cebus olivaceus. American Journal of Primatology*, 20, 43–50.

(1995). Fruit preferences of four sympatric primate species at Ketambe, Northern Sumatra, Indonesia. *International Journal of Primatology*, 16, 221–45.

(1996). Dental microwear of European Miocene catarrhines: evidence for diets and tooth use. *Journal of Human Evolution*, 31, 335–66.

(1998). Dental allometry, morphology, and wear as evidence for diet in fossil primates. *Evolutionary Anthropology*, 6, 205–17.

Ungar, P. S. & Kay, R. F. (1995). The dietary adaptations of European Miocene catarrhines. *Proceedings of the National Academy of Sciences USA*, **92**, 5479–81.

Ungar, P. S. & Teaford, M. F. (1996). Preliminary examination of non-occlusal dental microwear in anthropoids: implications for the study of fossil primates. *American Journal of Physical Anthropology*, **100**, 101–13.

Ungar, P. S. & Williamson, M. (2000). Exploring the effects of tooth wear on functional morphology: a preliminary study using dental topographic analysis. *Paleontologica Electronica*, **3**, Article 1, 18 pp.

van Schaik, C. P. & Knott, C. D. (2001). Geographic variation in tool use on Neesia fruits in orangutans. *American Journal of Physical Anthropology*, **114**, 331–42.

Walker, A. C. (1997). *Proconsul* – function and phylogeny. In *Function, Phylogeny and Fossils: Miocene Hominoid Evolution and Adaptation*, ed. D. R. Begun, C. V. Ward & M. D. Rose, pp. 209–24. New York: Plenum.

Walker, A. C., Teaford, M. F. & Ungar, P. S. (1994). Enamel microwear differences between species of *Proconsul* from the early Miocene of Kenya. *American*

Journal of Physical Anthropology, Supplement **18**, 202–3.

Ward, S. (1997). The taxonomy and phylogenetic relationships of *Sivapithecus* revisited. In *Function, Phylogeny and Fossils: Miocene Hominoid Evolution and Adaptation*, ed. D. R. Begun, C. V. Ward & M. D. Rose, pp. 269–90. New York: Plenum.

Ward, S., Beecher, R. & Kelley, J. (1991). Post-canine occlusal anatomy of *Sivapithecus*, *Lufengpithecus*, and *Pongo*. *American Journal of Physical Anthropology*, Supplement **12**, 181.

Ward, S., Brown, B., Hill, A., Kelley, J. & Downs, W. (1999). *Equatorius*: a new hominoid genus from the middle Miocene of Kenya. *Science*, **285**, 1382–6.

Wu, R. & Xu, Q. (1985). *Ramapithecus* and *Sivapithecus* from Lufeng, China. In *Palaeoanthropology and Palaeolithic Archaeology in the People's Republic of China*, ed. R. Wu & J. Olson, pp. 53–68. New York: Academic Press.

Wynn, J. G. & Retallack, G. J. (2001). Paleoenvironmental reconstruction of middle Miocene paleosols bearing *Kenyapithecus* and *Victoriapithecus*, Nyakach Formation, southwestern Kenya. *Journal of Human Evolution*, **40**, 263–88.

17 • Paleontology, terrestriality, and the intelligence of great apes

DANIEL L. GEBO

Department of Anthropology, Northern Illinois University, De Kalb

INTRODUCTION

The level of intelligence among great apes (orangutans, gorillas, chimpanzees, and bonobos) has produced an astonishing array of phenomena to explore. Great apes are believed to be self-aware, social manipulators, makers and users of tools, and generally good problem solvers (e.g., Boesch & Boesch 1984; Byrne 1995; Byrne & Whiten 1988, 1991; de Waal 1989; Gallup 1970, 1991; Goodall 1986; Kohler 1925; McGrew 1992; Parker, Mitchell & Boccia 1994; Premack 1988; Russon, Bard & Galdikas 1996, Russon *et al.* 1998). In contrast, other primates such as gibbons or monkeys show lesser abilities in these tasks (e.g., Anderson 1984; Byrne 1995; Cheney & Seyfarth 1990; Gallup 1991; Povinelli 1987; Povinelli & Cant 1995; Visalbergi & Trinca 1987). Two types of explanation, ecological and social, have been used to explain this dichotomy.

Ecological explanations have attempted to explain increased brain size among primates as a function of enhanced cognitive skills to increase foraging success (Clutton-Brock & Harvey 1980; Gibson 1986; Milton 1988; Parker & Gibson 1977; Povinelli & Cant 1995). For example, Clutton-Brock and Harvey (1980) and Milton (1988) have both shown that frugivory and increased brain size are correlated. Milton (1988) suggested that the complex mental–spatial maps used to find food among fruit-eating primates may play a significant role in increasing intellectual abilities. Parker and Gibson (1977: 37) have also discussed intelligence in cebus monkeys and great apes, arguing that their enhanced intelligence was favored "in situations of locally variable limited seasonal availability of embedded or encased high protein foods susceptible to extractive foraging and feeding."

The complex dynamic of group living has also been used to explain intellect development in primates (Byrne 1995; Byrne & Whiten 1988; Dunbar 1992; Jolly 1969; Kummer 1982). For example, Dunbar's 1992 analysis demonstrated that within the haplorhine primates, increased relative size of the neocortex is more strongly related to changes in group size rather than to ecological factors. Dunbar (1992) further states that in primates, species living in large groups, having large body sizes, and living terrestrially tend to have a relatively large neocortex.

A third explanation, locomotion, is a different type of ecological explanation. Povinelli and Cant (1995: 404) believe that crossing gaps "is the single most important problem of habitat structure for arboreal animals of moderate (10 kg) to large weight (40 kg and greater)." To illustrate, they compare *Macaca fascicularis*, *Hylobates syndactylus*, and *Pongo pygmaeus*. In their view, the locomotor problems faced by the larger orangutans are more severe than those of *Macaca* and *Hylobates*, particularly those associated with crossing gaps. "Orangutans maneuver through a highly deformable habitat in which the immediate structural elements available for use change in position relative to one another and to the animal as it moves" (Povinelli & Cant 1995: 404). Further, they note that "In observing orangutan locomotion, it often appears that the animal attempts one method of dealing with a problem, and if something goes wrong, it then changes its behavior." Correspondingly, they classified locomotor movements into "stereotypic" and "non-stereotypic." Thus, clambering, a non-stereotypic movement pattern common in orangutans, is different from quadrupedalism or brachiation, which are stereotypic movements common to *Macaca* and *Hylobates*. Povinelli and Cant (1995) argue that this distinction between stereotypic and non-stereotypic locomotor systems is the origin of great ape intelligence. The large increase in body size that characterized the ancestral great ape disrupted the regular or stereotypic

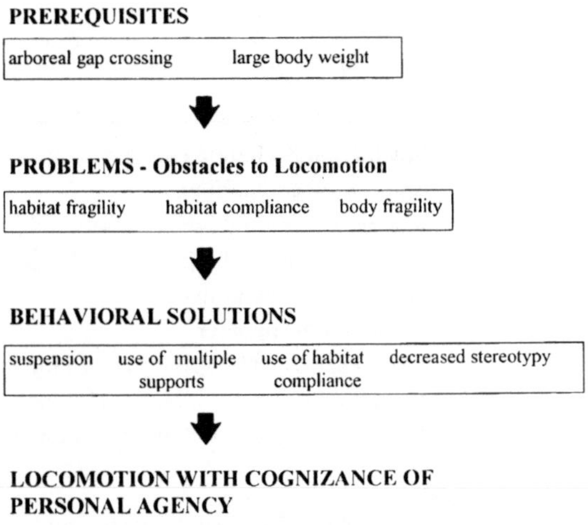

PREREQUISITES

arboreal gap crossing	large body weight

PROBLEMS - Obstacles to Locomotion

habitat fragility	habitat compliance	body fragility

BEHAVIORAL SOLUTIONS

suspension	use of multiple supports	use of habitat compliance	decreased stereotypy

LOCOMOTION WITH COGNIZANCE OF PERSONAL AGENCY

Figure 17.1. Povinelli and Cant's model for cognizance. They entitle their figure "Conceptual structure of the clambering model of the origins of self-conception" (redrawn from Povinelli & Cant, 1995, p. 408).

locomotion typical of smaller primates and forced it to utilize non-stereotypic locomotion, for example, cautious climbing or clambering. This change in locomotor behavior drove the evolution of a new governing mental system for dealing with the environment, one factor of which was the individual's accounting for "itself" in analyzing the structural problems it encountered when moving through the canopy. This new mental system allows "cognizance of one's actions – an ability to engage in a type of mental experimentation or simulation in which one is able to plan actions and predict their likely consequences before acting" (Povinelli & Cant 1995: 409). Figure 17.1 outlines their model. Here, spatial gaps in the canopy, along with large size, induced a decrease in locomotor stereotypy (increased clambering in their model), and this locomotor change led to a concept of self, a critical step in the evolution of higher cognitive abilities.

Although this model is very interesting, I believe several implications and alternative explanations need to be considered. This chapter will explore the uniqueness of "clambering" among primates and the evolution of large body size in living and fossil primates in light of a well-established phylogeny of great apes.

BODY SIZE AND APE LOCOMOTION

There is no doubt that the extant great apes (orangutans, gorillas, bonobos, and chimpanzees) are far larger than gibbons (Table 17.1). Thus, size is a particularly relevant factor for great ape locomotion relative to other extant primates. Brachiation (48% to 84%) and climbing (6% to 74%) dominate the movement patterns of gibbons (Cannon & Leighton 1994; Fleagle 1980; Gittins 1983; Srikosamatara 1984; see Hunt, Chapter 10, this volume for a quantitative review), the smallest of the living apes (5 to 12 kg, Table 17.1). Gibbons are also highly arboreal, utilizing the mid to upper levels of the canopy (Fleagle 1999).

For the large-sized African great apes, chimpanzees and gorillas, locomotion is primarily terrestrial. Hunt (1992) and Doran (1993b) have shown that terrestrial knuckle-walking represents at least 85% of all locomotor movements. Although gorillas and chimpanzees are quite capable climbers when using trees (Doran 1993a,b; Hunt 1991, 1992; Remis 1995), their lessened use of the high canopy and trees overall significantly reduces the risk of falling compared with Asian apes. Although bonobos are more arboreal than chimpanzees, they still engage in considerable terrestrial locomotion (see Hunt, Chapter 10, this volume).

Orangutan locomotion has been described in great detail (Cant 1987a,b; MacKinnon 1974; Povinelli & Cant 1995; Sugardjito 1982; Sugardjito & van Hooff 1986; Tuttle 1975, 1986; Tuttle & Cortright 1988). These studies document the great variety of cautious arboreal movements utilized by orangutans, especially when climbing, clambering (an orthograde body with varying combinations of four appendage grasping; Povinelli & Cant 1995) or quadrumanous scrambling, arm swinging, tree swaying, and bridging. Orangutans use the mid to upper levels of the canopy for feeding quite frequently (Cant 1987a), but at such heights their large size is highly problematic (Cant 1992). Cartmill (1985) has noted the problems of increased size for damage from falling and Biewiener's (1982) analysis has shown a decrease in safety factors in bone strength as size increases.[1] Thus, large size poses a severe problem for high-canopy mammals, and orangutans are the largest arboreal mammals.

However, few, if any, movements used by orangutans are truly unique to them. All primates climb, and most clamber, scramble, or use multiple supports. This is especially true when feeding in the terminal branches, a situation in which body size is always greater than twig diameter. Lorises, primates that are much smaller than orangutans, are also highly cautious climbers and

Table 17.1. *Body weights and locomotor preferences of living and fossil apes.*[1]

Living Apes	Size (kg)	Locomotion
Hylobates syndactylus	11–11.9	Brachiation & climbing
Hylobates concolor	7.6–7.8	Brachiation & climbing
Hylobates hoolock	6.9	Brachiation & climbing
Hylobates lar	5.3–5.9	Brachiation & climbing
Hylobates muelleri	5.4–5.7	Brachiation & climbing
Pongo pygmaeus	36–79	Suspensory clamberer
Pan troglodytes	34–60	Terrestrial knuckle-walking & climbing
Pan paniscus	33–45	Terrestrial knuckle-walking & climbing
Gorilla gorilla	71–175	Terrestrial knuckle-walking & climbing
Fossil Apes (above 10 kg)		
Afropithecus turkanensis	50	Arboreal quadrupedalism & climbing
Ankarapithecus meteai	82	Unknown
Dryopithecus fontani	35	Brachiation & climbing
Dryopithecus laietanus	20	Brachiation & climbing
Gigantopithecus giganteus	190	Unknown
Griphopithecus alpani	28	Unknown
Kenyapithecus africanus	30	Semiterrestrial quadruped
Lufengpithecus lufengensis	50	Unknown
Morotopithecus bishopi	40	Brachiation & climbing
Oreopithecus bambolii	30	Brachiation & climbing
Otavipithecus namibiensis	17.5	Unknown
Ouranopithecus macedonensis	110	Unknown
Proconsul heseloni	17	Arboreal quadrupedalism & climbing
Proconsul nyanzae	28	Arboreal quadrupedalism & climbing
Rangwapithecus gordoni	15	Arboreal quadrupedalism & climbing
Sivapithecus punjabicus	40	Semiterrestrial quadruped
Sivapithecus sivalensis	75	Semiterrestrial quadruped
Samburupithecus kiptalami	60	Unknown
Ugandapithecus major	50	Arboreal quadrupedalism & climbing
Fossil Apes (below 10 kg)		
Dendropithecus macinnesi	9	Arboreal quadrupedalism & climbing
Kalepithecus songhorensis	5	Unknown
Limnopithecus legetet	5	Arboreal quadrupedalism & climbing
Micropithecus clarki	3.5	Arboreal quadrupedalism & climbing
Nyanzapithecus vancouveringi	9	Arboreal quadrupedalism & climbing
Simiolus enjiessi	7	Arboreal quadrupedalism & climbing

Note: [1] Size data from Fleagle (1999).

clamberers but they do not show enhanced cognitive abilities. Like orangutans, lorises do not leap but they still must cross arboreal spatial gaps. *Nycticebus* and *Perodicticus* even prefer the high canopy (Bearder 1987; Charles-Dominique 1977), like orangutans. The problem of branch compliance, however, is certainly less critical for lorises than for orangutans and climbing/clambering in lorises is of the more pronograde (horizontal) variety, rather than orthograde (vertical) as in the living apes.

The spider monkey, *Ateles*, another primate studied by Cant (1986), is more orthograde and is one of the largest of the South American monkeys (7.2–9.1 kg; Fleagle 1999). Its locomotor pattern includes frequent use of climbing/clambering as well as brachiation and quadrupedalism, and these monkeys have often been referred to as ape-like in their movements (e.g., Cant 1986). In this comparative case, we see a monkey that is large relative to its ancestors and that approaches Povinelli and Cant's 10 kg rubicon, yet it climbs/clambers around in the mid to high canopy frequently and lives in a fragile arboreal environment (hence the evolution of a prehensile tail, an extra grasping organ, see Emmons & Gentry 1983). *Ateles* also lives in large social groups similar to those of chimpanzees and is highly frugivorous, like living apes. On the other hand, spider monkeys possess small brains, and have not been noted for their intellectual prowess like their smaller cousins the cebus monkeys.

To be fair, Povinelli and Cant (1995: 405) further define orangutan non-stereotyped locomotion as "consisting of nondiscrete and highly variable schemata in which limb maneuvers are not repeated very often, and there is a great deal of assimilation of changing structural contexts and accommodation to them." They discuss a macaque walking across a horizontal branch as their example of stereotypic locomotion with its discrete action schemata, repetition, and with few changes due to structural contexts. They emphasize the repetitiveness in stereotypic behavior and note that their "reconceptualization of locomotion emphasizes a continuous spectrum from the stereotyped to the non-stereotyped" (Povinelli & Cant 1995: 405).

Let's examine these claims. If brachiation in *Hylobates*, *Ateles*, and perhaps *Pongo*, is viewed as stereotypic, as is quadrupedalism within the loris and spider monkey locomotor spectrum, then both these latter movements would be repetitive, show little structural assimilation, and presumably show the use of discrete action schemata. In fact, the quadrupedalism of lorises and spider monkeys is quite distinct from these qualities. Movements by lorises blend much more smoothly into their climbing activities than this stereotypic characterization suggests. Both lorises and spider monkeys seem to move in unusual arboreal environments (often not walking along the tops of branches) that require deliberation, choices of substrates, and movement decisions (Cant 1986; Charles-Dominique 1977; Gebo 1987; Mittermeier 1978; Walker 1969, 1979). Thus, lorises and

spider monkeys appear to utilize both stereotypic and non-stereotypic movements, as do orangutans, and their movements span a range within the continuous spectrum proposed by Povinelli and Cant.

Habitat use and locomotor abilities of lorises, spider monkeys, and other primates would further suggest that clambering (and other movements involving decreased stereotypy) within the high canopy cannot simply be the product of increased size (>10 kg). The unusual factor in orangutan movements is their use of the high canopy given their great body size combined with their decidedly calculated approach to their environment as a whole. Certainly, their more frequent use of compliant supports (due to their great size) is one factor in which orangutans seem to be truly unique relative to other primates (Povinelli & Cant 1995; but see Schmitt 1999). The most important point to note here is that clambering and non-stereotypic movements are not exclusively orangutanian and that orangutans take cautious and calculated approaches to other facets of their behavior. Further, enhanced cognition has not been noted among other non-great-ape primates that clamber.

Large body size is also shared with other primates. For example, many Old World monkeys achieve body sizes above 10 kg but these taxa are not known for their clambering abilities. Likewise, other great apes climb trees and must make decisions about limb strength and compliance, but rarely "clamber." Thus, large body size does not necessarily predispose a primate toward using decreased stereotypy in arboreal locomotion, at least in the form of clambering. Further, orangutans have adapted to their unusual habitus by transforming their hands and feet into hooks, evolving hyper-mobile limb joints, and reducing the size of their foot bones for a suspensory rather than weight bearing role (Tuttle & Cortright 1988; Rose 1988; Tuttle 1970, 1975). This derived morphology suggests a very specialized mode of locomotion for orangutans. If the more deliberate nature of orangutan locomotion is what is truly making their locomotor style unique (cognitively non-stereotypic), relative to other primates (even the African apes), then how reasonable is an orangutan model for great ape evolution? In the end, the "clambering model of the origins of self-conception" (Figure 17.1) has several assumptions that require rethinking. Hunt (Chapter 10, this volume) re-examines this model from the perspective of living primates. Here I consider it from the perspective of the fossil record.

APE BODIES AND THE FOSSIL RECORD

There is a clear phylogenetic component to be considered in any assessment of sophisticated cognitive abilities in the great apes, if, as it now seems, within primates, only great apes and humans are capable of these cognitive abilities. Although phylogenetic factors have been noted in several papers on great ape intelligence, only Povinelli and Cant (1995) have seriously attempted to incorporate the paleontological record into their argument for the origin of increased intellect of great apes. In the early Miocene of Africa, there is an abundance of fossil hominoids; at least 12 genera and some 19 species have been named to date. Despite very similar dental and cranial anatomy, these fossil apes are diverse in size, ranging from the small-sized *Micropithecus* (3.5 kg) to the huge *Ugandapithecus major* (>50 kg) (Andrews 1978, 1985; Fleagle 1999; Harrison 1986; Szalay & Delson 1979). We know very little about the body anatomy of these taxa, with the exception of *Proconsul*. The consensus is that *Proconsul* is more monkey- than ape-like in its body plan (Figure 17.2; Fleagle 1983; Langdon 1984; Rose 1983, 1993; Walker & Pickford 1983; Ward 1993; Ward, Walker & Teaford 1991). It was probably a pronograde quadruped, a capable climber, but without the unique upper body and forelimb anatomy of the living apes (Rose 1993).

This suggests that *Proconsul*, and probably other related Miocene apes, are morphologically too primitive to be good models for the immediate common ancestor of great apes, humans, and gibbons. The common ancestor of all living apes and humans had a highly modified upper body with long and mobile arms. Anatomically, these features would include a wide and dorsally flattened thorax, a long clavicle, dorsally placed scapulae, a ball-like humeral head with small tubercles, humeral torsion, a long forelimb, enhanced mobility at the elbow and wrist, as well as features related to orthograde (upright) body posture (e.g., Cartmill & Milton 1977; Gebo 1996; Harrison 1986; Keith 1923; Tuttle 1975). This novel anatomy appears to be related to brachiating and arm-suspensory behavior (Gebo 1996; Gregory 1928; Keith 1923; Morton 1924; Washburn 1968; and see Hunt, Chapter 10, this volume). In other words, living apes and humans, our part of the larger Miocene ape radiation, possess an upper body and arm anatomy that is highly divergent from our original ape ancestors, like

Figure 17.2. Ape body plans. In living apes (for example, the gibbon, above), the upper body and forelimbs are highly modified. Note the long arms with shoulders pushed out to the side and their shorter, broader, and flatter chests. A reconstruction of the fossil ape *Proconsul* (below) shows a more monkey-like body plan, with limbs of about equal length and a long and deep chest. The shoulders are set close to the midline of the body like those of a baboon or a dog.

Proconsul. Some have argued that the unique upper body and forelimb anatomy of living apes could have evolved more than once in a variety of hominoids (Larson 1998; Napier 1963; Rose 1997; Simons 1962, 1967), but the great similarity of these forms makes this evolutionary interpretation highly unlikely. Harrison & Rook (1997: 331) state it this way: "the post cranial features and character complexes shared by extant hominoids are so detailed and so pervasive that they are extremely unlikely to be the product of convergent evolution." In sum, we have two fundamentally different types of body evolution among hominoids in the Miocene. First, we have the more monkey-like bodies of *Proconsul* and its relatives (Figure 17.2), and second, we have the specialized upper body and forelimb anatomy for a lineage of brachiators, the "dolichobrachiotherians" (long-armed beasts), a lineage more closely related to living apes and humans.

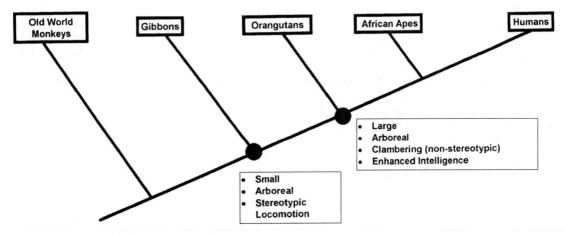

Figure 17.3. Arboreal model by Povinelli & Cant (1995).

The first discovered fossil ape to display the unique upper body and forelimb anatomy similar to living apes was named in 1872 (Gervais 1872). *Oreopithecus bambolii* from the late Miocene of Italy has had a controversial history (see Andrews 1992; Delson 1979,1987; Harrison 1987; Harrison & Rook 1997; Hürzeler 1960). Because of its unique dental adaptations, it has been interpreted as an Old World monkey (Delson 1979), confusing its important link to the living apes. Hürzeler's original claim that "not even by the boldest mental acrobatics can this ulnar fragment be interpreted as anything but a hominoid" (1958: 35) has been validated by the subsequent discovery of the complete skeleton of *Oreopithecus* in 1958 (Hürzeler 1958, 1960) and the more recent work by Harrison (1986), Sarmiento (1987), and Rose (1993). *Oreopithecus* was also much larger than gibbons with males being about 30 kg in size (Ward, Flinn & Begun, Chapter 18, this volume). Povinelli & Cant (1995) also believe that *Oreopithecus* (and perhaps *Dryopithecus*) had a body plan essential for the evolution of self-conception.

In 1996, new body parts of *Dryopithecus* (Köhler, Moyà-Soyà & Alba 2001; Moyà-Solà & Köhler 1996) were described from the late Miocene of Spain. This new evidence shows that *Dryopithecus* had a very long arm with very long fingers, a long clavicle, and lumbar vertebrae that suggest orthogrady. In short, this evidence as well as the older material (Begun 1992b, 1994; Morbeck 1983; Pilbeam & Simons 1971; Rose 1989) suggest that *Dryopithecus*, like *Oreopithecus*, does indeed possess the body plan of living apes and belongs to the clade of brachiators. *Dryopithecus* was also large bodied, ranging in size from 20 to 40 kg (Ward *et al.*, Chapter 18, this volume). In sum, I would argue that both *Dryopithecus* and *Oreopithecus* utilize orthogrady, forelimb suspension, and a flexible orientation of the body about fixed handholds, and had considerable hind limb mobility, the basic components proposed to be essential in the evolution of self-conception as noted by Povinelli and Cant (1995: 412) for *Oreopithecus*. These abilities are like those found in orangutans, and these fossil taxa fit well with Povinelli and Cant's ancestral condition for the great apes (see Figure 17.3).

In 1997, Gebo *et al.* (1997) found new body parts of a large-bodied hominoid from Uganda. This material was combined with the older material recovered by W. W. Bishop in the 1960s to name *Morotopithecus bishopi*. This taxon is much earlier than late Miocene *Oreopithecus* and *Dryopithecus*, and dates to 20.6 Ma in the early Miocene of Africa, overlapping in time with *Proconsul*. *Morotopithecus* is not known as well as the late Miocene European apes but it possesses lumbar vertebrae that suggest an orthograde body plan as well as a shoulder that is very similar to that of living apes, implying a high degree of arm mobility for brachiating and arm suspensory abilities (Gebo *et al.* 1997; MacLatchy *et al.* 2000). On the other hand, its femoral morphology is more primitive than those of all of the living apes, including gibbons, suggesting a phylogenetic position before the separation of the gibbon lineage. *Morotopithecus* is much larger than *Oreopithecus* and *Dryopithecus*, being some 40–50 kg in size.

Other Miocene hominoids have been linked to the living ape clade by dental or facial evidence (e.g.,

Afropithecus, Ankarapithecus, Heliopithecus, Lufeng-pithecus, Griphopithecus, Gigantopithecus, Graecopithecus (Ouranopithecus), and *Otavipithecus;* see Andrews 1992; Andrews & Alpagut 2001; Andrews *et al.* 1996; Begun 1995; Conroy 1994; de Bonis & Koufos 2001; and see Singleton, Chapter 16, this volume), but few body parts have been recovered to allow a better assessment of loco-motor ability. The recent fossil evidence for *Kenyapithe-cus* (McCrossin 1997; McCrossin *et al.* 1998) has been interpreted as indicative of a body form similar to that of living apes as well as showing arm and hand features like those of African apes (Benefit & McCrossin 1995; McCrossin & Benefit 1997; but see Rose 1997; C. V. Ward 1997). Species of *Kenyapithecus* are similar in size to *Dryopithecus.*

Curiously, one Miocene ape, *Sivapithecus,* seemed to be securely linked on the basis of facial features to the orangutan lineage (Andrews & Cronin 1982; Pilbeam *et al.* 1980; S. Ward 1997; Ward & Pilbeam 1983), but aspects of its humeral anatomy are possibly very prim-itive relative to the extant apes (Pilbeam *et al.* 1990; Richmond & Whalen 2001). If true, the arm anatomy of *Sivapithecus* could place this genus in a pre-hylobatid phyletic position (Benefit 2000) or more likely among the ancestral great apes. At present, no unequivocal evidence exists (i.e., no humeral heads are known) to remove *Sivapithecus* from the clade of living apes and we will have to wait and see about this evolutionary interpretation. One anatomical point is clear, however, the locomotor pattern of *Sivapithecus,* which could be characterized as a quadrupedal climber, is very different from that of orangutans (Madar *et al.* 2002; Rose 1993). Like the other fossil "brachiators," *Sivaptihecus* is large bodied (Table 17.1; See Ward *et al.,* Chapter 18, this volume).

The relationship between any of these Miocene fos-sils and the living great apes is still very unclear. This makes it quite difficult to use the fossil record to recon-struct the ancestral condition of the great apes, or the evolutionary pattern of locomotor behavior. Povinelli and Cant's assumption of "orangutan-like behavior" for the common ancestor cannot be supported (or rejected), although as noted above, the unique, derived morpho-logical features associated with orangutan behavior are not found in any Miocene hominoid. Povinelli and Cant (1995: 412) singled out *Oreopithecus* as best resembling the ancestral condition of the great ape/human clade with orangutans specializing "to deal with problems

Table 17.2. *Brain size for living and fossil apes*[1]

Taxa	Brain Size (cc)
Hylobates klossi	78–103
Hylobates lar	82–125
Hylobates concolor	82–136
Hylobates syndactylus	100–152
Pan paniscus	275–381
Pan troglodytes	282–454
Pongo pygmaeus	276–502
Gorilla gorilla	350–752
Proconsul heseloni (female)	167
Oreopithecus bambolii (female)	128

Note: [1] Brain size data from Harrison (1989), Tuttle (1986) and Walker *et al.* (1983).

of increased body weight while maintaining arboreal suspensory patterns." I agree with this evolutionary assessment but with one important exception. I would move the *Oreopithecus*-like ancestral condition back one evolutionary node to represent the ancestral condition for all living apes. This move has several implications for the Povinelli and Cant model (see below).

Finally, I add one last comment concerning brain size among the Miocene fossil hominoids and the living apes (Table 17.2). Great apes do indeed possess much larger brains, in absolute size, than gibbons. The lowest values for any of the living great apes is for female *Pan paniscus* and female *Pongo pygmaeus* at about 275 cc (see Tuttle 1986; and see MacLeod, Chapter 7, and Begun & Kordos, Chapter 14, this volume). Most mean values for the great apes are around 350 cc or above. For the fossil hominoids, the story is much more obscure. Few fossil hominoids are known with intact skulls to esti-mate their brain size and brain size estimates are prob-lematic (Conroy 1987; Begun & Kordos, Chapter 14, this volume). However, some evidence suggests that *Oreopithecus* had a relatively small brain compared with living great apes (Harrison, 1989; Harrison and Rook, 1997; Begun & Kordos, Chapter 14, this volume). In the context of the evolution of positional behavior and intel-ligence this suggests that if *Oreopithecus* was specialized in non-stereotyped arboreal locomotion, it did so with a smaller brain than that which supports self-concept in living great apes (Russon 1998).

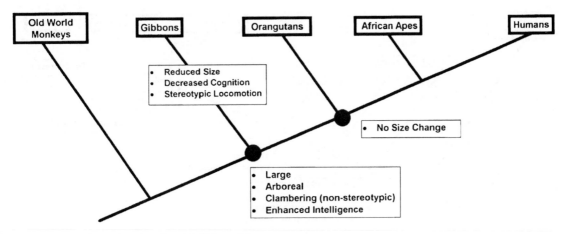

Figure 17.4. An alternative arboreal model. This model moves large body size back (one node from its position in Figure 17.3) to the ancestor of all living apes and humans.

ANCESTRAL CONDITION OF GREAT APES

Povinelli and Cant (1995) imply that the ancestral ape (of the living ape clade, i.e., hylobatids and great apes) was small and arboreal (Figure 17.3) and differed from the ancestral great ape condition. Increased body size in great apes, while retaining a primarily arboreal habitus, starts the chain of events leading to increased cognition (Figure 17.1). The large body size and morphology of *Morotopithecus* (as well as most other Miocene taxa) may force a re-evaluation of these assumptions and a re-appraisal of Povinelli and Cant's argument, as well as a reassessment of the gibbon lineage (Figure 17.4). First, the ancestral condition for the living ape clade had a body that was fundamentally changed from the monkey-like body plan of the ancestral hominoid. We all agree on this point. Was this ancestral ape large or small (Figures 17.3 and 17.4)? If the ancestor of all living apes was large (above 10 kg according to Povinelli and Cant, 1995), instead of small as implied by Povinelli and Cant, then all of the gap-crossing problems discussed by Cant (1992) and Povinelli and Cant (1995) would apply (Figure 17.1). This ancestral ape would be a capable climber and clamberer with an orthograde back, would use fore-limb suspension and brachiation, and would be capable of moving its body about a fixed handhold. This view would move the size increase and clambering connection, noted for great apes by Povinelli and Cant (1995),

back one evolutionary node, to the ancestral condition for all living apes and humans (Figure 17.4). A large-bodied living ape ancestral condition implies that the gibbon lineage is dwarfed (see Dunbar 1992; Groves 1972; Pilbeam 1996). According to Povinelli and Cant (1995), gibbons utilize a more stereotypic locomotor pattern and this would mean a reversal from the ancestral condition. Since gibbons have not to date shown the increased intellectual prowess of the extant great apes, they must have secondarily decreased their cognitive skills relative to the ancestral condition as well (Figure 17.4). Thus, gibbons would need to be viewed as "dwarfed idiots."

Lastly, under this evolutionary view no size increase is necessary for the origin of great apes, a critical part of the Povinelli and Cant cognition scheme. Thus, a large-sized ancestral condition for the common ancestor of all living apes necessarily implies a reversal in both size and cognition in gibbons. All of the fossil taxa that have been linked to living apes (e.g., *Oreopithecus*, *Dryopithecus*, *Morotopithecus*, *Sivapithecus*, *Kenyapithecus*) are estimated to be above 10 kg (Table 17.2); this suggests that the last common ancestor was, in fact, a large animal.

Another potential problem of the Povinelli and Cant model might also come from the future interpretation of fossil taxa linked to living apes, but which possess different types of body plans (e.g., *Sivapithecus*). If the body and limb anatomy of these fossil hominoids does

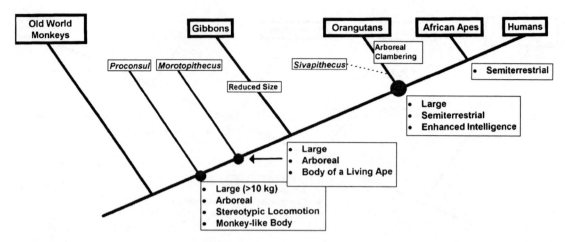

Figure 17.5. A terrestrial model of great ape evolution.

not display the *Pongo*- or *Oreopithecus*-like specializations, it would create several exceptions to the Povinelli and Cant model.

Finally, the common ancestor of great apes may not have been highly arboreal, making the locomotor pattern of orangutans a rather poor model (Figure 17.5). For example, *Kenyapithecus* is viewed as terrestrial (McCrossin & Benefit 1997). Likewise, the anatomical interpretation and phylogenetic position of *Sivapithecus* is critical for any assessment of the ancestral condition for great apes since several anatomical studies, particularly those of the face and palate, suggest a very close evolutionary relationship for *Sivapithecus* and orangutans (Ward & Kimbel 1983; Ward & Pilbeam 1983; S. Ward 1997). On the other hand, other studies have interpreted the dental anatomy of *Sivapithecus* as being rather hominid-like (e.g., Kay & Simons 1983) or the body as not especially orangutan-like, suggesting different phylogenetic positions for *Sivapithecus* (Pilbeam *et al.* 1990; Pilbeam & Young 2001). If the characters of the cranium signal the true phylogenetic position of *Sivapithecus* as the sister taxon to orangutans, then it suggests a very different ancestral morphological condition and locomotor pattern from that of orangutans. Here, we have an anatomical mosaic to interpret. For example, the big toe, thumb, hip, knee, elbow, and forelimb have been suggested to support an arboreal heritage for *Sivapithecus* (Madar *et al.* 2002; Rose 1993, 1994, 1997; C. V. Ward 1997), while aspects of the wrist (Spoor, Sondaar & Hussain 1991), the humeri (Pilbeam *et al.* 1990; Richmond & Whalen 2001; Rose 1993; Senut

1986), and the foot bones (Gebo 1996; Rose 1993) suggest that *Sivapithecus* was at least semiterrestrial. In fact, the most recent analysis of foot bones and phalanges of *Sivapithecus* stated that these elements most resembled African apes and that none of the newly described elements "precludes significant terrestriality" (Madar *et al.* 2002: 746). Smith and Pilbeam (1980) further discuss the consideration of a terrestrial ancestral condition for orangutans. Since chimpanzees and gorillas are also semiterrestrial, a terrestrial locomotor component for part of the ancestral condition for great apes is likely. This differs significantly from the arboreal/clambering ancestor envisioned by Povinelli and Cant (1995). On the other hand, Dunbar's (1992) linkage of terrestriality, large groups, and large body size with increased cognitive abilities fits well with this view of great ape evolution. Thus, if locomotion was a central pressure in the evolution of enhanced intelligence among great apes, then terrestriality, rather than arboreal clambering, would become the critical paleoenvironmental factor (Figure 17.5).

CONCLUSION

What can ape bodies tell us about the origin of great ape intelligence? The evolutionary node that unites all living apes and humans appears to have been a large-sized (at least 20 kg) arboreal primate, having an orthograde body, with highly mobile and long arms. It was a capable brachiator and could suspend by its arms for long periods of time, as well as climbing and clambering.

The ancestor for the clade of living apes and humans possessed a highly derived body plan compared with other Miocene hominoids. The diet would be primarily frugivorous (Kay 1977; Singleton, Chapter 16, this volume) and this evolutionary node would be very similar to the great ape node of Povinelli and Cant (1995).

Two contrasting views, an arboreal and a terrestrial model, have been proposed for the ancestral condition of great apes and humans (Figures 17.3 and 17.5). The arboreal model of Povinelli and Cant (1995) suggests an ancestor with increased body size compared with the common ancestor of gibbons, great apes, and humans. This ancestor would be very orangutan-like in its locomotor abilities (a clamberer), highly arboreal, and capable of conceiving of itself (Figure 17.3). Figure 17.4 modifies this view by shifting the body size increase back one evolutionary node from the Povinelli and Cant model. In this modified arboreal model, gibbons need to be viewed as reversals, with a decrease in body size and in cognitive abilities.

In contrast, a terrestrial model has been proposed as an alternative view of great ape evolution (Figure 17.5). This model is influenced by the large body sizes of early Miocene apes and the locomotor pattern of the ancestral orangutan clade, likely represented by *Sivapithecus*, a fossil ape strongly linked to orangutans but divergent in body form. If *Sivapithecus* mirrors the ancestral condition of great apes, then its mosaic anatomical structure needs to be fully understood. Is *Sivapithecus* primarily arboreal or is this form partly terrestrial? The many novel locomotor abilities and unique anatomical adaptations peculiar to orangutans are not likely to be part of the ancestral condition of great apes. If a terrestrial phase in great ape evolution has occurred, then Dunbar's interpretation for the origin of enhanced intellectual abilities is more likely than Povinelli and Cant's. Any evidence for terrestriality in the paleontological record at this evolutionary position would buttress the social intellect explanations for the origin of great ape cognition.

We all have to make assumptions about the past (ancestral conditions) and in the absence of time machines, the best test is a comparative one. In this light, Dunbar (1992) has performed the best effort to date. Terrestriality, group size, and large bodies correlate best with increased intellect within primates. Enhanced cognition in great apes and humans is indeed tied to their phylogenetic past, a past that fits well with Dunbar's analysis of the living species. The past can be explored in the paleontological record but it is a slow undertaking with evolutionary assumptions changing quickly with the discovery of new fossils. This makes our attempts to learn ultimate causes a very difficult process. In the end, I hope paleontology will provide a significant piece to the puzzle of great ape intelligence and perhaps the use of the terrestrial environment is a start, but broad based investigation of correlations among the living will likely provide faster results.

ACKNOWLEDGMENTS

I thank Anne Russon for inviting my participation in this volume as well as David Begun. I thank Marian Dagosto for all of her comments and Kim Reed for her time and skill (Figure 17.2). I also thank the reviewers, especially Mike Rose for his ever helpful and thoughtful suggestions.

ENDNOTE

1 Cartmill (1985) notes that bone strength is proportional to cross-sectional area (L^2) while body size is a measure of volume (L^3). As size increases, area/volume ratios decrease; thus, a limb will break more readily as size increases. For example, think of a cube where each side equals 1 centimeter. The area of one side will be 1 and the volume will be 1 (area/volume ratio = 1). On the other hand, if one side of a cube is 3 centimeters long then its area will be 9 and its volume 27. Thus, the 3 centimeter sided cube will possess an area/volume ratio one third that of our 1 centimeter cube. This relationship suggests that large animals are at greater risks with greater bodily harm when they fall. Biewiener (1982) has experimentally shown that the safety margin for the risk of bone fracture decreases as size increases. Thus, larger animals are more likely to break bones than small ones.

REFERENCES

Anderson, J. R. (1984). The development of self recognition: a review. *Developmental Psychobiology*, **17**, 35–49.

Andrews, P. (1978). A revision of the Miocene Hominoidea of East Africa. *Bulletin of the British Museum of Natural History: Geology Series*, **30**, 85–224.

(1985). Family group systematics and evolution among catarrhine primates. In *Ancestors: The Hard Evidence*, ed. E. Delson, pp. 14–22. New York: Liss.

(1992). Evolution and environment in the Hominoidea. *Nature*, **360**, 641–6.

(1996). The nature of the evidence. *Nature*, **379**, 123–4.

Andrews, P. & Alpagut, B. (2001). Functional morphology of *Ankarapithecus meteai*. In *Phylogeny of the Neogene Hominoid Primates of Eurasia*, ed. L. de Bonis, G. D. Koufos & P. Andrews, pp. 213–230. Cambridge, UK: Cambridge University Press.

Andrews, P. & Cronin, J. E. (1982). The relationships of *Sivapithecus* and *Ramapithecus* and the evolution of the orangutan. *Nature*, **297**, 541–6.

Andrews, P., Harrison, T., Delson, E., Bernor, R. L. & Martin, L. (1996). Distribution and biochronology of European and Southwest Asian Miocene catarrhines. In *The Evolution of Western Eurasian Neogene Mammal Faunas*, ed. R. L. Bernor, V. Fahlbusch & H.-W. Mittmann, pp. 168–206. New York: Columbia University Press.

Bearder, S. K. (1987). Lorises, bushbabies, and tarsiers: diverse societies in solitary foragers. In *Primate Societies*, ed. B. B. Smuts, D. L. Cheney, R. M. Seyfarth, R. W. Wrangham & T. T. Struhsaker, pp. 11–24. Chicago, IL: University of Chicago Press.

Begun, D. R. (1992). Phyletic diversity and locomotion in primitive European hominids. *American Journal of Physical Anthropology*, **87**, 311–40.

(1994). Relations among the great apes and humans: new interpretations based on the fossil great ape *Dryopithecus*. *Yearbook of Physical Anthropology*, **37**, 11–63.

(1995). Late Miocene European orang-utans, gorillas, humans or none of the above? *Journal of Human Evolution*, **29**, 169–80.

Benefit, B. R. (2000). Old World monkey origins and diversification: an evolutionary study of diet and dentition. In *Old World Monkeys*, ed. P. L. Whitehead & C. J. Jolly, pp. 133–79. Cambridge UK: Cambridge University Press.

Benefit, B. R. & McCrossin, M. L. (1995). Miocene hominoids and hominid origins. *Annual Review of Anthropology*, **24**, 237–56.

Biewiener, A. A. (1982). Bone strength in small mammals and bipedal birds: do safety factors change with body size? *Journal of Experimental Biology*, **98**, 289–301.

Boesch, C. & Boesch, H. (1984). Mental map in wild chimpanzees: an analysis of hammer transports for nut cracking. *Primates*, **25**, 160–70.

Byrne, R. W. (1995). *The Thinking Ape: Evolutionary Origins of Intelligence*. Oxford: Oxford University Press.

Byrne, R. W. & Whiten, A. (1988). *Machiavellian Intelligence: Social Expertise and the Evolution of Intellect in Monkeys, Apes and Humans*. Oxford: Clarendon Press.

(1991). Computation and mindreading in primate tactical deception. In *Natural Theories of Mind*, ed. A. Whiten, pp. 127–41. Oxford: Basil Blackwood.

Cannon, C. H. & Leighton, M. (1994). Comparative locomotory ecology of gibbons and macaques: selection of canopy elements for crossing gaps. *American Journal of Physical Anthropology*, **93**, 505–24.

Cant, J. G. H. (1986). Locomotion and feeding postures of spider and howling monkeys: field study and evolutionary interpretation. *Folia Primatologica*, **46**, 1–14.

(1987a). Positional behavior of female Bornean orangutans (*Pongo pygmaeus*). *American Journal of Primatology*, **12**, 71–90.

(1987b). Effects of sexual dimorphism in body size on feeding postural behavior of Sumatran orangutans (*Pongo pygmaeus*). *American Journal of Physical Anthropology*, **74**, 143–8.

(1992). Positional behavior and body size of arboreal primates: a theoretical framework for field studies and an illustration of its application. *American Journal of Physical Anthropology*, **88**, 273–83.

Cartmill, M. (1985). Climbing. In *Functional Vertebrate Morphology*, ed. M. Hildebrand, D. M. Bramble, K. Liem & D. B. Wake, pp. 73–88. Cambridge, MA: Belknap Press.

Cartmill, M. & Milton, K. (1977). The lorisiform wrist joint and evolution of "brachiating" adaptations in the Hominoidea. *American Journal of Physical Anthropology*, **47**, 249–72.

Charles-Dominique, P. (1977). *Ecology and Behaviour of Nocturnal Primates*. New York: Columbia University Press.

Cheney, D. L. & Seyfarth, R. M. (1990). *How Monkeys See the World: Inside the Mind of Another Species*. Chicago, IL: University of Chicago Press.

Clutton-Brock, T. H. & Harvey, P. H. (1980). Primates, brains, and ecology. *Journal of the Zoological Society of London*, **190**, 309–23.

Conroy, G. C. (1987). Problems of body-weight estimation in fossil primates. *International Journal of Primatology*, **8(2)**, 115–37.

(1994). *Otavipithecus*: or How to build a better hominid – Not. *Journal of Human Evolution*, **27**, 373–83.

de Bonis, L. & Koufos, K. D. (2001). Phylogenetic relationships of *Ouranopithecus macedoniensis* (Mammalia, Primates, Hominoidea, Hominidae) of the late Miocene deposits of Central Macedonia (Greece). In *Phylogeny of the Neogene Hominoid Primates of Eurasia*, ed. L. de Bonis, G. D. Koufos, & P. Andrews, pp. 254–68. Cambridge, UK: Cambridge University Press.

de Waal, F. B. M. (1989). *Peacemaking Among Primates*. Cambridge MA: Harvard University Press.

Delson, E. (1979). *Oreopithecus* is a cercopithecid after all. *American Journal of Physical Anthropology*, **50**, 431–2.

(1987). An anthropoid enigma: historical introduction to the study of *Oreopithecus bambolii*. *Journal of Human Evolution*, **15**, 523–31.

Doran, D. M. (1993a). Comparative locomotor behavior of chimpanzees and bonobos: the influence of morphology on locomotion. *American Journal of Physical Anthropology*, **91**, 83–98.

(1993b). Sex differences in adult chimpanzee positional behavior: the influence of body size on locomotion and posture. *American Journal of Physical Anthropology*, **91**, 99–116.

Dunbar, R. I. M. (1992). Neocortex size as a constraint on group size in primates. *Journal of Human Evolution*, **20**, 469–93.

Emmons, L. H. & Gentry, A. H. (1983). Tropical forest structure and the distribution of gliding and prehensile-tailed vertebrates. *American Naturalist*, **121**(4), 513–24.

Fleagle, J. G. (1980). Locomotion and posture. In *Malayan Forest Primates*, ed. D. J. Chivers, pp. 191–207. New York: Plenum Press.

(1983). Locomotor adaptations of Oligocene and Miocene hominoids and their phyletic implications. In *New Interpretations of Ape and Human Ancestry*, ed. R. L. Ciochon & R. S. Corruccini, pp. 301–25. New York: Plenum Press.

(1999). *Primate Adaptation and Evolution*, 2nd edn. New York: Academic Press.

Gallup, G. G. (1970). Chimpanzees: self recognition. *Science*, **167**, 86–7.

(1991). Toward a comparative psychology of self-awareness: species limitation and cognitive consequences. In *The Self: An Interdisciplinary Approach*, ed. G. R. Goethals & J. Strauss, pp. 121–35. New York: Springer-Verlag.

Gebo, D. L. (1987). Locomotor diversity in prosimian primates. *American Journal of Primatology*, **13**, 271–81.

(1996). Climbing, brachiation, and terrestrial quadrupedalism: historical precursors of hominid bipedalism. *American Journal of Physical Anthropology*, **101**, 55–92.

Gebo, D. L., MacLatchy, L., Kityo, R., Deino, A., Kingston, J. & Pilbeam, D. (1997). A hominoid genus from the early Miocene of Uganda. *Science*, **276**, 401–4.

Gervais, P. (1872). Sur un fossile, d'une espèce non encore décrite, qui a été découverte au Monte Bamboli, (Italic). *Comptes rendus de l' Académie des sciences de Paris*, **74**, 1217.

Gibson, K. R. (1986). Cognition, brain size, and the extraction of embedded food resources. In *Primate Ontogeny, Cognition, and Social Behaviour*, ed. J. G. Else & P. C. Lee, pp. 93–104. Cambridge UK: Cambridge University Press.

Gittins, S. P. (1983). The use of the forest canopy by the agile gibbon. *Folia Primatologica*, **40**, 134–44.

Goodall, J. (1986). *The Chimpanzees of Gombe: Patterns of Behavior*. Cambridge, MA: Harvard University Press.

Gregory, W. K. (1928). Were the ancestors of man primitive brachiators? *Proceedings of the American Philosophical Society*, **67**(2), 129–50.

Groves, C. P. (1972). Systematics and phylogeny of gibbons. In *Gibbon and Siamang*, Vol. 1, ed. D. M. Rumbaugh, pp. 1–89. Basel: Karger.

Harrison, T. (1986). A reassessment of the phylogenetic relationships of *Oreopithecus bambolii* Gervais. *Journal of Human Evolution*, **15**, 541–83.

(1987). The phylogenetic relationships of the early catarrhine primates: a review of the current evidence. *Journal of Human Evolution*, **16**, 41–80.

(1989). New estimates of cranial capacity, body size, and encephalization in *Oreopithecus bambolii*. *American Journal of Physical Anthropology*, **78**, 237 (abstract).

Harrison, T. & Rook, L. (1997). Enigmatic anthropoid or misunderstood ape? The phylogenetic status of *Oreopithecus bambolii* reconsidered. In *Function, Phylogeny, and Fossils: Miocene Hominoid Evolution and Adaptations*, ed. D. R. Begun, C. V. Ward & M. D. Rose, pp. 327–62. New York: Plenum Press.

Hunt, K. D. (1991). Positional behavior in the Hominoidea. *International Journal of Primatology*, **12**, 95–118.

(1992). Positional behavior of *Pan troglodytes* in the Mahale Mountains and Gombe Stream National Parks, Tanzania. *American Journal of Physical Anthropology*, **87**, 83–105.

Hürzeler, J. (1958). *Oreopithecus bambolii* Gervais: a preliminary report. *Verhandlungen der naturforschenden Gesellschaft*, **69**, 1–47.

(1960). The significance of *Oreopithecus* in the genealogy of man. *Triangle*, **4**, 164–74.

Jolly, A. (1969). Lemur social behavior and primate intelligence. *Science*, **153**, 501–6.

Kay, R. F. (1977). Diets of early Miocene African hominoids. *Nature*, **268**, 628–30.

Kay, R. F. & Simons, E. L. (1983). A reassessment of the relationship between later Miocene and subsequent Hominoidea. In *New Interpretations of Ape and Human Ancestry*, ed. R. L. Ciochon & R. S. Corruccini, pp. 577–624. New York: Plenum.

Keith, A. (1923). Man's posture: its evolution and disorders. *The British Medical Journal*, **1**, 451–4, 499–502, 545–8, 587–90, 624–6, 669–72.

Köhler, W. (1925). *The Mentality of Apes*. London: Routledge and Kegan.

Köhler, M., Moyà-Solà, S. & Alba, D. M. (2001). Eurasian hominoid evolution in the light of recent *Dryopithecus* findings. In *Phylogeny of the Neogene Hominoid Primates of Eurasia*, ed. L. de Bonis, G. D. Koufos & P. Andrews, pp. 192–212. Cambridge UK: Cambridge University Press.

Kummer, H. (1982). Social knowledge in free-ranging primates. In *Animal Mind, Human Mind*, ed. D. Griffin, pp. 113–30. Berlin: Springer.

Langdon, J. (1984). The Miocene hominoid foot. Ph.D. Dissertation, Yale University, New Haven, CT.

Larson, S. G. (1998). Parallel evolution in the hominoid trunk and forelimb. *Evolutionary Anthropology*, **6(3)**, 87–99.

MacKinnon, J. (1974). The behaviour and ecology of wild orang-utans (*Pongo pygmaeus*). *Animal Behaviour*, **22**, 3–74.

MacLatchy, L., Gebo, D., Kityo, R. & Pilbeam, D. (2000). Postcranial functional morphology of *Morotopithecus bishopi*, with implications for the evolution of modern ape locomotion. *Journal of Human Evolution*, **39**, 159–83.

Madar, S. I., Rose, M. D., Kelley, J., MacLatchy, L. & Pilbeam, D. (2002). New *Sivapithecus* postcranial specimens from the Siwaliks of Pakistan. *Journal of Human Evolution*, **42**, 705–52.

McCrossin, M. L. (1997). New postcranial remains of *Kenyapithecus* and their implications for understanding the origins of hominoid terrestriality. *American Journal of Physical Anthropology Supplement*, **24**, 164 (abstract).

McCrossin, M. L. & Benefit, B. R. (1997). On the relationships and adaptations of *Kenyapithecus*, a large-bodied hominoid from the middle Miocene of Eastern Africa. In *Function, Phylogeny, and Fossils: Miocene Hominoid Evolution and Adaptations*, ed. D. R. Begun, C. V. Ward & M. D. Rose, pp. 241–67. New York: Plenum Press.

McCrossin, M. L., Benefit, B. R., Fitau S. N., Palmer, A. K. & Blue, K. T. (1998). Fossil evidence for the origins of terrestriality among Old World higher primates. In *Primate Locomotion: Recent Advances*, ed. E. Strasser, J. Fleagle, A. Rosenberger & H. McHenry, pp. 353–96. New York: Plenum.

McGrew, W. C. (1992). *Chimpanzee Material Culture: Implications for Human Evolution*. Cambridge UK: Cambridge University Press.

Milton, K. (1988). Foraging behaviour and the evolution of intellect in monkeys, apes and humans. In *Machiavellian Intelligence: Social Expertise and the Evolution of Intellect in Monkeys, Apes and Humans*, ed. R. W. Byrne & A. Whiten, pp. 285–305. Oxford: Clarendon Press.

Mittermeier, R. A. (1978). Locomotion and posture in *Ateles geoffroyi* and *Ateles paniscus*. *Folia Primatologica*, **30**, 161–93.

Morbeck, M. E. (1983). Miocene hominoid discoveries from Rudàbanya: implications from the postcranial skeleton. In *New Interpretations of Ape and Human Ancestry*, ed. R. L. Ciochon & R. S. Corruccini, pp. 369–404. New York: Plenum Press.

Morton, D. J. (1924). Evolution of the human foot (II). *American Journal of Physical Anthropology*, **7**, 1–52.

Moyà-Solà, S. & Köhler, M. (1996). A *Dryopithecus* skeleton and the origins of great-ape locomotion. *Nature*, **379**, 156–9.

Napier, J. R. (1963). Brachiation and brachiators. In *The Primates*, ed. J. Napier & N. A. Barnicott, *Symposia of the Zoological Society of London*, **10**, 183–95.

Parker, S. T. & Gibson, K. R. (1977). Object manipulation, tool use, and sensorimotor intelligence as feeding adaptations in great apes and cebus monkeys. *Journal of Human Evolution*, **6**, 623–41.

Parker, S. T., Mitchell, R. W. & Boccia, M. L. (eds.) (1994). *Self-Awareness in Animals and Humans*. New York: Cambridge University Press.

Pilbeam, D. (1996). Genetic and morphological records of the Hominoidea and hominid origins: a synthesis. *Molecular Phylogenetics and Evolution*, **5**, 155–68.

Pilbeam, D. & Simons, E. L. (1971). Humerus of *Dryopithecus* from Saint Gaudens, France. *Nature*, **229**, 406–7.

Pilbeam, D. & Young, N. M. (2001). *Sivapithecus* and hominoid evolution: some brief comments. In *Phylogeny of the Neogene Hominoid Primates of Eurasia*, ed. L. de Bonis, G. D. Koufos & P. Andrews, pp. 349–64. Cambridge UK: Cambridge University Press.

Pilbeam, D., Rose, M. D., Badgley, C. & Lipschultz, B. (1980). Miocene hominoids from Pakistan. *Postilla*, **181**, 1–94.

Pilbeam, D., Rose, M. D., Barry, J. C. & Ibrahim Shah, S. M. (1990). New *Sivapithecus* humeri from Pakistan and the relationship of *Sivapithecus* and *Pongo*. *Nature*, **348**, 237–9.

Povinelli, D. J. (1987). Monkeys, apes, mirrors and minds: the evolution of self-awareness in primates. *Human Evolution*, **2**, 493–507.

Povinelli, D. J. & Cant, J. G. H. (1995). Arboreal clambering and the evolution of self-conception. *Quarterly Review of Biology*, **70**, 393–419.

Premarck, D. (1988). Does the chimpanzee have a theory of mind? In *Machiavellian Intelligence: Social Expertise and the Evolution of Intellect in Monkeys, Apes, and Humans*, ed. R. W. Byrne & A. Whiten, pp. 94–110. Oxford: Clarendon Press.

Remis, M. J. (1995). Effects of body size and social context on the arboreal activities of lowland gorillas in the Central African Republic. *American Journal of Physical Anthropology*, **97**, 413–33.

Richmond, B. G. & Whalen, M. (2001). Forelimb function, bone curvature and phylogeny of *Sivapithecus*. In *Phylogeny of the Neogene Hominoid Primates of Eurasia*, ed. L. de Bonis, G. D. Koufos & P. Andrews, pp. 326–48. Cambridge UK: Cambridge University Press.

Rose, M. D. (1983). Miocene hominoid postcranial morphology: monkey-like, ape-like, neither, or both? In *New Interpretations of Ape and Human ancestry*, ed. R. L. Ciochon & R. S. Corruccini, pp. 405–17. New York: Plenum.

(1988). Functional anatomy of the cheiridia. In *Orang-utan Biology*, ed. J. H. Schwartz, pp. 299–310. Oxford: Oxford University Press.

(1989). New postcranial specimens of catarrhines from the middle Miocene Chinji Formation, Pakistan: descriptions and a discussion of proximal humeral functional morphology in anthropoids. *Journal of Human Evolution*, **18**, 131–62.

(1993). Locomotor anatomy of Miocene hominoids. In *Postcranial Adaptation in Nonhuman Primates*, ed. D. L. Gebo, pp. 70–95. DeKalb, IL: Northern Illinois University Press.

(1994). Quadrupedalism in some Miocene catarrhines. *Journal of Human Evolution*, **26**, 387–411.

(1997). Functional and phylogenetic features of the forelimb in Miocene hominoids. In *Function, Phyogeny, and Fossils: Miocene Hominoid Evolution and Adaptations*, ed. D. R. Begun, C. V. Ward & M. D. Rose, pp. 79–100. New York: Plenum.

Russon, A. E. (1998). The nature and evolution of orangutan intelligence. *Primates*, **39**, 485–503.

Russon, A. E., Bard, K. A. & Galdikas, B. M. F. (eds.) (1996). *Reaching into Thought: The Minds of the Great Apes*. Cambridge UK: Cambridge University Press.

Russon, A. E., Mitchell, R. W., Lefebvre, L. & Abravanel, E. (1998). The comparative evolution of imitation. In *Piaget, Evolution, and Development*, ed. J. Langer & M. Killen, pp. 103–43. Hillsdale, NJ: Lawrence Erlbaum Associates.

Sarmiento, E. E. (1987). The phylogenetic position of *Oreopithecus* and its significance in the origin of the Hominoidea. *American Museum Novitates*, **2881**, 1–44.

Schmitt, D. (1999). Compliant walking in primates. *Journal of the Zoological Society of London*, **248**, 149–60.

Senut, B. (1986). New data on Miocene hominoid humeri from Pakistan and Kenya. In *Primate Evolution*, ed. J. G. Else & P. C. Lee, pp. 151–61. Cambridge UK: Cambridge University Press.

Simons, E. L. (1962). Fossil evidence relating to the early evolution of primate behavior. *Annals of the New York Academy of Sciences*, **102**, 282–95.

Simons, E. L. (1967). Fossil primates and the evolution of some primate locomotor systems. *American Journal of Physical Anthropology*, **26**, 241–54.

Smith, R. J. & Pilbeam, D. R. (1980). Evolution of the orang-utan. *Nature*, **284**, 447–8.

Spoor, C. F., Sondaar, P. Y. & Hussain, T. (1991). A new hominoid hamate and first metacarpal from the late Miocene Nagri Formation of Pakistan. *Journal of Human Evolution*, **21**, 413–24.

Srikosamatara, S. (1984). Ecology of pileated gibbons in South-East Thailand. In *The Lesser Apes*, ed. H. Preuschoft, D. J. Chivers, W. Y. Brockelman & N. Creel, pp. 242–57. Edinburgh: Edinburgh University Press.

Straus, W. L. & Schön, M. A. (1960). Cranial capacity of *Oreopithecus bambolii. Science*, **132**, 670–72.

Sugardjito, J. (1982). Locomotor behavior of the Sumatran Orang Utan (*Pongo pygmaeus abelii*) at Ketambe, Gunung Leuser National Park. *Malayan Nature Journal* **35**, 57–64.

Sugardjito, J. & van Hooff, J. A. R. A. M. (1986). Age–sex class differences in the positional behaviour of the Sumatran Orang-Utan (*Pongo pygmaeus abelii*) in the Gunung Leuser National Park, Indonesia. *Folia Primatologica*, **47**, 14–25.

Szalay, F. S. & Delson, E. (1979). *Evolutionary History of the Primates*. New York: Academic Press.

Tuttle, R. H. (1970). Postural, propulsive and prehensile capabilities in the cheiridia of chimpanzees and other great apes. In *The Chimpanzee*, ed. G. H. Bourne, vol. 2, pp. 167–253. Basel: Karger.

(1975). Parallelism, brachiation, and hominoid phylogeny. In *Phylogeny of the Primates: A Multidisciplinary Approach*, ed. W. P. Luckett & F. S. Szalay, pp. 447–80. New York: Plenum Press.

(1986). *Apes of the World: Their Social Behavior, Communication, Mentality, and Ecology*. New Jersey: Noyes.

Tuttle, R. H. & Cortright, G. W. (1988). Positional behavior adaptive complexes and evolution. In *Orang-utan Biology*, ed. J. H. Schwartz, pp. 311–30. Oxford: Oxford University Press.

Visalbergi, E. & Trinca, L. (1987). Tool use in capuchin monkeys: distinguishing between performing and understanding. *Primates*, **30**, 511–21.

Walker, A. C. (1969). The locomotion of the lorises, with special reference to the potto. *Journal of East African Wildlife*, **7**, 1–5.

(1979). Prosimian locomotor behavior. In *The Study of Prosimian Behavior*, ed. G. A. Doyle & R. D. Martin, pp. 543–65. New York: Academic Press.

Walker, A. & Pickford, M. (1983). New postcranial fossils of *Proconsul africanus* and *Proconsul nyanzae*. In *New Interpretations of Ape and Human Ancestry*, ed. R. L. Ciochon & R. S. Corruccini, pp. 325–51. New York: Plenum.

Walker, A., Falk, D., Smith, R. & Pickford, M. (1983). The skull of *Proconsul africanus*: reconstruction and cranial capacity. *Nature*, **305**, 525–7.

Ward, C. V. (1993). Torso morphology and locomotion in *Proconsul nyanzae*. *American Journal of Physical Anthropology*, **92**, 291–328.

(1997). Functional anatomy and phyletic implications of the hominoid trunk and hindlimb. In *Function, Phylogeny, and Fossils: Miocene Hominoid Evolution and Adaptations*, ed. D. R. Begun, C. V. Ward & M. D. Rose, pp. 101–30. New York: Plenum.

Ward, C. V., Walker, A. & Teaford, M. F. (1991). *Proconsul* did not have a tail. *Journal of Human Evolution*, **21**, 215–20.

Ward, S. (1997). The taxonomy and phylogenetic relationships of *Sivapithecus* revisited. In *Function, Phylogeny, and Fossils: Miocene Hominoid Evolution and Adaptations*, ed. D. R. Begun, C. V. Ward & M. D. Rose, pp. 269–90. New York: Plenum Press.

Ward, S. C. & Pilbeam, D. R. (1983). Maxillofacial morphology of Miocene hominoids from Africa and Indo-Pakistan. In *New Interpretations of Ape and Human Ancestry*, ed. R. L. Ciochon & R. S. Corruccini, pp. 211–38. New York: Plenum Press.

Ward, S. C. & Kimbel, W. H. (1983). Subnasal alveolar morphology and the systematic position of *Sivapithecus*. *American Journal of Physical Anthropology*, **61**, 157–71.

Washburn, S. L. (1968). *The Study of Human Evolution*. Condon Lectures, Oregon State System of Higher Education, Eugene, OR.

18 · Body size and intelligence in hominoid evolution

CAROL V. WARD,[1] MARK FLINN,[2] AND DAVID R. BEGUN[3]

[1]*Department of Anthropology – Department of Pathology and Anatomical Sciences, University of Missouri, Columbia*
[2]*Department of Anthropology, University of Missouri, Columbia;* [3]*Department of Anthropology, University of Toronto, Toronto*

INTRODUCTION

Great apes and humans are the largest-brained primates. Aside from a few extinct subfossil lemurs, they are also the largest in body mass. Body size is a key aspect of a species' biology, a large organism having different energetic, ecological, and physical constraints than a small one. Brain size, in so far as it determines abilities to acquire, process, and act on information, is also a key aspect of a species' biology and is linked to body size. Large animals have different informational problems to solve than do small ones, hence their respective sensory organs and nervous systems are sized and organized differently.

Mammalian body and brain size scale consistently with each other (Figure 18.1). This relation is generally described by allometric exponents that vary between 2/3 and 3/4 (e.g., Bauchot & Stephan 1966, 1969; Hofman 1982; Jerison 1973; Lande 1979; Martin 1981; Martin & Harvey 1985; Stephan 1972). From a paleontological perspective, the body–brain size relation offers an appealing way to evaluate the cognitive abilities of fossil taxa, a problem of particular importance for understanding hominid evolution.

A convincing theoretical basis for this general allometric statistical pattern, however, remains elusive (Deacon 1997; Harvey & Krebs 1990). Body mass is not a strict determinant of brain size, as species of similar size can have different brain sizes and cognitive abilities (Pagel & Harvey 1989). In addition, comparative analyses indicate considerable variation among taxa from general mammalian patterns (Pagel & Harvey 1989). Hominid brains, for example, are double or more their expected size as mammals. There are also phylogenetic differences in typical brain–body size relations within primates that reflect grade shifts in encephalization across taxa (Armstrong 1985a,b; Martin & Harvey 1985, Pagel & Harvey 1989).

One reason for the lack of a universal brain–body size correlation among mammalian species is that factors other than body mass or metabolism, such as locomotion, diet, predation risk, social structure, and life history, affect relations between body and brain size (see recent reviews in de Waal & Tyack 2003; other chapters in this volume). All of these factors and others may contribute to selective pressures for cognitive abilities. As such, allometric scaling models developed from analyses of relations between physical variables such as metabolic rate and body mass may not be appropriate models for relations between body and brain size.

Evolving a large brain depends upon a complex balance of costs and benefits, which vary from species to species. There are not likely to be simple explanations based upon simple physical principles. Observed correlations between brain size and body size, the variability in these relations, and the reasons underlying phylogenetic differences, require consideration of both direct and indirect influences. Direct influences include structural and metabolic constraints on encephalization and size-related needs: large-bodied animals are better able to support large crania and energetically expensive neural tissue than are small-bodied animals, and larger bodies may require more neurons to control. If this were all there was to the relation between body size and brain size, then we would expect simple and consistent statistical associations. However, additional indirect influences can independently affect both body size and intelligence, including the effects of selective pressures shaping other aspects of a species' biology, such as locomotion, diet, predation risk, social interactions, and life history. If these indirect influences are important for determining brain–body size relations, then we expect more

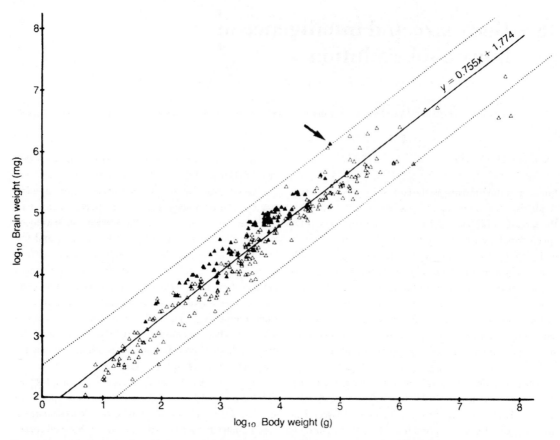

Figure 18.1. Logarithmic plot of brain weight on body weight for 309 placental mammal species. Black triangles are primates, open triangles are non-primates. The arrow points to humans. The best-fit (solid) line is the major axis for the entire sample. Dotted lines denote five-fold variation on either side of the major axis. Note the considerable scatter about the line, despite the linear nature of the data. Reprinted by permission from Martin, 1981; *Nature* vol. **293**, pp. 220–223. Copyright © Macmillan Journals Limited.

complex and variable statistical patterns, as indicated by analyses of inter-taxa differences (Martin & Harvey 1985).

Because of the apparent complexities of brain–body relations and neurobiological differences among extant mammals, simple consideration of relative brain size provides only an incomplete picture of the cognitive abilities of fossil species. To understand scaling relations among body size, brain size, cognition, and other aspects of a species' adaptation, we need first to understand the underlying selective pressures shaping cognitive function and related variables. Complex models involving ecological and social factors are required. Such models may provide new insights into the causal relations underlying statistical associations between body and brain size.

This chapter first examines the interrelations among multiple relevant variables and their relations with cognitive capacities and brain size that apply generally in primates, especially those linked with body size. This puts us in a stronger position to interpret the cognitive capabilities of extinct taxa, and therefore to understand the evolution of intelligence in the hominids.

PREVIOUS HYPOTHESES RELATING BODY MASS TO INTELLIGENCE

Logarithmic scaling between brain size and body size in mammals is often interpreted to suggest that increases in body size result in increases in brain mass in the absence of any selection for a particular brain function (Hofman

1983; Jerison 1973; Martin 1981, 1983) (Figure 18.1). The scatter about the line is interpreted as a change in brain mass that must be explained by some other factors. The questions to be answered are: (1) why do these observed scaling relations exist? and (2) why do some species depart from them?

Somatic factors: size, scaling and metabolism

A possible answer to the first question is that having a larger body with more sensory receptors sending input and more motor units to control requires greater processing power, hence larger brain size. This appears unlikely, however, because animals of similar size can vary dramatically in brain–body size relations. As an example among catarrhines, *Theropithecus oswaldi*, a large fossil papionin, was similar in body size to chimpanzees and female gorillas, with estimates ranging from about 20 to 128 kg (Delson *et al.* 2000; Martin 1993; and C. V. Ward unpublished data). Direct estimates of the cranial capacity of three specimens with body sizes ranging from 32 to 70 kg are 154, 155 and 200 cc (Martin 1993). Chimpanzees and gorillas of about the same body mass range, on the other hand, have brain sizes of roughly 275–580 cc (Tobias 1971). This variability, and the grade shifts in this relation evident among taxa (Begun & Kordos, Chapter 14, this volume), undermine the interpretation that brain size and body size are related by simple physical or physiological laws.

A related possibility is that somatic brain regions, which govern somatic and autonomic sensorimotor function, should scale with body size because larger bodies with their greater number of cells might need more neurons to receive input and send output (Aboitiz 1996; Fox & Wilczynski 1986). In contrast, extrasomatic regions of the brain concerned with higher cognitive processing, such as the neocortex, are expected not to follow any fixed scaling pattern. Somatic regions of the brain correlate weakly with body size ($r = 0.5$), suggesting that a larger body does not require a larger somatic brain (Rilling & Insel 1998). Non-somatic regions show even weaker correlations.

A frequently cited explanation for observed brain–body size correlations is metabolic rate of either the individual or its mother (Jerison 1973; Martin 1983). Smaller animals have higher metabolic rates, limiting the size of metabolically expensive brain tissue. There are significant problems with this hypothesis, however. Taxa do not always scale as predicted. Metabolic rate is not always correlated with adult brain size (Harvey & Krebs 1990) or neonatal brain size (Pagel & Harvey 1988). Furthermore, taxa with similar metabolic structures and body masses can have markedly different trajectories of postnatal brain growth (Periera & Leigh 2002). Maternal or individual metabolic rates do not seem to constrain brain size tightly.

A small-bodied organism faces stricter structural, metabolic, and other constraints on attaining large brain size than a large-bodied one. A large body is necessary for attaining large brain size (Dunbar 1993). Smaller animals are usually subject to higher extrinsic mortality rates than are larger ones, decreasing the selective advantages of growing a larger brain at the expense of rapid generational turnover times. They also tend to have relatively faster metabolisms than do larger animals (Kleiber 1932), so maintaining a large brain would pose a relatively greater burden on them. Large body size results in both a slower metabolism and less predation risk, decreasing costs associated with growing and maintaining a large brain. Therefore, one mechanism for being able to afford a large brain in the presence of cognitive selection pressures would be to increase body size (Dunbar 1993). This would alter the cost–benefit ratio of increasing brain size by decreasing metabolic costs, and accordingly facilitate brain expansion. In addition, selection for slower life history or increased body size would decrease constraints imposed by life history and metabolism on brain size (van Schaik & Deaner 2003; Kelley, Chapter 15, van Schaik, Preuschoft & Watts, Chapter 11, this volume), easing constraints on brain expansion in species facing selection for increased intelligence.

Because the brain is so metabolically expensive, consuming up to 10% of calories for most mammals and up to 20% for modern humans (Armstrong 1990), it should be as small as possible for a given body mass and set of species-specific cognitive demands (Geary & Huffman 2002). The only way for an expanded brain to be retained by selection is if the benefits to the individual of improved cognitive processing outweigh the metabolic and structural costs. The expensive nature of brain tissue may partially explain why brain regions expand differentially in taxa responding to different information-processing demands (e.g., Adolphs 2003; Armstrong 1985b; Barton & Harvey 2000; de Winter & Oxnard 2001; Purves 1994; Semendeferi & Damasio

2000; Whiting & Barton 2003; MacLeod, Chapter 7, this volume; *contra* Finlay & Darlington 1995, Finlay, Darlington & Nicastro 2001; Rakic 1988; 1995): it is too costly to sustain expansions that are not strictly necessary.

Another factor arguing against evolution of a large, unspecified cortex of the sort proposed by Finlay and Darlington (1995), Finlay *et al.* (2001) or Barton (1999) is that the energetic costs of maintaining a large brain would not necessarily be balanced by significant functional improvements (Aboitiz 1996; cf. La Cerra & Bingham 1998). To expand the brain, neurons must increase in number rather than size to maintain conduction speed, as dendrite breadth must increase with the square power of length to maintain conduction velocity (Kaas 2000). With more neurons, each neuron will communicate with absolutely more but proportionately fewer neurons than before. Clusters of specialized neurons should appear with cortical expansion to permit fine-tuned processing of information, or there can be relatively little improvement in cognitive sophistication (Geary & Huffman 2002; Kaas 2000; Nimchinsky *et al.* 1999; and see MacLeod, Chapter 7, this volume). For these reasons, areal specializations alongside greater interconnectedness both characterize the human and probably the great ape cortex (MacLeod, Chapter 7, this volume). Great apes and humans have larger neocortices, the area primarily responsible for flexible problem solving, than less socially complex species (Adolphs 2003; Barton 1996; Clark, Mitra & Wang 2001; de Winter & Oxnard 2001; Dunbar 1993; Dunbar & Bever 1998; Preuss 2001; Sawaguchi 1997), and also have augmented neocerebellar structures compared with other anthropoids that may be related to their especially complex behavioral challenges (MacLeod, Chapter 7, this volume).

Variation among mammal species in relative brain size and cognitive potential suggests that selection for overall or regional brain size increase affects metabolic rate or metabolic tradeoffs within an organism. A species under selective pressure to increase its cognitive complexity may experience selection to modify diet, altering calorie or nutrient intake to support brain expansion. Metabolic rate can also vary among mammalian species of similar body size, so it can also be modified by selection. For example, platyrrhines have higher rates of oxygen metabolism than do strepsirhines of similar sizes (Armstrong 1990). This appears to have happened in the evolution of *Homo*, which reduced its gut size, diverting more metabolic energy to the brain (Aiello & Wheeler 1995). That the extra energy from a reduced gut was devoted to the brain and not to increasing reproductive output or some other reproductively valuable function can only be explained if brain size, and by inference intelligence, was under strong selective pressure.

Locomotion

Povinelli and Cant (1995) argued that great apes, as large-bodied arboreal primates, face unique challenges in negotiating arboreal substrates due to increased substrate unpredictability and compliance, and face severe costs of failing to support their body weight high in the trees. These conditions would pose selective pressures for especially flexible and complex mental calculations during locomotion that would have survival and therefore reproductive consequences, and could have resulted in selection for negotiating safer movement in an arboreal setting. This led, they propose, to the evolution of self-concept and its supporting mental representation capabilities in the great ape lineage. However, as noted by Begun, Chapter 2, Gebo, Chapter 17, Hunt, Chapter 10, and Russon, Chapter 1, this volume, large bodied arboreal hominoids can be small brained (*Oreopithecus*), self-concepts may occur in mainly terrestrial hominids (*Gorilla*), and travel on highly compliant branches with deliberate, slow, non-stereotypical clambering occurs in small primates as well (some prosimians).

Diet

Diet and body size are associated in primates (Clutton-Brock & Harvey 1977; Milton & LeMay 1976). Because larger animals tend to have relatively slower metabolic rates than smaller ones (Kleiber 1932), however, body size can affect the types and amounts of food in which a species will specialize. Very small primates are insectivorous, large ones are folivorous, and frugivores are typically intermediate in size (Kay 1984). When size and phylogenetic factors are controlled for, there is no set relation between diet and metabolism in primates, with folivores and frugivores often having similar metabolic rates (Elgar & Harvey 1987). There is also no correlation between encephalization and dietary quality or challenge, as measured by percentage of fruit in the diet (Ross, Chapter 8, this volume) or seasonality (Parker &

Gibson 1977, 1979), or between extractive foraging and neocortex size in primates (Barton & Dunbar 1997; Dunbar 1992). Identifying dietary features related to intelligence, however, may require more specific dietary measures (Ross & Jones 1999). Neither of these diet measures considers the particular form of frugivory in which great apes specialize, which is extended to include foods higher in protein and fat and non-fruit fallback foods on a seasonal basis to survive recurrent periods of fruit scarcity (Yamagiwa, Chapter 12, this volume). Even so, dietary pressures alone are unlikely to explain the evolution of enhanced intelligence in the great apes.

Social complexity

As a consequence of selection to cope with ecological pressures, most primates live in social groups (Wrangham 1980). Resource distribution affects the cost–benefit equation of living in groups, so dietary specializations can affect grouping size and patterns (Alexander 1974). Body size also affects social systems by altering susceptibility to predators, conspecific competition, resource availability and distribution, and habitat use.

The social brain hypothesis proposes that cognitive enhancements in anthropoid primates are associated with social complexity and is supported broadly across primates by comparative analyses (Barton & Dunbar 1997; Dunbar 1992; review in van Schaik et al., Chapter 11, this volume). These analyses typically find that group size and proxy measures for brain size (e.g., cranial capacity, neocortex ratios) are associated in a wide range of primates (e.g., Kudo & Dunbar 2001; Pawlowski, Lowen & Dunbar 1998; van Schaik & Deaner 2003). The social brain hypothesis as initially presented, however, fails to explain why primates with great-ape-like social systems, such as capuchins and macaques (Preuschoft & van Schaik 2000; Thierry, Wunderlich & Gueth 1989; Perry 2003), are not as intelligent as great apes or why great apes, with group sizes typical of other anthropoids, consistently show more complex cognition than all other anthropoids (in this volume, see Russon, Chapter 1). Closer examination, however, reveals that despite apparent social similarities, living great apes face more dynamic social problems than other nonhuman primates (van Schaik et al., Chapter 11, this volume), and so ancestral hominids may have been under stronger selective pressure to become better equipped for flexible

social problem-solving abilities (Dunbar 1996; Whiten 1997; van Schaik et al., Chapter 11, this volume). Why great apes are more complex socially has not been made clear by the social brain model.

Social complexity may be related to patterns of sexual dimorphism in body size. Males are selected to grow to large size in taxa for which size is an advantage in male–male competitions that affect mating success, and this makes for rigid social structures. In monogamous primates where mating competition is minimal (e.g., gibbons), or where male–male coalitions are a significant component of their competition (e.g., *Pan*, hominins, capuchins), body size dimorphism is reduced. Gibbons are not relatively more intelligent than other primates so reduced size dimorphism alone is not directly correlated with greater intelligence. In species in which decreased body size dimorphism is related to coalitionary behavior, however, the situation may be different. The social complexities of building and maintaining effective kin and non-kin coalitions, as documented among humans, male chimpanzees, and capuchin monkeys (Pawlowski et al. 1998; Wrangham 1999), may have selected for increased cognitive capacities (van Schaik et al., Chapter 11, this volume).

This particular combination of body mass and social factors may in part explain encephalization in *Pan* and *Homo*, but it does not explain the roughly equal levels of encephalization in *Gorilla*, *Pongo*, and probably *Dryopithecus*, *Sivapithecus*, and *Australopithecus*, all of which were strongly sexually dimorphic (Begun 2002; Kelley 2002; McHenry 1982). The combination of unusually complex coalitionary behavior and reduction of body size dimorphism may have happened independently in *Pan* and *Homo*, since *Australopithecus*, which is more closely related to *Homo*, lacks at least some of these features (Ward et al. 1999). While complex coalitionary behavior represents an aspect of social complexity that may select for intelligence, it is not the sole factor influencing selection for enhanced intelligence in hominids because encephalization preceded reduction in sexual dimorphism in hominids.

Life history

Body size is related to brain size via life history in several ways, in addition to easing metabolic constraints on brain growth as outlined above. Large body size tends to decrease extrinsic mortality by reducing susceptibility to predators (Williams 1957; and see recent reviews in van Schaik & Deaner 2003; van Schaik *et al.* Preuschoft & Watts, Chapter 11, this volume). Large bodies take longer to grow, and a longer growth period may favor relatively larger brains by prolonging brain growth and programming (Barton 1993; Kelley, Chapter 15, this volume; Ross, Chapter 8, this volume; van Schaik & Deaner, 2003). Slow life histories have been hypothesized to allow longer time for brain growth or the learning involved to become a successful adult in humans (Dobzhansky 1962; Hallowell 1963; Mann 1975) and nonhumans (Joffe 1997; van Schaik & Deaner 2003; and see Ross, Chapter 8, this volume). However, time to reproductive maturity is not tightly related to rate or timing of brain development. Primates with the longest juvenile periods (humans), complete most of their brain size growth in infancy, well before the most complex learning tasks are tackled (Pereira & Leigh 2003; Ross, Chapter 8, this volume). Thus, prolonged juvenility may allow for brain growth, brain maintenance, experiential learning, or all three, depending on the species. While life history is an important correlate of intelligence and body size, slowing life history alone will not automatically result in increases in brain size and encephalization, but will only provide a conditions necessary for doing so when there is a fitness advantage to increased intelligence (see Kelley, Chapter 15, this volume).

A SYNTHESIS TO EXPLAIN BRAIN–BODY SIZE RELATIONS IN THE HOMINOIDS

We suggest that body size and brain size co-evolved in significant but complex ways during hominid evolutionary history. Observed correlations between body and brain size are real. Allometry, however, does not signify a single universal constraint or scaling law. Instead, observed relations reflect a multi-factorial and often mutually reinforcing set of selective pressures. The specific allometric relation for each taxon depends on its phylogenetic history and its particular ecological and social circumstances. Considering only one or a subset of these circumstances will contribute to unsatisfying explanations for the relation between body and brain sizes.

Observed brain–body size scaling relations in hominids, as in other primates and non-primate mammals, result from parallel selection on both brain size and body size. Because selection for body size is related

to selection for many other aspects of a species' biology, such as metabolism, diet, habitat, life history, and social behavior, selection can produce similar combinations of traits. Situations favoring increased intelligence are often similar to those favoring increased body size. This would produce correlations independent of direct causal relations. Because closely related taxa share other adaptations that can affect and be affected by size, and these sets of adaptations often co-evolve, common patterns across taxa could result in the general relations generated by allometric analyses. One would not expect all taxa to share exactly the same relations, given different selective pressures and adaptive constraints faced by each. Species therefore should vary about a statistically derived line (as in Figure 18.1). Only by elucidating patterns of selection shaping many parts of a species' biology and behavior can we hope to determine these relations and predict why and how variables are interrelated, and hence why observed scaling relations occur.

Selection pressures for enhancing cognition derive from situations that require increased flexibility and complexity in behavior and problem-solving (Geary & Huffman 2002). They concern biotic more than abiotic situations because the former are generally more variable, complex, and unpredictable. Broadly speaking, the most challenging may be predator–prey interactions and dynamic situations within social groups (Geary & Huffman 2002; West-Eberhard 2003). The more complex these become, the more complex and flexible cognition must be. Extant hominids face the most complex foraging challenges and the most sophisticated social interactions and relationships known in nonhuman primates (see many contributions in this volume)

Body size affects the cost–benefit ratio of evolving enhanced cognitive capacities by affecting susceptibility to predators and conspecific competitors, as well as diet, habitat use, the social system broadly, and life history, it also alters physical influences on brain size. Body size is associated with ecological dominance (Alexander 1989, 1990), a situation in which Darwin's traditional hostile forces of nature (predation risk, food shortages, disease, and climate) decrease in their effects on differential reproduction relative to competition with conspecifics. Ecological dominance is accomplished in different ways by different species, but large body size is a common avenue. It represents a gradient, with some taxa being more ecologically dominant than others. An increase in body size reduces susceptibility to predation and lowers metabolic rate, potentially increasing ecological dominance, as well as relaxing energetic constraints on encephalization. Increases in intelligence can also increase ecological dominance, as they render individuals better able to locate and obtain food resources, evade predators, and otherwise modify their environments. The relative reduction in differential reproduction due to decreased extra-specific costs also effectively increases the fitness value of sophisticated social problem-solving abilities, in species for which sociality is most relevant to reproductive success.

This spiral of ecological dominance and increased social competition may have contributed to the evolution of the human grade of cognitive abilities (Alexander 1990; Flinn, Ward & Geary in press). Examples of non-human species with relatively high ecological dominance include elephants, dolphins, orcas, sperm whales, lions, and the great apes. Intraspecific interactions have significant fitness effects on individuals in most primate species (Alexander 1990; Flinn *et al.* in press), providing an initial condition in which an increase in ecological dominance will increase social competition and lead to more intense intra-specific arms races in social intelligence.

When social competition has significant fitness effects, relatively intelligent individuals who are able to negotiate their social and environmental settings better then their less cognitively sophisticated conspecifics stand to achieve higher net fertility. If a species' social and physical environments are such that greater intelligence does not have significant fitness benefits, then large brains are not expected. Examples of long-lived, relatively large, relatively asocial, but not particularly encephalized species include Galapagos tortoises and rhinoceroses. One apparent exception to this rule, orangutans, who are often characterized as asocial yet highly intelligent, are actually more social than often supposed and show social complexity comparable to other great apes (see van Schaik *et al.*, Chapter 11, this volume); they also share other key cognitive challenges with other great apes, such as especially complex foraging problems (see Yamagiwa, Chapter 12, this volume).

In terms of the model proposed here, *Oreopithecus* may be an exception that proves the rule. *Oreopithecus* probably was highly folivorous (Singleton, Chapter 16, this volume) and insular, and probably experienced little ecological competition or predator pressure due to its island habitat (Harrison & Rook 1997). Although it fits the large size–low predation pattern, its folivorous diet would have made it difficult to obtain adequate caloric and other nutrient resources to maintain a large brain. This and its comparatively unchallenging ecology would have made a large brain an attribute that it neither needed nor could afford, resulting in selection for a smaller brain, and correspondingly reduced cognitive abilities. Outside of primates, river dolphins and male angler fish are other examples suggesting that evolution can act to diminish brain size in the absence of positive selective pressures.

Most anthropoid primates tend to be frugivorous and experience social competition, although some taxa have undergone stronger selective pressure to negotiate more complex social systems than others. Great apes, because of their size and largely frugivorous diets, live in societies that tend to especially flexible fission–fusion with relatively high subordinate leverage and complex non-kin social relations that can affect social and therefore reproductive success (review in van Schaik *et al.*, Chapter 11, this volume). This social complexity could favor enhanced cognitive abilities, and presumably brain size, until these increases are in turn constrained by other factors, and individuals are then selected to allocate energy to other efforts, such as parental effort. This arms race is species specific, because different ecological conditions and phylogenetic histories affect different species, and it explains phylogenetic differences in scaling patterns. Capuchin monkeys may share many aspects of their social system with chimpanzees, but a capuchin is only selected to out-compete other capuchin monkeys. It does not have to be as intelligent as a chimpanzee, reflecting its different phylogenetic heritage. The immediate ancestor of chimpanzees was already more encephalized than capuchins, and presumably more socially complex. Differences in such evolutionary starting points of intra-specific arms races, coupled with other constraints on different taxa, affects their ultimate trajectories.

The multiple covariates of selection may explain the lack of a tight correlation with social complexity and brain size. Because competition is relative to species, one should not predict equivalence in encephalization (i.e., EQ or neocortical index) or intelligence between taxa as mediated solely by social systems (e.g., Pawlowski *et al.* 1998; Preuschoft & van Schaik 2000; van Schaik *et al.*, Chapter 11, this volume). Instead, among close phylogenetic relatives, we should see more socially complex species having relatively larger brains (or neocortices and associated structures). Living catarrhines are generally more encephalized than platyrrhines and tend to have more complex social systems, though the most encephalized platyrrhines share some complex social features with cercopithecids. Among catarrhines, papionins are generally more encephalized than other cercopithecines, and hominids are more encephalized than hylobatids, after accounting for body mass (Gibson, Rumbaugh & Beran 2001; Begun & Kordos, Chapter 14, this volume). Generally, their higher encephalization levels are associated with greater social complexity, with levels of social complexity broadly tracking these encephalization differences (e.g., Dunbar 1996).

The neocortex is the primary site of learning and higher level cognitive processing, although other components such as the amygdala have supportive functions (Adolphs 2003; Siegal & Varley 2002). The cerebellum is also important, appearing to coordinate with the cortex to produce complex cognition (Rilling & Insel 1998; and see MacLeod, Chapter 7, this volume). The neocortex and the cerebellum are the two largest regions of the primate brain (MacLeod, Chapter 7, this volume); the cerebellum is disproportionately enlarged in apes over other nonhuman anthropoids and both enlarge at greater rates relative to the brain as a whole than more conservative components (see MacLeod, Chapter 7, this volume). Expansion of the brain to achieve enhanced cortical and cerebellar function would result in greater increases in overall brain size than would expansion driven by the functions of other regions. Doubling the neocortex results in a larger brain than doubling the hippocampus, for example. This is an important reason for the generally high association between behavioral complexities and brain size, with both social and ecological problems being important sources for these complexities.

In summary, particular combinations of diet, life history, social system, intelligence, and body size are likely to co-evolve, resulting in broad allometry between body and brain size. Some combinations appear unlikely. Large brains are costly for small-bodied primates, which are usually under selection for a high reproductive rate and fast life history due to high extrinsic mortality rates. Small primates are more likely to rely on insects for food, and coupled with high predation risks, this results in increased costs of grouping, and thus solitary life or small groups. Similarly, large primates are not expected to be relatively small brained. For primates, large size reduces predation risk, enabling flexibility in foraging party size. In great apes, even the comparatively solitary orangutan, it enables unusually flexible fission–fusion societies with high subordinate leverage and complex kin and non-kin interactions, all of which require exceptional cognitive sophistication (van Schaik *et al.*, Chapter 11, this volume). Foraging patterns in great apes also tend to be especially complex. During hominoid evolution, constraints lifted by increasing body mass, combined with concomitant increases in ecological dominance in inherently social species, contributed to selecting for increased social and cognitive complexity.

IMPLICATIONS FOR THE EVOLUTION OF HOMINID INTELLIGENCE

Our ability to infer the cognitive capacities of fossil primates depends on the assessment of brain size, body size, dimorphism, diet, life history, and social system. The evolution of body mass in fossil apes is somewhat difficult to assess given uncertainties in determining phylogenetic relations of some taxa, and the diverse range of sizes of Miocene apes. Extant great apes range from about 33 to 170 kg in body mass (Smith & Jungers 1997). Basal catarrhines were considerably smaller, with propliopithecids (including *Aegyptopithecus zeuxis*) ranging from 5 to 7 kg (Fleagle 1999). Thus, it is likely that hominoids evolved from fairly small-bodied ancestors.

Proconsul, a stem hominoid with no direct evolutionary relation with extant apes (Begun, Ward & Rose 1997), ranged in size from about 9 to 60 kg (Table 18.1). Other apparently stem hominoids (*Afropithecus*, *Morotopithecus*) are also within this range, though toward the upper end. While a few possible stem hominoids (e.g., *Micropithecus*) are as small or smaller than gibbons, most stem hominoids are larger than siamangs, and it is likely that hylobatids are phylogenetic dwarfs (Begun, Chapter 2, this volume). This range does not follow any temporal or spatial patterning, however, and no trends are readily apparent. Among extant hominoids and their fossil relatives, only hylobatids are less than 20 kg in body mass. *Dryopithecus*, suggested to share a particularly close phylogenetic relation with hominids, is known from four species that all tend to be slightly smaller than chimpanzees in size (Begun 2002). Their 25–45 kg range is the likely ancestral condition for African hominids, as australopithecine females also fall within this range. This is interesting, as *Pan* female body mass means range from 33.2 to 45.8 kg (Smith & Jungers 1997), suggesting that loss of significant body mass dimorphism in *Pan* may have involved females increasing size

Table 18.1. *Body mass estimates for fossil hominoids discussed in this chapter*

	Body mass (kg)		
	Males	Females	Evidence
Proconsul heseloni	?	10	Dental, cranial & postcranial
Proconsul nyanzae	35	15	Dental, palatal & postcranial
Micropithecus clarki	?	3.5	Dental and palatal
Afropithecus	35	?	Dental, facial & postcranial
Morotopithecus	54	?	Dental, facial & postcranial
Dryopithecus laietanus	35	20	Dental, cranial and postcranial
Dryopithecus brancoi	40	25	Dental, cranial and postcranial
Sivapithecus punjabicus	40	20	Dental, cranial and postcranial
Sivapithecus parvada	60	?	Dental, postcranial
Oreopithecus	30	15	Dental, cranial and postcranial
Australopithecus[a]	70	31	Postcranial

Note: [a] *Australopithecus afarensis.*
Sources: Based on estimates from Fleagle (1999), Gebo *et al.* (1997), Harrison (1989), Jungers (1987), Leakey & Walker (1997), McHenry (1988), Ruff *et al.* (1989), Walker *et al.* (1993) and personal observations (which authors).

in addition to or even instead of males decreasing in size.

Most hominoid taxa, living and fossil, are primarily frugivorous, although different species had relatively higher dependence on leaves (Kay & Ungar 1997) and some show use of hard foods (Singleton, Chapter 16, this volume). All extant hominids have anatomical and behavioral adaptations for processing especially challenging foods, often used as fallback resources in times of primary food scarcity (Bryne, Chapter 3, Russon & Begun, Chapter 19, Yamagiwa, Chapter 12, this volume). Most fossil hominids also have anatomical indications of an enhanced ability to exploit fallback foods, either in the form of large or specialized anterior teeth or large, thickly enameled molars and robust jaws (Russon & Begun, Chapter 19, Singleton, Chapter 16, this volume).

Early Miocene stem hominoids were not suspensory like extant hominoids, though a possible case has been made for *Morotopithecus* (Gebo, Chapter 17, this volume). Among middle and late Miocene hominids *Dryopithecus* had a clearly extant hominoid-like below-branch adaptation conceivably associated with the shift to a great-ape-sized brain and intelligence (i.e., Povinelli & Cant 1995). *Sivapithecus*, however, did not have the same type of below-branch positional behavior

characteristic of extant great apes (reviews in Rose 1997; Ward 1997) and *Oreopithecus* was highly suspensory but small brained. This diversity suggests that locomotor pattern alone is not correlated in a straightforward manner with the evolution of intelligence.

The prolonged life histories and periods of immaturity characteristic of modern apes first appeared in the Miocene. The only basal hominoid for which evidence is available is *Proconsul heseloni*, which appears to have had a developmental trajectory, defined using timing of the eruption of the first molar, more like that of a hylobatid than a hominid (Beynon *et al.* 1998; Kelley 1997). Life history evolution seems to parallel the evolution of encephalization in hominoids. *Proconsul heseloni*, the only basal hominoid for which data are available, had a relative cranial capacity roughly like that of a similarly sized cercopithecids (Begun & Kordos, Chapter 14, this volume; Walker *et al.* 1983). *Afropithecus*, larger than *P. heseloni* and close in size to *P. nyanzae*, appears to have a delayed age of first molar eruption (Kelley & Smith 2003; Kelley, Chapter 15, this volume). Its brain size is unknown, but if similar to the similarly sized *P. nyanzae*, which is possible given the anatomy of the cranium, it may provide evidence than an extended life is a necessary but not a sufficient factor to account for brain size increases (see also

Kelley, Chapter 15, Russon & Begun, Chapter 19, this volume). *Dryopithecus* has a further delayed age of first molar eruption, a life history change correlated to increased brain size, and is known to have had a great-ape-sized brain (Begun & Kordos, Chapter 14, Kelley, Chapter 15, this volume). *Sivapithecus* also had a delayed age at first molar eruption, though no direct evidence of brain size exists in this otherwise well-known taxon (Kelley 1997, Chapter 15, this volume).

All living and fossil hominoids for which there are data available are highly sexually dimorphic in body mass except for hylobatids, *Pan*, and *Homo*, implying intense mate competition and some level of group complexity (Plavcan 2001; Yamagiwa, Chapter 12, this volume). This suggests that polygynous mating systems with fairly high levels of male–male competition for access to females represent the ancestral hominoid condition. Reduced size body mass dimorphism is associated with monogamy in hylobatids. *Pan* and *Homo* have independently reduced body mass dimorphism levels yet increased (perhaps both) or at least maintained (in *Pan*) significant levels of encephalization, suggesting that their male–male coalitionary behavior is associated with the dimorphism changes.

In addition to coalitionary behavior, chimpanzees, orangutans, and *Homo* share the traits of tool use and manufacture. If chimpanzees and orangutans are more intelligent than other apes, this would involve some as yet undetected brain attribute other than mass to account for cognitive differences, because brain mass alone does not distinguish among great apes, and no significant cognitive differences have been documented. This has profound implications for interpreting fossil hominin behavior and for the suitability of chimpanzees as a source of behavioral models of human evolution.

If *Pongo* and *Gorilla* are as intelligent as *Pan*, it may be that the presence of coalitions maintain and even reinforce encephalization in *Pan* and *Homo* but that other factors achieve the same end in other fossil and living hominids. For *Pongo* and *Gorilla* it could be foraging challenges, other social problems or, at least in the case of *Pongo*, very slow reproductive turnover. All great apes appear to share fission–fusion tendencies rendered more complex by the effects of large body size (increased social leverage, less rigid dominance, enhanced social tolerance), so complex social problems may simply manifest themselves in other ways. It is also the case that *Pan* shares dietary complexities with the other great apes associated with seasonal fruit scarcities, so shared

ecological pressures may be among the forces behind their encephalization. Once achieved, encephalization is likely to be maintained if social interactions remain important, although there is no reason a priori to believe that only one mechanism is involved.

In summary, the evolution of hominoid intelligence can best be studied by examining a combination of many types of data. The last common ancestor of hominoids was likely the size of a large cercopithecid, perhaps a baboon, with a similar life history and frugivorous diet. The hominid last common ancestor increased its brain size and body size, extended periods of its life history, and altered its diet. It also may have begun further restructuring its brain to improve cognitive function internally, leading to the more complex cortical structure, both internally and externally, of extant great apes (Adolphs 2003; McLeod, Chapter 7, this volume; Nimchinsky *et al.* 1999; Semendeferi & Damasio 2000). The increased ecological dominance resulting from large body mass resulted in social interactions having increased relative roles in determining individual reproductive success, resulting in selection for increased intelligence. This process tapered off somewhat through the late Miocene and early hominin evolution, when other constraints on cognitive abilities appear to have been reached (see Potts, Chapter 13, and Begun & Kordos, Chapter 14, this volume). The process of encephalization later took off again in *Homo*.

CONCLUSIONS

Complexities in brain–body size relations make predictions of brain size from body size and assessment of cognitive capacities from brain–body size ratio more complicated than once supposed. To track the evolution of intelligence in the fossil record, one cannot simply calculate EQ and have the whole story. However, recognition of the interrelations between body size, metabolism, ecological dominance, sociality, life history, diet, and other factors help explain previously enigmatic aspects of brain size and scaling relations within primates. With more complex models incorporating these other adaptive links, we can better explain variations in brain size, body size, and cognitive abilities among extant animals. If we can identify some of these other aspects of species' biology in the fossil record, we can then more accurately track changes in intelligence over evolutionary time.

Many of these factors have been identified as correlates of intelligence. Here, we suggest that the concepts of ecological dominance and intra-specific arms races in cognitive capacities (Alexander 1989) are important, yet hitherto unrecognized, phenomena. Ecological dominance alters selective pressures in regard to predation and to sociality. Given possible associations between body size, longevity, and diet on the one hand, and ecological dominance on the other, increased selective pressure for mental adaptations to a complex social and ecological environment may result in increased brain size.

The recognition of the importance of social competition for sophisticated cognitive capacities may explain some broad intertaxic scaling patterns, such as why platyrrhines and catarrhines with similar social systems are not similarly encephalized. Social competition is relative within a species, with individuals competing against conspecifics and not against an external factor. If levels of intelligence are reached as a consequence of social arms races, they are necessarily dependent on lineage history and phylogenetic starting points. Most primates, particularly haplorhines, are inherently social, and when ecological dominance is increased by reducing predation, increasing dietary quality, or changing other factors such as locomotion, social competition increases in relative importance for individual reproductive success. This produces within-species arms races in social skills that will continue until capped by other constraints, whether ecological, metabolic, or structural.

The evolution of body size in great apes influenced the evolution of great ape intelligence. Size decreased metabolic constraints on encephalization as it increased ecological dominance by reducing predation risk. It also led to longer life histories, which in turn favored increased cognitive capacities. All of these factors are interrelated, and feed back on one another. It is in this context that we are in an improved position to study how and why intelligence evolved in great apes.

ACKNOWLEDGMENTS

Carol Ward and Mark Flinn would like to extend their gratitude to David Begun and Anne Russon for the invitation to contribute to this volume. We all thank Anne Russon for her many careful and useful comments on an earlier draft of this paper.

REFERENCES

Aboitiz, F. (1996). Does bigger mean better? Evolutionary determinants of brain size and structure. *Brain, Behavior and Evolution*, **47**, 225–45.

Adolphs, R. (2003). Cognitive neuroscience of human social behavior. *Nature Reviews Neuroscience*, **4**, 165–78.

Aiello, L. & Wheeler, P. (1995). The expensive tissue hypothesis. *Current Anthropology*, **36**, 199–211.

Alexander, R. (1974). The evolution of social behavior. *Annual Review of Ecology and Systematics*, **5**, 325–83.

(1989). The evolution of the human psyche. In *The Human Revolution: Behavioral and Biological Perspectives on the Origins of Modern Humans*, ed. P. Mellars & C. Stringer, pp. 455–513. Princeton, NJ: Princeton University Press.

(1990). How did humans evolve? Reflections on the uniquely unique species. *University of Michigan Special Publication*, **1**, 1–38.

Armstrong, E. (1985a). Relative brain size in monkeys and prosimians. *American Journal of Physical Anthropology*, **66**, 263–73.

(1985b). Allometric considerations of the adult mammalian brain, with special emphasis on primates. In *Size and Scaling in Primate Biology*, ed. W. L. Jungers, pp. 115–46. New York: Plenum Press.

(1990). Brains, bodies and metabolism. *Brain, Behavior and Evolution*, **36**, 166–76.

Barton, R. A. (1993). Independent contrasts analysis of neocortial size and socioecology in primates. *Behavioral and Brain Sciences*, **16**, 469–95.

(1996). Neocortex size and behavioural ecology in primates. *Proceedings of the Royal Society of London B*, **263**, 173–7.

(1999). The evolutionary ecology of the primate brain. In *Comparative Primate Socioecology*, ed. P. Lee, pp. 167–97. Cambridge, UK: Cambridge University Press.

Barton, R. & Dunbar, R. (1997). Evolution of the social brain. In *Machiavellian Intelligence*, ed. A. Whiten & R. W. Byrne, pp. 240–63. Cambridge, UK: Cambridge University Press.

Barton, R. & Harvey, P. (2000). Mosaic evolution of brain structure in mammals. *Nature*, **405**, 1055–7.

Bauchot, R. & Stephan, H. (1966). Donnés nouvelles sur l'encéphalisation des insectivores et des prosimiens. *Mammalia*, **30**, 160–96.

(1969). Encéphalisation et niveau évolutif chez les simians. *Mammalia*, **33**, 225–75.

Begun, D. R. (2002). European hominoids. In *The Primate Fossil Record*, ed. W. Hartwig, pp. 221–40. Cambridge, UK: Cambridge University Press.

Begun, D. R., Ward, C. & Rose, M. (eds.) (1997). *Function, Phylogeny, and Fossils: Miocene Hominoid Evolution and Adaptations*. New York: Plenum Press.

Beynon, A. D., Dean, M. C., Leakey, M. G., Reid, D. J. & Walker, A. (1998). Comparative dental development and microstructure of *Proconsul* teeth from Rusinga Island, Kenya. *Journal of Human Evolution*, **35**, 351–70.

Clark, D., Mitra, P. & Wang, S.-H. (2001). Scalable architecture in mammalian brains. *Nature*, **411**, 189–93.

Clutton-Brock, T. & Harvey, P. (1977). Species differences in feeding and ranging behaviour in primates. In *Primate Ecology*, ed. T. H. Clutton-Brock, pp. 557–79. London: Academic Press.

    (1980). Primates, brains and ecology. *Journal of the Zoological Society of London*, **190**, 309–23.

de Waal, F. & Tyack, P. (eds.) (2003). *Animal Social Complexity: Intelligence, Culture and Individualized Societies*. Cambridge, MA: Harvard University Press.

de Winter, W. & Oxnard, C. (2001). Evolutionary radiations and convergences in the structural organization of mammalian brains. *Nature*, **409**, 710–14.

Deacon, T. W., 1997. What makes the human brain different? *Annual Reviews of Anthropology*, **26**, 337–57.

Delson, E., Terranova, C., Jungers, W., Sargis, E., Jablonski, N. & Dechow, P. (2000). Body mass in Cercopithecidae (Primates, Mammalia): estimation and scaling in extinct and extant taxa. *American Museum of Natural History, Anthropological Papers*, **83**, 1–159.

Dobzhansky, T. (1962). *Mankind Evolving: The Evolution of the Human Species*. New Haven, CT: Yale University Press.

Dunbar, R. (1992). Neocortex size as a constraint on group size in primates. *Journal of Human Evolution*, **20**, 469–93.

    (1993). Coevolution of neocortical size, group and language in humans. *Behavioral and Brain Sciences*, **16**, 681–735.

    (1996). *Grooming, Gossip and the Evolution of Language*. Cambridge, MA: Harvard University Press.

Dunbar, R. & Bever, J. (1998). Neocortex size predicts group size in carnivores and some insectivores. *Ethology*, **104**, 695–708.

Elgar, M. A. & Harvey, P. H. (1987). Basal metabolic rates in mammals: allometry, phylogeny and ecology. *Functional Ecology*, **1**, 25–36.

Finlay, B. & Darlington, R. (1995). Linked regularities in the development and evolution of mammalian brains. *Science*, **268**, 1578–84.

Finlay, B., Darlington, R. & Nicastro, N. (2001). Developmental structure in brain evolution. *Brain and Behavioral Sciences*, **24**, 263–78.

Fleagle, J. (1999). *Primate Evolution and Adaptations*. New York: Academic Press.

Flinn, M., Geary, D. & Ward, C. (in press.). The social competition model of human evolution. *Evolution and Human Behavior*.

Fox, J. & Wilczynski, W. (1986). Allometry of major CNS divisions: towards a reevaluation of somatic brain–body scaling. *Brain, Behavior and Evolution*, **28**, 157–69.

Geary, D. & Huffman, K. (2002). Brain and cognitive evolution: forms of modularity and functions of mind. *Psychological Bulletin*, **128**, 667–98.

Gibson, K., Rumbaugh, D. & Beran, M. (2001). Bigger is better: primate brain size in relationship to cognition. In *Evolutionary Anatomy of the Primate Cerebral Cortex*, ed. D. Falk & K. Gibson, pp. 79–97. Cambridge, UK: Cambridge University Press.

Harrison, T. & Rook, L. (1997). Enigmatic anthropoid or misunderstood ape? The phylogenetic status of *Oreopithecus bambolii* reconsidered. In *Function, Phylogeny, and Fossils: Miocene Hominoid Evolution and Adaptations*, ed. D. R. Begun, C. Ward & M. Rose, pp. 327–62. New York: Plenum Press.

Hallowell, A. I. (1963). The protocultural foundations of human adaptation. In *Social Life of Early Man*, ed. S. L. Washburn, pp. 236–55. Chicago, IL: Aldine.

Harvey, P. & Clutton-Brock, T. (1985). Life history variation in primates. *Evolution*, **39**, 559–81.

Harvey, P. & Krebs, J. (1990). Comparing Brains. *Science*, **249**, 140–5.

Hofman, M. (1982). Encephalization in mammals in relation to the size of the cerebral cortex. *Brain, Behavior and Evolution*, **20**, 84–96.

    (1983). Evolution of the brain in neonatal and adult placental mammals: a theoretical approach. *Journal of Theoretical Biology*, **105**, 317–22.

Jerison, H. (1973). *Evolution of the Brain and Intelligence*. New York: Academic Press.

Joffe, T. (1997). Social pressures have selected for an extended juvenile period in primates. *Journal of Human Evolution*, **32**(6), 593–605.

Kaas, J. (2000). Why brain size is so important: design problems and solutions as neocortex gets bigger or smaller. *Brain and Mind*, **1**, 7–23.

Kay, R. (1984). On the use of anatomical features to infer foraging behaviour in extinct primates. In *Adaptations for Foraging in Primates*, ed. P. S. Rodman & J. G. H. Cant, pp. 21–53. New York: Columbia University Press.

Kay, R. & Ungar, P. (1997). Dental evidence for diet in some Miocene catarrhines with comments on the effects of phylogeny on the interpretation of adaptation. In *Function, Phylogeny, and Fossils: Miocene Hominoid Evolution and Adaptations*, ed. D. R. Begun, C. V. Ward & M. D. Rose, pp. 131–51. New York: Plenum Press.

Kelley, J. (1997). Paleobiological and phylogenetic significance of life history in Miocene hominoids. In *Function, Phylogeny, and Fossils: Miocene Hominoid Evolution and Adaptations*, ed. D. R. Begun, C. V. Ward & M. D. Rose, pp. 178–208. New York: Plenum Press.

(2002). The hominoid radiation in Asia. In *The Primate Fossil Record*, ed. W. Hartwig, pp. 369–84. Cambridge, UK: Cambridge University Press.

Kelley, J. & Smith, T. (2003). Age at first emergence in early Miocene *Afropithecus turkanensis* and the evolution of life history in the Hominoidea. *Journal of Human Evolution*, 44, 307–29.

Kleiber, M. (1932). Body and size and metabolism. *Hilgardia*, 6, 315–53.

Kudo, H. & Dunbar, R. (2001). Neocortex size and social network size in primates. *Animal Behaviour*, 62, 711–22.

La Cerra, P. & Bingham, R. (1998). The adaptive nature of the human neurocognitive architecture: an alternative model. *Proceedings of the National Academy of Sciences*, 95, 11 290–4.

Lande, R. (1979). Quantitative genetic analysis of multivariate evolution applied to brain:body size allometry. *Evolution*, 33, 402–16.

Mann, A. E. (1975). Paleodemographic aspects of the South African Australopithecines. *University of Pennsylvania Publications in Anthropology*, 1, 1–171.

Marino, L. (2002). Convergence of complex cognitive abilities in cetaceans and primates. *Brain, Behavior and Evolution*, 59, 21–32.

Marino, L., Rilling, J., Lin, S. & Ridgway, S. (2000). Relative volume of the cerebellum in dolphins and comparison with anthropoid primates. *Brain, Behavior and Evolution*, 56, 204–11.

Martin, R. (1981). Relative brain size and basal mebolic rate in terrestrial vertebrates. *Nature*, 293, 57–60.

(1982). Allometric approaches to the evolution of the primate nervous system. In *Primate Brain Evolution: Methods and Concepts*, ed. D. Falk & E. Armstrong, pp. 39–56. New York: Plenum Press.

(1983). *Human Brain Evolution in an Ecological Context*. 52nd James Arthur Lecture on the Evolution of the Human Brain. American Museum of Natural History.

(1993). Allometric aspects of skull morphology. In *Theropithecus: The Rise and Fall of a Primate Genus*, ed. N. Jablonski, pp. 273–98. Cambridge, UK: Cambridge University Press.

Martin, R. & Harvey, P. (1985). Brain size allometry: ontogeny and phylogeny. In *Size and Scaling in Primate Biology*, ed. W. L. Jungers, pp. 147–73. New York: Plenum Press.

Matano, S. (2001). Brief communication. Proportions of the ventral half of the cerebellar dentate nucleus in humans and great apes. *American Journal of Physical Anthropology*, 114, 163–5.

McHenry, H. M. (1982). How big were early hominids? *Evolutionary Anthropology*, 1, 15–20.

Milton, K. & LeMay, M. (1976). Body weight, diet and home range in primates. *Nature*, 259, 459–62.

Nimchinsky, E., Gilissen, E., Allman, J., Perl, D., Erwin, J. & Hof, P. (1999). A neuronal morphologic type unique to humans and great apes. *Proceedings of the National Academy of Sciences*, 96, 5268–73.

Ottoni, E. & Mannu, M. (2003). Spontaneous use of tools by semifree-ranging capuchin monkeys. In *Animal Social Complexity: Intelligence, Culture and Individualized Societies*, ed. F. de Waal & P. Tyack, pp. 440–3. Cambridge, MA: Harvard University Press.

Pagel, M. & Harvey, P. (1988). The taxon-level problem in the evolution of mammalian brain size: facts and artifacts. *American Naturalist*, 132, 344–59.

(1989). Taxonomic differences in the scaling of brain on body weight among mammals. *Science*, 244, 1589–93.

Parker, S. T. (1996). Apprenticeship in tool-mediated extractive foraging: the origins of imitation, teaching and self-awareness in great apes. In *Reaching into Thought: The Minds of the Great Apes*, ed. A. E. Russon, K. A. Bard & S. T. Parker, pp. 348–70. Cambridge, UK: Cambridge University Press.

Parker, S. T. & Gibson, K. R. (1977). Object manipulation, tool use and sensorimotor intelligence as feeding adaptations in cebus monkeys and great apes. *Journal of Human Evolution*, 6, 623–41.

(1979). A developmental model of the evolution of language and intelligence in early hominids. *Behavioral and Brain Sciences*, 2, 364–408.

Passingham, R. (1975). The brain and intelligence. *Brain, Behavior and Evolution*, 11, 1–15.

Paulin, M. (1993). The role of the cerebellum in motor control and perception. *Brain, Behavior and Evolution*, **41**, 39–50.

Pawlowski, B., Lowen, C. & Dunbar, R. (1998). Neocortex size, social skills and mating success in primates. *Behaviour*, **135**, 357–68.

Pereira, M. E. & Leigh, S. R. (2003). Modes of primate development. In *Primate Life History and Socioecology*, ed. P. M. Kappeler and M. E. Periera, pp. 149–76. Chicago, IL: University of Chicago Press.

Perry, S. (2003). Case Study 4a: coalitionary aggression in white-faced capuchins. In *Animal Social Complexity: Intelligence, Culture and Individualized Societies*, ed. F. de Waal & P. Tyack, pp. 111–14. Cambridge, MA: Harvard University Press.

Plavcan, J. (2001). Sexual dimorphism in primate evolution. *Yearbook of Physical Anthropology*, **44**, 25–53.

Povinelli, D. & Cant, J. (1995). Arboreal clambering and the evolution of self-conception. *Quarterly Review of Biology*, **70**, 393–421.

Preuschoft, S. & van Schaik, C. P. (2000). Dominance and communication: conflict management in various social settings. In *Natural Conflict Resolution*, ed. F. Aureli & F. B. M. de Waal, pp. 77–105. Berkeley, CA: University of California Press.

Preuss, T. (2001). The discovery of cerebral diversity: an unwelcome scientific revolution. In *Evolutionary Anatomy of the Primate Cerebral Cortex*, ed. D. Falk & K. Gibson, pp. 138–64. Cambridge, UK: Cambridge University Press.

Purves, T. M. (1994). *Neural Activity and the Growth of the Brain*. Cambridge, UK: Cambridge University Press.

Rakic, P. (1988). Specification of cerebral cortical areas. *Science*, **241**, 170–6.

　(1995). A small step for the cell, a giant leap for mankind: a hypothesis of neocortical expansion during evolution. *Trends in Neurosciences*, **18**, 383–8.

Read, A. & Harvey, P. (1989). Life history differences among the eutherian radiations. *Journal of Zoology, London*, **219**, 329–53.

Rilling, J. & Insel, T. (1998). Evolution of the cerebellum in primates: differences in relative volume among monkeys, apes and humans. *Brain, Behavior and Evolution*, **52**, 308–14.

　(1999). The primate neocortex in comparative perspective. *Journal of Human Evolution*, **37**, 191–223.

Rilling, J. & Seligman, R. (2002). A quantitative morphometric comparative analysis of the primate temporal lobe. *Journal of Human Evolution*, **42**, 505–33.

Rose, M. (1997). Functional and phylogenetic features of the forelimb in Miocene hominoids. In *Function, Phylogeny, and Fossils: Miocene Hominoid Evolution and Adaptations*, ed. D. R. Begun, C. V. Ward & M. D. Rose, pp. 79–100. New York: Plenum Press.

Ross, C. & Jones, K. E. (1999). The evolution of primate reproductive rates. In *Comparative Primate Socioecology*, ed. P. C. Lee, pp. 73–110. Cambridge, UK: Cambridge University Press.

Russon, A. (1998). The nature and evolution of intelligence in orangutans (*Pongo pygmaeus*). *Primates*, **39**, 485–503.

Sawaguchi, T. (1997). Possible involvement of sexual selection in neocortical evolution of monkeys and apes. *Folia Primatologica*, **68**, 95–9.

Semendeferi, K. & Damasio, H. (2000). The brain and its main anatomical subdivisions in living hominoids using magnetic resonance imaging. *Journal of Human Evolution*, **38**, 317–22.

Siegal, M. & Varley, R. (2002). Neural systems involved in "Theory of Mind." *Nature Reviews Neuroscience*, **3**, 463–71.

Smith, R. J. & Jungers, W. L. (1997). Body mass in comparative primatology. *Journal of Human Evolution*, **32**, 523–59.

Stephan, H. (1972). Evolution of primate brains: a comparative anatomical investigation. In *The Functional and Evolutionary Biology of Primates*, ed. R. H. Tuttle, pp. 155–74. Chicago, IL: Aldine-Atherton.

Stephan, H., Frahm, H. & Baron, G. (1981). New and revised data on volumes of brain structures in insectivores and primates. *Folia Primatologica*, **35**, 1–29.

Thierry, B., Wunderlich, D. & Gueth, C. (1989). Possession and transfer of objects in a group of brown capuchins (*Cebus apella*). *Behaviour*, **110**, 294–305.

Tobias, P. (1971). The distribution of cranial capacity values among living hominoids. *Proceedings of the 3rd International Congress of Primatology, Zurich 1970*, **1**, 18–35.

van Schaik, C. & Deaner, R. (2003). Life history and cognitive evolution in primates. In *Animal Social Complexity: Intelligence, Culture and Individualized Societies*, ed. F. de Waal & P. Tyack, pp. 5–25. Cambridge, MA: Harvard University Press.

Vokaer, M., Bier, J., Elincx, S., Claes, T., Paquier, P., Goldman, S., Bartholome, E. & Pandolfo, M. (2002).

The cerebellum may be directly involved in cognitive functions. *Neurology*, **58**, 967–70.

Ward, C. (1997). Functional anatomy and phyletic implications of the hominoid trunk and forelimb. In *Function, Phylogeny, and Fossils: Miocene Hominoid Evolution and Adaptations*, ed. D. R. Begun, C. V. Ward & M. D. Rose, pp. 101–30. New York: Plenum Press.

Walker, A., Falk, D., Smith, R. & Pickford, M. (1983). The skull of *Proconsul africanus*: reconstruction and cranial capacity. *Nature*, **305**, 525–7.

Ward, C. V., Walker, A. & Leakey, M. G. (1999). The new species. *Australopithecus anamensis. Evolutionary Anthropology*, **70**, 197–205.

West-Eberhard, M. J. (2003). *Developmental Plasticity and Evolution*. Oxford: Oxford University Press.

Whiten, A. (1997). The Machiavellian mindreader. In *Machiavellian Intelligence II: Extensions and Evaluations*, ed. A. Whiten & R. W. Byrne, pp. 144–73. Cambridge, UK: Cambridge University Press.

Whiting, B. & Barton, R. (2003). The evolution of the cortico-cerebellar complex in primates: anatomical connections predict patterns of correlated evolution. *Journal of Human Evolution*, **44**, 3–12.

Williams, G. C. (1957). Pleiotropy, natural selection, and the evolution of senescence. *Evolution*, **11**, 398–411.

Wrangham, R. (1980). An ecological model of female-bonded primate groups. *Behaviour*, **75**, 262–300.

Wrangham, R. W., Jones, J. H., Laden, B., Pilbeam, D. & Conklin-Brittain, N. (1999). The raw and the stolen: cooking and the ecology of human origins. *Current Anthropology*, **40**(5), 567–94.

Zhang, J. & Sejnowski, T. (2000). A universal scaling law between gray matter and white matter of cerebral cortex. *Proceedings of the National Academy of Sciences*, **97**, 5621–6

Part IV
Integration

19 • Evolutionary origins of great ape intelligence: an integrated view

ANNE E. RUSSON[1] AND DAVID R. BEGUN[2]

[1]*Psychology Department, Glendon College of York University, Toronto*
[2]*Anthropology Department, University of Toronto, Toronto*

Among the great apes that once ranged the forests of the Old World, only four species survive. Their evolutionary history reveals a huge range of morphological and behavioral diversity, all of which must be considered successful adaptations in their own time. Some of these attributes (large brains, sclerocarp and hard-object feeding, frugivory, folivory, gigantism, terrestriality, and suspensory positional behavior) survive in modern great apes. Our questions are: what combination of behaviors and attributes characterized the ancestor of living great apes? what was the significance of this suite of features for cognition? and how did it arise in evolution? To that end, we offer our model of a distinct great ape cognition along with its biological underpinnings and environmental challenges, then attempt to trace the evolutionary origins of this ensemble of features.

COGNITION

All living great apes express a distinctive grade of cognition intermediate between other nonhuman primates and humans. Their cognition normally reaches rudimentary symbolic levels, where symbolic means using internal signs like mental images to stand for referents or solving problems mentally. It supports rudimentary cognitive hierarchization or metarepresentation to levels of complexity in the range of human 2 to 3.5 year olds, but not beyond (in this volume, see Blake, Chapter 5, Byrne, Chapter 3, Parker, Chapter 4, Russon, Chapter 6, Yamakoshi, Chapter 9).

Great apes' high-level cognitive achievements are generalized in that they manifest system wide and relatively evenly across cognitive domains (Russon, Chapter 6, this volume). Evolutionary reconstructions, however, have typically fixed on specific high-level abilities, singly or in combination, such as self-concept or intelligent tool use (see Russon, Chapter 1, this volume).

While the challenges these abilities address may have provided the evolutionary impetus to enhancing great ape cognition, evolutionary reconstructions have more to explain than these. No single ability, combination of abilities, or cognitive domain encompasses what sets great ape cognition apart. In the physical domain, great apes do use tools in ways that require their grade of cognition (Yamakoshi, Chapter 9, this volume) but they devise equally complex manual techniques (Byrne, Chapter 3, this volume) and solve equally complex spatial problems (Hunt, Chapter 10, Russon, Chapter 6, this volume). They show exceptionally complex social cognition in social routines, scripts, and fission–fusion flexibility, as well as in imitation, teaching, self-concept, perspective-taking, deception, and pretense (in this volume see Blake, Chapter 5, Parker, Chapter 4, Russon, Chapter 6, van Schaik et al., Chapter 14, Yamagiwa, Chapter 12). Their communication reaches rudimentary symbolic levels, even considering only strictly defined gestures and language (Blake, Chapter 5, this volume), as does their logico-mathematical cognition (e.g., analogical reasoning, classification, quantification) (e.g., Langer 2000; Thompson & Oden 2000). The latter has not figured in evolutionary reconstructions but perhaps it should. Enhanced logico-mathematical capacities offer important advantages; classification and quantification, for example, may aid in managing great apes' broad diets and social exchange (Russon 2002), and analogical reasoning may support limited cognitive interconnections (see below). Others have also emphasized generalized features of great apes' cognitive enhancements (in this volume, Byrne, Chapter 3, Parker, Chapter 4, Russon, Chapter 6, van Schaik et al., Chapter 11, Yamakoshi, Chapter 9). Features our contributors identify include regular, sequential plans of many actions, hierarchical organization, bimanual role differentiation, complex event representations, scripts and routines, and coordinating more components in solving a task.

Individual great apes can also interconnect abilities from different domains to solve a single problem or use one ability to facilitate another (Russon, Chapter 6, this volume). This is an important source of cognitive power because it enables solving multifaceted problems and boosts problem-specific abilities. It is not commonly recognized in great ape cognition but evidence for its role in exceptionally complex achievements in the wild, for example Taï chimpanzees' cooperative hunting, suggests that it should. It is important in evolutionary perspective because it is a plausible source of the "fluidity of thought" or "multiple intelligences working together" that stands out in humans. Its appearance in great apes ties well with evidence that some of their species-typical problems require coordinating abilities across cognitive domains, for example adjusting foraging strategies as social needs, feeding needs, and their interactions fluctuate (Yamagiwa, Chapter 12, this volume). It also speaks to claims that only humans have this capacity.

Great apes' cognitive achievements appear to be products of generative systems, i.e., systems that construct problem-specific cognitive structures to suit the particular challenges encountered. Their skills in stone nut cracking (Inoue-Nakamura & Matsuzawa 1997), language (Miles 1991; Miles, Mitchell & Harper 1996), and classification (Langer 1996) all show constructive processes. Models characterizing great ape cognition in terms of centralized constructive processes like hierarchization or hierarchical mental construction take this position (Byrne 1995; Gibson 1993). Hierarchization is especially important because hierarchical cognitive systems may be intrinsically generative (Gibson 1990; Rumbaugh, Washburn & Hillix 1996). Generativity helps explain several ostensibly anomalous features of great ape cognition that have incited debate – notably, achievement variability across individuals, tasks, rearing/testing conditions, and communities, and "atypical" abilities that emerge with special rearing. If great apes' cognitive systems are generative, these "anomalies" may simply be normal expressions of generative cognitive systems.

Development is a defining feature of primate cognition. Distinctive in great apes is prolonging cognitive development beyond infancy and emergence of their distinctively complex achievements during juvenility (Parker & McKinney 1999). Prolonged cognitive development probably relates to their more complex social and ecological challenges compared with other anthropoid primates (Byrne, Chapter 3, Parker, Chapter 4, van Schaik et al., Chapter 11, Yamagiwa, Chapter 12, Yamakoshi, Chapter 9, this volume) coupled with the longer time they need to grow their exceptionally large brains (Ross, Chapter 8, this volume). Great apes' enhanced cultural potential (e.g., more powerful social learning, greater social tolerance) is considered essential to their cognitive development, underlining how difficult these challenges must be. Even with larger brains and more time to learn, immature great apes need more sophisticated and extensive social support than other anthropoid primates.

Many cognitive features believed critical to hominin evolution are then shared by great apes, including symbolism, generativity, and cognitive fluidity as well as specific abilities like complex tool use and manufacture, mental representation of absent items, perspective taking, cooperative hunting, food sharing, and symbolic communication. While great apes share these features only to rudimentary symbolic levels, these achievements are significant comparatively. Rudimentary symbolism in particular has been taken as an exclusively human leap forward in cognitive evolution. If great apes share this capacity, however, it must have evolved with ancestral hominids.

BIOLOGICAL BASES OF GREAT APE COGNITION

The brain

Efforts to establish what in the brain confers high cognitive potential have focused on brain size because it predicts many other brain features (e.g., structures, gyrification, organization). The picture for great apes remains unclear because all available size measures are problematic as indices of cognitive potential and samples of great ape brains have typically been very small (see Begun & Kordos, Chapter 14, MacLeod, Chapter 7, Ross, Chapter 8, this volume). As larger samples are becoming available, within-species variation is appearing to be extensive, so the many published findings based on small samples must now be treated as suggestive. These limitations in mind, modern great ape brains suggest the following cognitive characterization.

Great apes brains appear to follow a distinctively "ape" design (MacLeod, Chapter 7, this volume). All apes, compared with other nonhuman anthropoids, show more complex cerebral convolutions and an aug-mented neocerebellum. The neocerebellum connects extensively with the cerebral cortex, and primarily through it the cerebellum contributes to cognitive pro-cesses such as planning complex motor patterns, visuo-spatial problem solving, and procedural learning. These cognitive processes support skills apes need as suspen-sory frugivores, for example spatial memory, mapping, and complex manipulation. A large sample of primate brains also suggests that apes may have disproportion-ately larger brains for their body size than other anthro-poids; this finding is tentative and runs counter to standard views, but it is consistent with these structural distinctions (see MacLeod, Chapter 7, Ross, Chapter 8, this volume). A distinctive ape brain is also consistent with apes' distinctive life histories: living apes have dis-proportionately prolonged immaturity with delay con-centrated in the juvenile period (Ross, Chapter 8, this volume); fossil hominoids may have shared this pattern (Kelley 1997, Chapter 15, this volume).

Great apes' higher cognitive potential over lesser apes, system wide, may well be a function of absolutely large brain size and its allometric effects on morphology. Large brains provide more "extra" neurons for cognition (Gibson, Rumbaugh & Beran 2001; Rumbaugh 1995). Lesser apes' brains resemble great ape brains morpho-logically but resemble typical anthropoid brains in abso-lute size (Begun, Chapter 2, this volume), and do not show these cognitive enhancements. Large brains are also more extensively interconnected; this may enable more complex cortical processing by enabling parallel processing and distributed networks, and so enhance problem solving via simultaneous processing in mul-tiple areas of the cortex and their connecting structures (Gibson 1990). This fits well with great apes' capacity for solving complex problems by interconnecting multiple cognitive structures.

Many specific brain features that distinguish great apes can also be explained by their brains' absolutely large size (e.g., greater lateralization, neocortex expan-sion, specialized areas). Even if these features owe prin-cipally to larger brain size, they can translate into impor-tant differences in cognitive potential. Brain structures that increase in size with increases in overall brain size do so at differential rates. Structures implicated in cog-nition (e.g., neocortex, cerebellum) typically increase at higher rates, so they come to represent a larger per-centage of the brain in larger-brained species. For this reason great apes have relatively larger neocerebellar

structures, magnifying the cognitive advantages of an ape cerebellum. This cerebellar advantage may con-tribute to handling the more severe tasks that great apes face as extremely large-bodied suspensory pri-mates. Large brain size also increases demands on cere-bral cortical connectivity that, in humans, may have favored neocortical reorganization towards lateraliza-tion and locally specialized functional units (Deacon 1990; Hopkins & Rilling 2000). Great ape brains, all weighing over 250 g, appear to be large enough to experi-ence similar effects: they show two specialized structures implicated in sophisticated communication, a planum temporale and spindle neurons of the anterior cingulate cortex, which are otherwise found only in humans. That the allometric effects of large brain size likely brought specialized structures along with greater interconnect-edness may be related to the co-occurrence of problem-specific and interconnected cognitive structures in great apes and humans.

Life histories

Life history traits are fundamental attributes of a species' biology that govern the pattern of maturation from con-ception to death (e.g., gestation period, age at weaning, maturation rate – age of female first reproduction, inter-birth interval, longevity). These traits typically occur in packages that fall roughly along a continuum of fast–slow rates of life. They correlate highly with body and brain size, but some taxa depart dramatically from the predicted life history–body size relationship. For their body sizes, primates have greatly protracted life his-tories with notably delayed maturation compared with most other mammals. Links between the brain and life histories may suggest broader biological factors associ-ated with high cognitive potential. Reasons for specific scaling factors are typically explored by assessing links among ecological, brain, and life-history features.

Anthropoid brain size is linked with delayed matu-ration, in particular prolonged juvenility. Anthropoids may then make tradeoffs against juvenile growth rates to support their large brains, diverting energy away from body growth to support the brain. Even after removing body size effects, juvenility appears to be further pro-longed relative to body size in apes. Great apes may do the same thing to a greater degree. Slower body growth probably affects juveniles, even though most primate brain growth occurs in infancy, because caregivers with-draw support at weaning (Ross, Chapter 8, this volume).

Juveniles' immature foraging skills and the slow rate at which great apes learn, added to withdrawal of caregiver nutritional subsidies, can only prolong the period in which their energy intake does not meet the energetic needs of supporting the brain and body growth. Especially in apes, prolonged juvenility may be best explained as an unavoidable but bearable cost imposed by large brains, rather than as directly adaptive (Ross, Chapter 8, van Schaik *et al.*, Chapter 11, this volume). No clear links occur between the brain and life history in great apes as a distinct group (Ross, Chapter 8, this volume).

Body size

There is no question about great ape body sizes – all are exceptionally large for primates – or about correlations between their large body size and their large brain size (Ward *et al.*, Chapter 18, this volume). Yet the reasons for this relationship are unresolved: direct cause–effect in one direction or the other, parallel adaptations to other selection pressures, or byproducts of selection on related factors.

Because brains scale to body size, ratios between the two have been used to index a species' "encephalization," the extent to which its brain has increased in cognitive potential, by assessing its enlargement beyond the size predicted by its body size. By these measures, great apes appear no more encephalized than other anthropoids: their brains are not relatively larger given their body size, even if they are absolutely larger (but see MacLeod, Chapter 7, this volume). This has prompted some to suggest that body size is the driving evolutionary adaptation and that great apes' large brains are mere side effects of their large bodies (e.g., see MacLeod, Chapter 7, this volume). Analyses that simply seek to "remove body size effects" implicitly take this view. Brain–body mass relationships are much more complex than such corrections suggest (Begun & Kordos, Chapter 14, Ward *et al.*, Chapter 18, this volume) and no acceptable method has yet been developed to apportion relative percentages of brain mass related directly to body mass and to selection for absolutely bigger brains.

ENVIRONMENTAL PRESSURES ON COGNITION

Establishing the function and evolution of complex cognition and its biological underpinnings involves exploring related behavioral challenges. Behavioral challenges affecting modern great apes are often used to suggest evolutionary selection pressures that may have shaped their cognitive enhancement. Their counterparts in evolutionary history are inferred from indirect indices, for example diet from dental morphology. Ecological challenges that primarily tap physical cognition include diet/foraging (Parker & Gibson 1979), diverse "technical" difficulties (Byrne 1997), and arboreality (Povinelli & Cant 1995). Social challenges, which tap both social and communicative cognition, involve both competition and cooperation (e.g., Byrne & Whiten 1988; Parker 1996; van Schaik *et al.*, Chapter 11, this volume). In light of our characterization of great ape cognition and contributions to this volume, we reconsider these challenges.

Ecological challenges

Food is considered a primary limiting ecological factor of primate populations because of its sparse distribution and anti-predator defenses (Yamagiwa, Chapter 12, this volume). Features considered to challenge cognition include eclectic frugivory, very large dietary repertoires and correspondingly large ranges, and essential "technically difficult" foods. Interest in difficult foods has focused on embedded foods, especially those that elicit tool use, but foods protected by other defenses such as barbs or noxious chemicals and obtained manually present comparable cognitive challenges (e.g., Byrne & Byrne 1991, 1993; Russon 1998; Stokes & Byrne 2001). The distribution of tool use in the wild (chimpanzees and orangutans) probably reflects opportunity and not differential hominid cognitive potential. Bonobos and gorillas can both use tools when opportunities arise.

Fallback foods on which great apes rely during fruit scarcities are often difficult to obtain. This may be especially true of the fallback foods on which orangutans and chimpanzees rely, some of which elicit use of foraging tools in the wild (Yamakoshi 1998; Yamagiwa, Chapter 12, this volume). Seasonal fruit scarcities also probably contribute to great apes' extremely broad dietary repertoires and their flexibility in using individual foods. Cognitively, the latter may require interpreting local indices of change to detect the availability of particular foods, given that great apes inhabit the tropics where seasonal change can be irregular. The last common ancestor (LCA) was also a generalized frugivore that may also have consumed hard foods needing preparation prior to ingestion and inhabited seasonal forest habitats that probably imposed periodic fruit scarcities. By implication, the same dietary pressures affecting modern great apes also affected the LCA: seasonality, dietary breadth, and the need for fallback foods.

Arboreal locomotion and navigation, two spatial problems, present extreme cognitive challenges to great apes because of their extremely large bodies and forest habitats. Navigating large ranges effectively and efficiently may require mapping skills sophisticated enough to calculate routes and distances mentally. Povinelli and Cant (1995) hypothesized that the great apes' work-it-out-as-you-go, non-stereotypic modes of arboreal locomotion, for example cautious clambering and gap crossing, require minds with the representational capacity to figure in the self. These "cognitive" positional modes are neither shared among nor unique to all living great apes, however (Gebo, Chapter 17, Hunt, Chapter 10, this volume). They are prominent in orangutans and lesser apes but not African great apes. They could have influenced great ape cognitive evolution if the LCA was a large arboreal clamberer but this is uncertain, perhaps even unlikely (Gebo, Chapter 17, this volume). Povinelli and Cant suggested *Oreopithecus* as a model of that ancestor, with the requisite large size and body plan for arboreal clambering. *Oreopithecus* was otherwise very unlike other hominids, however (e.g., folivorous versus frugivorous, unusually small brained), and probably represents an isolated adaptation to a refugium rather than great apes' ancestral condition (see Begun & Kordos, Chapter 14, Gebo, Chapter 17, Potts, Chapter 13, Singleton, Chapter 16, this volume). Even if arboreal locomotion demands complex cognition in orangutans, there is little to indicate that it does, or did, in the great ape lineage.

Social challenges

Primate social life is recognized as having high potential for cognitive complexity. It is puzzling about great apes that they use more complex cognition than other anthropoids to solve social problems, but the problems themselves are not obviously more complex. Their social unit sizes are well within the range of other anthropoids, their demographic composition is no more complex, and few if any more complex social phenomena are known (van Schaik *et al.*, Chapter 11, this volume). To add to the puzzle, great ape species differ widely in their social systems but are very similar in cognitive potential.

Van Schaik *et al.* propose social challenges in great apes that may help explain their enhanced social cognition: fission–fusion tendencies with individuals out of contact with conspecifics for lengthy periods and foraging females solitary; relatively high subordinate leverage leading to less rigid dominance and enhanced social tolerance; greater intrasexual bonds with non-kin, and extensive flexibility in social organization and affiliation. These are clearly shared by chimpanzees and orangutans, and perhaps by the other species. Most are consequences of large size and exceptionally slow life histories, which reduce vulnerability to predators, increase vulnerability to hostile conspecifics, increase the potential for contest competition (especially for females and in species unable to switch to high-fiber fallback foods), and favor non-kin bonding. They require more complex cognition to handle greater flexibility in social relations and interactions and in the interplay among a more complex array of labile factors (e.g., balance rivalry with interdependence, or social with predation or foraging pressures). Rejoining conspecifics after lengthy absences increases needs for sophisticated navigation, distance communication, and renegotiating relationships. Two examples of complex communication in wild great apes concern rejoining companions: tree drumming (Boesch & Boesch-Achermann 2000) and placing indicators of travel direction (Savage-Rumbaugh *et al.* 1996). Higher subordinate leverage, less rigid dominance, and enhanced social tolerance are likely to improve opportunities for social learning, cultural transmission, and more flexible use of eye contact (Yamagiwa, Chapter 12, this volume). Similar social complexities also occur in some monkeys (capuchins, some macaques), however, so alone they cannot explain the enhanced cognition seen in great apes.

Great ape sociality should be affected by diet because social groups must adjust to ecological conditions. Effects probably differ more in great apes than in other anthropoids because of great apes' broad, technically difficult, and seasonally varying diet (Yamagiwa, Chapter 12, this volume). Social foraging strategies during fruit scarcities, when dietary and social competition pressures are at their worst, expose these effects. Significant to cognition is that great ape foraging groups change as a function of food availability, although patterns differ between species depending in part on the preferred type of fallback food. This is consistent with suggestions that fission–fusion in *Pan* functions to allow flexibility in handling challenges that vary over time and space (Boesch & Boesch-Achermann 2000), great ape life allows and requires facultative switches between solitary and gregarious foraging (van Schaik *et al.*, Chapter 11, this volume), and ephemeral activity subgroups show exceptional flexibility relative to ecological conditions (Parker, Chapter 4, this volume). All great apes then share the challenge, as a normal circumstance, of complex problems wherein pressures from two distinct cognitive domains interact.

EVOLUTIONARY RECONSTRUCTIONS

The origin of great ape cognitive capabilities is to be found in the Miocene, when the great apes originated and diversified. Here, we examine the evidence of brain size and morphology, life history, body size, positional behavior, diet, and environment in ancestral hominoids as they relate to the evolution of great ape intelligence. Patterns are summarized in Table 19.1.

Ecology: habitat and diet

The local habitats of early Miocene hominoids were most likely warm, moist forests in tropical and subtropical zones that enjoyed low seasonality and climatic stability (Andrews, Begun & Zylstra 1997; Potts, Chapter 13, this volume). Soft fruit, their dietary mainstay (Singleton, Chapter 16, this volume), would have been available year-round, albeit patchily distributed spatially and temporally.

Hominid emergence in the late middle Miocene, 14–12 Ma, coincides with increasing climatic fluctuation, especially increasing seasonality (Potts, Chapter 13, this volume). This may have restricted soft fruit availability for several months annually, at least in some regions. The earliest Eurasian hominoid, *Griphopithecus*, which is more modern in dental anatomy than *Proconsul*, shows for the first time a fully developed suite of masticatory characters indicative of hard-object feeding (Güleç & Begun 2003; Heizmann & Begun 2001; Singleton, Chapter 16, this volume). The ability of the ancestors of hominids to exploit hard objects may have allowed their expansion into Eurasia at the end of the early Miocene, as a way of avoiding competition with the many frugivores making the same trip northward (Heizmann & Begun

2001). In later hominids, the ability to exploit these resources may have served as an important parachute during times of scarcity in more seasonal environments when soft fruits, generally preferred by hominids, are more difficult to find. Greater seasonality is indicated in both Europe and Asia in the late Miocene, suggesting fruit scarcities with hard objects serving as fallback foods in some taxa. *Sivapithecus* is often reconstructed as having had an essentially soft fruit diet based on microwear (Teaford & Walker 1984), although morphologically it shared many features with hard-object feeders (thick enamel, low, rounded cusps, large molars, thick, massive mandibles), suggesting an ability to exploit hard objects when needed. *Dryopithecus* was not a hard-object feeder and may have lived in less seasonal environments than *Sivapithecus* (Andrews *et al.* 1997; Begun 1994; Singleton, Chapter 16, this volume; Potts, Chapter 13, this volume). However, seasonality was probably greater in environments inhabited by *Dryopithecus* than in most early Miocene hominoid environments, and evidence of the anterior dentition suggests enhanced abilities for pre-ingestive processing of embedded foods (Begun 1992). Either way, late Miocene hominids probably extended their frugivory with fallback foods during fruit scarcities. Their large body size may also represent a response to increased seasonality because it enhances energy-storing capacities for surviving periods of fruit scarcity (Knott 1998; Yamagiwa, Chapter 12, this volume). Living great apes show similar dietary breadth. Species differ in how they adjust to fruit scarcities, but all share the overall pattern of relying on fallback foods. Orangutans and chimpanzees use "hard" fallback foods (e.g., embedded, barks, pith), perhaps analogous to *Sivapithecus*, and gorillas and bonobos lean to folivory (although bonobos appear to enjoy especially rich habitats abundant with THV, which may or may not serve as fallback foods), possibly more similar to the *Dryopithecus* strategy. These environmental pressures and species traits imply considerable cognitive–behavioral adaptation, all in the direction of increased flexibility or adaptability (Potts, Chapter 13, this volume).

The latest Miocene experienced cooling, drying, and more pronounced seasonality, causing a worldwide shift from moist, warm forest to drier, open grassland and a corresponding shift in vegetation (Cerling *et al.* 1997; Potts, Chapter 13, this volume). Effects on hominids' preferred habitats, moist warm forests, include shrinkage, fragmentation, and retreat. Preferred

Table 19.1. *Selection pressures and biological traits in modern great apes and ancestral hominoids*

Epoch	Ma	Species	Brain[4,5,8] Size (g)	Morphology	Life history	Body Size (kg-f)	Anatomy[3,7]	Diet[2]	Environments Local	Regional & global	Sociality Sexual dimorph[8]	Organization
Modern												
		Chimpanzee	325	GA	DM PJ	34–60	Semi-terrestrial knucklewalk & climb	Fruit + hard fallback + omnivory	F–W to O, S+, T	High seasonal fruit scarcity	Low	(f–f+)
		Bonobo	314	GA	DM PJ	33–45	Mixed knucklewalk & climb	Fruit + THV	F, S–, T		Low	(f–f+)
		Gorilla	426	GA	DM PJ	71–175	Semi-terrestrial knucklewalk & climb	Fruit + folivory fallback, folivory	F, S+, T		High	(f–f+)
		Orangutan	288	GA	DM PJ	36–79	Suspensory clamber & climb	Fruit + hard fallback	F, S+, T		High	(f–f+)
Early Miocene	20.6	*Moratopithecus* (E. Africa)				30–35?	?Suspensory primitive brachiate & climb	Soft fruit		Less seasonal fluctuation	High	
	20–18	*Proconsul* (E. Africa)	146	H[4]	(DM)	10–15	Pronograde arboreal quadruped & climb	Soft fruit	F–W, S–		High	
	18–17	*Afropithecus* (E. Africa)			(DM)	25–30	Pronograde arboreal quadruped & climb	Hard object	F–W, S–		High	

Table 19.1. (cont.)

Epoch	Ma	Species	Brain[4,5,8] Size (g)	Brain[4,5,8] Morphology	Life history	Body Size (kg-f)	Body Anatomy[3,7]	Diet[2]	Environments Local	Environments Regional & global	Sociality Sexual dimorph[8]	Sociality Organization
Middle to Late Miocene												
16.5–15		Griphopithecus* (W Asia, C Europe, E Africa)				20–25	Pronograde arboreal quadruped & climb	Hard object	W–OP, S–/S+, ST	Increasing fluctuation	High	
15.5–14		Kenyapithecus* (E Africa)				30	Pronograde arboreal quadruped & climb	Hard object	W+GP, S+, T, dry	Increasing fluctuation + cooling	High	
12–9		Dryopithecus (Europe)	289	GA?	DM[5,8]	20–30	Suspensory brachiate & climb	Soft fruit, eclectic	F+W/SW, S+, ST	Seasonal fruit scarcity	High	
12.3–7		Sivapithecus (S Asia)			DM[5]	20–40	Suspensory arboreal or semi-terrestrial quadruped	Fruit + hard object	F+W, S+, warm	Seasonal fruit scarcity	High	
10–9.5		Ouranopithecus (S Europe)				35?	Partly terrestrial?	Hard object specialist	F to O, S+	Seasonal fruit scarcity	High	
8–7		Oreopithecus (S Europe)	112			15	Suspensory brachiate & climb	Folivory	F–SW, S+, moist, ST	Insular	High	

Notes:

Epoch: Early Miocene (22–17 Ma), Mid Miocene (17–11.5 Ma), Late Miocene (11.5–5.5 Ma).

Species: underlined, good hominid candidate; *, possible hominid candidate.

Brain: Size (for females, where available): weight in grams; (GA), estimated weight in the modern great ape range; *Morphology:* H, hominoid; GA, great ape (hominid); ?, unknown.

Body size: size (for females, where available): size > 25 kg is interpreted as within the hominid range.

Life history: DM, delayed maturation; PJ, prolonged juvenile period; bracketed values are estimates.

Habitat: *Type:* F, forest; FP, forest patches; W, Woodland; WG, wooded, more grassy; GP, grassy patches; SW, swamp; O, open; OP, open patches; *Seasonality:* S–, low; S+, higher; *Zone:* T, tropical; ST, subtropical.

Social: *Sexual dimorphism:* High (roughly 2:1 M:F); Low (closer to 1:1 M:F); *Organization:* f–f+ = flexible fission–fusion.

Shading highlights significant traits in living great apes and their appearance in the fossil hominoid record.

Sources: 1, Potts, Chapter 13; 2, Singleton, Chapter 16; 3, Gebo, Chapter 17; 4, Begun & Kordos, Chapter 14; 5, Kelley 1997, Chapter 15; 6, MacLeod, Chapter 7; 7, Hunt, Chapter 10; 8, Begun, Chapter 2; 9, Ward, Flinn & Begun, Chapter 18 this volume.

foods would have been available in smaller patches and more dispersed in distribution, an ecological situation less suitable to large-bodied hominid foragers – especially in groups. Hominid presence is indicated in isolated moist forest and swamp refuges and forest–open woodland mosaics, suggesting they tracked their preferred habitats where possible. Overall, their presence was increasingly restricted southward (Begun 2001, 2002; Harrison & Rook 1997; Potts, Chapter 13, this volume).

Fewer hominids are known in the fossil record of the late Miocene after the last occurrence of *Dryopithecus* and *Sivapithecus* between about 9.5 and 7 Ma. They appear to have become extinct locally while their descendants may have moved south at this time (Begun 2001). In Europe, the most ecologically specialized hominids are known from this time. *Oreopithecus* from Tuscany had an exceptionally small brain, well-developed suspensory positional behavior and highly folivorous diet, while *Ouranopithecus* was among the largest of the Miocene hominids and had a specialized hard-food diet. It is also most likely during this time that the ancestors of the African apes and humans arrived in Africa and that gorillas shortly thereafter diverged from the chimpanzee–human clade. Recent evidence from Thailand suggests that orangutan ancestors may have first appeared in Southeast Asia at this time as well (Chaimanee *et al.* 2003). These patterns overall also suggest habitat tracking, i.e., maintaining established habitat and fruit preferences (Potts, Chapter 13, this volume).

In the Plio-Pleistocene, worldwide climate was marked by strong arid–moist and temperature oscillations, wider climatic fluctuation, instability, profound habitat variability, and arid–monsoon seasonality (Potts, Chapter 13, this volume). Hominids would have experienced increased episodic disturbance, intra-annual variability in food availability, and repeated forest contraction–expansion and fragmentation–coalescence. Predictable effects include impoverished habitat in size and quality, even greater variability in food availability and abundance, changing species communities, changing competitor and predator patterns, and variable population densities (Potts, Chapter 13, this volume). Plio-Pleistocene pressures likely led to further diversification of strategies to augment capacities for handling unpredictable habitat instabilities. Gorillas shifted towards folivory, especially for fallback foods, smaller ranges, and reduced foraging complexity. Chimpanzees shifted to greater omnivory, including increased meat consumption, and use of savanna habitats. Bonobos maintained forest habitats and increased THV consumption. Orangutans maintained earlier diets and remained in tropical moist forests of southeast Asia, which persisted in large blocks on Borneo and Sumatra until this century. Hominins became increasingly dependent on terrestrial resources and developed a variety of approaches (megadontia, tools) to maximize dietary breadth and ecological flexibility.

The brain

Great apes' distinctive brains seem be defined by their large absolute size and hominoid morphology. Reconstructing their evolutionary origins comes down to when and why these features evolved. This exercise remains hampered by the dearth of fossil material on ancestral apes, especially crania.

Proconsulids, early Miocene stem hominoids, were relatively unspecialized pronograde quadrupeds but were distinguished from primitive anthropoids by their large size, taillessness, powerful appendages, and brains with a few hominoid features (Begun & Kordos, Chapter 14, this volume; Kelley 1997, Chapter 15, this volume; Ward *et al.* 1991, Chapter 18, this volume). The early hominids, *Dryopithecus* in Europe and *Sivapithecus* in South Asia, are either known or supposed on indirect but solid grounds to have had brain sizes in the range of modern great apes; where known, their endocasts show greater resemblances to modern hominids than do *Proconsul* endocasts (Begun & Kordos, Chapter 14, this volume). *Sivapithecus* neurocrania are not known. For *Dryopithecus*, partial neurocrania yield brain size estimates at the low end in absolute size but at the high end relative to body mass compared with the ranges for modern great apes. The fact that great apes with brains ranging from 280 to 700 cc, or humans with brains ranging from 1000 to 2000 cc, have not been shown to differ in cognitive capacity could be taken to indicate that there is a loose causal relationship between brain mass and cognitive capacity (Kelley, Chapter 15, this volume). On the other hand, the fact that there is no overlap in brain mass between monkeys and great apes or between great apes and humans suggests that normal brain mass minima in each taxon represent thresholds for cognitive change beyond which cognition is not affected, until the next threshold is attained. If this is the case, and absolutely large brains are what generate great apes' grade of cognition, then the rubicon represented by *Dryopithecus* and the smallest extant great apes (280–350 cc) evolved in the late middle Miocene with *Dryopithecus* and *Sivapithecus*. The emergence of the hominid-sized brain is associated with increasing seasonality, seasonal fruit scarcities, and frugivorous diet enhanced with hard foods. Though there are indications of hominid-like cerebral reorganization in *Dryopithecus*, its endocast is distinct from that of extant hominids so it is not clear whether their brains provided equivalent cognitive potential. At a minimum however, the cognitive potential of late Miocene hominids spans the considerable gap between great apes and other nonhuman primates, probably coming closer to the former.

The Plio-Pleistocene is likely to have exerted further selection pressures on hominid cognition given its negative effects of great ape habitats. Brain size has not changed, but organizational differences between extant and Miocene hominids probably occurred at this time (Begun & Kordos, Chapter 14, Potts, Chapter 13, this volume). The most telling findings from the fossil record may be that (1) partial de-coupling of size and morphology is a common feature in the evolution of catarrhine brains, and (2) hominoid brain evolution is highly diverse, with reduction in some lineages and increases in others. Some lineages experience brain mass loss in connection with body mass reduction (e.g., *Hylobates*) or independent of body mass change (e.g., *Oreopithecus*). The pattern of brain size diversity in fossil great apes more closely matches broad patterns of diet than of size (Begun & Kordos, Chapter 14, this volume), especially frugivory extended (seasonally) with challenging fallback foods. Brain size has been surprisingly stable in hominid evolution until *Homo*, despite dramatic changes and diversity in body mass, diet, positional behavior, and ecological conditions. It may be that a hominoid brain size at least 250 g represents a rubicon that generates hominid levels of cognitive and behavioral complexity. Conversely, although large bodies do not always imply large brains in hominoids, large brains always co-occur with large bodies.

Body size and life history

Fossil hominids were predominantly large bodied but somewhat smaller than living great apes. The smallest *Dryopithecus* (female *D. laietanus* and *D. brancoi*) was probably smaller on average than the smallest living great apes, the smallest females possibly weighing about 20 kg (Begun, Chapter 2, Ward *et al.*, Chapter 18, this volume). The smallest *Sivapithecus*, female *S. punjabicus*, probably ranged from close to *Dryopithecus* in body size to as large as the smallest living hominids. Other clearly hominid taxa such as *Ouranopithecus*, other species of *Sivapithecus*, and *Lufengpithecus* are in the size range of large chimpanzees and small gorillas; so is *Morotopithecus*, though it is less clearly a great ape. The LCA was therefore almost certainly large compared with most primates. Hylobatids are small bodied, but this is probably a result of secondary reduction in size compared with the common hominoid ancestor (Begun, Chapter 2, this volume). The range of body sizes in the proconsulids is broad and overlaps with the hominids.

In addition to being the size of an extant great ape, *Sivapithecus* and *Dryopithecus* M1 emergence age estimates suggest life history prolongation roughly equivalent to that of modern great apes (Kelley 1997, Chapter 15, this volume). The stem hominoids *Proconsul* and *Afropithecus* may show the first signs of life history prolongation. *Proconsul* may have been intermediate between hominids and non-hominids in M1 emergence age (Kelley 1997, Chapter 15, this volume), although Begun & Kordos (Chapter 14, this volume) and Kelley (Chapter 15, this volume) both also find *Proconsul* to be equivalent to *Papio* in M1 emergence and brain size. *Afropithecus* may have been within the great ape ranges for M1 emergence and brain size (Kelley & Smith, 2003; Kelley, Chapter 15, this volume). However, in our view the poorly preserved neurocranium of *Afropithecus* tentatively suggests a somewhat smaller brain than in a similarly sized chimpanzee, which would be consistent with the lower end of the range of estimates of M1 emergence and brain size provided by Kelley (Chapter 15, this volume). Either way, the implication is that prolonged immaturity emerged with the hominoids but became more clearly prolonged as brain size increased with the first hominids, because of energetic constraints, social constraints, or both. There is likely a complex interrelationship among life history, body mass, and body size that has yet to be fully understood in vertebrates in general (Ward *et al.*, Chapter 18, this volume).

Sociality

Characterizing sociality in the LCA is a highly speculative exercise resting exclusively on indirect indices.

Several features of great ape sociality result from large size and exceptionally slow life histories, both of which characterize early hominids (many Chapters in this volume). If, as argued for living great apes, large size and slow life histories give impetus to these social features, then hominids should share them. All great apes but no lesser apes also share fission–fusion tendencies that are affected by fruit scarcities and fallback foods; early hominids likely experienced similar dietary pressures, so they too may have had a fission–fusion form of sociality.

The main influence on female sociality, food availability, depends on fallback foods in great apes (Yamagiwa, Chapter 12, this volume). In chimpanzees and orangutans, which rely on similar hard fallback foods, females restrict their social grouping during fruit scarcities and increase it during periods of abundance. In gorillas and bonobos, which have more folivorous fallback patterns, females grouping patterns remain more stable. The main influence on male sociality, access to females, is primarily shaped by sexual dimorphism in great apes. In highly dimorphic orangutans and gorillas, males tend to be solitary and corral females for mating. In less dimorphic chimpanzees and bonobos, males associate with one another via dominance ranking systems. In early hominids, challenging fallback foods co-occur with high sexual dimorphism (Begun, Chapter 2, Singleton, Chapter 16, Ward et al., Chapter 18, this volume), so their social systems may have resembled the orangutan's, perhaps in less dispersed form. Attributes include polygynous mating systems with solitary males attempting to monopolize multiple females or female ranges, male dominance based on size, and female associations waxing and waning with the seasons.

DISCUSSION

Stem hominoids lived in moist tropical forest habitat with low seasonality, and probably exhibited dedicated frugivory, social complexity commensurate with frugivory, polygynous social structures with relatively high male–male competition, life histories with somewhat prolonged immaturity, brains mostly of anthropoid size and design, and body mass somewhere in the range between monkeys and great apes (10–25 kg). From this starting point and considering the many factors discussed in this book, we suggest the following patterns and processes in the evolution of great ape intelligence (Figure 19.1).

Ecology

Compared with the first hominoids, the first well-known fossil hominids, *Dryopithecus* and *Sivapithecus*, inhabited middle to late Miocene moist tropical forests with greater seasonality, frugivory extended in the direction of challenging foods, polygyny/high male–male competition, life histories with prolonged immaturity and prolonged juvenility, and larger bodies and brains reaching into the modern great ape range. The greater seasonality combined with incorporation of hard or otherwise challenging foods in the diet suggests a dietary shift towards adding fallback foods requiring pre-ingestive preparation as diet supplements during fruit scarcities. Increased absolute brain size indicates increased cognitive potential. Altogether, this suggests that seasonality resulted in a more cognitively challenging diet that favored larger brains. Ecological pressures on hominids intensified under the increasingly seasonal and unpredictable conditions of the latest Miocene and Plio-Pleistocene. Their effects on cognitive evolution were perhaps constrained by habitat tracking, with the great apes adopting a more conservative ecological approach and the hominins exploiting more radically different environments.

Brain–Body–Sociality

In the anthropoid/hominoid phylogenetic context, hominid large brain and body size likely co-occurred with slow life histories, prolonged immaturity, lower predation risk, higher vulnerability to hostile conspecifics, stronger relations with non-kin, high subordinate leverage, and relaxed dominance. Which came first is neither interesting intellectually nor a useful question processually. We will never know, and these variables were probably a package as soon as they appeared in early hominids.

Socially, this package is consistent with unusually flexible fission–fusion tendencies and enhanced social tolerance (van Schaik et al., Chapter 11, this volume). The former would have favored larger brains for more complex social problem-solving; the latter may have further boosted cognition by enhancing conditions for socio-cultural learning. Some social intelligence models argue for an "arms race" in cognition, once cognitive solutions to social problems take hold, because competing successfully depends on outwitting increasingly savvy conspecifics (e.g., Ward et al., Chapter 18,

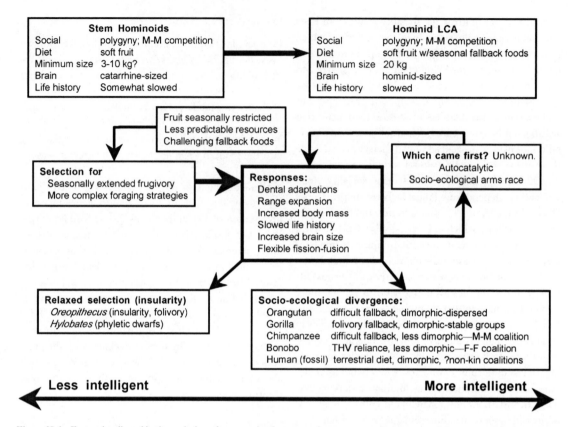

Figure 19.1. Factors implicated in the evolution of great ape intelligence. Early hominids are distinguished from early hominoids mostly by body and brain size and slowed growth. Ecological changes may have been the catalyst for a feedback reaction between larger bodies and slower growth on the one hand and ecological challenges on the other. Which response typical of extant hominoids came first may never be known, and may not even be important. The combination of characters is unique to hominids. While autocatalytic, directionality is not inevitable, as we see in the examples of hominoids that have smaller brains and presumably less intelligence.

this volume). Biological and ecological variables exert similar dynamic effects, and in concert with social pressures they feed back and contribute to further cognitive evolution.

Additional pressures between the brain and sociality may have arisen through prolonging juvenility, which has been linked with their large brains' higher energy demands (Kelley, Chapter 15, Ross, Chapter 8, this volume). Prolongation increases vulnerability for juveniles, who are handicapped by poor foraging skills and small size. Learning foraging skills is exceptionally slow and difficult because of great apes' difficult diets; complex skills for obtaining their most difficult foods, some of them fallback foods, may not be mastered until near adulthood. Juveniles' poor foraging skills and slow learning essentially extend their dependency, aggravating

pressures on caregivers, especially mothers. These pressures have been linked with enhancing apprenticeship (e.g., imitation, teaching) as a means of speeding their skill acquisition (e.g., Parker 1996).

Body–diet–brain

Brain size correlates with diet more closely than with body size (Begun & Kordos, Chapter 14, this volume). Large bodies are none the less linked with diet. The hominid combination of body size, diet, and brain size probably aggravated cognitive challenges.

Hominids, exceptionally large bodied, would have required more and/or better food than smaller-bodied hominids, although not proportional to their greater size because of their lower metabolic rates. Fruit specialists' diets are typically diversified because fruits are energy rich but poor in important nutrients like proteins and fat; hominids in particular are too large to be dedicated frugivores, and at some point they diversified their diets to include foods richer in protein and fat (Waterman 1984; Yamagiwa, Chapter 12, this volume). Whenever large body size appeared between stem hominoids and early hominids, broadening the diet was one probable avenue of obtaining more food. Compared with stem hominoids, early hominid dentition indicates expanding beyond soft fruits to eclectic frugivory or additional hard foods. If modern great apes are any index, their broader diets increased cognitive challenges by increasing foraging complexity, which increases memory load and the range and complexity of skills needed to locate and obtain food.

Large brains, with their high energetic costs, favor better-quality diets (e.g., meat in hominins). Non-fruit foods are generally differently distributed and more highly defended against predators than fruits. Effects on behavior include broadening and/or shifting foraging ranges and foraging skill repertoires; this increases the variety and especially the complexity of foraging skills, which translates into greater cognitive challenges. In hominids, then, improving diets to support large brains likely generated new pressures to enlarge the brain even more. In other words, hominid diets and large brains may have generated their own dietary cognitive arms race.

Diet–Sociality

Hominid diets and sociality mutually affect one another, as shown by great apes' foraging strategies during seasonal fruit scarcities. Foraging strategies are affected by both fruit scarcities (through females) and social pressures (through male competition). For cognition, this is the sort of intertwined tangle of complex social and ecological demands that requires interconnected cognition, that is, handling diverse demands in one integrated solution; it is a recurrent feature of normal great ape life. Potts, Chapter 13, and Ward et al., Chapter 18, this volume also recognize this situation.

This myriad of interdependent biological, social, and ecological factors affecting intelligence in hominids is complicated beyond our ability to discern first causes or prime movers. We do know, however, that these attributes co-occur only in hominids. Some of them occur in other mammals, but never all together and never

to the degree expressed in hominids. First causes may then be less important to present day outcomes than changes induced by multiple interdependencies among these factors.

It is also probable that in the evolution of hominid brains, this attribute package entailed "arms races" involving both ecological (dietary) and social pressures. Arms races are always constrained by initial conditions. As Ward et al., Chapter 18, this volume note, within a taxon individuals compete mainly with conspecifics. Pressures on a hominid come from other hominids in their ecological and social context. Given their different evolutionary trajectories, arms races in different social, biological, and evolutionary contexts should produce different outcomes. This is the reason we do not see monkeys, even capuchins and baboons, as intelligent as great apes and humans. Monkeys experience different ecological conditions and do not need to be as intelligent as great apes to compete with other monkeys. For the hominids, diet and moist tropical forests are good candidates for constraints. The great apes never really got out of the fruit market and that may have limited their capacity to take in enough energy to enlarge their brains beyond some ceiling. Their persistent tracking of moist tropical forests would impose other constraints on their adaptation, especially given ever-dwindling forest size and productivity. The possibility that some sort of systemic equilibrium sets in is suggested by the distinct "grades" of intelligence and brain–body size scaling patterns that are evident within the primates, as opposed to continuous gradation.

CONCLUSIONS

Our interpretation of available evidence is that the evolution of a great ape grade of intelligence involved a web of factors, causally interrelated and mutually adjusted. Constituent pressures and traits may have affected one another in spiraling or arms race fashion before reaching the particular combination seen in the hominids. Great ape adaptation constitutes an integrated package of cognitive–behavioral–social–morphological traits dovetailed to a particular constellation of ecological and social pressures and possibilities, rather than an assemblage of individual traits adapted independently to specific pressures. Their cognitive system, one component of this package, was shaped by all these traits and shaped all these traits in turn.

Many cognitive enhancements taken as key hominin adaptations are now recognized in great apes, and were probably present in the common ancestor of all hominids. While these cognitive enhancements do not reach human levels in any great ape, they none the less point to the ancestral condition of hominid cognition. These include enhancements to individual cognitive abilities (e.g., distance communication, mental representation of distant entities, spatio-temporal mapping, adaptability to novel and variable situations, attributing others' perspectives, tool manufacture and use, food sharing, cooperative hunting) as well as to centralized processes (e.g., rudimentary symbolism, generativity, multiple intelligences working together). Evidence offered here indicates that these cognitive enhancements are part and parcel of a biological package that evolved with the great apes, including larger brains, larger bodies, and extended life histories, in concert with the package of socio-ecological pressures they faced and created. This is consistent with other recent findings, for example that cultures in orangutans and chimpanzees show complexities previously thought possible only in humans (van Schaik *et al.* 2003; Whiten *et al.* 1999).

The cognitive achievements of humans originated as cognitive responses in fossil great apes to increasingly difficult life in the evolving sub-tropical forests of Eurasia. The unique cognitive adaptations of hominins evolved in response to the more severe challenges (for an ape) of more open forested or grassland ecological settings, and are mere elaborations of the cognitive adaptations of their great ape ancestors. In other words, the origin of the human cognitive capacity makes sense only in the light of great ape cognitive evolution.

REFERENCES

Andrews, P., Begun, D. R. & Zylstra, M. (1997). Interrelationships between functional morphology and paleoenvironments in Miocene hominoids. In *Function, Phylogeny, and Fossils: Miocene Hominoid Evolution and Adaptations*, ed. D. R. Begun, C. V. Ward & M. D. Rose, pp. 29–58. New York: Plenum Publishing Co.

Begun, D. R. (1992). Miocene fossil hominids and the chimp–human clade. *Science* **257**, 1929–33.

Begun, D. R. (1994). Relations among the great apes and humans: new interpretations based on the fossil great ape *Dryopithecus*. *Yearbook of Physical Anthropology*, **37**, 11–63.

Begun, D. R. (2001). African and Eurasian Miocene hominoids and the origins of the Hominidae. In *Hominoid Evolution and Environmental Change in the Neogene of Europe. Vol. 2. Phylogeny of the Neogene Hominoid Primates of Eurasia*, ed. L. de Bonis, G. Koufos & P. Andrews, pp. 231–53. Cambridge, UK: Cambridge University Press.

Begun, D. R. (2002). European hominoids. In *The Primate Fossil Record*, ed. W. Hartwig, pp. 339–68. Cambridge, UK: Cambridge University Press.

Boesch, C. & Boesch-Achermann, H. (2000). *The Chimpanzees of the Taï Forest: Behavioural Ecology and Evolution*. Oxford: Oxford University Press.

Byrne, R. W. (1995). *The Thinking Ape*. Oxford: Oxford University Press.

Byrne, R. W. (1997). The technical intelligence hypothesis: an alternative evolutionary stimulus to intelligence? In *Machiavellian Intelligence II: Extensions and Evaluations*, ed. R. W. Byrne & A. Whiten, pp. 289–311. Cambridge, UK: Cambridge University Press.

Byrne, R. W. & Byrne, J. M. E. (1991). Hand preferences in the skilled gathering tasks of mountain gorillas (*Gorilla gorilla berengei*). *Cortex*, **27**, 521–46.

Byrne, R. W. & Byrne, J. M. E. (1993). Complex leaf-gathering skills of mountain gorillas (*Gorilla g. beringei*): variability and standardization. *American Journal of Primatology*, **31**, 241–61.

Byrne, R. W. & Whiten, A. (ed.) (1988). *Machiavellian Intelligence: Social Expertise and the Evolution of Intellect in Monkeys, Apes, and Humans*. Oxford: Clarendon Press.

Cerling, T., Harris, J. R. MacFadden, B.J., *et al.* (1997). Global vegetation change through the Miocene/Pliocene boundary. *Nature*, **389**, 153–8.

Chaimanee, Y., Jolly, D. Benammi, M., *et al.* (2003). A middle Miocene hominoid from Thailand and orangutan origins. *Nature*, **422**, 61–5.

Deacon, T. W. (1990). Rethinking mammalian brain evolution. *American Zoologist*, **30**, 629–705.

Gibson, K. R. (1990). New perspectives on instincts and intelligence: brain size and the emergence of hierarchical mental construction skills. In, *"Language" and Intelligence in Monkeys and Apes: Comparative Developmental Perspectives*, ed. S. T. Parker & K. R. Gibson, pp. 97–128. New York: Cambridge University Press.

Gibson, K. R. (1993). Beyond neoteny and recapitulation: new approaches to the evolution of cognitive development. In *Tools, Language and Cognition in Human Evolution*, ed. K. R. Gibson & T. Ingold, pp. 273–8. Cambridge, UK: Cambridge University Press.

Gibson, K. R., Rumbaugh, D. & Beran, M. (2001). Bigger is better: primate brain size in relationship to cognition. In *Evolutionary Anatomy of the Primate Cerebral Cortex*, ed. D. Falk & K. R. Gibson, pp. 79–97. Cambridge, UK: Cambridge University Press.

Güleç, E. & Begun, D. R. (2003). Functional morphology and affinities of the hominoid mandible from Çandir. *Courier Forschungsinstitut Senckenberg*, **240**, 89–112.

Harrison, T. & Rook, L. (1997). Enigmatic anthropoid or misunderstood ape: the phylogenetic status of *Oreopithecus bambolii* reconsidered. In *Function, Phylogeny, and Fossils: Miocene Hominoid Origins and Adaptations*, ed. D. R. Begun, C. V. Ward & M. D. Rose, pp. 327–62. New York: Plenum Press.

Heizmann, E. & D. R. Begun (2001). The oldest European hominoid. *Journal of Human Evolution*, **41**, 465–81.

Hopkins, W. D. & Rilling, J. K. (2000). A comparative MRI study of the relationship between neuroanatomical asymmetry and interhemispheric connectivity in primates: implication for the evolution of functional asymmetries. *Behavioral Neuroscience*, **114**(4), 739–48.

Inoue-Nakamura, N. & Matsuzawa, T. (1997). Development of stone tool use by wild chimpanzees (*Pan troglodytes*). *Journal of Comparative Psychology*, **111**, 159–73.

Kelley, J. (1997). Paleobiological and phylogenetic significance of life history in Miocene hominoids. In *Function, Phylogeny, and Fossils: Miocene Hominoid Evolution and Adaptations*, ed. D. R. Begun, C. V. Ward & M. D. Rose, pp. 173–208. New York: Plenum Press.

Kelley, J. & Smith, T. (2003). Age at first molar emergence in early Miocene *Afropithecus turkanensis* and the evolution of life history in the Hominoidea. *Journal of Human Evolution*, **44**, 307–29.

Knott, C. D. (1998). Changes in orangutan caloric intake, energy balance, and ketones in response to fluctuating fruit availability. *International Journal of Primatology*, **19**(6), 1061–79.

Langer, J. (1996). Heterochrony and the evolution of primate cognitive development. In *Reaching into Thought: The Minds of the Great Apes*, ed. A. E. Russon, K. A. Bard & S. T. Parker, pp. 257–77. Cambridge, UK: Cambridge University Press.

Langer, J. (2000). The descent of cognitive development (with peer commentaries). *Developmental Science*, **3**(4), 361–88.

Miles, H. L. (1991). The development of symbolic communication in apes and early hominids. In *Studies in Language Origins*, Vol. 2, ed. J. W. von Raffler-Engel, J. Wind, & A. Jonker, pp. 9–20. Amsterdam: John Benjamins.

Miles, H. L., Mitchell, R. W. & Harper, S. (1996). Imitation, pretense and self-awareness in a signing orangutan. In *Reaching into Thought: The Minds of the Great Apes*, ed. A. E. Russon, K. A. Bard, & S. T. Parker, pp. 278–99. Cambridge, UK: Cambridge University Press.

Parker, S. T. (1996). Apprenticeship in tool-mediated extractive foraging: the origins of imitation, teaching and self-awareness in great apes. In *Reaching into Thought*, ed. A. E. Russon, K. A. Bard & S. T. Parker, pp. 348–70. Cambridge, UK: Cambridge University Press.

Parker, S. T. & Gibson, K. R. (1979). A developmental model of the evolution of language and intelligence in early hominids. *Behavioral and Brain Sciences*, **2**, 364–408.

Parker, S. T. & McKinney, M. (1999). *Origins of Intelligence: The Evolution of Cognitive Development in Monkeys, Apes, and Humans*. Baltimore, MD: The Johns Hopkins Press.

Povinelli, D. J. & Cant, J. G. H. (1995). Arboreal clambering and the evolution of self-conception. *Quarterly Review of Biology*, **70**(4), 393–421.

Rumbaugh, D. (1995). Primate language and cognition: common ground. *Social Research*, **62**, 711–30.

Rumbaugh, D. M., Washburn, D. A., & Hillix, W. A. (1996). Respondents, operants, and *emergents*: toward an integrated perspective on behavior. In *Learning as a Self-Organizing Process*, ed. K. Pribram & J. King, pp. 57–73. Hillsdale, NJ: Lawrence Erlbaum Associates.

Russon A. E. (1998). The nature and evolution of intelligence in orangutans (*Pongo pygmaeus*). *Primates*, **39**, 485–503.

Russon, A. E. (2002). Return of the native: cognition and site specific expertise in orangutan rehabilitation. *International Journal of Primatology*, **23**(3), 461–78.

Savage-Rumbaugh, E. S., Williams, S. L., Furuichi, T. & Kano, T. (1996). Language perceived: *paniscus* branches out. In *Great Ape Societies*, ed. W. C. McGrew, L. F. Marchant & T. Nishida, pp. 173–84. Cambridge, UK: Cambridge University Press.

Stokes, E. J. & Byrne, R. W. (2001). Cognitive abilities for behavioral flexibility in wild chimpanzees

(*Pan troglodytes*): the effect of snare injury on complex manual food processing. *Animal Cognition*, **4**, 11–28.

Teaford, M. F. & Walker, A. C. (1984). Quantitative differences in dental microwear between primate species with different diets and a comment on the presumed diet of *Sivapithecus*. *American Journal of Physical Anthropology*, **64**, 191–200.

Thompson, R. K. R. & Oden, D. L. (2000). Categorical perception and conceptual judgments by nonhuman primates: the paleological monkey and the analogical ape. *Cognitive Science*, **24**(3), 363–96.

Van Schaik, C., Ancrenaz, M., Borgen, G., Galdikas, G., Knott, C., Singleton, I., Suzuki, A., Utami, S. & Merrill, M. (2003). Orangutan cultures and the evolution of material culture. *Science*, **299**, 102–5.

Ward, C. V., Walker, A. C. & Teaford, M. F. (1991). *Proconsul* did not have a tail. *Journal of Human Evolution*, **21**: 215–20.

Waterman, P. G. (1984). Food acquisition and processing as a function of plant chemistry. In *Food Acquisition and Processing in Primates*, ed. D. J. Chivers, B. A. Wood & A. Bilsborough, pp. 177–211. New York: Plenum.

Whiten, A., Goodall, J., McGrew, W., Nishida, T., Reynolds, V., Sugiyama, Y., Tutin, C., Wrangham, R., & Boesch, C. (1999). Cultures in chimpanzees. *Nature*, **399**, 682–5.

Yamakoshi, G. (1998). Dietary responses to fruit scarcity of wild chimpanzees at Bossou, Guinea: possible implications for ecological importance of tool use. *American Journal of Physical Anthropology*, **106**, 283–95.

Author index

Species index

Aegyptopithecus zeuxis 260–262, 266–270, 342–344
African great apes: *see* hominines
African hominids 342–344
Afropithecus 238, 287–289, 293, 303–305, 307–313, 342–344, 362
Alaska sea otter (*Enhydra lutris nereis*) 32–33
Alouatta 260–263
anatomically modern humans 2–3
anthropoid primates 22, 88, 105, 107–108, 110–113, 114–116, 117, 122, 133–134, 136, 260–265, 266–269, 338, 339, 342
Atelines 5, 6, 47, 128, 143–156, 185, 199–201, 210–211, 321–323
Australopithecus 165, 260, 264–265, 339, 342–344

baboon (*Papio*) 7, 17–18, 45, 47, 141, 159, 181, 182–184, 199–201, 342–344, 362, 365
bonobo (*Pan paniscus*) 15–22, 35–37, 47, 48, 63–64, 66–70, 105, 113, 126–127, 128, 135, 141, 173–174, 177, 179–181, 193, 197–201, 210–212, 217–218, 219, 221–223, 249–253, 263–264, 305–307, 321–323, 358–361, 362–363

California sea otter (*Enhydra lutris*) 32–33
callitrichids 267–269
capuchin (*Cebus*) 113, 141, 142–159, 160, 161–162, 163
catarrhines 266–270, 282, 283–284, 292–293, 342–344
catarrhines, fossil 260–262
Cebinae 6–7, 143–156, 185, 267–269, 312–313, 320, 321–323, 339, 365
Cercocebus 267–269
cercopithecoids 17–19, 143–156, 199–201, 243–247, 249–253, 262–263, 266–267
cetaceans 190
chimpanzee (*Pan troglodytes*) 2, 3, 7, 8, 9, 15–22, 32–34, 35–37, 39, 45, 46–47,

48, 49–50, 51, 52, 53–54, 55–56, 61–70, 88–89, 90, 92, 93, 108–110, 113, 117, 126–127, 128, 134–135, 136, 141–159, 160, 172, 173–174, 177, 181, 193, 196, 198–201, 210–212, 213, 217, 218–219, 221–223, 242–243, 248–253, 263–264, 283–284, 287–292, 298–299, 304–307, 312–313, 321–323, 326, 328, 337–338, 339, 342–344, 356–361, 362–363
Chiroptera (bats) 283–284, 285–286
colobinae 143–156, 159
Colobus polykomos 262–263

Darwin's finch 32–33
dolphins 21
Dryopithecus 17–22, 237–243, 249–253, 260, 263–264, 266–270, 271–272, 287–289, 293, 305–312, 327–328, 339, 342–344, 358–362, 363
Dryopithecus brancoi 17–18, 260, 263–264
Dryopithecus laietanus 17–18

Eastern lowland gorilla (*Gorilla gorilla graueri*) 210–211, 215–217
Egyptian vulture (*Neophron percnopterus*) 32–33
Equatorius 242–243, 304–305, 307–313

Gelada baboon (*Theropithecus gelada*) 199–201
gibbon (*Hylobates* sp.) 15–19, 21, 107–108, 113, 114–116, 117, 134–135, 136, 143–156, 173–175, 176, 177, 179, 246–247, 251–252, 262–264, 266–270, 305–312, 320–323, 324–325, 326, 327–328, 339, 342–344, 357, 362
Gigantopithecus 246–247
gorilla (*Gorilla gorilla*) 6, 7, 8, 17–22, 37–38, 49, 63–64, 65–66, 67–70, 105–106, 108–111, 113, 116, 134–135, 136, 141–142, 172, 173–174, 178–179, 181, 183–184,

193, 194–196, 199–201, 210–211, 212–213, 215–217, 219, 221–223, 246–247, 248–253, 266–267, 269–270, 287–289, 305–307, 321–323, 326, 328, 337–338, 339, 342–344, 356–357, 358–361, 362–363
Griphopithecus 238, 243–246, 304–305, 358

Hamadryas baboons (*Papio hamadryas*) 45, 46–48, 49, 51, 52–53, 54, 55–56, 199–201
hominids, fossil 15–21, 22, 49, 107, 109, 114–116, 238, 249–253, 260, 266–267, 269–270, 287–289, 335–336, 339, 342–344, 358, 361, 362
hominines 3, 21–22, 267–270, 309–311, 339, 342–344, 358–361, 362, 366
hominins, fossil 3, 32
hominoids 15, 17–22, 105, 110–116, 117, 249–253, 266–267, 342, 358, 362, 363
hominoids, fossil 114–116, 165, 237–253, 260, 262–263, 266–267, 287–289, 324–325, 327–328, 342–344
Homo 342–344
Homo erectus 2, 3
howler monkey (*Alouatta palliata*) 5, 6
humans 15–22, 31–32, 34–35, 52, 54–55, 63–71, 106–116, 117, 122, 134–135, 136, 161–162, 223–224, 249–253, 260, 269–271, 280–282, 283–284, 289–290, 292–293, 342–344
humans, fossil (pre *Homo*) 18
hybrid baboons 48

Japanese macaques (*Macaca fuscata*) 66–67, 262–263

Kenyapithecus 238–243, 304–305, 307–312, 326, 328

LCA hominin
LCA hominoid 342–344

Subject index

Lightning Source UK Ltd.
Milton Keynes UK
22 April 2010

153138UK00004B/18/A